METALS IN WATER

Woodhead Publishing
Advances in Pollution Research

METALS IN WATER

Global Sources, Significance, and Treatment

Edited by

Sushil Kumar Shukla
Department of Environmental Sciences, Central University of Jharkhand, Ranchi, Jharkhand, India

Sunil Kumar
Waste Reprocessing Division, CSIR-National Environmental Engineering Research Institute, Nagpur, Maharashtra, India

Sughosh Madhav
Department of Civil Engineering, Jamia Millia Islamia, New Delhi, India

Pradeep Kumar Mishra
Dr APJ Abdul Kalam Technical University, Lucknow, Uttar Pradesh, India

Woodhead Publishing is an imprint of Elsevier
50 Hampshire Street, 5th Floor, Cambridge, MA 02139, United States
The Boulevard, Langford Lane, Kidlington, OX5 1GB, United Kingdom

ISBN: 978-0-323-95919-3 (print)
ISBN: 978-0-323-95920-9 (online)

For Information on all Woodhead Publishing publications
visit our website at https://www.elsevier.com/books-and-journals

Publisher: Candice G. Janco
Acquisitions Editor: Maria Elekidou
Editorial Project Manager: Ali Afzal-Khan
Production Project Manager: Kumar Anbazhagan
Cover Designer: Mark Rogers

Typeset by MPS Limited, Chennai, India

Contents

3. Heavy metal contamination in water: consequences on human health and environment 39

Anjali Sharma, Ajmer Singh Grewal, Devkant Sharma and Arun Lal Srivastav

4. Metal pollution in the aquatic environment and impact on flora and fauna

Sweta and Bhaskar Singh

5. Metal in water: an assessment of toxicity with its biogeochemistry

Bipradeep Mondal and Nirmali Bordoloi

6. Assessment and impact of metal toxicity on wildlife and human health

Nitin Verma, Mahesh Rachamalla, P. Sravan Kumar and Kamal Dua

7. Metal contamination in water resources due to various anthropogenic activities

Amrita Daripa, Lal Chand Malav, Dinesh K. Yadav and Sudipta Chattaraj

8. Impact assessment of heavy metal pollution in surface water bodies

Soumya Pandey and Neeta Kumari

9. Airborne heavy metals deposition and contamination to water resources

Harshbardhan Kumar, Gurudatta Singh, Virendra Kumar Mishra, Ravindra Pratap Singh and Pardeep Singh

10. Metal pollution in marine environment: sources and impact assessment

Rahul Mishra, Ekta Singh, Aman Kumar, Akshay Kumar Singh, Sughosh Madhav, Sushil Kumar Shukla and Sunil Kumar

11. Marine environmental chemistry and ecotoxicology of heavy metals

Shriya Garg and Mangesh Gauns

12. Various methods for the recovery of metals from the wastewater

Priya Mukherjee, Uttkarshni Sharma, Ankita Rani, Priyanka Mishra and Pichiah Saravanan

13. Processes of decontamination and elimination of toxic metals from water and wastewaters

Sylvester Chibueze Izah, Clement Takon Ngun, Paschal Okiroro Iniaghe, Ayobami Omozemoje Aigberua and Tamaraukepreye Catherine Odubo

List of contributors

Ritu Aggrawal Department of Botany, Maharaj Singh College, Saharanpur, Uttar Pradesh, India

Ayobami Omozemoje Aigberua Department of Environment, Research and Development, Anal Concept Limited, Elelenwo, Rivers State, Nigeria

Sweety Nath Barbhuiya Laboratory of Cell and Molecular Biology, Department of Life Science and Bioinformatics, Assam University, Silchar, Assam, India; Department of Zoology, Patharkandi College, Patharkandi, Assam, India

Dharmeswar Barhoi Laboratory of Cell and Molecular Biology, Department of Life Science and Bioinformatics, Assam University, Silchar, Assam, India; The Assam Royal Global University, Guwahati, Assam, India

Kuldeep Bauddh Department of Environmental Sciences, Central University of Jharkhand, Ranchi, India

Preetismita Borah CSIO-Central Scientific Instruments Organization, Chandigarh, Chandigarh, India

Nirmali Bordoloi Department of Environmental Sciences, Central University of Jharkhand, Ranchi, Jharkhand, India

Sudipta Chattaraj ICAR-National Bureau of Soil Survey and Land Use Planning, Nagpur, Maharashtra, India

Amrita Daripa ICAR-National Bureau of Soil Survey and Land Use Planning, Nagpur, Maharashtra, India

Kamal Dua Discipline of Pharmacy, Graduate School of Health, University of Technology Sydney, Broadway, NSW, Australia

Himanshu Dwivedi Department of Botany, Maharaj Singh College, Saharanpur, Uttar Pradesh, India

Shriya Garg Academy of Scientific and Innovative Research (AcSIR), Gaziabad, India; CSIR- National Institute of Oceanography, Goa, India

Mangesh Gauns Academy of Scientific and Innovative Research (AcSIR), Gaziabad, India; CSIR- National Institute of Oceanography, Goa, India

Ajay Giri Department of Basic Educations, Ghazipur, Uttar Pradesh, India

Deen Dayal Giri Department of Botany, Maharaj Singh College, Saharanpur, Uttar Pradesh, India

Sarbani Giri Laboratory of Cell and Molecular Biology, Department of Life Science and Bioinformatics, Assam University, Silchar, Assam, India

Ajmer Singh Grewal Guru Gobind Singh College of Pharmacy, Yamunanagar, Haryana, India

Paschal Okiroro Iniaghe Department of Chemistry, Faculty of Science, Federal University Otuoke, Bayelsa State, Nigeria

Sylvester Chibueze Izah Department of Microbiology, Faculty of Science, Bayelsa Medical University, Yenagoa, Bayelsa State, Nigeria

Dharmendra Kumar Jigyasu Department of Geology, University of Lucknow, Lucknow, India; Central Muga Eri Research and Training Institute, Jorhat, India

Deepak Kashyap CSIO-Central Scientific Instruments Organization, Chandigarh, Chandigarh, India

Juhi Khan Department of Botany, IFTM University, Moradabad, Uttar Pradesh, India

Sushil Kumar Kharia College of Agriculture, SKRAU, Bikaner, Rajasthan, India

Aman Kumar Waste Reprocessing Division, CSIR-National Environmental Engineering Research Institute, Nagpur, Maharashtra, India

Harshbardhan Kumar Academy of Scientific and Innovative Research (AcSIR), Ghaziabad, Uttar Pradesh, India; CSIR-National Institute of Oceanography, Dona Paula, Goa, India

Lawrence Kumar Department of Nanoscience and Technology, Central University of Jharkhand, Ranchi, Jharkhand, India

Manish Kumar CSIO-Central Scientific Instruments Organization, Chandigarh, Chandigarh, India

P. Sravan Kumar Department of Biology, University of Saskatchewan, Saskatoon, SK, Canada

Pawan Kumar Department of Physics, Mahatma Gandhi Central University, Motihari, Bihar, India

Sandeep Kumar Division of Environment Science, ICAR-IARI, New Delhi, India

Sunil Kumar Waste Reprocessing Division, CSIR-National Environmental Engineering Research Institute, Nagpur, Maharashtra, India

Khushbu Kumari Department of Environmental Sciences, Central University of Jharkhand, Ranchi, India

Neeta Kumari Department of Civil and Environmental Engineering, Birla Institute of Technology, Mesra, Ranchi, Jharkhand, India

Sughosh Madhav Department of Civil Engineering, Jamia Millia Islamia, New Delhi, India

Lal Chand Malav ICAR-National Bureau of Soil Survey and Land Use Planning, Regional Center, Udaipur, Rajasthan, India

Priyanka Mishra Environmental Nanotechnology Laboratory, Department of Environmental Science and Engineering, Indian Institute of Technology (ISM), Dhanbad, Jharkhand, India

Rahul Mishra Waste Reprocessing Division, CSIR-National Environmental Engineering Research Institute, Nagpur, Maharashtra, India

Virendra Kumar Mishra Institute of Environment and sustainable Development, Banaras Hindu University, Varanasi, Uttar Pradesh, India

Bipradeep Mondal Department of Environmental Sciences, Central University of Jharkhand, Ranchi, Jharkhand, India

Mritunjay Department of Civil Engineering, National Institute of Technology Patna, Patna, India

Priya Mukherjee Environmental Nanotechnology Laboratory, Department of Environmental Science and Engineering, Indian Institute of Technology (ISM), Dhanbad, Jharkhand, India

Clement Takon Ngun Department of Biochemistry and Biophysics, Saratov State University, Saratov, Saratov Region, Russian Federation

Tamaraukepreye Catherine Odubo Department of Microbiology, Faculty of Science, Bayelsa Medical University, Yenagoa, Bayelsa State, Nigeria

Soumya Pandey Department of Civil and Environmental Engineering, Birla Institute of Technology, Mesra, Ranchi, Jharkhand, India

Sanjeet Kumar Paswan Department of Nanoscience and Technology, Central University of Jharkhand, Ranchi, Jharkhand, India

Majid Peyravi Department of Chemical Engineering, Babol Noshirvani University of Technology, Babol, Iran

Abdur Quaff (Rahman) Department of Civil Engineering, National Institute of Technology Patna, Patna, India

Mahesh Rachamalla Department of Biology, University of Saskatchewan, Saskatoon, SK, Canada

Ankita Rani Environmental Nanotechnology Laboratory, Department of Environmental Science and Engineering, Indian Institute of Technology (ISM), Dhanbad, Jharkhand, India

Hossein Rezaei Department of Chemical Engineering, Babol Noshirvani University of Technology, Babol, Iran

Biswa Mohan Sahoo Roland Institute of Pharmaceutical Sciences, Berhampur, Odisha, India

Pichiah Saravanan Environmental Nanotechnology Laboratory, Department of Environmental Science and Engineering, Indian Institute of Technology (ISM), Dhanbad, Jharkhand, India

Anjali Sharma Guru Gobind Singh College of Pharmacy, Yamunanagar, Haryana, India

Devkant Sharma Ch. Devi Lal College of Pharmacy, Jagadhari, Haryana, India

Uttkarshni Sharma Environmental Nanotechnology Laboratory, Department of Environmental Science and Engineering, Indian Institute of Technology (ISM), Dhanbad, Jharkhand, India

Vaishali Sharma Panjab University, Chandigarh, Chandigarh, India

Sushil Kumar Shukla Department of Environmental Sciences, Central University of Jharkhand, Ranchi, Jharkhand, India

Akshay Kumar Singh Department of Transport Science and Technology, Central University of Jharkhand, Ranchi, Jharkhand, India

Bhaskar Singh Department of Environmental Sciences, Central University of Jharkhand, Ranchi, Jharkhand, India

Ekta Singh Waste Reprocessing Division, CSIR-National Environmental Engineering Research Institute, Nagpur, Maharashtra, India

Gurudatta Singh Institute of Environment and sustainable Development, Banaras Hindu University, Varanasi, Uttar Pradesh, India

Munendra Singh Department of Geology, University of Lucknow, Lucknow, India

Pardeep Singh Department of Environmental Science, PGDAV College University of Delhi, New Delhi, Delhi, India

Priyanka Singh Department of Geology, Babasaheb Bhimrao Ambedkar University, Lucknow, India

Ram Kishore Singh Department of Nanoscience and Technology, Central University of Jharkhand, Ranchi, Jharkhand, India

Ravindra Pratap Singh Central Public Works Department, New Delhi, India

Sandeep Singh Department of Earth Sciences, Indian Institute of Technology, Roorkee, India

Shobha Singh Department of Nanoscience and Technology, Central University of Jharkhand, Ranchi, Jharkhand, India

Arun Lal Srivastav School of Engineering and Technology, Chitkara University, Baddi, Himachal Pradesh, India

B.H. Sunil ICAR-National Bureau of Soil Survey and Land Use Planning, Nagpur, Maharashtra, India

Sweta Department of Environmental Sciences, Central University of Jharkhand, Ranchi, Jharkhand, India

Rinkey Tiwari Departments of Botany, Meerut College, Meerut, Uttar Pradesh, India

Nitin Verma Chitkara University School of Pharmacy, Chitkara University, Solan, Himachal Pradesh, India

Brijesh Yadav ICAR-National Bureau of Soil Survey and Land Use Planning, Regional Center, Udaipur, Rajasthan, India

Dinesh K. Yadav ICAR-Indian Institute of Soil Science, Bhopal, Madhya Pradesh, India

Preface

Heavy metals are found in the earth's crust naturally and get into the water and food web by diverse geochemical cycles. Various anthropogenic activities such as agricultural, industrial, transportation, and mining operations contribute to a significant proportion of heavy metal pollution in water resources. Some heavy metals are essential for human life in small quantities, but their large amounts in water and food chain are toxic for humans and plants. Heavy metals are nonbiodegradable, and thus, they persist for a long time in the environment and have the ability of biomagnification and bioaccumulation. Heavy metals adversely impact human health. Hence, remediation and removal of heavy metals are necessary prior to water consumption. Adequate knowledge of heavy metal detection sensors and toxicity indices is required to deal with heavy metal contamination in water resources.

This book elaborates on different techniques for the detection and measurement of metal contamination in water resources. It enumerates various toxicity indices of heavy metals. This book discusses natural and anthropogenic sources of metals in the water resources. It also discusses geogenic and airborne deposition of heavy metals in water resources. It then discusses anthropogenic sources of heavy metal contamination like farming, mining, industrial operation, and transportation. It discusses heavy metal pollution in different aquatic ecosystems like surface water, groundwater, and marine environment. This book also elaborates on the impact of heavy metals on human health. Furthermore, it discusses recovery of precious heavy metals from industrial effluents and wastewater. It also includes various traditional and recent remediation techniques for wastewater treatment. It then discusses heavy metal remediation using low-cost adsorbent, membrane technology, phytoremediation, and nanotechnology. It includes time series analysis of dissolved trace metals in the river ecosystems.

This book contains both latest practical and theoretical aspects of heavy metal pollution in the water resources and remediation measures. It covers holistic heavy metal pollution issues in water resources. The authors have laid an emphasis on the recent research on heavy metal contamination in water resources and its impact across the globe. This book is an honest effort to showcase the most recent research work in heavy metal remediation drafted by leading researchers, academicians, and scientists. This book will be helpful for undergraduate or university students, teachers, and researchers, especially those working in the areas of hydrochemistry, water pollution, and remediation and health aspects. Primary audiences are graduate students, environmental engineers and scientists, scholars working on the issues related to water resources and environmental studies, and scientists working in the field of water resources, heavy metal pollution, water treatment, and social issues.

We are grateful to all the authors whose contributions have made this book possible. We are confident that the chapters presented in this book will act as a good source of knowledge and reference material for the scientists and researchers in their research work. We sincerely welcome feedback from all the valuable readers and critics.

Editorial team

CHAPTER

1

Measurement techniques for detection of metals in water resources

Ekta Singh[1], Aman Kumar[1], Rahul Mishra[1], Akshay Kumar Singh[2], Sughosh Madhav[3], Sushil Kumar Shukla[4] and Sunil Kumar[1]

[1]Waste Reprocessing Division, CSIR-National Environmental Engineering Research Institute, Nagpur, Maharashtra, India [2]Department of Transport Science and Technology, Central University of Jharkhand, Ranchi, Jharkhand, India [3]Department of Civil Engineering, Jamia Millia Islamia, New Delhi, India [4]Department of Environmental Sciences, Central University of Jharkhand, Ranchi, Jharkhand, India

1.1 Introduction

Metals are present naturally in the Earth's crust, and their composition varies between different places with different concentrations. Anthropogenic activities are majorly accountable for the fast growth of heavy metals in the environment during the last few decades (Buledi et al., 2020). Heavy metal pollution has created a surge in sudden environmental contamination occurrences, resulting in significant concentrations of heavy metals entering the environment in a short period of time, causing severe damage (Vareda et al., 2019). The anthropogenic releases of heavy metals into water sources as a result of industrial activity, household effluent, and wastewater discharge have become a worldwide problem (Fu and Wang, 2011; Mishra et al., 2022). The primary cause of their emission comprises residential and industrial waste as well as metal-containing herbicides and fertilizers. The natural sources of heavy metals include erosion, precipitation and weathering from soil and rock that can introduce them into the ecosystem (Fig. 1.1) (Poornima et al., 2016). Heavy metals have a density of 4 g/cm^3 and can be found as elements, ions or complexes (Ugulu, 2015). Heavy metals such as cobalt (Co), copper (Cu), magnesium (Mg), mercury (Hg), nickel (Ni), cadmium (Cd), chromium (Cr), lead (Pb), and arsenic (As) can pollute water, air, and soil

Metals in Water
DOI: https://doi.org/10.1016/B978-0-323-95919-3.00009-4

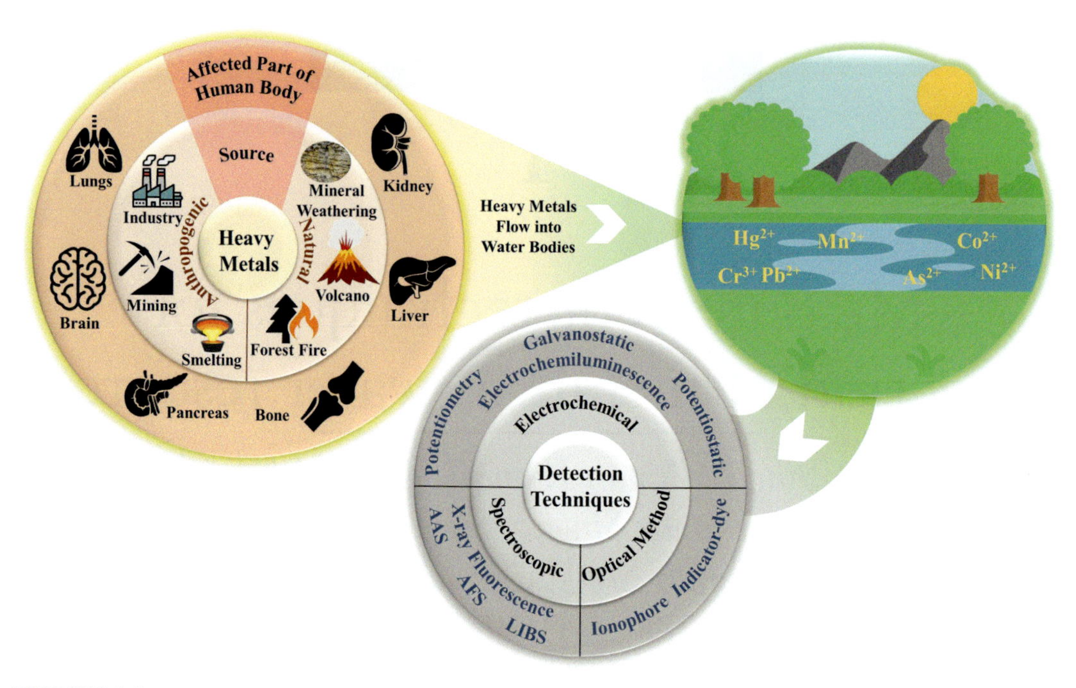

FIGURE 1.1 Heavy metal circulation cycle, its impact on human bodies and available detection techniques.

which results in harmful consequences and health problems (Fig. 1.1) (Jaishankar et al., 2014; Poornima et al., 2016; Lin et al., 2017; Mishra et al., 2022; Arora et al., 2008).

A variety of heavy metals are toxic to humans. Lead poisoning has also been related to developmental issues in infants' brain systems, negative behavioral impacts, reduced renal and fertility function and a higher risk of ischemic heart disease and stroke (Carrington and Bolger, 2014; Mishra et al., 2022). Because inorganic arsenic competes with phosphorus in the oxidative phosphorylation process and inhibits the succinate and pyruvate pathways, it disrupts metabolic activities (Zang and Bolger, 2014). Another carcinogen, cadmium, affects gastrointestinal symptoms such as vomiting, stomach aches, nausea, bone abnormalities, as well as kidney failure (Zang and Bolger, 2014). Although elemental mercury is not easily absorbed by the body, its vapors are poisonous and have a significant influence on the nervous system. The reproductive organs, kidneys, liver and blood can be affected. Multiple organ systems can be affected by the most poisonous form of mercury is methylmercury (Zang and Bolger, 2014). At levels of 50–150 mg/kg body weight, chromium, specifically Cr (VI), can cause mortality (Zang and Bolger, 2014). Lung and nasal cancers can be caused by nickel compounds particularly nickel sulfate when inhaled (Arora et al., 2008). Liver, kidney damage, anemia, along with damage to immune and developmental systems is caused by high copper concentrations (Arora et al., 2008). It's worth noting that some heavy metals, like zinc, copper and iron are biologically important to organisms and only become hazardous when found in excess concentrations. As a result, cost-effective, small, rapid, precise, and sensitive heavy metal monitoring in the drinking water, food, biological fluids and environment is extremely desired (Kaur et al., 2015). Diverse traditional analytical approaches for the identification of heavy metals at ultra-low levels have been developed over time using various techniques such as electrochemical sensing, spectroscopic detection and optical method of detection. Fig. 1.1 shows various heavy metal ion detection techniques, as well as the sources and health effects of these heavy metals. This chapter provides a brief explanation of the various methods for detecting heavy metals in water. It also highlights the source and distribution of heavy metals pollution in water and the consecutive impact that it poses on the environment.

1.2 Heavy metal distribution in water

Heavy metals originate from geologic weathering, direct atmospheric deposition or via industrial waste products' discharge, which is eventually deposited into water bodies like a sink (Mishra et al., 2022). With the characteristics of high toxicity as well as difficulty to decompose, heavy metals serve as major contaminants in surface water (Gao et al., 2015). The sources and distribution of heavy metal ions in the environment are shows in Fig. 1.2. Sediments, interstitial water and overlying water are the most distributed portion of heavy metal contaminants in water. In comparison to upstream rivers and surface water, the concentration of heavy metals is higher in the river's sediments and downstream (Kumar et al.,

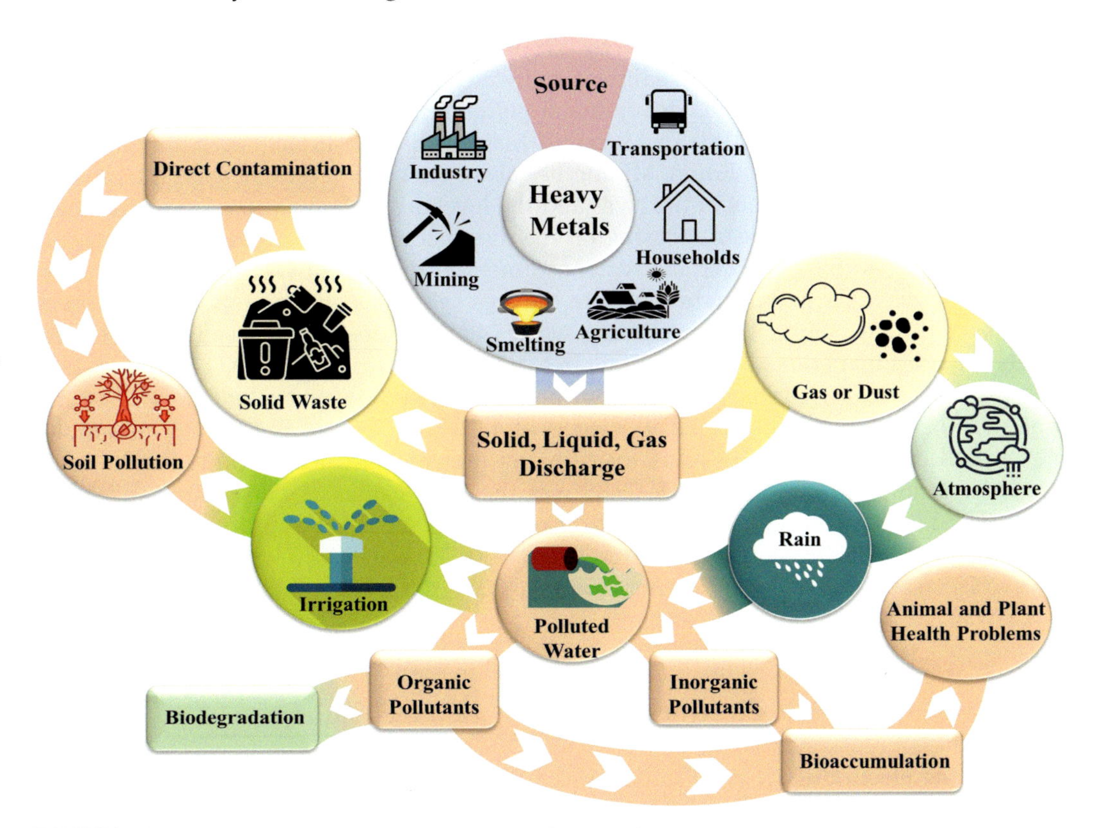

FIGURE 1.2 Sources and proliferation of heavy metal ions in the environment.

2013). River water is a potential source of heavy metals as it finally deposits (Banerjee et al., 2016). High heavy metal concentration and dispersion in sediments and water are important for the movement of transmitting energy and matter in downstream regions (Lin et al., 2020). To reduce human exposure and potential risks, it is necessary to utilize systematic sampling for understanding heavy metals source, characteristics, spatial distribution and possible health effects. About 800 surface waters were studied by Rothwell et al. (2010). It was observed that heavy metals concentration was greater in southeast Scotland mostly due to nonpoint and point source contamination. Different sources generate heavy metals that enter into rivers, reservoirs and lakes that eventually cause hazards to human health and the environment. Thus, presently, it is necessary to study heavy metals in water like estuaries, rivers and lakes. Many researchers have examined the amount of heavy metal pollution in water bodies such as lakes and rivers around the world. For this, concentrations of about 12 dissolved heavy metals were collected by Li et al. (2019a,b) from literatures globally. Findings confirmed that the concentration of Fe, Mn, Ni, Cu, Cr and Cd showed an upward trend in the global lakes and rivers whereas, the concentration of Zn and Pb showed a downward trend. In addition, one of the most relevant sources of metal pollution in lakes and rivers are the manufacturing and mining industries. Heavy metals accumulating in sediments, as a result, pose a serious risk to aquatic lives. Lin et al. (2020) conducted research on Three Gorges Projects and found significant changes in its condition of water conservation and the characteristics of industrial pollution. The indoor space and regular changes have been significantly noted by Pb, Cr, Cd, Zn and Cu metals' concentration values in groundwater. Moreover, sediment and water quality have been deteriorating from mining activities for a long time. Heavy metal pollution and spatial distribution have been studied by Omwene et al. (2018) in Mustafaq Malpas River's surface sediments and confirmed possible sources of pollution. Mining has also been observed to increase the presence of Cd, As and Cr in sediments of the river and the level of Cr shows several possible effects on biota. Researchers have been focusing on the dispersion and migratory pattern of heavy metal ions in the surface of water and water sediment during recent years to obtain information into possible risks to the environment in the area as well as offer a foundation for future governance of that pollutant source. Using water quality index (WQI), evaluation of water quality in Maddhapara Granite Mining area, Dinajpur, Bangladesh, Xi'an area, China, Saraydüzü Dam Lake, Turkey and Dinajpur basin, Bangladesh was evaluated by Howladar et al. (2018), Kükrer and Mutlu (2019), Gao et al. (2020) and Yu et al. (2020). The sample's WQI varied between 10.23 and 63.64 for drinking purposes according to the results of the Howladar et al. (2018). About 17.62–29.88 was the WQI values as per Kükrer and Mutlu (2019). The WQI results reported that 28.82% of samples were not appropriate for drinking as per Gao et al. (2020). Samples of water collected in 2011 and 2017 were compared by Yu et al. (2020). They reported that in comparison to 2011, the WQI values were lower in 2017, which was 72.8 and 59.2, respectively.

1.3 Electrochemical detection techniques

Biosensing electrodes are used in heavy metal's electrochemical sensing to flow current through an aqueous solution and create a meaningful and detectable electrical signal that corresponds to electrochemical processes occurring within the solution owing to the existence of metal ions (Kudr et al., 2015; Locatelli and Melucci, 2013). Count electrode, reference electrode and the working electrode (WE) are generally used in these procedures (Cui et al., 2015). The WE may be customized with various constituents to determine particular heavy metal ions (Pan et al., 2009). For determining heavy metals in a selective and sensitive manner, these interface materials are critical. Previously, carbon and mercury were the most common interface materials which are used in the manufacture of these electrodes. Hanging mercury drop electrodes and dropping mercury electrodes, on the other hand, were inappropriate for automated heavy metal analysis due to their toxicity and mechanical instability. Chemically modified electrodes are being developed by employing a variety of interface materials, such as carbon nanotubes, metal oxides, polymers, nanomaterials, electrochemical biosensors and others. Various researchers have detected heavy metal ions using different electrochemical techniques, which are shown in Table 1.1. Several electrochemical methods are described below.

1.3.1 Potentiometry

Some heavy metals can be determined using potentiometric methods, which are highly selective. At zero current, the emf (E) is measured in this procedure. This signifies that no current is used in these measurements. By employing selective electrodes, potentiometric technologies are commonly used to do quantitative analysis of heavy metals in aqueous solutions. In contrast to other electroanalytical techniques, these analyze quantitative and qualitative data by measuring activity in a solution instead of concentration. Because of their wide-ranging response, good selectivity, low cost and quick response time, these approaches have been extensively utilized in heavy metal analysis in the complicated environment (Aragay and Merkoçi, 2012). Nevertheless, these approaches have some drawbacks such as challenges in electrode downsizing, limited sensitivity and high detection limits. In this regard, these approaches in conjunction with nanomaterials as the interface material are viable research areas (Bakker and Pretsch, 2008; Düzgün et al., 2011). Increased sensitivity and decreased detection limits for heavy metals have been shown by these potentiometric nanoelectrodes in a variety of environments. In potentiometric technologies, field-effect transistors (FETs) and ion-selective electrodes (IESs) are two major types of devices used (Aragay and Merkoçi, 2012; Gumpu et al., 2015). To decrease the matrix interferences, a selective polymeric membrane is present in ISEs. This is a fascinating field of analytical research because of its nondestructive nature, quick reaction and precision (Radu and Diamond, 2007). Due to excellent physiochemical properties, receptors replace this polymeric membrane by using carbon nanostructure materials as transducers in ISEs directly connected to the transducer surface. Electrically conducting polysulfoaminoantraquinone were used for developing ISEs in the sensing membrane to detect Pb^{2+} (1.6×10^{-7} M of detection limits) (Huang et al., 2014). Pb^{2+} ions were increased in the solution with increasing potential. Another study utilized carbon paste electrode in conjunction with multiwalled carbon nanotubes to detect Hg^{2+} ions' (Khani et al., 2010). For Hg^{2+} ions, a detection limit of 2.5×10^{-9} M has been confirmed for minimal time (5 s) and the benefits of long utilization of up to 55 days. With the coupling of

TABLE 1.1 Several electrochemical techniques for heavy metal ions detection.

Heavy metal ions	Technique	Detection range	Detection limit	Reference
Zn^{2+}, Cd^{2+} and Pb^{2+}	SIA-ASV	• Zn^{2+}: 12–100 μg/L • Cd^{2+} and Pb^{2+}: 100 μg/L	• 11, 0.8 and 0.2 μg/L	Injang et al. (2010)
Hg^{2+}	SWASV	• 6–35 nM	• 2.4 nM	Nguyen et al. (2017)
Pb^{2+}	CV	• 1.0×10^{-5} to 10.0 ng/mL	• 0.5 nM	Li et al. (2013)
Pb^{2+} and Cd^{2+}	SWASV	• Pb^{2+}: 5–100 μg/L • Cd^{2+}: 5–500 μg/L	• 1.8 and 1.2 μg/L	Shen et al. (2017)
Pb^{2+}	SWASV	• 0.01–1.0 μM	• 7.8 pM	Tang et al. (2013)
Pb^{2+} and Cd^{2+}	ASV	• 0.5 –70 ng/mL	• 0.1 ng/mL	Chaiyo et al. (2016)
Hg^{2+}	Amperometry	• 0–80 nM	• 0.2 nM	Xuan et al. (2013)
Pb^{2+} and Cd^{2+}	SWASV	• 10–100 ppb	• 7 and 11 ppb	Medina-Sánchez et al. (2015)
Cu^{2+}	DPV and DSV	• 1.0–1000 nM	• 0.48 nM	Cui et al. (2014)
Pb^{2+}	SWASV	• 1–80 μg/L	• 0.3 μg/L	Hong et al. (2017)
Cu^{2+} and Pb^{2+}	SWASV	• 3.31–22.29 ppb	• 0.613 and 0.546 μg/L	Bui et al. (2012)
Cu^{2+}, Zn^{2+}, Pb^{2+} and Cd^{2+}	ASV	• 20–120 mg/L	• 26, 17, 0.55 and 2.8 μg/L	Sahoo et al. (2013)
Pb^{2+}	DPV	• 1.0×10^{-9} to 8.1×10^{-7} M	• 0.6 nM	Alizadeh and Amjadi (2011)
Pb^{2+} and Cd^{2+}	SWASV	–	• 1 ppb	Nantaphol et al. (2017)
Cd^{2+} and Pb^{2+}	SWASV	• 0.01–10 nM	• 0.027 and 0.00423 nM	Zhang et al. (2011)
Pb^{2+}	CV	–	• 0.8 ppb	Wang et al. (2015a,b)
Zn^{2+}, Cu^{2+}, Cd^{2+} and Pb^{2+}	SWASV	• Zn^{2+}:4.2–16.8 μM • Cd^{2+}, Pb^{2+} and Cu^{2+}: 2–8 μM	• 67, 44, 25 and 12 nM	Guo et al. (2011)
Hg^{2+}	Conductometry	• 0.1–100 μM	• 25 nM	Soldatkin et al. (2012)
As^{3+}	DPASV	• 20–100 μM	• 0.48 μM	Thotiyl et al. (2012)
Zn^{2+}	LSASV	• 10^{-6}–10^{-4} M	• 50 nM	Feier et al. (2012)
Hg^{2+}, Cu^{2+}, Pb^{2+} and Cd^{2+}	CC Stripping method	• 1 ng/mL–10 μg/mL	• 2.6 ± 0.9 ng/mL, 0.8 ± 0.2 ng/mL, 2.8 ± 0.6 ng/mL and 1.9 ± 0.4 ng/mL	Choi et al. (2015)
Hg^{2+}	Amperometry	• 0.05–14.77 ppm	• 5 nM	Arduini et al. (2011)

potentiometric sensors, some researchers reported the organic polymers application for heavy metals detection. For assessing Fe^{3+} using ion-selective potentiometery, a coated-wire ISE was reported by Gupta et al. (2011) based on the iron-cyclam complex. Solid-contact ISE was used in another study with 1,2-di-(o-salicylaldiminophenylthio) ethane for detection of Cu^{2+} (Brinić et al., 2012). Very low detection limits were achieved by applying some nano-FETs for heavy metal detection (Zhou and Wei, 2008). For detecting Hg^{2+} ions, single-walled carbon nanotubes were executed by Kim et al. (2009) to FETs construction with the 10×10^{-9} M detection limits. With the 1×10^{-9} M detection limit, Cu^{2+} has also been detected by using FETs and silicon nanowires.

1.3.2 Potentiostatic techniques

A potentiostat apparatus helps adjust the potential difference between the working and recorded reference electrodes. This manages the voltage between both the electrodes. To assess the concentration of analytes, the resulting current is recorded. These tests are also known as controlled potential techniques. Based on resultant measured current waveforms and the type of voltage signal applied, several categories are formed from these controlled potential techniques. Amperometry, Chronocoulometry, and Voltammetry/Polarography are three basic subdivisions of potentiostatic technologies. These three techniques are described below.

1.3.2.1 *Amperometry*

Amperometry is a type of supervised potential technology wherein a nonmercury WE is used to measure and regulate very tiny currents at a set potential. In the solution that contains electroactive substances, between the working and reference electrode, a potential step signal is typically delivered. At the electrode surface, the large current flow is led by resultant reduction that is equal to the gradient concentration. Amperometric techniques are experiments in which the current is noted as a function of time. Because of the fixed voltage of the WE, this approach identifies only single component from electrochemically reducible type. At some chosen magnitude and polarity of the applied voltage, a faradaic reaction is measured by the analyte. But this faradaic reaction is incomplete because of the WE's less surface area and only a portion of analyte reacting. Using three polysaccharides such as dextran, sodium alginate and chitosan, an amperometric biosensor was used by Dabhade et al. (2021) for detecting Cr(VI) in water with a detection limit of 0.05 ppm. Using glucose, mutarotase, and invertase oxidase, an amperometric biosensor was used by Mohammadi et al. (2005) for detecting Hg^{2+} with a detection limit of 1 nM. Chronoamperometery (CA) is an amperometric method, which utilizes a potential step to the WE and detects the resulting current as a function of time. In comparison to other amperometric methods, CA has a shorter time scale. This process necessitates the use of complicated and expensive equipment as well as specialized employees to operate these machines. These methods are also time and labor intensive. Though, it offers extremely precise findings.

1.3.2.2 *Chronocoulometry*

Integral analogs of amperometric techniques are chronocoulometric techniques. These methods entail measuring the charge amount that has transferred after applying a regulated potential, which is computed by taking the integral of current vs voltage or time. These methods are typically used to conduct exhaustive electrolysis for quantitative analysis, however, they provide very little information about the analyte type. Chronocoulometry methods are commonly used to measure the amount of species adsorption that go through a reaction at the electrode surface in electroactive materials. In comparison to continuous potential in voltammertric and other amperometric technologies, at constant current, this electrolysis is performed. Chronocoulometry employs larger surface area electrodes compared to amerometric approaches, resulting in increased performance due to the analyte's complete reaction. This method requires a high current efficiency despite the advantages of simplicity and great efficiency; therefore, it is not generally employed for electrochemical measurement (Bansod et al., 2017).

1.3.2.3 *Voltammetry*

Voltammetric technologies are extensively utilized for the assessment of heavy metals in several environments. In comparison to the fixed potential point in the amperometric approach, at numerous potential points, these approaches measure current in a current-voltage curve. Because of voltammetry's great sensitivity and accuracy, it is an extensively utilized technique for determining heavy metals. These approaches are well fitted for improving the detection limit and suppressing the background current. The reactions reversibility in the electrochemical cell configuration can provide qualitative information. It is an excellent approach for analyzing colored and turbid solutions which would not be evaluated using any other electroanalytical methods.

In voltammetric measurement, pulse voltammetry utilizes a voltage signal pulse with various amplitudes and shapes. Square wave voltammetry (SWV), reverse pulse voltammetry or polarography (RPV), differential pulse voltammetry or polarography (DPV), tast polarography or staircase voltammetry and normal pulse voltammetry (NPV) or normal pulse polarography (NPP) are subcategories of pulse voltammetry. The detection limit and background current are improved and suppressed, respectively by these techniques. Owing to greater sensitivity, which is appropriate for trace-level analysis, square wave and DPV are most commonly utilized.

1.3.3 Galvanostatic techniques

The current between the counter and WE is controlled using a current source (galvanostat). Chronopotentiometric or galvanostatic approaches are used to characterize such investigations.

Simple instrumentation is used in galvanostatic procedures, unlike potentiostatic techniques, because no feedback from the reference electrode is necessary. Nevertheless, massive double layer charging impacts arise during the test which makes these strategies ineffective. Current cyclic and reversal chronopotentiometry are two types of chronopotentiometry. Galvanostatic stripping chronopotentiometry (SCP) is another widely used method for detecting heavy metal ions. SCP is a good alternate to anodic stripping voltammetry (ASV) as this is less sensitive to the existence of substantial amounts of organic materials in the environment (Gozzo et al., 1999; Estela et al., 1995). SCP has been widely employed to identify trace metals in drinks, food as well as biological materials due to these advantages (Serrano et al., 2003). For heavy metal, researchers used continuous current SCP in a handling mercury drop electrode (Town and van Leeuwen, 2002). Szłyk and Szydłowska-Czerniak (2004) introduced a galvanostatic stripping chronopotentiometric technique to measure copper, lead and cadmium in commercial butter and margarine.

AC voltammetry and electrochemical impedance spectroscopy (EIS) are two of the most extensively utilized impedance measuring technologies for detecting the concentration of analytes in a solution. Among these two technologies, EIS has been used by a number of experts to determine metal ions from a variety of samples. The EIS technology is commonly used for investigating the altered electrode's interfacial characteristics, particularly multilayer films. It was also demonstrated as a useful instrument to identify relevant interface qualities, which may be used in biosensing (Jovanović et al., 2013). The circuit response to a modifying voltage or current as a frequency function is described by EIS which is a very well part of ac theory (Research, 1987). In comparison to other electroanalytical technologies, this technology is cheap and simple to identify hazardous metal ions in chemical and biological media. The concentration of metal ions can be forecasted in an electrolytic solution by measuring resistive-capacitive (RC) and impedance parameters of the electrical equivalent circuit. For detecting the concentration of analytes among various impedance assessments, the frequency response analyzer has been shown as the most extensively utilized approach. This is a single-sine approach that involves overlaying a short sine wave (5–15 mV) of a particular frequency on a dc bias voltage and measuring the resultant ac current. By scanning the desired frequency, various impedance values are recorded for various currents and voltages. Although, this method necessitates a lengthy data acquisition period.

1.3.4 Electrochemiluminescence techniques

Chemiluminiscence is affected by several homogeneous electron transfer processes including radical ions that occur in chemical solutions. Electrochemiluminescence (ECL) is a type of chemiluminescence that is caused by the electrolytic generation of free radicals in chemical processes. These methods are frequently utilized for identifying a specific metal ion in a liquid using fluorescence detection, which is inexpensive, easy to use and very sensitive (parts per billion/ trillion). Hg^{2+} ions detection was done by Gao et al. (2013) using modified DNA strands. Hg^{2+} ions were detected by using the electrochemical platform over a range of 2×10^{-10} M to 2×10^{-8} M with 1.05×10^{-10} M of the detection limit. In ECL sensing, because of narrow Gaussian emission and broad excitation spectra, quantum dots (QDs) are used. The quantum size effect allows for a large frequency band from ultraviolet to near infrared (Gill et al., 2008). In 2004, the first ECL sensor was produced by Zou and Ju (2004) based on CdSe QDs. Presently, Cu^{2+} ions were detected using CdTe QDs which utilized a low potential ECL technique (Cheng et al., 2010). Although, due to their toxicity, QDs are no longer used for detecting heavy metal ions.

1.4 Spectroscopic detection

Heavy metal is detected using highly sensitive techniques such as X-ray fluorescence spectrometry (XRF), inductively coupled plasma-optical emission spectrometry, neutron activation analysis, inductively coupled plasma mass spectroscopy, and atomic absorption spectroscopy (AAS) (Wang et al., 2015a,b; Losev et al., 2015; Sitko et al., 2015; Gong et al., 2016). With low limits of detection, they can simultaneously determine the concentration of heavy metal for a wide elements range. Though this equipment is highly costly and they necessitate the use of qualified staff to operate the sophisticated equipment. Table 1.2

TABLE 1.2 Studies on detection of heavy metal ions from various sources using different spectroscopy techniques.

Heavy metal ions	Technique	Source/sample	Reference
Fe	AAS	Natural and mineral waters	Tautkus et al. (2004)
Ca, Mg, Fe, Mn, Cu, Zn, Cd	AAS	Chinese taponin tablet recipe	Dong and Zhu, 2002
Ca, Mg, Fe, Cu, Zn	AAS	Chinese tablets	Dong and Zhu, 2002
Cd, Cu, Fe, Ni, Pb, Zn, Tl	AAS	Seawater	Kojuncu et al. (2004)
Pb, Cd, Zn, Ni, Cr, Mn, Fe	AAS	Water and therapeutic mud	Radulescu et al. (2014)
Pb, Sb, Al, As	AAS	Tube wells	Tareen et al. (2014)
Mn, Pb, Cr, Cd	AAS	River	Tsade (2016)
Pb	GF-AAS	Water sample	Chen et al. (2005)
Hg	GF-AAS	Muscle samples of fish	Moraes et al. (2013)
Cr	GF-AAS	French wine and grapes	Cabrera-Vique et al. (1997)
Cd	GF-AAS	Wine	Cvetković et al. (2006)
Pt	GF-AAS	Rocks	Odonchimeg et al. (2016)
Co, Ni	GF-AAS	Water	Minami et al. (2003)
Hg	AFS	Water	Logar et al. (2002)
Sb	AFS	Marine algae, molluscs	de Gregori et al. (2007)
Se	AFS	Cow milk	Muniz-Naveiro et al. (2007)
As	AFS	Seafood	Schaeffer et al. (2005)
Pb	AFS	Water	Beltrán et al. (2015)
Hg	XRFS	Wastewater	Marguí et al. (2010)
Zn, Cu, As, Pb	XRFS	Soil	Radu and Diamond (2009)
Pb, Zn, Ni, Cu, Cr, Cd, As	XRFS	Soil	Taha (2017)
Pb, Zn, Cu, Mn, Ni, Fe	XRFS	Coastal seawaters	Yuan-Zhen et al. (2012)
Pb, Cu	XRFS	Aqueous solution	Hutton et al. (2014)

presents the studies conducted on the detection of heavy metal ions using different spectroscopy detection techniques.

1.4.1 Atomic absorption spectroscopy

For the chemical elements quantitative assessment, AAS is one of the most broadly utilized techniques. The basic idea of AAS is that when light is irradiated, electrons in atoms are excited to higher energy levels and when they deexcite, they release light with frequencies unique to that atom. The process' sensitivity results from the concept that a transition line's width is relatively little, and so it can be regarded as "elemental" sensitivity. The material is firstly atomized before it is examined at the atomic level. Flame and electro-thermal atomizers are used to do this, although other atomization techniques are also available. The radiation flow is then evaluated after the atomized sample has been subjected to radiation. The radiation flux is also evaluated without the sample before this. Absorbance is the ratio of fluxes without and with the sample, which may be determined and converted into a concentration of an analyte using the Lambert–Beer law (Sikdar and Kundu, 2018). Inductively coupled plasma-optical emission spectrometry (ICP-OES), flame atomic absorption spectrometry (FAAS) and graphite furnace (or electro-thermal) atomic absorption spectrometry (GF-AAS/ETAAS) are examples of spectrophotometers, which are currently used.

1.4.2 Atomic fluorescence spectrometry

Atomic fluorescence spectrometry (AFS) is a fast-evolving heavy metal detection technology, which relies on the atom's excitation in the elements' vapor state which is examined by specific wavelength radiation that produces atomic fluorescence. The substance concentration is directly proportional to the atomic fluorescence. Using the intensity of fluorescence, the heavy metals in the water and soil may be accurately assessed.

Except that the light source and other elements are not in a linear path, but rather at a 90° right angle, to prevent the excitation light source's radiation from influencing the atomic fluorescence detection signal, AFS is similar to AAS. It offers both atomic absorption and atomic emission benefits. Nevertheless, its drawbacks include the utilization of only eleven elements, like Se, Hg and As as well as a limited application scope (Li et al., 2019a,b). Chen et al. (2016) established an isolation-free, marker-free, super sensitive analysis of homologous DNA in terms of enzyme chain replacement's signal amplification that is employed to target protein and DNA detection using AFS for detecting amplify DNA and metal ions. After the formation of AFS, there have been a few examples of when AFS has been combined with other methodologies, the most famous of this is Tsuji and Kga's HG-AFS technology (Junior et al., 2017).

1.4.3 Laser-induced breakdown spectrometry

Laser-induced breakdown spectrometry (LIBS) is a noncontact, rapid and universal atomic spectroscopy technique that can offer quantitative as well as qualitative analytical data for any material without any significant preparation of sample (Galbács, 2015). On the other hand, in comparison to gas or solid analysis, LIBS has lower accuracy and worse LODs for direct analysis of liquids. It is because of the laser-induced plasma tendency which is produced in the bulk liquid's center to burst out (Lee et al., 2012). In the harsh environment, LIBS can also be utilized for the detection of noncontact optical fields (Randall et al., 2013). To stimulate the material to form transient plasma, the method helps generate a high-energy laser that effectively focuses on the sample's surface via a focusing lens. Analysis of emission line wavelength and the distribution intensity of the plasma can obtain the composition and content of elements (Wang et al., 2014). LIBS necessitates to be enhanced in sensitivity and detection accuracy in comparison to spectral analysis techniques like ICP-Ms and AAS. One can focus on improving detection limit, decreasing matrix effect, enhancing the signal-to-noise ratio, collecting intensity of spectral signal, improving the source of laser light and other aspects to resolve the exiting issues. multipulse LIBS (MP-LIBS), double-pulse LIBS (DP-LIBS) and single-pulse LIBS (SP-LIBS) were compared by Jedlinszki and Galbács (2011). It was confirmed that MP-LIBS' sensitive values substantially increased in comparison to SP-LIBS. LIBS develops towards advancement after decades of development, with the introduction of multipulse LIBS, double pulse LIBS, polarization-resolved LIBS, femtosecond filament LIBS, femtosecond LIBS, nanosecond LIBS and other techniques (Galbács, 2015). Without any treatment of sample, excellent results of LIBS carry out from some types of samples. By the utilization of sample preparation technology, the usage of LIBS can be extended. When compared to ICP-Ms, ICP-OES and XRF, the analytical performance seems equivalent (Jantzi et al., 2016). In the environmental monitoring field, LIBS has gained recognition as an excellent technology. It has shown significant findings for heavy metals detection in soil, water and atmosphere (El Haddad et al., 2013). There are two methods for the detection of aerosols in the atmosphere. One option is to aim the laser at the enrichment filter that can detect materials at lower concentrations and another is to aim the laser directly on the aerosol for identification that has a greater limit of identification in comparison to the former one (Gallou et al., 2011). For detecting metal element s like Zn, Pb, Ni in water, LIBS was utilized earlier by Järvinen et al. (2014). With greater sensitivity, this technique can even monitor industrial wastewater online.

1.4.4 X-ray fluorescence spectrometry

In this technique, the material is ionized when gamma rays or X-rays irradiate it. Electrons from even the inner orbitals of K or L shells can be ejected by this kind of bombardment with high energy radiation. Electrons from higher energy shells fulfill these vacancies that are accompanied by fluorescence or photon emission called XRF that is further assessed. Each element generates its own fluorescence spectrum as each element has a particular set of energy levels with a unique set of energies accompanied by several peaks of different intensities. Therefore, to determine the samples' elemental composition easily, XRF is a very beneficial technology (Malik et al., 2019). The block diagram of the X-ray fluorescence spectrometer comprises a display system, data processing, fluorescence detector, sample chamber and an X-ray source. In numerous FDA-regulated items, the XRF approach has been utilized to screen and detect harmful components (Palmer et al., 2009).

1.5 Optical methods of detection

Conventional methods such as luminescence spectrometry, reflection, or absorption can be used for the detection of optical effects. For heavy metals ions' optical detection, ionophores, specific indicator dyes, capillary-type devices, integrated optics and optical fiber, etc., are commonly executed. Some limitations also arise when optical ion sensing is used to identify heavy metals (Wolfbeis, 2008). More than one metal ion is interacted by several nonselective optical indicators as well as hydrogen ions are interacted by several metal ion indicators. Therefore, pH must be carefully monitored in order to apply a suitable correction factor. Masking agents can be used to compensate for indicator dyes' poor selectivity (Oehme and Wolfbeis, 1997).

1.5.1 Ionophore-based sensors

Ionophores have been used to detect heavy metals by optical techniques due to the shortcomings of indicator reagents. These ionophores are efficient for the detection of heavy metal ions due to their capacity to interact with selective ions, their complexing characteristics and ion carrier. There are numerous sensing methods that may be used to make ionophores sensors effective for metal ions, such as combining ionophores with appropriate dyes or adding chromogenic or fuorogenic constituents to ionophores. Extraction of ions into membranes by employing ion carriers is another method that is best for detecting heavy metal ions (Hisamoto et al., 1995; Lerchi et al., 1994). The heavy metals' detection in these kinds of optical sensors is based on heavy metals' selective binding with ionophores, whereas the optical signal is generated by a proton-selective chromo-ionophore. Protons corresponding to the metal ion's charge are emitted from the chromo-ionophore upon heavy metal ion interaction, leading to changes in the chromo-color ionophore's or fluorescence. Heavy metal interacts to ionophores via covalent bonds with several heteroatoms (sulfur, oxygen and nitrogen) found in the ionophores.

1.5.2 Indicator dye-based sensors

Indicator dye-based sensors focus on the heavy metals' reaction with the indicator dye, which results in an alteration for the absorbance or fluorescence of the interaction reagents. In these kinds of sensors, the indicator acts as a transducer in which it is impossible to direct optical detection. "Quenchers" are heavy metal ions in another set of indicators, causing both dynamic and static quenching of indicator dye luminescence when these two mixes (Lakowicz, 1994). The quencher interacts with the excited state in dynamic quenching, whereas in the ground state, a quencher reacts with a fluorophore in static quenching. The dye is not absorbed during the dynamic quenching process; hence it is reversible. Fluorescent indicators offer the benefits of increased selectivity and sensitivity, but they also have drawbacks that must be discounted. Several indicators are nonselective and they can interact with multiple metal ions. Due to nonavailability in the pure form necessary for sensing application, the requirement of extra reagents, poor stability and unfavorable analytical wavelengths, certain indicators cannot be employed in heavy metals optical sensing. Several indicators attach permanently or just at extremes of pH. Different dye requires for each heavy metal ion which is the key issue with indicator dye-based sensors and thus a different analytical wavelength must be employed each time, complicating the overall detection technique (Oehme and Wolfbeis, 1997).

1.6 Conclusion

Heavy metal pollution is a very serious issue for the environment and human health. In comparison to natural activities, anthropogenic activities increase the heavy metals concentration in the environment. Different anthropogenic emission sources affect the distribution of heavy metals in the water. Heavy metal from various points eventually goes into the water through compost, solid waste, irrigation and runoff, atmospheric sedimentation and parent materials. Many studies found the severe health impact of heavy metals ion toxicity to living beings, and this attracts the researchers to develop several kinds of technologies for detection of heavy metals in water to deem it safe for usage. Due to the scarcity of freshwater, it is necessary to recycle or reuse contaminated water after proper treatment. Several detection techniques require identifying heavy metals in water before treatment. The common detection techniques are present which are mentioned in the above sections. Among them, spectroscopic detection and electrochemical techniques have high sensitivity, low LODs and are appropriate for identifying a larger sample size. LIBS can be utilized for real-time identification of trace elements in the environment and is fast, simple and spatially resolved technique. Although, the rapid detection of heavy metals is limited by their low stability of analysis. AFS is one of the most important techniques for detecting heavy metals and has outstanding selectivity and sensitivity. But its limitation is that its usage is not widespread. By using conventional techniques, the studies on identifying heavy metals have become progressively mature. However, more research is still required for the commercialization of these techniques.

References

Alizadeh, T., Amjadi, S., 2011. Preparation of nano-sized Pb^{2+} imprinted polymer and its application as the chemical interface of an electrochemical sensor for toxic lead determination in different real samples. Journal of Hazardous Materials 190 (1–3), 451–459.

Aragay, G., Merkoçi, A., 2012. Nanomaterials application in electrochemical detection of heavy metals. Electrochimica Acta 84, 49–61.

Arduini, F., Majorani, C., Amine, A., Moscone, D., Palleschi, G., 2011. Hg^{2+} detection by measuring thiol groups with a highly sensitive screen-printed electrode modified with a nanostructured carbon black film. Electrochimica Acta 56 (11), 4209–4215.

Arora, M., Kiran, B., Rani, S., Rani, A., Kaur, B., Mittal, N., 2008. Heavy metal accumulation in vegetables irrigated with water from different sources. Food Chemistry 111 (4), 811–815.

Bakker, E., Pretsch, E., 2008. Nanoscale potentiometry. TrAC Trends in Analytical Chemistry 27 (7), 612–618.

Banerjee, S., Kumar, A., Maiti, S.K., Chowdhury, A., 2016. Seasonal variation in heavy metal contaminations in water and sediments of Jamshedpur stretch of Subarnarekha river, India. Environmental Earth Sciences 75 (3), 265.

Bansod, B., Kumar, T., Thakur, R., Rana, S., Singh, I., 2017. A review on various electrochemical techniques for heavy metal ions detection with different sensing platforms. Biosensors and Bioelectronics 94, 443–455.

Beltrán, B., Leal, L.O., Ferrer, L., Cerdà, V., 2015. Determination of lead by atomic fluorescence spectrometry using an automated extraction/pre-concentration flow system. Journal of Analytical Atomic Spectrometry 30 (5), 1072–1079.

Brinić, S., Buzuk, M., Bralić, M., Generalić, E., 2012. Solid-contact Cu (II) ion-selective electrode based on 1, 2-di-(o-salicylaldiminophenylthio) ethane. Journal of Solid State Electrochemistry 16 (4), 1333–1341.

Bui, M.P.N., Li, C.A., Han, K.N., Pham, X.H., Seong, G.H., 2012. Simultaneous detection of ultratrace lead and copper with gold nanoparticles patterned on carbon nanotube thin film. Analyst 137 (8), 1888–1894.

Buledi, J.A., Amin, S., Haider, S.I., Bhanger, M.I., Solangi, A.R., 2020. A review on detection of heavy metals from aqueous media using nanomaterial-based sensors. Environmental Science and Pollution Research 1–9.

Cabrera-Vique, C., Teissedre, P.L., Cabanis, M.T., Cabanis, J.C., 1997. Determination and levels of chromium in French wine and grapes by graphite furnace atomic absorption spectrometry. Journal of Agricultural and Food Chemistry 45 (5), 1808–1811.

Carrington, C.D., Bolger, P.M., 2014. Toxic metals: lead. Encyclopedia of Food Safety 2, 349–351. Available from: https://doi.org/10.1016/B978-0-12-378612-8.00203-1.

Chaiyo, S., Apiluk, A., Siangproh, W., Chailapakul, O., 2016. High sensitivity and specificity simultaneous determination of lead, cadmium and copper using μPAD with dual electrochemical and colorimetric detection. Sensors and Actuators B: Chemical 233, 540–549.

Chen, J., Xiao, S., Wu, X., Fang, K., Liu, W., 2005. Determination of lead in water samples by graphite furnace atomic absorption spectrometry after cloud point extraction. Talanta 67 (5), 992–996.

Chen, P., Wu, P., Chen, J., Yang, P., Zhang, X., Zheng, C., et al., 2016. Label-free and separation-free atomic fluorescence spectrometry-based bioassay: sensitive determination of single-strand DNA, protein, and double-strand DNA. Analytical Chemistry 88 (4), 2065–2071.

Cheng, L., Liu, X., Lei, J., Ju, H., 2010. Low-potential electrochemiluminescent sensing based on surface unpassivation of CdTe quantum dots and competition of analyte cation to stabilizer. Analytical Chemistry 82 (8), 3359–3364.

Choi, S.M., Kim, D.M., Jung, O.S., Shim, Y.B., 2015. A disposable chronocoulometric sensor for heavy metal ions using a diaminoterthiophene-modified electrode doped with graphene oxide. Analytica Chimica Acta 892, 77–84.

Cui, L., Wu, J., Li, J., Ge, Y., Ju, H., 2014. Electrochemical detection of Cu^{2+} through Ag nanoparticle assembly regulated by copper-catalyzed oxidation of cysteamine. Biosensors and Bioelectronics 55, 272–277.

Cui, L., Wu, J., Ju, H., 2015. Electrochemical sensing of heavy metal ions with inorganic, organic and biomaterials. Biosensors and Bioelectronics 63, 276–286.

Cvetković, J., Arpadjan, S., Karadjova, I., Stafilov, T., 2006. Determination of cadmium in wine by electrothermal atomic absorption spectrometry. Acta Pharmaceutica 56 (1), 69–77.

Dabhade, A., Jayaraman, S., Paramasivan, B., 2021. Development of glucose oxidase-chitosan immobilized paper biosensor using screen-printed electrode for amperometric detection of Cr (VI) in water. 3 Biotech 11 (4), 1–11.

de Gregori, I., Quiroz, W., Pinochet, H., Pannier, F., Potin-Gautier, M., 2007. Speciation analysis of antimony in marine biota by HPLC-(UV)-HG-AFS: extraction procedures and stability of antimony species. Talanta 73 (3), 458–465.

Dong, S.F., Zhu, Z.G., 2002. Determination of the contents of Ca, Mg, Fe, Cu and Zn in suxiao jiuxin pill and the analysis of Ca/Mg and Cu/Zn values. Guang pu xue yu Guang pu fen xi = Guang pu 22 (3), 478–479.

Düzgün, A., Zelada-Guillén, G.A., Crespo, G.A., Macho, S., Riu, J., Rius, F.X., 2011. Nanostructured materials in potentiometry. Analytical and Bioanalytical Chemistry 399 (1), 171–181.

El Haddad, J., Villot-Kadri, M., Ismaël, A., Gallou, G., Michel, K., Bruyère, D., et al., 2013. Artificial neural network for on-site quantitative analysis of soils using laser induced breakdown spectroscopy. Spectrochimica Acta. Part B: Atomic Spectroscopy 79, 51–57.

Estela, J.M., Tomás, C., Cladera, A., Cerda, V., 1995. Potentiometric stripping analysis: a review. Critical Reviews in Analytical Chemistry 25 (2), 91–141.

Feier, B., Floner, D., Cristea, C., Bodoki, E., Sandulescu, R., Geneste, F., 2012. Flow electrochemical analyses of zinc by stripping voltammetry on graphite felt electrode. Talanta 98, 152–156.

Fu, F., Wang, Q., 2011. Removal of heavy metal ions from wastewaters: a review. Journal of Environmental Management 92 (3), 407–418.

Galbács, G., 2015. A critical review of recent progress in analytical laser-induced breakdown spectroscopy. Analytical and Bioanalytical Chemistry 407 (25), 7537–7562.

Gallou, G., Sirven, J.B., Dutouquet, C., Bihan, O.L., Frejafon, E., 2011. Aerosols analysis by LIBS for monitoring of air pollution by industrial sources. Aerosol Science and Technology 45 (8), 918–926.

Gao, A., Tang, C.X., He, X.W., Yin, X.B., 2013. Electrochemiluminescent lead biosensor based on GR-5 lead-dependent DNAzyme for Ru $(phen)_3^{2+}$ intercalation and lead recognition. Analyst 138 (1), 263–268.

Gao, X., Zhuang, W., Chen, C.T.A., Zhang, Y., 2015. Sediment quality of the SW coastal Laizhou Bay, Bohai Sea, China: a comprehensive assessment based on the analysis of heavy metals. PLoS One 10 (3), e0122190.

Gao, Y., Qian, H., Ren, W., Wang, H., Liu, F., Yang, F., 2020. Hydrogeochemical characterization and quality assessment of groundwater based on integrated-weight water quality index in a concentrated urban area. Journal of Cleaner Production 260, 121006.

Gill, R., Zayats, M., Willner, I., 2008. Semiconductor quantum dots for bioanalysis. Angewandte Chemie International Edition 47 (40), 7602–7625.

Gong, T., Liu, J., Liu, X., Liu, J., Xiang, J., Wu, Y., 2016. A sensitive and selective sensing platform based on CdTe QDs in the presence of L-cysteine for detection of silver, mercury and copper ions in water and various drinks. Food Chemistry 213, 306–312.

Gozzo, M.L., Colacicco, L., Callà, C., Barbaresi, G., Parroni, R., Giardina, B., et al., 1999. Determination of copper, zinc, and selenium in human plasma and urine samples by potentiometric stripping analysis and constant current stripping analysis. Clinica Chimica Acta 285 (1–2), 53–68.

Gumpu, M.B., Sethuraman, S., Krishnan, U.M., Rayappan, J.B.B., 2015. A review on detection of heavy metal ions in water—an electrochemical approach. Sensors and Actuators B: Chemical 213, 515–533.

Guo, X., Yun, Y., Shanov, V.N., Halsall, H.B., Heineman, W.R., 2011. Determination of trace metals by anodic stripping voltammetry using a carbon nanotube tower electrode. Electroanalysis 23 (5), 1252–1259.

Gupta, V.K., Sethi, B., Upadhyay, N., Kumar, S., Singh, R., Singh, L.P., 2011. Iron (III) selective electrode based on S-methyl N-(methylcarbamoyloxy) thioacetimidate as a sensing material. International Journal of Electrochemical Science 6, 650–663.

Hisamoto, H., Nakagawa, E., Nagatsuka, K., Abe, Y., Sato, S., Siswanta, D., et al., 1995. Silver ion selective optodes based on novel thia ether compounds. Analytical Chemistry 67 (8), 1315–1321.

Hong, Y., Zou, J., Ge, G., Xiao, W., Gao, L., Shao, J., et al., 2017. Finite element modeling simulation-assisted design of integrated microfluidic chips for heavy metal ion stripping analysis. Journal of Physics D: Applied Physics 50 (41), 415303.

Howladar, M.F., Al Numanbakth, M., Faruque, M.O., 2018. An application of Water Quality Index (WQI) and multivariate statistics to evaluate the water quality around Maddhapara Granite Mining Industrial Area, Dinajpur, Bangladesh. Environmental Systems Research 6 (1), 1–18.

Huang, M.R., Ding, Y.B., Li, X.G., 2014. Combinatorial screening of potentiometric Pb (II) sensors from polysulfoaminoanthraquinone solid ionophore. ACS Combinatorial Science 16 (3), 128–138.

Hutton, L.A., O'Neil, G.D., Read, T.L., Ayres, Z.J., Newton, M.E., Macpherson, J.V., 2014. Electrochemical X-ray fluorescence spectroscopy for trace heavy metal analysis: enhancing X-ray fluorescence detection capabilities by four orders of magnitude. Analytical Chemistry 86 (9), 4566–4572.

Injang, U., Noyrod, P., Siangproh, W., Dungchai, W., Motomizu, S., Chailapakul, O., 2010. Determination of trace heavy metals in herbs by sequential injection analysis-anodic stripping voltammetry using screen-printed carbon nanotubes electrodes. Analytica Chimica Acta 668 (1), 54–60.

Jaishankar, M., Tseten, T., Anbalagan, N., Mathew, B.B., Beeregowda, K.N., 2014. Toxicity, mechanism and health effects of some heavy metals. Interdisciplinary Toxicology 7 (2), 60.

Jantzi, S.C., Motto-Ros, V., Trichard, F., Markushin, Y., Melikechi, N., De Giacomo, A., 2016. Sample treatment and preparation for laser-induced breakdown spectroscopy. Spectrochimica Acta. Part B: Atomic Spectroscopy 115, 52–63.

Järvinen, S.T., Saari, S., Keskinen, J., Toivonen, J., 2014. Detection of Ni, Pb and Zn in water using electrodynamic single-particle levitation and laser-induced breakdown spectroscopy. Spectrochimica Acta. Part B: Atomic Spectroscopy 99, 9–14.

Jedlinszki, N., Galbács, G., 2011. An evaluation of the analytical performance of collinear multi-pulse laser induced breakdown spectroscopy. Microchemical Journal 97 (2), 255–263.

Jovanović, Z., Buică, G.O., Mišković-Stanković, V., Ungureanu, E.M., Amarandei, C.A., 2013. Electrochemical impedance spectroscopy investigations on glassy carbon electrodes modified with poly (4-azulen-1-yl-2, 6-bis (2-thienyl) pyridine). UPB Scientific Bulletin, Series B: Chemistry and Materials Science 75 (1), 125–134.

Junior, M.M.S., Portugal, L.A., Serra, A.M., Ferrer, L., Cerdà, V., Ferreira, S.L., 2017. On line automated system for the determination of Sb (V), Sb (III), thrimethyl antimony (v) and total antimony in soil employing multisyringe flow injection analysis coupled to HG-AFS. Talanta 165, 502–507.

Kaur, B., Srivastava, R., Satpati, B., 2015. Ultratrace detection of toxic heavy metal ions found in water bodies using hydroxyapatite supported nanocrystalline ZSM-5 modified electrodes. New Journal of Chemistry 39 (7), 5137–5149.

Khani, H., Rofouei, M.K., Arab, P., Gupta, V.K., Vafaei, Z., 2010. Multi-walled carbon nanotubes-ionic liquid-carbon paste electrode as a super selectivity sensor: application to potentiometric monitoring of mercury ion (II). Journal of Hazardous Materials 183 (1–3), 402–409.

Kim, T.H., Lee, J., Hong, S., 2009. Highly selective environmental nanosensors based on anomalous response of carbon nanotube conductance to mercury ions. The Journal of Physical Chemistry C 113 (45), 19393–19396.

Kojuncu, Ý., Bundalevska, J.M., Ay, Ü., Čundeva, K., Stafilov, T., Akçin, G., 2004. Atomic absorption spectrometry determination of Cd, Cu, Fe, Ni, Pb, Zn, and TI traces in seawater following flotation separation. Separation Science and Technology 39 (11), 2751–2765.

Kudr, J., Nguyen, H.V., Gumulec, J., Nejdl, L., Blazkova, I., Ruttkay-Nedecky, B., et al., 2015. Simultaneous automatic electrochemical detection of zinc, cadmium, copper and lead ions in environmental samples using a thin-film mercury electrode and an artificial neural network. Sensors 15 (1), 592–610.

Kükrer, S., Mutlu, E., 2019. Assessment of surface water quality using water quality index and multivariate statistical analyses in Saraydüzü Dam Lake, Turkey. Environmental Monitoring and Assessment 191 (2), 1–16.

Kumar, R.N., Solanki, R., Kumar, J.N., 2013. Seasonal variation in heavy metal contamination in water and sediments of river Sabarmati and Kharicut canal at Ahmedabad, Gujarat. Environmental Monitoring and Assessment 185 (1), 359–368.

Lakowicz, J.R. (Ed.), 1994. Topics in Fluorescence Spectroscopy: Volume 4: Probe Design and Chemical Sensing, 4. Springer Science & Business Media.

Lee, Y., Oh, S.W., Han, S.H., 2012. Laser-induced breakdown spectroscopy (LIBS) of heavy metal ions at the sub-parts per million level in water. Applied Spectroscopy 66 (12), 1385–1396.

Lerchi, M., Reitter, E., Simon, W., Pretsch, E., Chowdhury, D.A., Kamata, S., 1994. Bulk optodes based on neutral dithiocarbamate ionophores with high selectivity and sensitivity for silver and mercury cations. Analytical Chemistry 66 (10), 1713–1717.

Li, F., Yang, L., Chen, M., Qian, Y., Tang, B., 2013. A novel and versatile sensing platform based on HRP-mimicking DNAzyme-catalyzed template-guided deposition of polyaniline. Biosensors and Bioelectronics 41, 903–906.

Li, X., Yang, D.Z., Yuan, H., Liang, J.P., Xu, T., Zhao, Z.L., et al., 2019a. Detection of trace heavy metals using atmospheric pressure glow discharge by optical emission spectra. High Voltage 4 (3), 228–233.

Li, Y., Zhou, Q., Ren, B., Luo, J., Yuan, J., Ding, X., et al., 2019b. Trends and health risks of dissolved heavy metal pollution in global river and lake water from 1970 to 2017. Reviews of Environmental Contamination and Toxicology 251, 1–24.

Lin, W.C., Li, Z., Burns, M.A., 2017. A drinking water sensor for lead and other heavy metals. Analytical Chemistry 89 (17), 8748–8756.

Lin, L., Li, C., Yang, W., Zhao, L., Liu, M., Li, Q., et al., 2020. Spatial variations and periodic changes in heavy metals in surface water and sediments of the Three Gorges Reservoir, China. Chemosphere 240, 124837.

Locatelli, C., Melucci, D., 2013. Voltammetric method for ultra-trace determination of total mercury and toxic metals in vegetables. Comparison with spectroscopy. Central European Journal of Chemistry 11 (5), 790–800.

Logar, M., Horvat, M., Akagi, H., Pihlar, B., 2002. Simultaneous determination of inorganic mercury and methylmercury compounds in natural waters. Analytical and Bioanalytical Chemistry 374 (6), 1015–1021.

Losev, V.N., Buyko, O.V., Trofimchuk, A.K., Zuy, O.N., 2015. Silica sequentially modified with polyhexamethylene guanidine and Arsenazo I for preconcentration and ICP–OES determination of metals in natural waters. Microchemical Journal 123, 84–89.

Malik, L.A., Bashir, A., Qureashi, A., Pandith, A.H., 2019. Detection and removal of heavy metal ions: a review. Environmental Chemistry Letters 17 (4), 1495–1521.

Marguí, E., Kregsamer, P., Hidalgo, M., Tapias, J., Queralt, I., Streli, C., 2010. Analytical approaches for Hg determination in wastewater samples by means of total reflection X-ray fluorescence spectrometry. Talanta 82 (2), 821–827.

Medina-Sánchez, M., Cadevall, M., Ros, J., Merkoçi, A., 2015. Eco-friendly electrochemical lab-on-paper for heavy metal detection. Analytical and Bioanalytical Chemistry 407 (28), 8445–8449.

Minami, T., Atsumi, K., Ueda, J., 2003. Determination of cobalt and nickel by graphite-furnace atomic absorption spectrometry after coprecipitation with scandium hydroxide. Analytical Sciences 19 (2), 313–315.

Mishra, R., Kumar, A., Singh, E., Kumar, S., Tripathi, V.K., Jha, S.K., et al., 2022. Current status of available techniques for removal of heavy metal contamination in the river ecosystem. Ecological Significance of River Ecosystems. Elsevier, pp. 217–234.

Mohammadi, H., Amine, A., Cosnier, S., Mousty, C., 2005. Mercury–enzyme inhibition assays with an amperometric sucrose biosensor based on a trienzymatic-clay matrix. Analytica Chimica Acta 543 (1–2), 143–149.

Moraes, P.M., Santos, F.A., Cavecci, B., Padilha, C.C., Vieira, J.C., Roldan, P.S., et al., 2013. GFAAS determination of mercury in muscle samples of fish from Amazon, Brazil. Food Chemistry 141 (3), 2614–2617.

Muniz-Naveiro, O., Domínguez-González, R., Bermejo-Barrera, A., Bermejo-Barrera, P., Cocho, J.A., Fraga, J.M., 2007. Selenium speciation in cow milk obtained after supplementation with different selenium forms to the cow feed using liquid chromatography coupled with hydride generation-atomic fluorescence spectrometry. Talanta 71 (4), 1587–1593.

Nantaphol, S., Channon, R.B., Kondo, T., Siangproh, W., Chailapakul, O., Henry, C.S., 2017. Boron doped diamond paste electrodes for microfluidic paper-based analytical devices. Analytical Chemistry 89 (7), 4100–4107.

Nguyen, H.L., Cao, H.H., Nguyen, D.T., Nguyen, V.A., 2017. Sodium dodecyl sulfate doped polyaniline for enhancing the electrochemical sensitivity of mercury ions. Electroanalysis 29 (2), 595–601.

Odonchimeg, S., Oyun, J., Javkhlantugs, N., 2016. Determination of plantinum in rocks by graphite furnace atomic absorption spectrometry after separation on sorbent. International Research Journal of Engineering and Technology 3, 753–757.

Oehme, I., Wolfbeis, O.S., 1997. Optical sensors for determination of heavy metal ions. Microchimica Acta 126 (3–4), 177–192.

Omwene, P.I., Öncel, M.S., Çelen, M., Kobya, M., 2018. Heavy metal pollution and spatial distribution in surface sediments of Mustafakemalpaşa stream located in the world's largest borate basin (Turkey). Chemosphere 208, 782–792.

Palmer, P.T., Jacobs, R., Baker, P.E., Ferguson, K., Webber, S., 2009. Use of field-portable XRF analyzers for rapid screening of toxic elements in FDA-regulated products. Journal of Agricultural and Food Chemistry 57 (7), 2605–2613.

Pan, D., Wang, Y., Chen, Z., Lou, T., Qin, W., 2009. Nanomaterial/ionophore-based electrode for anodic stripping voltammetric determination of lead: an electrochemical sensing platform toward heavy metals. Analytical Chemistry 81 (12), 5088–5094.

Poornima, V., Alexandar, V., Iswariya, S., Perumal, P.T., Uma, T.S., 2016. Gold nanoparticle-based nanosystems for the colorimetric detection of Hg^{2+} ion contamination in the environment. RSC Advances 6 (52), 46711–46722.

Radu, A., Diamond, D., 2007. Ion-selective electrodes in trace level analysis of heavy metals: potentiometry for the XXI century. Comprehensive Analytical Chemistry 49, 25–52.

Radu, T., Diamond, D., 2009. Comparison of soil pollution concentrations determined using AAS and portable XRF techniques. Journal of Hazardous Materials 171 (1–3), 1168–1171.

Radulescu, C., Dulama, I.D., Stihi, C., Ionita, I., Chilian, A., Necula, C., et al., 2014. Determination of heavy metal levels in water and therapeutic mud by atomic absorption spectrometry. Romanian Journal of Physics 59 (9–10), 1057–1066.

Randall, D.W., Hayes, R.T., Wong, P.A., 2013. A simple laser induced breakdown spectroscopy (LIBS) system for use at multiple levels in the undergraduate chemistry curriculum. Journal of Chemical Education 90 (4), 456–462.

Research, P.A., 1987. Application note AC-1 subject: basics of electrochemical impedance.

Rothwell, J.J., Dise, N.B., Taylor, K.G., Allott, T.E.H., Scholefield, P., Davies, H., et al., 2010. A spatial and seasonal assessment of river water chemistry across North West England. Science of the Total Environment 408 (4), 841–855.

Sahoo, P.K., Panigrahy, B., Sahoo, S., Satpati, A.K., Li, D., Bahadur, D., 2013. In situ synthesis and properties of reduced graphene oxide/Bi nanocomposites: as an electroactive material for analysis of heavy metals. Biosensors and Bioelectronics 43, 293–296.

Schaeffer, R., Soeroes, C., Ipolyi, I., Fodor, P., Thomaidis, N.S., 2005. Determination of arsenic species in seafood samples from the Aegean Sea by liquid chromatography–(photo-oxidation)–hydride generation–atomic fluorescence spectrometry. Analytica Chimica Acta 547 (1), 109–118.

Serrano, N., Díaz-Cruz, J.M., Ariño, C., Esteban, M., 2003. Comparison of constant-current stripping chronopotentiometry and anodic stripping voltammetry in metal speciation studies using mercury drop and film electrodes. Journal of Electroanalytical Chemistry 560 (2), 105–116.

Shen, L.L., Zhang, G.R., Li, W., Biesalski, M., Etzold, B.J., 2017. Modifier-free microfluidic electrochemical sensor for heavy-metal detection. ACS Omega 2 (8), 4593–4603.

Sikdar, S., Kundu, M., 2018. A review on detection and abatement of heavy metals. Chembioeng Reviews 5 (1), 18–29.

Sitko, R., Janik, P., Zawisza, B., Talik, E., Margui, E., Queralt, I., 2015. Green approach for ultratrace determination of divalent metal ions and arsenic species using total-reflection X-ray fluorescence spectrometry and mercapto-modified graphene oxide nanosheets as a novel adsorbent. Analytical Chemistry 87 (6), 3535–3542.

Soldatkin, O.O., Kucherenko, I.S., Pyeshkova, V.M., Kukla, A.L., Jaffrezic-Renault, N., El'Skaya, A.V., et al., 2012. Novel conductometric biosensor based on three-enzyme system for selective determination of heavy metal ions. Bioelectrochemistry (Amsterdam, Netherlands) 83, 25–30.

Szłyk, E., Szydłowska-Czerniak, A., 2004. Determination of cadmium, lead, and copper in margarines and butters by galvanostatic stripping chronopotentiometry. Journal of Agricultural and Food Chemistry 52 (13), 4064–4071.

Taha, K., 2017. Heavy elements analyses in the soil using X-ray fluorescence and inductively coupled plasma-atomic emission spectroscopy. International Journal on Advanced Science, Engineering and Information Technology 5, 118–120.

Tang, S., Tong, P., Li, H., Tang, J., Zhang, L., 2013. Ultrasensitive electrochemical detection of Pb^{2+} based on rolling circle amplification and quantum dotstagging. Biosensors and Bioelectronics 42, 608–611.

Tareen, A.K., Sultan, I.N., Parakulsuksatid, P., Shafi, M., Khan, A., Khan, M.W., et al., 2014. Detection of heavy metals (Pb, Sb, Al, As) through atomic absorption spectroscopy from drinking water of District Pishin, Balochistan, Pakistan. International Journal of Current Microbiology and Applied Sciences 3 (1), 299–308.

Tautkus, S., Steponeniene, L., Kazlauskas, R., 2004. Determination of iron in natural and mineral waters by flame atomic absorption spectrometry. Journal of the Serbian Chemical Society 69 (5), 393–402.

Thotiyl, M.O., Basit, H., Sánchez, J.A., Goyer, C., Coche-Guerente, L., Dumy, P., et al., 2012. Multilayer assemblies of polyelectrolyte–gold nanoparticles for the electrocatalytic oxidation and detection of arsenic (III). Journal of Colloid and Interface Science 383 (1), 130–139.

Town, R.M., van Leeuwen, H.P., 2002. Effects of adsorption in stripping chronopotentiometric metal speciation analysis. Journal of Electroanalytical Chemistry 523 (1–2), 1–15.

Tsade, H.K., 2016. Atomic absorption spectroscopic determination of heavy metal concentrations in Kulufo River, Arbaminch, Gamo Gofa, Ethiopia. International Journal of Environmental Analytical Chemistry 3 (177), 2.

Ugulu, I., 2015. Determination of heavy metal accumulation in plant samples by spectrometric techniques in Turkey. Applied Spectroscopy Reviews 50 (2), 113–151.

Vareda, J.P., Valente, A.J., Durães, L., 2019. Assessment of heavy metal pollution from anthropogenic activities and remediation strategies: a review. Journal of Environmental Management 246, 101–118.

Wang, Z., Yuan, T.B., Hou, Z.Y., Zhou, W.D., Lu, J.D., Ding, H.B., et al., 2014. Laser-induced breakdown spectroscopy in China. Frontiers of Physics 9 (4), 419–438.

Wang, H., Wu, Z., Chen, B., He, M., Hu, B., 2015a. Chip-based array magnetic solid phase microextraction on-line coupled with inductively coupled plasma mass spectrometry for the determination of trace heavy metals in cells. Analyst 140 (16), 5619–5626.

Wang, N., Kanhere, E., Triantafyllou, M.S., Miao, J.M., 2015b. Shark-inspired MEMS chemical sensor with epithelium-like micropillar electrode array for lead detection. In 2015 Transducers-2015 18th International Conference on Solid-State Sensors, Actuators and Microsystems (TRANSDUCERS) (pp. 1464–1467). IEEE.

Wolfbeis, O.S., 2008. Fiber-optic chemical sensors and biosensors. Analytical Chemistry 80 (12), 4269–4283.

Xuan, F., Luo, X., Hsing, I.M., 2013. Conformation-dependent exonuclease III activity mediated by metal ions reshuffling on thymine-rich DNA duplexes for an ultrasensitive electrochemical method for Hg^{2+} detection. Analytical Chemistry 85 (9), 4586–4593.

Yu, C., Yin, X., Li, H., Yang, Z., 2020. A hybrid water-quality-index and grey water footprint assessment approach for comprehensively evaluating water resources utilization considering multiple pollutants. Journal of Cleaner Production 248, 119225.

Yuan-Zhen, P.E.N.G., Huang, Y.M., Dong-Xing, Y.U.A.N., Yan, L.I., Zhen-Bin, G.O.N.G., 2012. Rapid analysis of heavy metals in coastal seawater using preconcentration with precipitation/co-precipitation on membrane and detection with X-ray fluorescence. Chinese Journal of Analytical Chemistry 40 (6), 877–882.

Zang, Y., Bolger, P.M., 2014. Toxic metals: cadmium, pp. 346–348.

Zhang, Y., Liu, Y., Ji, X., Banks, C.E., Zhang, W., 2011. Sea cucumber-like hydroxyapatite: cation exchange membrane-assisted synthesis and its application in ultra-sensitive heavy metal detection. Chemical Communications 47 (14), 4126–4128.

Zhou, F.S., Wei, Q.H., 2008. Scaling laws for nanoFET sensors. Nanotechnology 19 (1)015504.

Zou, G., Ju, H., 2004. Electrogenerated chemiluminescence from a CdSe nanocrystal film and its sensing application in aqueous solution. Analytical Chemistry 76 (23), 6871–6876.

CHAPTER

2

Various indices to find out pollution and toxicity impact of metals

Lal Chand Malav[1], Amrita Daripa[2], Sushil Kumar Kharia[3], Sandeep Kumar[4], Brijesh Yadav[1], B.H. Sunil[2] and Sudipta Chattaraj[2]

[1]ICAR-National Bureau of Soil Survey and Land Use Planning, Regional Center, Udaipur, Rajasthan, India [2]ICAR-National Bureau of Soil Survey and Land Use Planning, Nagpur, Maharashtra, India [3]College of Agriculture, SKRAU, Bikaner, Rajasthan, India [4]Division of Environment Science, ICAR-IARI, New Delhi, India

2.1 Introduction

Metals are generally defined as "elements, which conduct electricity, have a metallic luster, are malleable and ductile, form cations, and have basic oxides". Heavy metals are generally added to the soil environment from natural sources such as parent rocks and volcanic eruptions processes. Rapid industrial development and agricultural operations such as pesticides and phosphate fertilizers, on the other hand, are primary anthropogenic causes. Heavy metals are recognized environmental contaminants because of their toxic effects, endurance in the soil, and upsurge in organisms (Ali et al., 2019; El Zokm et al., 2020). Adding more heavy metals by anthropogenic activities and natural phenomena creates significant challenges in modern human society to protect the environment. Heavy metals can persist for a more extended period in the soil. They can enter food chains and contaminate them. Consequently, it causes health issues due to toxicity and chronic effects, a real threat for animals and humans (Madhav et al., 2020). The content of toxic metals above critical level reduces microbial population of soils and degrades the soil quality and health. Higher amounts of hazardous metals in soil organisms significantly harm the soil quality and lead to population decreases. In fish species, heavy metals are significant neurotoxins (Rani et al., 2022). Our capacity to recognize typical contaminant reaction patterns in the soil environment is critical for positive soil health assessments. As a result, the degree of human change in the soil ecosystem has been determined to be significant.

Metals in Water
DOI: https://doi.org/10.1016/B978-0-323-95919-3.00013-6

To investigate the effect of toxicity of heavy metal on the soil environment, we need to evaluate different ecological risk indices. Various pollution indices are useful instruments for assessing the overall degree of soil pollution. Data on toxic heavy metal bioaccumulation in biotas, such as ocean's organisms and cereals, can be utilized to determine healthiness risks in the inhabitants. El Zokm et al. (2020) used various single and integrated ecological risk indices to assess heavy metals' role in the Jeddah Coast's marine pollution in the Red Sea. They reported that the Ecological contamination and Contamination severity index is handy in assessing pollution in sediments by heavy metals.

2.2 Various techniques for toxicity estimation of the metals

Heavy metal is the most vital and crucial genre in our daily lives. Heavy metal concentrations in water, soil, air, and food that above the WHO recommended limit cause a variety of disorders in humans. As a result, the estimation and assessment of each metal's effect on human bodies is very necessary. To evaluate the heavy metal toxicity, various classical analytical techniques are used such as titrimetric and, more recently, advanced instrumental techniques such as UV-Vis spectroscopy, atomic absorption spectrometry (AAS), and inductively coupled plasma mass spectrometry (ICP-MS) (Soylak et al., 2002).

2.3 Titrimetric method

The titration approach is generally used to determine the total mercury concentration in water. The detection limit of this approach is stated to be as low as 0.05 mM. Similarity, titration method is used for chromium and nickel determination.

2.4 UV-Vis spectrophotometry

Over the last few years, UV-Vis spectrophotometer has been widely used for the detection of heavy metals because it is a readily available and inexpensive (Andruch et al., 2012). UV-Vis spectroscopy is based on the principle that specific wavelengths of UV-Vis light can absorb by a specific pollutant molecules. This absorbed light cause electron migration from the ground state to an excited state which reduce the amount of light transmitted. According to the Lambert–Beer law, the absorption and concentration have a strong correlation which is used to detect water pollutants.

2.5 Atomic absorption spectrometry

The determination of many heavy metals is done using atomic absorption spectroscopy. Flame emission is used to determine sodium and potassium levels (Boettner and Grunder, 1968). Atomic vapors formed when the sample's atoms breathed in the flame of the AAS. Some elements remained at ground state, absorbing the radiation produced by a lamp customized for each metal. The concentration of the element is proportional to the light absorbed.

2.6 Inductively coupled plasma—atomic emission spectrometry

It's the emission spectrophotometric technique, in which excited atoms emit radiation when they revert to a normal state. Each element emits a signal in a distinctive wavelength of radiation that indicates the properties of that element. The intensity of energy radiated at a given wavelength increases with the concentration of an element. The inductively coupled plasma—atomic emission spectrometry (ICP-AES) can measure the quantity of an atom in a sample corresponding to the content of the atom present in the soil to its standard content (Amjad et al., 2020).

2.7 Estimation of heavy metal in plant samples

Heavy metal analysis in plant samples involved two primary steps: (1) acid digestion in a di-acid mixture of materials (Aqua Regia, HNO_3: HCl = 1:3), it converts plant tissue's organic chemical integrity into inorganic and molecular form; (2) metals in an acid digested sample were estimated using ICP-OES. [Plasma-Optical Emission Spectrometer with Inductively Coupled Plasma (Feist et al., 2008)].

2.8 Estimation of heavy metals in soil

After collection of representative samples, soil samples are air-dried and crushed with the help of mortar and pestle to run through 2 mm sieve. There are two different methods can be used for estimation of metals in the samples.

1. Total metal estimation
   Total heavy metal concentration in the soil samples can be determined by Aqua Regia extraction method (HCl-HNO_3 in 3:1 ratio) using ICP-OES (Vercoutere et al., 1995).
2. Available metal content

The DTPA-$CaCl_2$-TEA method can be used to extract accessible metals from soil, and ICP-OES can be used to estimate them (Kashem and Singh, 1999).

Detection of heavy metal ions and organic contaminants is still difficult due to low levels in samples and the great complexity of sample matrices, despite significant improvements in current instrumental analysis. Therefore, for the assessment of heavy metals, many computational tools have been applied.

2.9 Role of pollution indices in evaluation of metal toxicity

There are lots of heavy metal pollution assessment index given by different researchers. These indexes are generally classified into two categories-individual and complex index. The individual index is a collection of methods for assessing environmental pollution with particular heavy metals on a unitary basis. Knowing the geochemical background or data calculated from other studies may be required to understand level of metals in soil. The complex indices group enables the thorough specification of heavy metal pollution levels. Each complex index was calculated using the total content of all considered toxic metals in the study area and (in some instances) particular resulting index values. Pollution indices are the key to a realistic appraisal of heavy metals in the soil environment. Muller (1969) and Hakanson (1980) established the first pollution indices/index to assess the potential of heavy metals to pollute the environment. The contamination evaluation indices of toxic metals help identify leading cause of toxic metals accumulation in the soil environment (Kowalska et al., 2018). It may be due to natural processes or anthropogenic activities (Caeiro et al., 2005). Furthermore, pollution indices are critical for determining soil quality and guaranteeing long-term viability, particularly in agro-ecosystems (Ripin et al., 2014).

2.10 Pollution indices

2.10.1 Various indices to assess toxic metal contamination in the soil environment

2.10.1.1 Geoaccumulation index

The geometrical mean of all metal concentrations in the veggies was calculated to produce this index (I_{geo}), which was established by Muller (1969). In the late 1960s, this strategy became increasingly popular.

$$I_{geo} = \log_2(C_n/1.5B_n)$$

The concentration of each heavy metal identified in the mine dirt is denoted by C_n (mg/kg), and B_n: soil's geochemical background number for toxic metals (mg/kg). Because of possible differences in the baseline data, the constant 1.5 is employed (Loska et al., 2004).

I_{geo} values classified in 7 classes. Category 0 ($I_{geo} = 0$): uncontaminated; Category 1 (I_{geo} value, 0–1): uncontaminated level to medium polluted level; Category 2 (1–2): medium polluted; Category 3 (2–3) medium polluted to high polluted; Category 4 (3–4): high polluted level; Category 5 (4–5): high to very high polluted; Category 6 (>5): very high polluted.

2.10.1.2 Enrichment factor

To estimate pollutants, Covelli and Fontolan (1997), favoured enrichment factor (EF) over I_{geo}. It's a new method for normalizing toxic metal composition in comparison to a reference metal. Any contaminant's concentration in the environment is measured by the EF. It's commonly used to determine the amount of heavy metal in topsoil as a result of anthropogenic effect. In two phases, the normalized EF value is determined by dividing the sample's heavy metal and heavy metal content, which is considered as reference value ratio, by the exact ratio of unpolluted concentration (Rubio et al., 2000; Ahamad et al., 2021).

$$EF = \frac{(C_i/C_{fe})_{(sample)}}{(C_i/C_{fe})_{(background)}}$$

Where C_i denotes metal concentration and C_{fe} denotes iron concentration (reference metal). To determine EF, Fe or Al is frequently used as a reference (normalizer) metal. Iron's geochemical composition is more akin to several important heavy metals for the environment in both oxic and anoxic environments, making its use more practical (Rubio et al., 2000). Its natural concentration is likewise rather consistent. When EF is less than 2, it appears to be crustal, and when it is greater than 2, it implies anthropogenic causes (Liaghati et al., 2004).

2.11 Contamination factor

This determines the level of pollution in the soil at a certain location by individual element (Halim et al., 2014). It's the ratio of each metals detected concentration in the sample to the provincial background content weight, expressed as:

$$CF_i = \frac{C_{metal\ (sample)}}{C_{metal\ (background)}}$$

The degree of pollution by different metals in the soils surrounding the copper mining area, as measured by contamination factor (CF), indicated that Cu had the highest contamination.

2.12 Pollution load index

In 1980, Tomlinson et al. presented a pollution load index (PLI). PLI determines the level of contamination load considering all metals at various locations. It's a straightforward geometric mean of the CF. (CF) and can be expressed as:

$$PLI = (CF_1 \times CF_2 \times CF_3 \times \ldots \times CF_n)^{1/n}$$

Tomlinson et al. (1980) recommended that PLI 1 indicates that background amounts of contaminants are present, while PLI > 1 indicates that soil quality is deteriorating.

2.13 Ecological risk factor

Hakanson (1980) calculated the ecological risk (Er) for each pollutant based on its unique toxicity response. This is calculated as:

$$ER_i = T_r \times CF_i$$

Where T_r is a hazardous metal's toxic response, and CF is a toxic metal's pollutant (CF). Metal's designated toxic response (Ti = Zn = 1; Cr = 2; Ni = Cu = Pb = 5; and As = 10) following Hakanson (1980).

2.14 Potential ecological risk index

The risk index (RI) is a modified version of Ecological Risk Factor that can thoroughly assess the level of environmental degradation sensitivity induced by hazardous metals (Sahoo et al., 2016). This index determined by following equation (Hakanson, 1980):

$$RI = \sum_{i=1}^{n=7} ER_i$$

ER_i is the ecological risk factor for particular heavy metal. Jiao et al. (2015) categorized RI values as <150 (small); between 150 and 300 (medium); between 300 and 600 (large); and >600 (very large).

2.15 Biogeochemical index

In the literature, there is no standard index for assessing the level of toxic metal content in the organic layer of soils in silvopasture ecosystem. Biogeochemical index (BGI) is developed to understand this. The computations require understanding the toxic metal concentration in O layer and directly beneath surface soil (Mazurek et al., 2017). BGI is calculated by:

$$BGI = CnO/CnA$$

where CnO represents heavy metal content in the O horizon and CnA represents heavy metal content in the A horizon. BGI is useful for determining the O horizon's ability to sorb contaminants. As a result, values greater than 1.0 indicate that the O layers of soil have a greater potential to absorb heavy metals.

2.16 Nemerow pollution index (PINemerow)

The NPI is a tool for determining the total extent of soil contamination and incorporates all heavy metals tested (Gong et al., 2008). It is determined using the formula below for both the O and A horizons:

$$PINemerow = \sqrt{\left(1/n \sum_{i-1}^{n} PI\right)}$$

PI represents computed PI weights, PI (SPI) max values represent highest PI value of all toxic metals, and n represents total numeral of hazardous heavy metals.

2.17 Various indices to assess toxic metal contamination in the aquatic environment

2.17.1 Contamination index

The degree of contamination is determined by the number of parameters that exceed the upper permitted limit or advice values of potentially toxic components, as well as the concentration exceeding these limit values (Kowalska et al., 2018). To estimate the degree of contamination index (Cd), the total of the water contaminant factor of the constituent components above the upper allowed levels is computed separately for each water sample evaluated. The Cd is a summation of various quality parameters' combined influence. that are considered hazardous to home water in general. The Cd was calculated using all identified values in this investigation, it is computed using the following:

$$\mathrm{Cd} = \sum_{i=1}^{n} \mathrm{Cf}_i$$

Cf_i = represent the contaminant factor for the i-th component and is calculated from the equation

$$\mathrm{Cf}_i = \frac{\mathrm{CA}_i}{\mathrm{CN}_i} - 1$$

Where CA_i is the i-th component's analytical value, and CN_i is the i-th component's maximum permitted concentration (N denotes the normative value). The Cd levels in groundwater can be classified into three pollution levels: small (1), medium (1–3), and large ($>$3).

2.18 Heavy metal evaluation index

The heavy metal evaluation index (HEI) technique, like the heavy metal pollution index (HPI), determines the overall quality of the water in terms of heavy metals (WHO, 2011). The following equation is used to compute HEI.

$$\mathrm{HPI} = \sum_{i=1}^{n} \mathrm{Hc}/\mathrm{Hmac}$$

Hc is the it h parameter's monitored value, and Hmac is the ith parameter's minimal acceptable concentration.

2.19 Pollution load index

It is an ecological risk indexandfor each contaminant the PLI is calculated using the formula:

$$\mathrm{PLI} = \text{anti } \log_{10}\left(1 - \frac{\mathrm{C} - \mathrm{B}}{\mathrm{T} - \mathrm{B}}\right)$$

B: uncontaminated standard number.

Minimum concentrations linked to estuarine system deterioration or changes in quality are referred to as T the threshold. Wilson and Jeffrey (1987) define B and T as distinguishable contaminants and C as pollutant content. The pollution load index calculation takes into consideration all n pollutants for each location:

$$\mathrm{PLI} = (\mathrm{PLI}_1, \mathrm{PLI}_2, \ldots, \mathrm{PLI}_n)^{1/n}$$

PLI values varies from 0 (highly polluted) to 10 (unpolluted)

This index allows you to compare different estuary systems. Simple to put into practice. We were able to achieve decent results by combining these indices with additional assembly techniques such as arithmetic mean and lowest subindices.

2.20 Metal pollution index

It is also known as Cd

$$\mathrm{MPI} = (\mathrm{M}_1, \mathrm{M}_2, \ldots, \mathrm{M}_n)^{1/n}$$

M_n is the metal n concentration in mg per kg of dry weight.

However, no benchmarks or criteria are used to assess pollution quantities. There is no differentiation between places that are uncontaminated and those that are heavily contaminated. When compared to other aggregation methods, the geometric average offers advantages, as Ott (1978) points out, because it reveals concentration differences.

2.21 New index of geoaccumulation

It is also known as background enrichment index

$$\mathrm{NIgeo} = \log_2 = \frac{C_n}{1.5 \times B_n}$$

A list of geographic backgrounds for various grain sizes was compiled. B_n is the content of the metal n in nonpolluted particles (soil separates);

C_n is the metal's concentration.
Unpolluted sediments New index of geoaccumulation (NIgeo) <1;
Very small polluted sediments $1 < \mathrm{NIgeo} < 2$;
Small polluted sediments $2 < \mathrm{NIgeo} < 3$;
Medium polluted sediments $3 < \mathrm{NIgeo} < 4$;
Large polluted sediments $4 < \mathrm{NIgeo} < 5$;
Very large polluted sediments $\mathrm{NIgeo} > 5$

The actual interpretation of these indices for rivers was given by Muller (1969), however this updated version has been applied to estuaries. The sediment must be classified by grain size. The ability to use a varied backdrop level depending ongrain size of sediments is a huge plus. Cr, Cu, Zn, and Pb were the only metals for which C_n was developed. It does not combine all pollutants into a single number.

2.22 Heavy metal pollution index

Taking into consideration the heavy metal concentration in the groundwater, HPI utilized to establish the suitability of drinking purposes. The methods of Mohan et al. (1996) were used to calculate it. HPI is calculated using the equation below.

$$\mathrm{HPI} = \frac{\sum_{i=1}^{n} W_i Q_i}{\sum_{i=1}^{n} W_i}$$

Where W_i is the heavy metal's unit weightage and Q_i is the heavy metal's subindex.

$$Q_i = \sum_{i=1}^{n} \frac{|M_i - I_i|}{(S_i - I_i)}$$

According to WHO, M_i signifies the highest permissible amount of the ith heavy metal identified during that season, and S_i denotes the maximum desired value of that particular (ith) heavy metal (Madhav et al., 2021). The modulus is used to calculate the quantitative midpoint recommended range and the heavy metal concentration of groundwater. The equation below is used to calculate the subindex Q_i. $W_i = \frac{1}{S_i}$

2.23 Entropy-weight based HM contamination index

The entropy-weight based HM contamination index (EHCI) uses the concept of information entropy to deliver water quality (Singha et al., 2019). Entropy weights (W_i) for heavy metals are calculated using the Shannon entropy knowledge-based approach, and subindices (Q_i) are then combined with the importances. The five categories of EHCI values are very good (50), good (50–100), medium (100–150), bad (150–200), and unsuited (>200). EHCI has a crucial value of 200 when it comes to drinking water (Singha et al., 2019; Dash et al., 2019).

$$\mathrm{EHCI} = \sum_{i=1}^{n} W_i Q_i Q_i = \frac{O_i}{S_i} \times 100$$

$$y_{ij} \frac{\begin{matrix} xij - \min(xij) \\ i = 1, 2, 3, \ldots, m \end{matrix}}{\begin{matrix} \max xij - \min(xij) \\ i = 1, 2, 3, \ldots, m \qquad\qquad i = 1, 2, 3, \ldots, m \end{matrix}}$$

$$P_{ij} = \frac{y_{ij}}{\sum_{i=1}^{m} y_{ij}}$$

$$E_j = -\frac{1}{\ln m} \sum_{i=1}^{m} \{p_{ij/n} p_{ij})$$

$$W_i = (1 - E_j) / \sum_{i=1}^{m} (1 - E_j)$$

x_{ij} is the jth HM concentration measured in the ith groundwater sample.
The chance of finding the jth HM in the ith groundwater sample is P_{ij}.
The entropy of the jth HM is E_j.

2.24 Heavy metal index

The heavy metal index (HMI) is a multidimensional technique that represents an area's water quality using principal component analysis (PCA) (Dash et al., 2019). PCA calculates factor loadings by taking into account factors with eigenvalues larger than 1. The associated heavy metal weight (p_i), which is then used to generate the HMI index, is calculated using PCA based on relative eigenvalues and factor loading. The five categories of HMI readings are very good (50), good (50–100), medium (100–200), poor (200–300), and very poor (>300).

$$\text{HMI} = \sum_{i=1}^{n} \left(p_i \times \frac{O_i}{I_i} \right) \times 100$$

2.25 Principal component analysis–based index

According to the authors, the principal component analysis–based index (PMI) is a multidimensional metal categorization approach based on PCA that does not require any allowed constraints in the computing process (Giri and Singh, 2019). The nonstandardized PCA metal index (NSPMI) at each sampling point is calculated using the factor's first extracted factor score (FS_i), the variance explained by the important index (V_i), and the total variance explained by the important index (V_i). The NSPMIs of all the derived factors are summed together, with both favorable and unfavorable values possible. As a result, a standardized PMI is calculated to make data interpretation more straightforward. The third quantile value, on the other hand, determines the essential value of PMI indices, which is not fixed.

$$\text{NSPMI} = \sum_{i=1}^{n} \frac{V_i}{V_t} \times \text{FS}_i$$

$$\text{PMI}_{\text{location}_n} = \frac{\text{NSPMI}_{\text{location}_n} - \text{NSPMI}_{\text{min}}}{\text{NSPMI}_{\text{max}} - \text{NSPMI}_{\text{min}}}$$

The NSPMI (nonstandardized PCA metal index) is a nonstandardized PCA metal index.
The variance explained by the ith factor is V_i.
The total variance explained is denoted by V_t.
The ith factor score is FS_i.
The PMI is a standardised metal index based on PCA.

The total number of groundwater sampling locations is given by n.

2.26 Modified heavy metal pollution index

For n number of heavy metals as water quality indicators, the m-HPI is calculated as follows:

$$\mathrm{m-HMPI} = \sum_{i=1}^{n} (\mathrm{m-HMPI})^{i}$$

where m-HMPI$_i$ is the modified HPI, which corresponds to the i th heavy metal indicator and can be calculated as follows:

$$\mathrm{m-HMPI} = w_i Q_{si}$$

where w_i is the relative weight of the associated ith heavy metal ion and Q_{si} is the ith heavy metal indicator's subindex.

2.27 Indices for determining the health risk posed by consumption of heavy metal-contaminated vegetables

To measure the health hazards associated with eating heavy metal-contaminated vegetables, utilize the indices below.

2.27.1 Hazard quotient

The proportion of a chemical's potential exposure to the amount where no harmful effects are hoped, or the proportion of estimated dosage to reference quantity, is known as the hazard quotient (HQ). It is largely employed by the USEPA to analyze the healthiness risks of harmful substances. A HQ of less than or equal to one indicates that bad outcomes are unlikely, and hence the risk is low. The statistical probability of danger is not represented by HQs bigger than one. Instead, they simply declare if an exposure concentration is higher than the reference concentration (RfC). HQ utilized to assess the risk of metal-polluted veggies on healthiness. The following equation is used to calculate the HQ (USEPA US Environmental Protection Agency, 1989).

$$\mathrm{HQ} = \mathrm{Exposure\ Concentration\ (ppm)/Reference\ Concentration\ (ppm)}$$

Or

$$\mathrm{HQ} = [W_{\mathrm{plant}}] \times [M_{\mathrm{plant}}]/\mathrm{RfD} \times B$$

Where,

W_{plant} is the dry weight of contaminated plant material consumed (mg/day),
M_{plant} is the metal concentration in vegetables (mg/kg),
B is the body mass of consumer (kg)
RfD is a reference dose of metal

2.27.2 Hazard index

The Hazard Index (HI) is another important HQ metric. It's the total of HQs for compounds that have the same effect on the same organ or organ system. Aggregate doses below an HI of 1.0 determined using target organ-specific HQs, like the HQ, are unlikely to cause adverse noncancer health impacts after a lifetime of exposure. Chromium, Nickel, Copper, Lead, Cadmium, Manganese, and Zinc have Rfd values of 1.5, 0.02, 0.04, 0.001, 0.033, and 0.30 (mg/kg/bw/day), respectively (USEPA IRIS, 2006).

2.27.3 Daily intake of metals

Consumers' daily vegetable intake was determined using data collected through a questionnaire during the study. The following equation was used to determine daily intake of metals (DIM) (Chary et al., 2008);

$$\text{DIM} = C_{\text{metal}} \times C_{\text{factor}} \times D_{\text{food intake}} / B_{\text{average weight}}$$

Where,

C_{metal} = metals conc. in crops (mg/kg)
C_{factor} = transformation factor
$D_{\text{food intake}}$ = daily intake of vegetable (g)
$B_{\text{average body}}$ = average body weight of consumer (kg)

The conversion factor of 0.085 is to convert fresh vegetable weight to dry weight.

2.28 Health risk index

Health risk index (HRI) is required to quantify the value of mortal vulnerability to heavy metals by determining the risk of toxic metals to health by following the course of pollution exposure to the human body. There are numerous heavy metal exposure routes that rely on a polluted media of consumables on the recipients. Plants with a higher level of heavy metals are consumed by the receptor population; it penetrates the living organism and creates medical issues. In this study, vegetables grown in contaminated wastewater were collected and utilized for determine the HRI. The daily metal intake and the oral reference dose are used to calculate the HRI value. A daily metal exposure to the human body that has no detrimental consequences over the course of a lifetime is referred to as an oral reference dose(USEPA IRIS, 2006). If the HRI score is less than one, the exposed population is considered safe. The given equation was used to construct the health risk index.

$$\text{Health risk index} = \text{DCM}/\text{RfD}$$

DCM is the daily consumption of metals, Rfd: oral reference amount

2.29 Daily dietary intake

As heavy metals contaminate vegetable crops, as a result, daily consumption will be assessed to compare with the United States' recommendations. The daily food intake was calculated using the formula below. (Wang et al., 2005);

$$\mathrm{DDI} = A \times B \times C/B_{\mathrm{W}}$$

Where,

A is the amount of metal in a vegetable (mg/kg)
B = the vegetable's dry weight (kg)
C = daily vegetable intake (approximate) (kg/day)
B_{W} stands for "average human body mass" (kg)

2.30 Metal pollution index/pollution load index

The MPI was used to specify the general heavy metal content in samples tested. The geometrical mean of all metal levels in the veggies was used to compute this. The following equation was used to calculate the metal Cd (Ghosh et al., 2013);

$$\mathrm{PLI} = \sqrt[n]{\mathrm{Cf}_1 \quad \times \mathrm{Cf}_2 \quad \times \ldots \times \mathrm{Cf}_n}$$

Where,

Cf_n = Concentration of metal n in sample

2.31 Conclusion

Heavy metal contamination is a threat to living beings. For that purpose, various assessment methods are available like spectroscopy and titration methods, etc. But pollution indices give one useful decision through combining all the characteristics in the soil-water-plant systems. Calculation indices through the help of various variables, helps in appropriate soil condition's interpretation. In this book chapter, several indices of pollution have been reviewed in three broader categories. Furthermore, a comparison of strengths and weaknesses of each index has been tabulated for a better understanding at the time of selection of the suitable indices. Out of all indices geoacculmulation index, EF, HQ, and pollution load index are the most useful and widely accepted. Other than these, complex indices like Nemerow Pollution Index and biogeochemical indices are the most important but less useful now. Based on their utility, advantages, and limitations, a specific set of pollution indicators could be used to evaluate soil contamination status for urban green space (Table 2.1).

TABLE 2.1 Strength and weakness of different pollution indices.

Criteria	Pollution indices	Advantages	Disadvantages
Indices to assess the health risk due to the consumption of heavy metals contaminated vegetables	Hazard Quotient (HQ)	Easy to calculate Less complex indices Widely used Simple quantity method	The key is choice of appropriate reference concentration of heavy metals
	Hazard Index (HI)	Combination of HQ Target organ-specific hazard Precise scale	Just based on HQ values Not so popular
	Daily Intake of Metals (DIM)	Based on survey Widely used	Data availability based on survey
	Health Risk Index (HRI)	Using regular metal consumption and an ingested standard dosage	Accuracy based on value of DIM The key is choice of appropriate reference concentration of heavy metals
	Daily Dietary Intake (DDI)	Allows comparison of the different metals accumulation in the human body Represent the actual risk	Not suitable for overall impact of the metals on human beings
	Metal Pollution Index (MPI)	Combines all heavy metals that have been assessed. Allows the comparison of contamination levels in various soil ecosystems. According to the PI values	It is not necessary for natural processes to vary. Heavy metals' available abilities aren't taken into account. The secret is to choose the right GB.
Various indices to assess toxic metal contamination in the soil environment	Geoaccumulation index	Allows you to compare current and prior contamination. Frequently used Quantity calculation in a straightforward manner Application in GB (Geochemical Background) The 1.5 multiplication factor decreases the potential for lithogenic effects to vary. Scale that is precise	Erroneous outcomes are caused by incorrect GB selection. Natural fluctuations may occur throughout the United Kingdom. The available abilities of heavy metal bands are ignored. There is no consideration for biogeochemical fluctuation.
	Enrichment factor (EF)	A good product for comparing the content of metal ions. Trying to figure out where heavy metal came from.	Anthropogenic influence is estimated. Evaluation of heavy metal's origins Metal variability is reduced. The element measures should be compared to one that is utilized for elements with low occurrence variability. Individual heavy metals are used to assess contamination. Scale that is precise

Contamination factor (CF)	Method that is simple and straightforward Each metal has its own unique design. The disparity between the sample and reference values is what this term refers to. The concentration of each metal is divided to get the result. Scale that is precise	It is not necessary for natural processes to vary. Heavy metals' available abilities aren't taken into account. GB is not included. It is vital to have a preindustrial reference value.
Pollution load index (PLI)	Combines any number of heavy metals that have been examined. Simple to use It's possible to compare pollution levels at different soil sites.	In terms of the United Kingdom, It is not necessary for natural processes to vary. Heavy Metals aren't included because they're not readily available.
Ecological risk factor (Er)	Easy to calculate Precise scale Simple and indirect method	Only for single contaminant
Potential Ecological Risk Index (RI)	Sum of all risk factors for considered metals Based on Er	GB is not included. It is vital to have a preindustrial reference value
Biogeochemical Index (BGI)	Heavy metals' vertical mobility is demonstrated. Precise scale is simple to compute.	It is not necessary for natural processes to vary. Heavy metals' available abilities aren't taken into account.
Nemerow Pollution Index (PINemerow)	Directly reflects contamination in the soil environment. The most polluted components are highlighted. It's possible to use GB, threshold, and baseline values. Frequently used Considers all of the distinct components Scale that is precise Based on PI values	The absence of a specific scale The weight factor is not included. Elements must be ranked.

(*Continued*)

TABLE 2.1 (Continued)

Criteria	Pollution indices	Advantages	Disadvantages
Various indices to assess toxic metal contamination in the aquatic environment	Contamination index (Cd)	Represents combined effects of the several quality parameters Easy to calculate Widely accepted	The absence of a specific scale Factor weighing is not included.
	Heavy metal evaluation index (HEI)	Easy to apply Widely used	Only specific to heavy metals only Does not include weighing Factor
	Metal pollution index (MPI)	Simple Easy to apply Widely accepted	The pollutant concentration is not compared to any baseline
	New index of geoaccumulation (NIgeo)	Easy to calculate Precise scale Simple and direct method	Used only for the assessment of river and estuaries water quality The sediment must be classified by grain size. Does not combine all pollutants into a single number
	Heavy metal pollution index (HPI)	Simple to use (determined using the ratio of topsoil concentration to GB values) GB program that is often utilized Scale that is precise	It is not necessary for natural processes to vary. Heavy metals' available abilities aren't taken into account. The secret is to choose the right GB.

References

Ahamad, A., Raju, N.J., Madhav, S., Gossel, W., Ram, P., Wycisk, P., 2021. Potentially toxic elements in soil and road dust around Sonbhadra industrial region, Uttar Pradesh, India: source apportionment and health risk assessment, Environmental Research, 202. p. 111685.

Ali, H., Khan, E., Ilahi, I., 2019. Environmental chemistry and ecotoxicology of hazardous heavy metals: environmental persistence, toxicity, and bioaccumulation. Journal of Chemistry 2019.

Amjad, M., Hussain, S., Javed, K., Khan, A.R., Shahjahan, M., 2020. The sources, toxicity, determination of heavy metals and their removal techniques from drinking water. World (Oakland, Calif.: 1993) 5 (2), 34–40.

Andruch, V., Kocúrová, L., Balogh, I.S., Škrlíková, J., 2012. Recent advances in coupling single-drop and dispersive liquid–liquid microextraction with UV–vis spectrophotometry and related detection techniques. Microchemical Journal 102, 1–10.

Boettner, E.A., Grunder, F.I., 1968. Water analysis by atomic absorption and flame emission spectroscopy.

Caeiro, S., Costa, M.H., Ramos, T.B., Fernandes, F., Silveira, N., Coimbra, A., et al., 2005. Assessing heavy metal contamination in Sado Estuary sediment: an index analysis approach. Ecological Indicators 5 (2), 151–169.

Chary, N.S., Kamala, C.T., Raj, D.S.S., 2008. Assessing risk of heavy metals from consuming food grown on sewage irrigated soils and food chain transfer. Ecotoxicology and Environmental Safety 69 (3), 513–524.

Covelli, S., Fontolan, G., 1997. Application of a normalization procedure in determining regional geochemical baselines. Environmental Geology 30 (1–2), 34–45.

Dash, S., Borah, S.S., Kalamdhad, A., 2019. A modified indexing approach for assessment of heavy metal contamination in Deepor Beel, India. Ecological Indicators 106, 105444.

El Zokm, G.M., Al-Mur, B.A., Okbah, M.A., 2020. Ecological risk indices for heavy metal pollution assessment in marine sediments of Jeddah Coast in the Red Sea. International Journal of Environmental Analytical Chemistry 1–22.

Feist, B., Mikula, B., Pytlakowska, K., Puzio, B., Buhl, F., 2008. Determination of heavy metals by ICP-OES and F-AAS after preconcentration with 2, 2′-bipyridyl and erythrosine. Journal of Hazardous Materials 152 (3), 1122–1129.

Ghosh, R., Xalxo, R., Ghosh, M., 2013. Estimation of heavy metal in vegetables from different market sites of tribal based Ranchi city through ICP-OES and to assess health risk. Current World Environment 8 (3), 435.

Giri, S., Singh, A.K., 2019. Assessment of metal pollution in groundwater using a novel multivariate metal pollution index in the mining areas of the Singhbhum copper belt. Environmental Earth Sciences 78 (6), 1–11.

Gong, Q., Deng, J., Xiang, Y., Wang, Q., Yang, L., 2008. Calculating pollution indices by heavy metals in ecological geochemistry assessment and a case study in parks of Beijing. Journal of China University of Geosciences 19, 230–241.

Hakanson, L., 1980. An ecological risk index for aquatic pollution control. A sedimentological approach. Water Research 14 (8), 975–1001.

Jiao, X., Teng, Y., Zhan, Y., Wu, J., Lin, X., 2015. Soil heavy metal pollution and risk assessment in Shenyang industrial district, Northeast China. PLoS One 10 (5), e0127736.

Kashem, M.A., Singh, B.R., 1999. Heavy metal contamination of soil and vegetation in the vicinity of industries in Bangladesh. Water, Air, and Soil Pollution 115 (1), 347–361.

Kowalska, J.B., Mazurek, R., Gąsiorek, M., Zaleski, T., 2018. Pollution indices as useful tools for the comprehensive evaluation of the degree of soil contamination–a review. Environmental Geochemistry and Health 40 (6), 2395–2420.

Liaghati, T., Preda, M., Cox, M., 2004. Heavy metal distribution and controlling factors within coastal plain sediments, Bells Creek catchment, southeast Queensland, Australia. Environment International 29 (7), 935–948.

Loska, K., Wiechuła, D., Korus, I., 2004. Metal contamination of farming soils affected by industry. Environment International 30 (2), 159–165.

Madhav, S., Ahamad, A., Singh, A.K., Kushawaha, J., Chauhan, J.S., Sharma, S., et al., 2020. Water pollutants: sources and impact on the environment and human health. Sensors in Water Pollutants Monitoring: Role of Material. Springer, Singapore, pp. 43–62.

Madhav, S., Raju, N.J., Ahamad, A., Singh, A.K., Ram, P., Gossel, W., 2021. Hydrogeochemical assessment of groundwater quality and associated potential human health risk in Bhadohi environs, India. Environmental Earth Sciences 80 (17), 1–14.

Mazurek, R., Kowalska, J., Gąsiorek, M., Zadrożny, P., Józefowska, A., Zaleski, T., et al., 2017. Assessment of heavy metals contamination in surface layers of Roztocze National Park forest soils (SE Poland) by indices of pollution. Chemosphere 168, 839–850.

Mohan, S.V., Nithila, P., Reddy, S.J., 1996. Estimation of heavy metals in drinking water and development of heavy metal pollution index. Journal of Environmental Science & Health Part A 31 (2), 283–289.

Muller, G., 1969. Index of geoaccumulation in sediments of the Rhine River. GeoJournal 2, 108–118.

Ott, W.R., 1978. Environmental Indices—Theory and Practice. Ann Arbor Science, Michigan, USA, 371 pp.

Rani, L., Srivastav, A.L., Kaushal, J., Grewal, A.S., Madhav, S., 2022. Heavy metal contamination in the river ecosystem. Ecological Significance of River Ecosystems. Elsevier, pp. 37–50.

Ripin, S.N.M., Hasan, S., Kamal, M.L., Hashim, N.M., 2014. Analysis and pollution assessment of heavy metal in soil, Perlis. The Malaysian Journal of Analytical Sciences 18 (1), 155–161.

Rubio, B., Nombela, M.A., Vilas, F., 2000. Geochemistry of major and trace elements in sediments of the Ria de Vigo (NW Spain): an assessment of metal pollution. Marine Pollution Bulletin 40 (11), 968–980.

Sahoo, P.K., Equeenuddin, S.M., Powell, M.A., 2016. Trace elements in soils around coal mines: current scenario, impact and available techniques for management. Current Pollution Reports 2 (1), 1–14.

Singha, S., Pasupuleti, S., Durbha, K.S., Singha, S.S., Singh, R., Venkatesh, A.S., 2019. An analytical hierarchy process-based geospatial modeling for delineation of potential anthropogenic contamination zones of groundwater from Arang block of Raipur district, Chhattisgarh, Central India. Environmental Earth Sciences 78 (24), 1–19.

Soylak, M., Aydin, F.A., Saracoglu, S., Elci, L., Dogan, M., 2002. Chemical analysis of drinking water samples from Yozgat, Turkey. Polish Journal of Environmental Studies 11 (2), 151–156.

Tomlinson, D.L., Wilson, J.G., Harris, C.R., Jeffrey, D.W., 1980. Problems in the assessment of heavy-metal levels in estuaries and the formation of a pollution index. Helgoländermeeresuntersuchungen 33 (1–4), 566–575.

USEPA (US Environmental Protection Agency), 1989. Risk Assessment Guidance for Superfund: Human Health Evaluation Manual [Part A]: Interim Final. U.S. Environmental Protection Agency, Washington, DC, USA [EPA/540/1–89/002].

USEPA IRIS, 2006. United States, Environmental Protection Agency, Integrated Risk Information System. http://www.epa.gov/iris/substS.

Vercoutere, K., Fortunati, U., Muntau, H., Griepink, B., Maier, E.A., 1995. The certified reference materials CRM 142 R light sandy soil, CRM 143 R sewage sludge amended soil and CRM 145 R sewage sludge for quality control in monitoring environmental and soil pollution. Fresenius' Journal of Analytical Chemistry 352 (1), 197–202.

Wang, X., Sato, T., Xing, B., Tao, S., 2005. Health risks of heavy metals to the general public in Tianjin, China via consumption of vegetables and fish. Science of the Total Environment 350 (1–3), 28–37.

WHO, 2011. Revision of Guidelines for Drinking Water Quality, fourth ed. World health organization, Geneva, Switzerland.

Wilson, J.G., Jeffrey, D.W., 1987. Europe-wide indices for monitoring estuarine quality. In: Kramer, K.J.M. (Ed.), Biological Indicators of Pollution. Royal Irish Academy, Dublin, Ireland, pp. 225–242.

CHAPTER

3

Heavy metal contamination in water: consequences on human health and environment

Anjali Sharma[1], *Ajmer Singh Grewal*[1], *Devkant Sharma*[2] *and Arun Lal Srivastav*[3]

[1]Guru Gobind Singh College of Pharmacy, Yamunanagar, Haryana, India [2]Ch. Devi Lal College of Pharmacy, Jagadhari, Haryana, India [3]School of Engineering and Technology, Chitkara University, Baddi, Himachal Pradesh, India

3.1 Introduction

Heavy metals include those metals which are toxic in nature in trace amounts (Maitra, 2016). Due to their hazardous nature, they pose a significant harmful effect on the environment including human health and hence they have become ta crucial area of interest to be investigated (Carlos et al., 2016). Owing to their ability to accumulate and toxic nature they are considered as the major source of environmental pollution (Hesse et al., 2018). Different factors like urbanization, agricultural activities along with Industrialization, had led to the increased concentration of heavy metal in comparison to their natural background level (He et al., 2016). Various atmospheric activities led to the mobility of heavy metals. The activities include blowing of winds, run off water which ultimately leads to enhanced concentration of heavy metals in the soil's top layer that ultimately pollutes water and air which possess life threatening effects on the living organism residing in these habitats. Plants and trees which grows on roadside receives large number of toxic gases including heavy metals like nickel are emitted through transport vehicles. Lead is the prominent hazardous metal contaminant which is found in aquatic ecosystem and soils available in nearby industrial region (Liu et al., 2017; Wu et al., 2018).

In the crust of earth heavy metals are conventionally accessible in trace amounts but are employed differently on the daily basis say for example self-cleanable cars and ovens, use

DOI: https://doi.org/10.1016/B978-0-323-95919-3.00015-X

of different plastic containers, smart mobile phones along with antiseptics and solar panel, etc. (Karthik et al., 2016). Even the presence of heavy metal trace is helpful in managing specific process of biological origin including iron and copper as they are being used in electron transport systems, for the synthesis of complex ones cobalt is utilized zinc helps in hydroxylation, regulation of enzyme is done by manganese and vanadium, glucose utilization process utilizes chromium, metabolic growth of animals is done by arsenic. selenium led to hormone production and antioxidant (Alsbou and Al-Khashman, 2018). Different heavy metals like mercury, cadmium, arsenic is widely available in environment (Zhang et al., 2018). Few metals are required in trace amount but if they are present in slightly higher concentration, they become hazardous extensive use and toxic nature of these are potentially life threatening to living beings (Sayel et al., 2014). When some of the metal got bounded via thiol group (−SH) some metal shows better affinity for sulfur inside the human body. When different combines in the presence of enzymes sulfur metal bond controls the metabolic reaction rates (Tepanosyan et al., 2017). The presence of −SH bonds they alter the proper functioning of present enzymes and ultimately deteriorates the human health which on prolongation becomes lethal (Lukina et al., 2016). High concentration of cadmium led to degenerative disorder of borne disease and highly concentrated lead items causes destruction of some parts of the nervous system in living things.

3.2 Anthropogenic sources of heavy metals

During the formation of Earth's crust, these metals are naturally occurred; and due to enhanced use of heavy metals in different aspects they had shown enhanced concentration in both the aquatic as well as the terrestrial ecosystem (Masindi and Muedi, 2018). Anthropogenic activity such as smelting process, mining of metals, foundries are the main cause of heavy metal pollution. leaching of heavy metals from varied sources including excretion, landfills, dumps of waste, livestock and manure of chicken, automobiles oil and roadworks are other sources of heavy metal pollution. Other sources include excessive use of insecticide, pesticide, fertilizers, etc. Natural processes like volcanic activity, corrosion, evaporation of metal from water and soil, weathering, soil erosion also leads to heavy metal pollution.

In the aquatic ecosystem, the heavy metals are entered from natural as well as anthropogenic sources. Entry of these may be through direct discharge of these metals in fresh as well as marine ecosystems or they can have entry through indirect approach like wet deposition and dry deposition along with land running-off process (Biney et al., 1994). Natural sources that led to the occurrence of heavy metals include volcanic activities during volcanic eruption, continental weathering along with fires in the forest. The contribution which are obtained from volcanoes are large and sporadic emissions subsists due to the occurrence activity explosive in nature or even other low sources emitting continuously plus geothermal activity along with degassing of magma (Khallaf et al., 2003). Apart from these anthropogenic factors affecting environment are fossil fuels burning and activities related to petroleum industry activities as shown in (Table 3.1).

Promising source of heavy metal pollution in aquatic ecosystem is caused through different anthropogenic inputs like geochemical structuring of compounds along with mining

TABLE 3.1 Toxic metal in industrial effluents.

Metal	Manufacturing industries	References
Copper	Plating, rayon and electrical	Biney et al. (1994)
Nickel	Electroplating, iron steel	Biney et al. (1994)
Lead	Paints, battery	Biney et al. (1994)
Chromium	Textile, metal plating, tanning, rubber and photography	Al Naggar et al. (2018)
Mercury	Chlor-Alkali, scientific instruments, chemicals	Al Naggar et al. (2018)
Arsenic	Phosphate and fertilizer, metal hardening, paints and textile	Biney et al. (1994)
Cadmium	Phosphate fertilizer, electronics, pigments and paints	Biney et al. (1994)
Zinc	Galvanizing, plating iron and steel	Biney et al. (1994)

activities. As a result of soil erosion and weathering heavy metals enter the environment as heavy metal are prime and natural constituents of rocks (Förstner, 1987). When heavy metals accumulate in a concentration higher than required concentration, they are capable of causing ecological damage. These five metals including As, Cd, Cu, Hg, and Zn have large potential and are capable of entering into the environment in a comparatively high concentration from the storms and discharge of water waste which result from agriculture waste and industrialization. Zinc and copper are used as fertilizers in small amounts while arsenic, cadmium, and copper are incorporated as algaecides and fungicides. As we know, the world is facing the problem of lead pollution which is caused due to combustion of heavy leaded petrol in the automobiles (Fifield and Haines, 2000).

3.2.1 Agricultural activities

As the population is growing and in order to accommodate the growing needs large amount of land is required to produce food with high concentration of pesticides, fertilizers along with soil amendments. For enhanced nutrition to the crop fertilizers are added to the soil and for increasing the bioavailability of the nutrients pH of the soil could be a better option. Soils are amended through manure of animals, sewage sludge along with sediments of rivers and harbors. Dredging process is involved in mobilization of heavy metals from the sediments. As the upper surface, soil, surface and under-ground waters are interconnected systems, and hence metals ions could be transfer from soil to surface and ground water through the infiltration. Various irrigation techniques are involved in triggering heavy metals release (Khallaf et al., 2003).

3.2.2 Mining

Metals obtained from ores have relatively very low ion concentration. Most metals occurring in ore deposits have only low concentration. When extraction from ore is done there occurred production of huge amount of rock waste containing heavy metals traces which were not removed during the process. The rock waste got disposed vin mines leading to spoilage of mine. Acid mine drainage is caused when the pyrite ions got weathered in tailing due to oxidation process in the environment and ultimately leading to the drainage of acid mining. The produced acidic environment is responsible for the mobilization of heavy metal ions from the waste. Various fatal health and environment problems are caused due to entry of these metal during respiration, through contaminated drinking water and contaminated food grown on the land containing contaminated soil with heavy metals. These metals include Hg, Pb, Cu, As, Cd, etc. (Ngo et al., 2011).

3.2.3 Transportation

Another factor responsible for environmental pollution includes transportation of metal ions from indoor as well as outdoor means. The metal ions from the fabric gases, dyes, heating, refrigerants, coverings of floor, burning of volatile compounds of organic nature includes the indoor means. In urban areas paints and leaded pipes are the major source of lead ion contamination. Potable water and dust of the household contain concentration of lead ions. Different heavy metals like Pb, Cd, Ni, Zn are released from catalytic converters and from automobiles exhaust which ultimately causes the contamination of soil and greenery on the roadside. If we see the atmospheric heavy metal contamination the incinerators which are used for the high temperature solid waste management are one of the major factors responsible for heavy metal contamination (Mahboob et al., 2014).

3.3 Ecotoxicology of metals

Heavy metals are regarded as hazardous chemicals available in the environment. Nonessential metals are also highly toxic to plants growth, animals along with humans even at a very low concentration. essential heavy metals at high concentrations are capable of causing adverse effects. For the identification of contaminants available in aquatic environment different procedures are used which incorporates three characteristics features which includes bioaccumulation, toxicity and persistence. Toxic substances capable of being persistent as well as bio accumulative are considered more hazardous (DeForest et al., 2007). Toxicity is defined as the characteristic property of a chemical substance capable of affecting the survival, proper growth, and regular reproduction process of an organism. Certain heavy metal showed carcinogenic effect, mutagenic effect along with, teratogenic effect depending upon the availability of dose along with the exposure time. Heavy metals also effect wildlife sanctuary incorporating human health. Some species of animals are comparatively more sensitive to heavy metals when compared to other ones. Complex process is involved in the induction of effect of heavy metals on the life of animals and human beings. Carbonic anhydrase in tissue of Bivalve Anodonta anatine is affected adversely when exposed to cadmium which in return affects the calcium metabolism and osmoregulation process in the tissues (Ngo et al., 2011). Cd is regarded as one the most hazardous metals which is highly responsible for the declining of aquatic ecosystem due to its high toxic effect, high potential of bioaccumulation and rapid transfer through food chains (Ngo, 2008).

3.4 Bioaccumulation and biomagnification

As we all know that biota of Aquatic life is well exposed to hazardous heavy metals constantly by different passages including water sources, sediments of the soil, and through food (Youssef and Tayel, 2004). Different anthropogenic as well as natural sources are responsible for toxicity of freshwater fishes which ultimately leads to toxic effect in the tissues of the aquatic ecosystem. Now a day's contamination of fish through heavy metals has emerged as a global issue which are capable of threating the life of fishes along with the consumer of fishes (Rahman et al., 2012) It has become major issue of concern to assess bioaccumulation of different traces of heavy metals in aquatic fauna including fishes. Assessing the level of heavy metal in the tissues of the fishes has become quite essential for the management and consumption of marine ecosystem by humans (Yousafzai et al., 2017). Fish possess low amount of cholesterol level and high amounts of unsaturated fatty acids and are an important protein source (Malakootian et al., 2016). It is recommended in balanced diet to consume an edible fish by humans. However, if the fish is contaminated through heavy metals is regarded as a major risk for the health of the human beings along with this high concern regarding intake in sensitive groups like women, toddlers, children are posed for a higher risk. Different factors are responsible for bioaccumulation of metals in freshwater fish which includes characteristics of fishes and the external environmental factors. Factors affecting the fish are age of the fish, size including length and breadth of the fish, physiology of the body fish along with the feeding habits of the fish. On the other hand, external environmental factors are bioavailability and concentration of metals in the column of water, other climatic factors and physicochemical properties of water, are high accumulation of metals in the varied tissues of fish which is generally different depending on the available structures and function of different tissues. Generally, the tissues of liver, gills and kidneys are metabolically active and possess high concentration of heavy metal accumulation when compared to other parts such as muscles and skin. Fish possess metal binding with proteins called metallothioneins. Fish gills are found to be the main target tissue for elimination as well as the bioaccumulation of nickel which is accumulated in the gills of fish (Mansouri et al., 2012). The muscles of fish do not accumulate heavy metals and are comparatively important to be consumed by the human beings (Khaled, 2009). Fish capable of accumulating traces of heavy metal are specific in nature. Many studies showed that upon bioaccumulation of traces of heavy metals in fishes had investigated the concentration of metal in muscles of fish as these are edible tissue and shows relevance with human health (Kumar et al., 2010). Different consequences including ecological, environmental and social are related with the accumulation of traces of heavy metals. As they show their implication with fish consuming wildlife along with the humans (Ali et al., 2017; Ali and Khan, 2018a,b). Heavy metals enter the human body through food chain proved to be fatal for the human health. It is investigated and surveyed that river pollution with heavy metals-contaminants along with wastewater bearing stress in the freshwater fish including Channa punctatus, makes it very weak and comparatively more vulnerable to the fatal diseases (Dwivedi et al., 2015). Pollution through Heavy metal is considered as one the most critical aspect which is capable of declining the population of the fishes of freshwater and other species of aquatic ecosystem (Javed and Usmani, 2015).

Generally, toxicity of metal is caused in mammals due to the chemical reaction of the ions with the structure of cell, different membrane system along with the structures of protein and enzymes present. Specific metal toxicity is usually caused when the target organ accumulates high concentration of heavy metals in in vivo. This mainly depends on the exposure route along with chemical nature of the compound such as its valency, volatility, lipid solubility profile, etc. (Al-Ghanim et al., 2016).

Cadmium is a naturally, occurring heavy metal available with the ores of zinc and lead. It does not show any significant role in the cell biology and hence it is not considered as an essential metal. In the normal atmosphere, cadmium occurs naturally from the fires occurred during volcano in the forest and finally transported through soil particles through the movement of wind. Cadmium entered into the human body through the diet, plain drinking water along with the wind and starts accumulating in body of the human beings. Cadmium start accumulating in the liver, kidney and vascular system (Ali, 2019).

Cadmium compounds are released from industrialization of plastic stabilizer, pigment of different colors, welders, and different types of batteries which are rechargeable in nature (Gil et al., 2011).Contamination due to chromium takes place due to inhalation of contaminated air, spoiled food and contaminated water which ultimately leads to pulmonary problems, irritation along with the ulceration and lesions in the wall of stomach and intestine, it also causes anemia and attacks the male reproductive system and thus led to declining of sperm count (WHO, 2000).

Cobalt is found naturally in the crust of earth in the form of these compounds which includes [CoAsS], [Co_3 $(AsO_4)_2$] and [$CoAs_2$]. This is an essential heavy material required by different mammals used in the preparation of various types of enzymes and vitamins. Cobalt has both genotoxic as well as carcinogenic effects capable of inhibiting the repair of DNA, alteration in the gene expression pattern, induces apoptosis, make changes in the chromosomal structure and leads to the defects in mitotic apparatus. Cobalt causes lung disease, asthmatic effect along with the disorders of central nervous system defects (Dayan and Paine, 2001). Lead is available naturally in the environment and theses are examined through different methods from air routes, water, and soils, etc. Contamination of environment through lead from nature is extremely low, but further leads significant in versatile industries, including paints, plastic, ceramics, insecticides, and petrol, etc. The lead is entered through ingestion, inhalation and through skin into the human body. It is distributed in the body through blood, soft tissues, and through bones and ultimately causes destruction to the kidneys. Lead has carcinogenetic as well as mutagenic effects on human lesions occurred gastrointestinal tract, in the immune system, endocrine glands along with reproductive system (Wendling et al., 2009). Manganese is very nutritious element for the growth of animals, plants, and human beings. This element is required for the proper growth and development and are recognized in different oxidation forms and exist in high concentration in earth's crust which can be further examined through food, soil and water and food. For the production of different chemicals, they are used in metal industry. A variety of diseases are caused due to this heavy metal especially disorder of the nervous system, and "manganism" (Deveci, 2006). It was recognized in 18 centuries that nickel exists in extremely small amounts in different vegetables including spinach. Naturally, the origin sources of nickel found in atmosphere are from dust, drastic eruptions of volcano and through the weathering of soils. aqueous nickel occurs naturally through biological cycles and different compounds form soils. In case of industry, nickel is used as a catalytic converter for automobiles and in electroplating purposes. In the year 2000 it was recommended by WHO that there is no safe concentration of nickel which could be consumed. The person who are working in nickel refineries and mines of nickel they possess higher risk of being contaminated which ultimately leads to the cancer of nose as well as abnormalities in the upper respiratory tract. Other problems associated with elevated nickel concentration includes problem related to heart and dysfunction of kidneys (Santamaria and Sulsky, 2010). It is reported around the world that traces of few heavy metals showed considerable deleterious effects on our safe environment. They are regarded injurious to human health to a great extent. Due to agricultural activities as well as globalization due to industrialization have elevated the level of these metals in air, soil and in water also which ultimately makes trace of heavy metal as the more complicated contaminant of the atmosphere (Hostýnek, 2002; Soares et al., 2003; Ramzan et al., 2011).

3.5 Impact on aquatic flora and fauna and human health

Fishes are rich in omega-3 and protein that human body needs in order to stay healthy. Certain heavy metals are absorbed by the tissues of fish which when transferred to humans can cause certain deadly diseases (Bryan et al., 1985). Good quality of fishes can be produced in a pollution free environment. Fisheries are the backbone of economy and are immensely affected by the heavy metal pollution, some pollutants are directly absorbed by the fishes through consumption like lead, cadmium, etc., whereas there are certain other pollutants like nitrates, oil which are indirectly absorbed through diffusion of cell membranes in the tissues. The presence of heavy metals has twofold consequences on the fisheries all over the world (Stolzenberg and Draeger, 1988; Nriagu, 1989). Firstly, it leads to the low productivity of the fishes and secondly it leads to bioaccumulation of fat-soluble contaminants which causes diseases. Labeorohita, for instance, is a freshwater fish of high value as it is the most commonly consumed all over the world. Thus, it is well suited for studying the model of heavy metal contamination. The structural changes in the tissues of the fish due to contamination of heavy metal can be best studied through histopathological studies (Purves, 1990). These structural changes depend on various factors like size of fish, nature of contaminant, solubility of the contaminant, that is, whether it fat soluble or water soluble, duration of exposure of the contaminant, part of the fish exposed, concentration of the pollutant, nature of the medium like salinity (Smith and Williamson, 1986). Certain other factors like water quality and the carcinogenicity of the pollutant also affects the histopathological studies. The change in the structural configuration of the fish tissues at the microscopic levels leads to the change in the functional activity of the pollutant (Ullrich et al., 2001; Wilken and Hintelmann, 1991). Thus, the main objective of this study is to get a clear glimpse of the extent of the pollutant level and the harmful effects of the heavy metals on the marine environment and their related health risks on humans (Kelly et al., 2012). Since mid-nineties Japan is facing a major health disability called Minamata disease which is caused due to heavy mercury pollution and cadmium pollution in the coastal water of the country. The main problem which has led to such a big problem in the country is the release of mineral mercury effluents through mining related activities in the country (Stumm and Morgan, 2012). The mercury pollution in the country is due to the bioaccumulation and biomagnification of mercury in the fishes which form the major diet for the local populace. Evidently mercury has also reached the people's plate through absorption in crops caused by the irrigation of the crops by effluent rich water (Brewster and Passmore, 1994). In India too, all our water bodies particularly freshwater bodies are under the stress of heavy metal pollution. For instance, the ground water is under stress due to excess burden of tube well irrigation and exposure of arsenic in the low stick water table. Inland fisheries in India are exposed to heavy metals through run offs (Salt et al., 1998). The rivers in our country carries urban sewage which is loaded with lead and other harmful metals. This combined with the pollution of other water bodies like lakes, wetlands, backwaters, lagoons, estuaries, etc., has led to the metal pollution of the inland fisheries. When the ions enter the human body, the chemical activity is propagated leading to toxicity as soon as they got combined with the structural proteins of the cell and the cell membrane. The metal toxicity mainly targets those organs of the human body which accumulates maximum concentration of ions in vivo. The toxicity basically depends on the chemical nature of the metal including its solubility in lipids, valency along with the route and intensity of chemical exposure (Marcus and Kertes, 1969). Apart from general toxicity there are different potential carcinogenic toxicities associated with metals. Metals like Ni, Cr when got exposed to human beings proved to be carcinogenic in nature.

3.6 Utility of wetland plants in heavy metal removal

Wetlands are known as natural purifiers for different environmental benefits they have. They act as buffer zones for metal polluted water and can cause breakdown of complex inorganic compounds into simpler complexes. This activity is termed as bio remediation and leads to eco-friendly removal of harmful metals from the water bodies. Mangrove wetlands, for instance, serve as the ecotone for the purification of heavy metals from the water bodies. These forests serve as natural purifier of contaminated water. Avicenna a variety of mangroves found along the wetlands of eastern ghats serve as the remover of cadmium from the backwaters of coastal lagoons. These plants capture the heavy metals from the contaminated water and use it or stores it for biochemical reactions thereby removing it from cycles of consumption, reconsumption and accumulation. This is how wetlands play an important role in the Phytoremediation of heavy metals in the environmental science. Both natural and manmade wetlands have almost similar roles to play in the treatment of heavy metals (Brix, 1997). Wetlands have high productivity apart from being low-cost and good-value clean up systems. The primary use of wetlands as natural filters for the removal of pollutants that will otherwise get transported in freshwater sources like rivers and lakes is now considered to be low-cost, cleaning option to upgrade the quality of waters. Moreover, wetlands in the last decades have been used to purify the contaminated waters all over the world. Vegetation of wetland plants mainly comprise of common reed (*Phragmites* spp.), cattail (*Typha* spp.), rush (*Juncus* spp.), and bulrush (*Scirpus* spp.) the most common being *Phagmites australis* (Cav.) The foremost biological constituents of the vegetation of wetland are macophytes. These macophytes entrap and assimilate the heavy metal contaminants into the active tissues of themselves and further act as reaction catalysts for purification process by elevating the root system zone with environmental diversity and ultimately leads to the promotion of a numerous biochemical reactions which in return enhances process. Water hyacinth (*Eichhornia crassipes*) is one of another plant which is grown in the artificial constructed wetland s as it has a faster growth time and has greater scope for the assimilation of wide variety of pollutants. Water hyacinths have the ability to carry out phytoremediation by root absorption, concentration in tissues, and metabolic degradation. However, after few years of research in 1970s and 1980s, it was found that such system is difficult to operate and sustain as there is a limit on the quality and quantity of purification process. Further, water hyacinth has a tremendous growth and poses a problem as an invasive species in the functioning of ecosystem due to its exotic invasive nature. It is also known for its rapid decomposition in comparison with other plants. the plant is also responsible for the excessive growth of biomass and waste which is difficult to handle. Apart from the above stated drawbacks it has also been found in the studies that growth of water hyacinth leads to the loss of biodiversity in the region and also leads to competition among the different species which affects the wetlands ecosystem in a negative way. The seasonality of the plant is another issue which affects the phytoremediation capability. also raised an ecological concern about growth seasonality and its resulting phytoremediation seasonality. Water hyacinth-based artificial wetlands are not very useful at full-scale size. Macrophytes can be other potential option as it absorbs pollutants in their superficial tissues and also provides a surface layer and an environment for microorganisms to grow. Moreover, macrophytes are capable of creating superior conditions for the sedimentation process of suspended solids and also helps in prevention of erosion by reduction of the velocity of water in the areas of wetland. The optimum growth of roots got aggregated decomposes the organic matter and hence, prevents the clogging by creating pores and channels for the transportation of water through the vertical-flow system which is loaded intermittently.

Macrophytes helps in transporting about 90% of the available oxygen of the rhizosphere, which enhances the growth of nitrifying bacterial strains and also stimulates decomposition of aerobic organic matter (Scholz and Lee, 2006). However, if compared with other microorganisms, macrophytes play only a secondary role, that is, in wetland degradation of left-over organic matter (Stottmeister et al., 2003). The organic matter gets accumulated in wetlands over the period of time by the annual turnover of macrophytes. Organic matter combines and bind directly with heavy metals and supply source of carbon and energy for the metabolic reactions of microbial metabolism. macrophytes are considered to be an indispensable part for the functioning of wetlands in the long run. A number of comprehensive treatments for the structure and function of natural aquatic ecosystem exist (Wetzel and Likens, 2000).

Phytoremediation along with wetland plants is a useful passive technique used for cleaning environmental pollutants having low to high level of contaminations. This technique is an eco-friendly, cost-effective esthetically pleasing driven through solar system. Now a days Phytoremediation is one of the most promising pathways for removal of heavy metal contaminants from aquatic ecosystems. Plants and their rhizosphere bacteria have different actions on the contaminants which includes Phyto stabilization, phytoextraction, rhizodegradation, phytovolatization, and rhizofiltration, etc. The vegetation which covers the wetland areas, plays a vital role in sequestering huge amount of nutrients (Knight, 1997) along with metals from the existing environment by keep them storing in the shoots and roots of the plant. Wetland plants possess high remediation potential for the macronutrients as they have high biomass production and high growth rate. From the environment Wetland plants takes up high concentration of heavy metals and they are ultimately accumulated in belowground tissues (Weis and Weis, 2004). For the nonhyperaccumulators the Restriction of translocation of shoot translocation is promising strategy for the metal tolerance. Performing this activity, the plants become capable of avoiding the potentially higher ill effects of heavy metal concentrations inside the tissues responsible for the process photosynthesis. It is suggested that accumulated amount of heavy metal is totally accountable for reduced chlorophyll content level and also it adversely affects the ratio of Chlorophyll A to Chlorophyll B (Manios et al., 2003). However, the ability to accumulate large concentration of heavy metals in the plant tissues above the ground represented a center point which represents the appropriateness of plants which are available for phytoextraction of the metal. The number of metals varies which is piled up in the top aerial part of plant which shows variations during the favorable season and ultimately leads to the internal growth advances in the plant, as well as they also to respond to variations that takes place in the extent of heavy metal along with their accessibility in the ground water and also in the soil they reported the versatile effect of plant species on the biogeochemistry of the heavy metal including sulfur species on the mobility of these in sediments of wetland (Hardej and Ozimek, 2002).From the results it was concluded that, in the existence of plants, sediments had increased the concentration of sulfate in the rhizosphere during the season of the growth (0.2–6.20 mmol/L). This finding gave an idea that during senescence when comparison is done between the vegetated as well as the nonvegetated sediments the difference in the level of sulfides was reported. It was suggested by Batty in the year 2003 that *P. australis* is capable of removing away almost 100% of supplied iron contents which exists in the concentration of 1 mg/L. Woolley Colliery reported that iron contents removal efficiency raised from 70% to 95% and when the reported system was placed with wetland plants. In 2002 it was reported by Cheng et al. that approximately 30% of the manganese and available copper were piled up in *Cyperus alternifolius* when examined experimental vertical-flow systems which is a complex procedure to assess the potential of phytoremediation of wetlands (Angier et al., 2002) as the conditions are variable such as plant-species diversity, the growing season, hydrology, soil/sediment types, and the ecological succession process in wetlands (Williams et al., 1999). In case of some ecosystem of some wetlands the biomass of different species biomass which dominate are capable of being remain stable for decades. However, in case of upstream wetlands they are subjected to variability in flow of freshwater, shift in dramatic community can takes place? Additionally, due to water-logging in soil even the Spartina alterniflora marshes inland zones are experiencing. The presence of expanded rhizosphere in the wetland of herbaceous shrub along with species of tree are capable of providing rich culture zone for those microbes which are involved in degradation process (Macek et al., 2000). The tides of marshy sediments of freshwater are reported to depict moderate to high level reduction and are considered that salty and marshy sediments are strong reducing agents (Odum, 1984). An integrated picture is created by through interactions in the community of microorganism microbial and the constitution of the characteristic properties of the of soil and the sediments. The process that determines rates of occurring remediation reactions is called as hydrology.

3.7 Biogeochemical processes in wetlands

The most customary process in wetland, which are used for removing the heavy metals from the effluents of the industrial wastes are,

1. Proper binding with soils as well as sediments along with the particulate matter
2. Heavy metal Precipitation by converting them into salts which are insoluble in nature
3. Removing biomass and through proper harvesting process

In the wetland the major concentration of heavy metal is removed basically through binding processes (Kadlec and Keoleian, 1986). As heavy metal possesses positive charge, these are readily complexed, got adsorbed and finally get bounded with the particles which are suspended, which frequently got settle on the floor of the substrate. Heavy metals tend to precipitate as salts which are insoluble includes carbonates and bicarbonates salts, followed by sulfides and the hydroxides are another method that helps in their long-term removal. These salts formation takes place through the reaction of traces of heavy metals with chemical constituents that are available in water column and are insoluble in nature; which ultimately leads to salts precipitation at the bottom and got fixed within the substrate of the wetland (Sheoran and Sheoran, 2006). During the initial stages when wetlands were established the binding processes are less and uptake by the biota was more dominant. Algae and other microorganisms they take up heavy metals that were available in the dissolved form, whereas macrophytes were capable of taking up them from sediments.

3.8 Conclusion

Heavy metals are required in very small concentrations but in larger concentrations their effect can be life threatening. The industries which release heavy metals and the worker who are working in that particular area should take comparatively extra care and should wear an extra layer of gear in order to protect from inhaling, digesting, as well as contacting the area contaminated with heavy metals. Diagnoses of the disease must be made appropriately as these symptoms resembles with the symptoms of neurological disorder. The foremost thing is to prevent the heavy metals entering the human body. In order to work on this industrial pollution, less exhausted areas should be taken into consideration. Another way through which heavy metals enter the human body includes the food we consume. The present study focuses on the heavy metal contamination, various anthropogenic sources including industrial effluent, mining, agricultural activities which leads to the heavy metal pollution are covered along with their consequences. The effect of the heavy metal pollution on the flora and fauna of the aquatic ecosystem along with its ill effects on mammals are discussed. Special emphasis on the removal of heavy metals through different techniques along with the utility of wetland for the removal of heavy metals in detail have been explained.

References

Al Naggar, Y., Khalil, M.S., Ghorab, M.A., 2018. Environmental pollution by heavy metals in the aquatic ecosystems of Egypt. Open Access Journal of Toxicology 3, 555603.

Al-Ghanim, K.A., Mahboob, S., Seemab, S., Sultana, S., Sultana, T., Al-Misned, F., et al., 2016. Monitoring of trace metals in tissues of *Wallago attu* (lanchi) from the Indus River as an indicator of environmental pollution. Saudi Journal of Biological Sciences 23 (1), 72–78.

Ali, H., Khan, E., 2018a. Assessment of potentially toxic heavy metals and health risk in water, sediments, and different fish species of River Kabul, Pakistan. Human and Ecological Risk Assessment: An International Journal 24 (8), 2101–2118.

Ali, H., Khan, E., 2018b. Bioaccumulation of non-essential hazardous heavy metals and metalloids in freshwater fish. Risk to human health. Environmental Chemistry Letters 16 (3), 903–917.

Ali, H., Ali, W., Ullah, K., Akbar, F., Ahrar, S., Ullah, I., et al., 2017. Bioaccumulation of Cu and Zn in *Schizothorax plagiostomus* and *Mastacembelus armatus* from river swat, river panjkora and river barandu in malakand division, Pakistan. Pakistan Journal of Zoology 49 (5), 1555–1561.

Alsbou, E.M.E., Al-Khashman, O.A., 2018. Heavy metal concentrations in roadside soil and street dust from Petra region, Jordan. Environmental Monitoring and Assessment 190 (1), 48.

Angier, J.T., McCarty, G.W., Rice, C.P., Bialek, K., 2002. Influence of a riparian wetland on nitrate and herbicides exported from an agricultural field. Journal of Agricultural and Food Chemistry 50 (15), 4424–4429.

Biney, C.A.A.T., Amuzu, A.T., Calamari, D., Kaba, N., Mbome, I.L., Naeve, H., et al., 1994. Review of heavy metals in the African aquatic environment. Ecotoxicology and Environmental Safety 28 (2), 134–159.

Brewster, M.D., Passmore, R.J., 1994. Use of electrochemical iron generation for removing heavy metals from contaminated groundwater. Environmental Progress 13 (2), 143–148.

Brix, H., 1997. Do macrophytes play a role in constructed treatment wetlands? Water Science and Technology 35 (5), 11–17.

Bryan, G.W., Langston, W.J., Hummerstone, L.G., Burt, G.R., 1985. A guide to the assessment of heavy metal contamination in estuaries using biological indicators. Occasional Publication of the Marine Biological Association 4.

Carlos, M.H.J., Stefani, P.V.Y., Janette, A.M., Melani, M.S.S., Gabriela, P.O., 2016. Assessing the effects of heavy metals in ACC deaminase and IAA production on plant growth-promoting bacteria. Microbiological Research 188, 53–61.

Dayan, A.D., Paine, A.J., 2001. Mechanisms of chromium toxicity, carcinogenicity and allergenicity: review of the literature from 1985 to 2000. Human & Experimental Toxicology 20 (9), 439–451.

DeForest, D.K., Brix, K.V., Adams, W.J., 2007. Assessing metal bioaccumulation in aquatic environments: the inverse relationship between bioaccumulation factors, trophic transfer factors and exposure concentration. Aquatic Toxicology 84 (2), 236–246.

Deveci, E., 2006. Ultrastructural effects of lead acetate on brain of rats. Toxicology and Industrial Health 22 (10), 419–422.

Dwivedi, A.C., Tiwari, A., Mayank, P., 2015. Seasonal determination of heavy metals in muscle, gill and liver tissues of *Nile tilapia, Oreochromis niloticus* (Linnaeus, 1758) from the tributary of the Ganga River, India. Zoology and Ecology 25 (2), 166–171.

Fifield, F.W., Haines, P.J. (Eds.), 2000. Environmental Analytical Chemistry. Wiley-Blackwell.

Förstner, U., 1987. Metal speciation in solid wastes—factors affecting mobility. Speciation of Metals in Water, Sediment and Soil Systems. Springer, Berlin, Heidelberg, pp. 11–41.

Gil, H.W., Kang, E.J., Lee, K.H., Yang, J.O., Lee, E.Y., Hong, S.Y., 2011. Effect of glutathione on the cadmium chelation of EDTA in a patient with cadmium intoxication. Human & Experimental Toxicology 30 (1), 79–83.

Hardej, M., Ozimek, T., 2002. The effect of sewage sludge flooding on growth and morphometric parameters of *Phragmites australis* (Cav.) Trin. ex Steudel. Ecological Engineering 18 (3), 343–350.

He, Z., Hu, Y., Yin, Z., Hu, Y., Zhong, H., 2016. Microbial diversity of chromium-contaminated soils and characterization of six chromium-removing bacteria. Environmental Management 57 (6), 1319–1328.

Hesse, E., O'Brien, S., Tromas, N., Bayer, F., Luján, A.M., van Veen, E.M., et al., 2018. Ecological selection of siderophore-producing microbial taxa in response to heavy metal contamination. Ecology Letters 21 (1), 117–127.

Hostýnek, J.J., 2002. Nickel-induced hypersensitivity: etiology, immune reactions, prevention and therapy. Archives of Dermatological Research 294 (6), 249–267.

Javed, M., Usmani, N., 2015. Stress response of biomolecules (carbohydrate, protein and lipid profiles) in fish *Channa punctatus* inhabiting river polluted by Thermal Power Plant effluent. Saudi Journal of Biological Sciences 22 (2), 237–242.

Kadlec, R.H., Keoleian, G.A., 1986. Metal ion exchange on peat. Peat and Water 61–93.

Karthik, C., Oves, M., Thangabalu, R., Sharma, R., Santhosh, S.B., Arulselvi, P.I., 2016. *Cellulosimicrobium funkei*-like enhances the growth of *Phaseolus vulgaris* by modulating oxidative damage under Chromium (VI) toxicity. Journal of Advanced Research 7 (6), 839–850.

Kelly, M., Allison, W.J., Garman, A.R., Symon, C.J., 2012. Mining and the Freshwater Environment. Springer Science & Business Media.

Khaled, A., 2009. Trace metals in fish of economic interest from the west of Alexandria, Egypt. Chemistry and Ecology 25 (4), 229–246.

Khallaf, E.A., Galal, M., Authman, M., 2003. The biology of *Oreochromis niloticus* in a polluted canal. Ecotoxicology 12 (5), 405–416.

Knight, R.L., 1997. Wildlife habitat and public use benefits of treatment wetlands. Water Science and Technology 35 (5), 35–43.

Kumar, B., Senthil Kumar, K., Priya, M., Mukhopadhyay, D., Shah, R., 2010. Distribution, partitioning, bioaccumulation of trace elements in water, sediment and fish from sewage fed fish ponds in eastern Kolkata, India. Toxicological & Environ Chemistry 92 (2), 243–260.

Liu, S.H., Zeng, G.M., Niu, Q.Y., Liu, Y., Zhou, L., Jiang, L.H., et al., 2017. Bioremediation mechanisms of combined pollution of PAHs and heavy metals by bacteria and fungi: a mini review. Bioresource Technology 224, 25–33.

Lukina, A.O., Boutin, C., Rowland, O., Carpenter, D.J., 2016. Evaluating trivalent chromium toxicity on wild terrestrial and wetland plants. Chemosphere 162, 355–364.

Macek, T., Mackova, M., Káš, J., 2000. Exploitation of plants for the removal of organics in environmental remediation. Biotechnology Advances 18 (1), 23–34.

Mahboob, S., Alkkahem Al-Balwai, H.F., Al-Misned, F., Al-Ghanim, K.A., Ahmad, Z., 2014. A study on the accumulation of nine heavy metals in some important fish species from a natural reservoir in Riyadh, Saudi Arabia. Toxicological & Environmental Chemistry 96 (5), 783–798.

Maitra, S., 2016. Study of genetic determinants of nickel and cadmium resistance in bacteria—a review. International Journal of Current Microbiology and Applied Sciences 5 (11), 459–471.

Malakootian, M., Mortazavi, M.S., Ahmadi, A., 2016. Heavy metals bioaccumulation in fish of southern Iran and risk assessment of fish consumption.

Manios, T., Stentiford, E.I., Millner, P.A., 2003. The effect of heavy metals accumulation on the chlorophyll concentration of *Typha latifolia* plants, growing in a substrate containing sewage sludge compost and watered with metaliferus water. Ecological Engineering 20 (1), 65–74.

Mansouri, B., Ebrahimpour, M., Babaei, H., 2012. Bioaccumulation and elimination of nickel in the organs of black fish (*Capoeta fusca*). Toxicology and Industrial Health 28 (4), 361–368.

Marcus, Y., Kertes, A.S., 1969. Ion exchange and solvent extraction of metal complexes.

Masindi, V., Muedi, K.L., 2018. Environmental contamination by heavy metals. Heavy Metals 10, 115–132.

Ngo, T.T.H., 2008. Effects of cadmium on calcium homeostasis and physiological conditions of the freshwater mussel Anodonta anatina (Doctoral dissertation).

Ngo, H.T.T., Gerstmann, S., Frank, H., 2011. Subchronic effects of environment-like cadmium levels on the bivalve *Anodonta anatina* (Linnaeus 1758): III. Effects on carbonic anhydrase activity in relation to calcium metabolism. Toxicological & Environmental Chemistry 93 (9), 1815–1825.

Nriagu, J.O., 1989. A global assessment of natural sources of atmospheric trace metals. Nature 338 (6210), 47–49.

Odum, W.E., 1984. The ecology of tidal freshwater marshes of the United States east coast: a community profile. The Team.

Purves, D., 1990. Toxic sludge. Nature 346 (6285), 617–618.
Rahman, M.S., Molla, A.H., Saha, N., Rahman, A., 2012. Study on heavy metals levels and its risk assessment in some edible fishes from Bangshi River, Savar, Dhaka, Bangladesh. Food Chemistry 134 (4), 1847–1854.
Ramzan, M., Malik, M.A., Iqbal, Z., Arshad, N., Khan, S.Y., Arshad, M., 2011. Study of hematological indices in tannery workers exposed to chromium in Sheikhupura (Pakistan). Toxicology and Industrial Health 27 (9), 857–864.
Salt, D.E., Smith, R.D., Raskin, I., 1998. Phytoremediation. Annual Review of Plant Biology 49 (1), 643–668.
Santamaria, A.B., Sulsky, S.I., 2010. Risk assessment of an essential element: manganese. Journal of Toxicology and Environmental Health, Part A 73 (2–3), 128–155.
Sayel, H., Joutey, N.T., Ghachtouli, N.E., 2014. Chromium resistant bacteria: impact on plant growth in soil microcosm. Archives of Environmental Protection 40 (2).
Scholz, M., Lee, B., 2006. Critical review of wetland systems to control urban runoff pollution. In Proceedings of the 10th International Conference on Wetland Systems for Water Pollution Control (pp. 1869–1877). International Water Association.
Sheoran, A.S., Sheoran, V., 2006. Heavy metal removal mechanism of acid mine drainage in wetlands: a critical review. Mineral's Engineering 19 (2), 105–116.
Smith, D.G., Williamson, R.B., 1986. Heavy metals in the New Zealand aquatic environment.
Soares, S.R.C., Bueno-Guimaraes, H.M., Ferreira, C.M., Rivero, D.H.R.F., De Castro, I., Garcia, M.L.B., et al., 2003. Urban air pollution induces micronuclei in peripheral erythrocytes of mice in vivo. Environmental Research 92 (3), 191–196.
Stolzenberg, H.C., Draeger, S., 1988. The macrophytic periphyton-algae in the severely metal-contaminated River Oker (eastern Lower Saxony). Braunschweiger Naturkundliche Schriften. Braunschweig 3 (1), 243–254.
Stottmeister, U., Wießner, A., Kuschk, P., Kappelmeyer, U., Kästner, M., Bederski, O., et al., 2003. Effects of plants and microorganisms in constructed wetlands for wastewater treatment. Biotechnology Advances 22 (1–2), 93–117.
Stumm, W., Morgan, J.J., 2012. Aquatic Chemistry: Chemical Equilibria and Rates in Natural Waters, vol. 126. John Wiley & Sons.
Tepanosyan, G., Maghakyan, N., Sahakyan, L., Saghatelyan, A., 2017. Heavy metals pollution levels and children health risk assessment of Yerevan kindergartens soils. Ecotoxicology and Environmental Safety 142, 257–265.
Ullrich, S.M., Tanton, T.W., Abdrashitova, S.A., 2001. Mercury in the aquatic environment: a review of factors affecting methylation. Critical Reviews in Environmental Science and Technology 31 (3), 241–293.
Weis, J.S., Weis, P., 2004. Metal uptake, transport and release by wetland plants: implications for phytoremediation and restoration. Environment International 30 (5), 685–700.
Wendling, L.A., Kirby, J.K., McLaughlin, M.J., 2009. Aging effects on cobalt availability in soils. Environmental Toxicology and Chemistry: An International Journal 28 (8), 1609–1617.
Wetzel, R.G., Likens, G.E., 2000. The heat budget of lakes. Limnological Analyses. Springer, New York, NY, pp. 45–56.
Wilken, R.D., Hintelmann, H., 1991. Mercury and methylmercury in sediments and suspended particles from the river Elbe, North Germany. Water Air & Soil Pollution 56 (1), 427–437.
Williams, J.B., Coleman, H.V., Pearson, J., 1999. Implications of pH effects and succession for phytoremediation in wetlands. In Proceedings of the National Conference on Environmental Remediation Science and Technology. Battelle, Columbia (pp. 243–248).
World Health Organization, 2000. Air Quality Guidelines for Europe. WHO Regional Office for Europe, Copenhagen.
Wu, X., Chen, S., Guo, J., Gao, G., 2018. Effect of air pollution on the stock yield of heavy pollution enterprises in China's key control cities. Journal of Cleaner Production 170, 399–406.
Yousafzai, A.M., Ullah, F., Bari, F., Raziq, S., Riaz, M., Khan, K., et al., 2017. Bioaccumulation of some heavy metals: analysis and comparison of *Cyprinus carpio* and *Labeo rohita* from Sardaryab, Khyber Pakhtunkhwa. BioMed Research International 2017.
Youssef, D.H., Tayel, F.T., 2004. Metal accumulation by three Tilapia spp. from some Egyptian inland waters. Chemistry and Ecology 20 (1), 61–71.
Zhang, J., Li, H., Zhou, Y., Dou, L., Cai, L., Mo, L., et al., 2018. Bioavailability and soil-to-crop transfer of heavy metals in farmland soils: a case study in the Pearl River Delta, South China. Environmental Pollution 235, 710–719.

CHAPTER

4

Metal pollution in the aquatic environment and impact on flora and fauna

Sweta and Bhaskar Singh

Department of Environmental Sciences, Central University of Jharkhand, Ranchi, Jharkhand, India

4.1 Introduction

The aquatic ecosystem plays a pivotal role in several ecosystem services like precipitation, recharging groundwater, hydrological cycle, and atmospheric circulation. These ecosystems have enormous economic importance either as direct or indirect benefits. Agriculture mainly depends on freshwater availability for irrigation. Due to uncontrolled industrial and other anthropogenic activities, the environment including aquatic ecosystems is adversely impacted. Industrial wastewater discharge into the surface water bodies disturbs the entire ecosystem posing adverse impacts on the living organisms present there (Hong et al., 2020). Dumping of wastewater severely changes the natural characteristics of the ecosystem by altering the temperature, pH, electrical conductivity (EC), nutrient composition, and excess availability of heavy metals (HMs) which ultimately caused deleterious effects to the living organisms (Fig. 4.1).

The presence of toxic HMs and metalloids in the municipal and industrial wastewater is well documented (Naggar et al., 2018; Hong et al., 2020; Hussain et al., 2021). HMs are released into the surface water bodies and adversely affect the aquatic flora and fauna (Naggar et al., 2018; Hong et al., 2020; Hussain et al., 2021). The presence of HMs like Cr, Cd, Ni, Cu, Co, Zn and some metalloids like As and Hg in the surface water bodies cause deleterious effects on the aquatic plants like macrophytes, micro- and macroalgae (Marinova et al., 2018; Naorbe and Serrano, 2018) and aquatic fauna like fish, frog, and larvae (Sfakianakis et al., 2015; Elbeshti et al., 2018; Hussain et al., 2021; Yousif et al., 2021). Common symptoms caused due to HMs exposure in the aquatic plants are lipid peroxidation, development of reactive oxygen species (ROS), reduction in photosynthetic rate, DNA

Metals in Water
DOI: https://doi.org/10.1016/B978-0-323-95919-3.00005-7

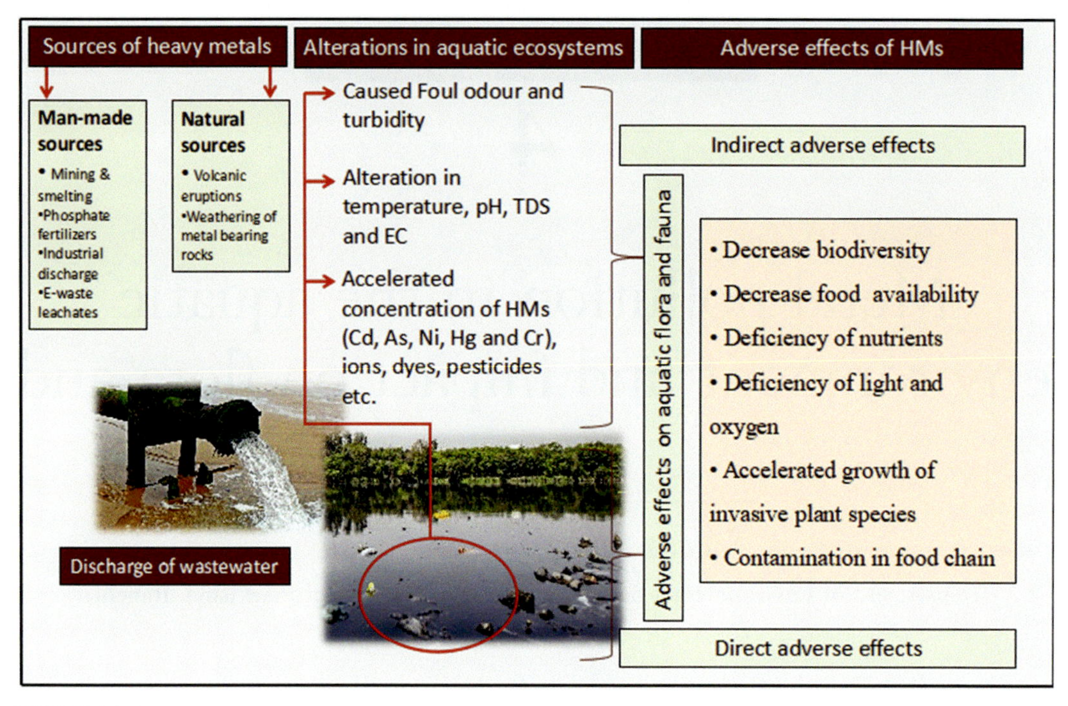

FIGURE 4.1 Sources of heavy metals, alterations in nature's natural characteristics of water bodies and associated adverse effects on aquatic flora and fauna.

damage, reduction in growth, compromised respiratory efficiency, etc. (Singh and Kalamdhad, 2011; Govind and Madhuri, 2014; Ates et al., 2015; Han et al., 2020; Hussain et al., 2021). Protein denaturation, generation of ROS, damage in cell organ, interruption in central nervous system, etc., are some examples of toxic effects of heavy metal and metalloids (HMMs) developed in aquatic animals. The presence of HMs causes nutrient deficiency in the water bodies. HMs are widely reported to cause severe damages to both the plants and animals starting from mild abnormality and culminating up to death of the exposed organism (Lee et al., 2018; Bakulski et al., 2020; Angulo-Bejarano et al., 2021; Balali-Mood et al., 2021; Ma et al., 2021a,b; Naeem et al., 2021).

This chapter explores the potential sources of HMMs to contaminate the water bodies. Further, adverse impacts of HMMs on the aquatic flora and fauna have been discussed thoroughly.

4.2 Heavy metal as a potential contaminant in aquatic environment

Surface water bodies, especially lotic ecosystems, are the major sink for the discharge of municipal and industrial discharge. The presence of variety of contaminants both of inorganic and organic substances in the aquatic ecosystems has been well documented in the published scholarly articles (Odubanjo et al., 2021; Pandiyan et al., 2021; Perumal et al., 2021). The common water contaminants are synthetic pesticides, pharmaceutical wastes, dyes, hydrocarbons, ionic compounds like nitrate, sulfate, chloride, etc., HMs like Ni, Cd, Cu, Pb, metalloids like As, Hg, etc. Due to release of wastewater from industries, municipality, mining and agricultural runoff, majority of water bodies have been identified to bear significant level of HMs (Perumal et al., 2021). Volcanic eruptions and weathering of rocks that contain metals are also considered a source of HMs pollution (He et al., 2005). Industrial discharge, acid mine drainage, agriculture runoff, atmospheric depositions, etc. are the common routes for the HM contamination in the water bodies. Once entered in the water bodies, HMs get accumulated into the phytoplankton that are the first component of a food chain. Being nondegradable, HMs can get bioaccumulated and biomagnified in the higher trophic levels and their elevated levels may cause severe illness and sometimes results in the development of noncurable diseases like cancer in human beings. Accumulation of HMs in the plants can also cause diverse adverse impacts that have been described below in detail.

4.3 Bioaccumulation of HMMs

The toxicity of HMMs primarily depends on their accumulation in the plants and animals. The bioaccumulation of HMMs may depends on the several factors including types and concentration of HMMs, plant and animal species, nutrients availability in the growing medium, etc. The bioaccumulation potential of HMMs in the aquatic plants and animals have been discussed in detail in this section.

4.3.1 Accumulation of HMMs in aquatic plants

Aquatic flora, a pivotal component in an aquatic ecosystem, is also the first organism that gets exposed to the toxicants present in the water bodies. HMMs are considered as potentially toxic substance due to their nondegradable nature. HMMs can easily be absorbed by the plants and also get accumulated in their tissues. Although, accumulation of HMMs in the plants depends on their types and concentration, different plant species have different efficiency for their bioaccumulation. Studies have reported the efficiency of accumulation of HMMs in aquatic plants (Engin et al., 2015; Hassanzadeh et al., 2021; Ma et al., 2021a,b).

Hassanzadeh et al. (2021) reported the bioaccumulation of Pb, Hg, Cr, Cu, Cd, Ag, Zn and Ti in an exotic fern *Azolla filiculoides*. The level of all the studied HMs was found in higher concentration in *A. filiculoides* than in the water except Ag and Cd. Accumulation of Cu, Mn, Cr, Zn, Ni, Pb, Co, and V in four commonly occurring aquatic plant species *Potamogeton malaianus*, *Eichhornia crassipes*, *Hydrilla verticillata* and *Nymphoides peltata* were reported by Bai et al. (2018).

Accumulation of Fe, Cr, Ni, Pb, Zn and Cd in six aquatic plant species was reported in *Potamogeton pectinatus*, *Ceratophyllum demersum*, *Najas marina*, *A. filiculoides*, *Typha domingensis* and *Phragmites australis* and the level of these metals in the plants were higher than the levels given by WHO (Al-Abbawy et al., 2021). Three aquatic plants *Myriophylhum aquaticum*, *Ludwigina palustris*, and *Mentha aquatica* were examined by Kamal et al. (2004)

to check the Fe, Cu, Zn, and Hg accumulation potential and found that all three studied plants have excellent ability to accumulate the metals in their tissues. Although the findings extracted from this study were to assess the phytoremediation potential of these plants, it may also be helpful to assess the risk to contaminate the food chain in aquatic ecosystem. Similarly, Branković et al. (2012) reported significant Fe, Mn, Cu, and Pb accumulation potential in eight aquatic plants (*M. aquatica, Alisma plantago-aquatica, Myriophyllum spicatum, Polygonum amphibium, Bidens tripartitus, Lycopus europaeus, Roripa amphibian* and *Typha angustifolia*) in comparison with the metal's concentration in water and sediment.

A food chain in a sewage-fed wetland situated in West Bengal, India showed organisms to accumulate substantial amount of Pb (Kundu et al., 2016). Yozukmaz et al. (2018) reported accumulation of Cr, Al, Ni, Mn, Zn, Cu, As, Hg, Cd, Pb in algae *Enteromorpha intestinalis*. The levels of Hg and Cd were found higher than the permissible limits.

4.3.2 Accumulation of HMMs in aquatic animals

Accumulation of HMMs in the animal tissues is the first step to cause the adverse health impact. Bioaccumulation of HMMs depends on their type and concentration and animal species and their body organ get exposed. Fishes accumulate various HMs and the gills of the fish are the sites for the direct metal uptake from the contaminated water (Rajeshkumar and Li, 2018; Mehana et al., 2020; Albu et al., 2021). Nwabunike (2016) reported significant bioaccumulation of Cd, Ni, Hg, Cr and Pb in Fin of *Clarias albopunctatus*. Recently, Onita (Mladin) et al. (2021) reported the accumulation of Cu, Cr, Cd, Pb, and Zn in liver, gills and kidney in a significant amount in three fish species *Barbus barbus, Squalius cephalus*, and *Chondrostoma nasus*.

Bioaccumulation of eight HMs, that is, Cd, Cu, Cr, Co, Pb, Hg, Ni and Zn in benthic species (molluscs, polychaetes, and crustaceans), prawn and fishes were observed by Pandiyan et al. (2021) at Point Calimere Wildlife Sanctuary, southern India. The level of Cd and Ni were reported to be higher in molluscs and polychaetes, respectively. The crabs accumulated higher levels of Pb, however, the level of Cu and Zn was found greater in the fishes. The levels of different HMs studied varied substantially among different sources examined. Rakib et al. (2021) reported the presence of Mn, Cr, Co, Fe, Zn, Rb, Cu, Hg, Se, and Pb in edible dried fish species *Trichiurus lepturus, Pampus chinensis, Harpodon neherius, Amblypharyngodon mola, Penaeus affinis, Ilisha megaloptera, Panna microdon, Lates calcarifer, Coilia dussumieri*, and *Gudusia chapra*. Bioaccumulation of Hg, Cd and Pb in the muscles of *Sander lucioperca, Micropterus salmoides, Esox lucius, Scardinius erythrophthalmus*, and *Lepomis macrochirus* collected from the Mechraâ-Hammadi Dam in Morocco were reported by Mahjoub et al. (2021).

Thanomsangad et al. (2020) in Thailand, reported accumulation of As, Cd, Cr, and Pb in three species of frogs *Hoplobatrachus rugulosus, Fejervarya limnocharis*, and *Occidozyga lima* near an e-waste dump site higher than the standards. Accumulation of As, Cr, Cd, Ni, Zn, Pb, Mn, Fe, and Cu in the fish species *Brachydanio albolineata, Rasbora tornieri*, and *Systomus rubripinnis* were found by Intamat et al. (2016) near the gold-mine site of Loei province, Thailand. Nasir et al. (2020) reported the accumulation of $27.56 \pm 9.92\ \mu g/g$ Cu in the liver of *Rana tigrina* sampled from industrial site and 6.56 ± 3.06 μg/g in *R. tigrina* taken from the nonindustrial sites. Rajeshkumar and Li (2018) reported the bioaccumulation of Cu, Pb, Cd and Cr in *Cyprinus carpio* and *Pelteobagrus fluvidrac* sampled from Meiliang Bay, Lake Taihu, China. All the studied HMs accumulated in both the fish species but the levels were lower than the Chinese Food Health Criterion. The total metal accumulation was higher in the gills and liver, and the lowest in the muscle.

4.4 Adverse effects of HMMs

Once HMs entered into the water bodies, it may be absorbed and get accumulated in the aquatic organism, that is, in zooplanktons and phytoplanktons (Fig. 4.2). Zooplanktons also feed on phytoplanktons, therefore, the level of HMs in zooplanktons gets enriched. HMs are passed from one trophic level to another as they are not easily excreted from the body. The bioaccumulation of HMs and concerned adverse effects in both the aquatic flora and fauna has been discussed below.

4.4.1 Adverse effects of HMMs on the aquatic plants

Several studies depict the severe adverse effects of HMMs on aquatic plants including algae as reported in Table 4.1 (Kapkov et al., 2011; Clément and Lamonica, 2018). Malar et al. (2016) reported that Pb induced adverse effects on the growth and biochemical

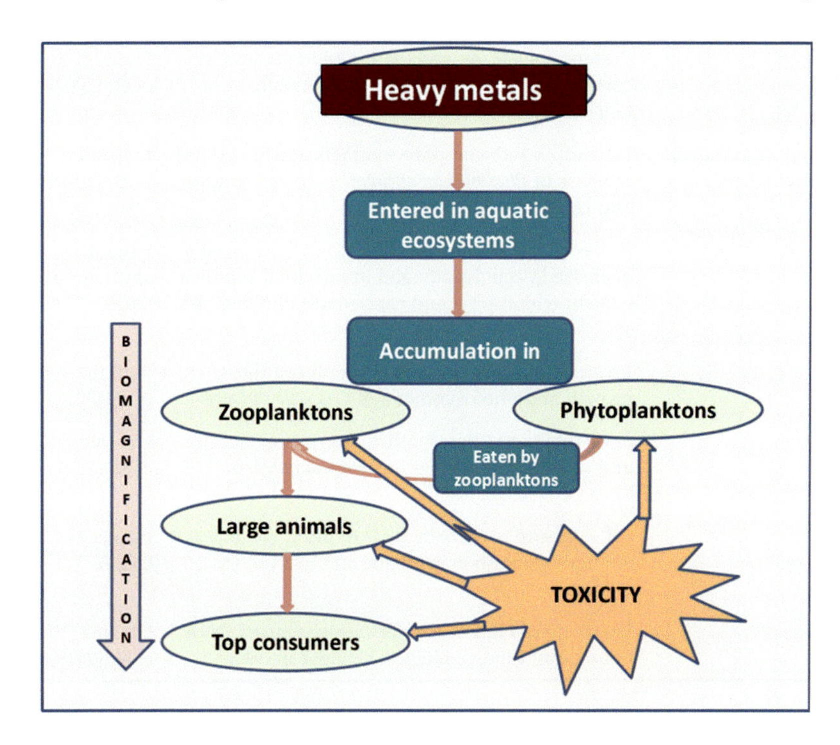

FIGURE 4.2 Adverse effects of HMs to aquatic life.

TABLE 4.1 Effect of HMs on aquatic flora.

Heavy metal(s)	Plant species	Effects	Reference
Cd	*Eichhornia crassipes*	The growth of plant was diminished by 50% at concentration of 1000 mg/L Pb	Melignani et al. (2019)
Pb	*Eichhornia crassipes*	The growth of plant was diminished by 50% at concentration of 1000 mg/L Pb	Malar et al. (2016)
Cd	*Eichhornia crassipes*	Cadmium exposure caused phytotoxicity in *Eichhornia crassipes* at 25.0 μg/mL	Mishra et al. (2007)
Hg, Cd and Pb	*Ipomoea aquatica*	Presence of Hg, Cd and Pb in the water bodies adversely affected the yield of *Ipomoea aquatica*	Göthberg et al. (2004)
Cu, Zn and Cd	*Prorocentrum minimum, Dunaliella tertiolecta, Thalassiosira weissflogii* and *Synechococcus* sp.	Metal caused reduction in chlorophyll content and cell-specific growth	Miao et al. (2005)
Cd and Cu	*Chlorolobion braunii*	Cd and Cu adversely affected the growth, chlorophyll a content, viability and photochemical efficiency of *Chlorolobion braunii*	Echeveste et al. (2017)
Cu and Cd	*Chlorella vulgaris*	Cu and Cd decreased chlorophyll content and growth along of *Chlorella vulgaris*	Qian et al. (2009)
Cd	*Sarcodia suiae*	Decrease in the chlorophyll a content, photosynthetic efficiency, respiratory efficiency with enhanced ambient cadmium levels	Han et al. (2020)
Cu, Zn and Cd	*Chlorella* sp.	Presence of Cu, Zn and Cd individually and in combinations casued growth inhibition to *Chlorella* sp	Franklin et al. (2002)
Cu and Zn	*Phormidium autumnale, Nitzschia palea* and *Uronema confervicolum*	Cu and Zn reduced photosynthetic efficiency and biomass of *Phormidium autumnale, Nitzschia palea* and *Uronema confervicolum*	Loustau et al. (2019)
Cu and Cd	*Nitzschia* sp.	Iincreasing concentration of Cu and Cd caused a decrease in cell density and intracellular pigment content (chlorophyl and carotenoid) of *Nitzschia* sp	Hindarti and Larasati (2019)
Cu, Zn, Cd, Cr and Pb	*Chlorella vulgaris*	Presence of HMs substantially reduced the growth and photosynthesis of *Chlorella vulgaris*	HuiLing et al. (2012)
Hg, CN, Ag, Cd, Cu and Cr	*Chlorella vulgaris*	HMs inhibited photosynthetic oxygen evolution and chlorophyll content in *Chlorella vulgaris*	Hussain et al. (2021)
Cd	*Sarcodia suiae*	Cd exposure to *Sarcodia suiae* reduced content of chlorophyll a, photosynthetic and respiratory efficiency	Han et al. (2020)
Hg^{2+}	*Chlorella vulgaris*	Exposure of Hg to *Chlorella vulgaris* caused DNA damage and morphological changes in cells	Hazlina et al. (2019)

parameters in *E. crassipes* (water hyacinth). They observed a 50% reduction in plant growth in the presence of 1000 mg/L Pb. A reduction in the chlorophyll content with increased level of Pb was also reported. The presence of Cd in water caused phytotoxic effects in *E. crassipes* at concentration of 25.0 μg/mL (Mishra et al., 2007). Melignani et al. (2019) studied the exposure of Cd on *E. crassipes* and reported stress and toxic effects on the species. They found a slight enhanced growth in *E. crassipes* at low Cd concentration but at higher dose it caused severe reduction in the plant growth. Miao et al. (2005) examined toxic effects of Cu, Zn, and Cd on marine phytoplankton, *Prorocentrum minimum, Dunaliella tertiolecta, Thalassiosira weissflogii* and *Synechococcus* sp. Cellular metal bioaccumulation was reported in all the studied species which resulted in reduction in chlorophyll content and cell-specific growth. Further, the bioaccumulation of Cu, Zn, and Cd and toxic effects were metal specific and species dependent.

In a study conducted by Echeveste et al. (2017), the presence of Cd and Cu in the water bodies reduced cell growth of *Chlorolobion braunii*. A significant increase in average cell volume was found when metal concentrations were increased. Although the concentrations of Cu did not affect the photosystem II, Cd caused negative alterations. It was noticed that Cu exhibited lesser toxic effects to *C. braunii* than that of Cd. Exposure of Cu and Cd to *Chlorella vulgaris* decreased chlorophyll content and growth along with enhanced ROS (Qian et al., 2009). ROS was found to be reduced by 9.15 fold times in comparison with the control when exposed to Cd (2.0) Cu (1.5). Han et al. (2020) have reported that the ambient cadmium levels decreased the chlorophyll a content, oxygen evolution rate in terms of photosynthetic efficiency, and oxygen consumption rate in terms of respiratory efficiency of *Sarcodia suiae*. Zeng et al. (2009) assessed the toxicity of Cd and Zn to a freshwater cyanobacteria *Microcystis aeruginosa* and reported that Cd caused significant toxic effects to two studied strains of *M. aeruginosa*. Exposure of Zn was less toxic than that of Cd.

Loustau et al. (2019) studied the physiological response on phototrophic biofilms by exposing Cu and Zn (individually and in combination) to *Phormidium autumnale, Nitzschia palea*, and *Uronema confervicolum*. The higher concentrations of both the metals were found to cause adverse effects on algae in terms of their photosynthetic efficiency and biomass. Exposure of 5 μmol/L of Zn, Cr, Cd, Cu and Pb reduced the growth and photosynthesis (chlorophyll fluorescence) of *C. vulgaris*, and the effects became weaker with enhanced exposure duration (HuiLing et al., 2012). It was also noticed that there was a significant reduction in biomass in the presence of Pb, but the inhibition rate was declined when exposure duration was increased. Further, it was observed that the percentage of inhibitions of dose 5 μmol/L exposure, the toxic order was Cu > Cd > Pb > Cr > Zn after 24 h and the order got changed to Cu > Cr > Cd > Zn > Pb after 96 h which indicates that the effect observed were also dependent on exposure duration along with the metal types and their concentrations. The responses of Pb and Cd exposure to *Chlorella sorokiniana* were studied to assess the photosynthesis, growth, enzyme activities, and respiration by Carfagna et al. (2013). Although the toxic effects were metal-specific, overall photosynthesis got reduced significantly because of a reduction in total chlorophyll content along with a significant decline in the soluble protein levels. Recently, Hussain et al. (2021) reported adverse effects of six HMs, that is, Hg, CN, Ag, Cd, Cu and Cr to a microalgae, *C. vulgaris*. They observed the negative effects of HMs on photosynthetic oxygen evolution and chlorophyll content in *C. vulgaris*. Exposure of Hg^{2+} to *C. vulgaris* caused DNA damages and adverse morphological changes in the cells (Hazlina et al., 2019).

4.4.2 Adverse effects on the aquatic animals

The adverse effect of HMMs on the aquatic animals largely depends on the type and concentration of HMMs and also on the exposed animal species. Once HMMs enter in the food chain, their concentration at higher tropic levels get increased and ultimately top consumers express high degree of toxic effects (El-Moselhy et al., 2014; Boldrocchi et al., 2019). HMs namely Cd, Co, Pb, and Hg were found to cause adverse effects on the oocyte maturation and ovulation in catfish (*Heteropneustes fossilis*) and the effect was found in the order Pb > Hg > Cd > Co (Gautam and Chaube, 2018). Huang et al. (2017) observed necrosis along with congestion and fibrosis congestion in *Pelophylax nigromaculata* in the presence of high concentration of Pb (at concentration of 100 and 1000 μg/L). Further, androgen receptors on the gonad cells were also adversely affected by Pb. Kumar et al. (2011) reported an enhanced production of total cholesterol, serum glucose, glutamic oxaloacetic acid transaminase (GOT), and glutamic pyruvic acid transaminase (GPT) due to the exposure of Cr in *C. carpio* (common carp). They also observed a reduced level of the serum total protein. The presence of HMMs in aquatic ecosystem adversely affects the aquatic animals (Table 4.2) (Tamele and Loureiro, 2020).

In a study conducted by Shaheen and Akhtar (2012), it was found that Cr(VI) caused adverse effects on biochemical and hematological parameters of *C. carpio*. The level of several parameters like mean corpuscular volume, erythrocyte sedimentation rate, white blood cells, aspartate aminotransferase, alkaline phosphatase, alanine aminotransferase, and acid phosphatase were increased significantly indicating anemia in Cr(VI) exposed *C. carpio*.

Exposure of Cd, Cu, Cr, Zn, and Pb to an Indian green frog *Euphlyctis hexadactylus* caused damage to kidney, liver, skin and lung (Jayawardena et al., 2017). Injuries observed in the liver were in the form of elevated biochemical components like aspartate transaminase, γ-glutamyltransferase, alanine transaminase and alkaline phosphatases in frog. The results also confirmed loss of functional and structural integrity and hepatocellular leakage. Ahmad et al. (2006) studied the effects of Cr on *Anguilla anguilla*. They assessed genotoxic effects and oxidative stress in gill and kidney of *A. anguilla*. It was reported that Cr caused decrease of DNA integrity and responses regarding oxidative stress in the gill and kidney of Cr exposed *A. anguilla*.

Ni exposure adversely affected the immune system in *Carassius auratus* (Kubrak et al., 2012). Kubrak et al. (2012) further reported that increased Ni accumulation enhanced renal Fe content that resulted a significant increase in lipid peroxide and protein carbonyl content. The suppression of enzymatic activities like superoxide dismutase by 50%–53%, glutathione peroxidase by 15%–45%, glutathione reductase by 31%–37% and glucose-6-phosphate dehydrogenase by 20%–44%, and these values are an indicative of oxidative stress development in the kidney of Ni exposed *C. auratus*. Mai et al. (2012) assessed the effect of Cu and Cd to *Crassostrea gigas*. They found that the exposure of Cu and Cd

TABLE 4.2 Effect of HMs on aquatic fauna.

Heavy metal(s)	Animal species	Effects	Reference
Cd, Co, Pb and Hg	Catfish (*Heteropneustes fossilis*)	The heavy metals adversely affected oocyte maturation and ovulation. The effect was found in the order of Pb > Hg > Cd > Co	Gautam and Chaube (2018)
Pb	Adult *Pelophylax nigromaculata*	At the level of 100 and 1000 μg/L, Pb caused germ cell necrosis, congestion and fibrosis congestion in *Pelophylax nigromaculata*	Huang et al. (2017)
Cr	*Cyprinus carpio*	Exposure of Cr to *Cyprinus carpio* enhanced the levels of total cholesterol, serum glucose, GOT and GPT significantly, however, the serum total protein was found decreased	Kumar et al. (2011)
Cr	*Cyprinus carpio*	Cr(VI) exposure caused adverse effects on biochemical and hematological parameters	Shaheen and Akhtar (2012)
Cr	*Cyprinus carpio*	Cr (III) reduced the amounts of Hb, RBC, Hct, WBC, MCH, and MCHC, however, albumin and glucose significantly level was increased	Abedi et al. (2013)
Cd, Cu, Cr, Zn and Pb	*Euphlyctis hexadactylus* (Indian green frog)	Metals caused damages to kidney, lung, liver and skin.	Jayawardena et al. (2017)
Cd	*Rana cyanophlyctis* (Indian Skipper Frog)	The lethal concentration of Cd was found at 23.048 mg/L after 96 h exposure	Srivastav et al. (2016)
Cu, Pb, Fe, Cd, Mn and Zn	*Clarias gariepinus* fish	Presence of Cu, Pb, Fe, Cd, Mn and Zn in water bodies increased GOT and GPT activities along with different biochemical parameters of *Clarias gariepinus*	Authman et al. (2013)
Ni	*Carassius auratus*	Caused oxidative stress in kidney	Kubrak et al. (2012)
$ZnSO_4$	*Clarias batrachus*	Caused histopathological alternation in the liver	Joshi (2011)
Cu and Cd	*Crassostrea gigas*	Exposure of both the metals caused embryotoxicity and genotoxicity	Mai et al. (2012)
Cd, Cu, Mn. Zn, Pb, Cr, Sn, and Hg	*Cirrhinus mrigala* and *Catla catla*	Metal specific genotoxicity was observed in the form of DNA damage in *Cirrhinus mrigala* and *Catla catla*. The impacts were more prolonged tin *Cirrhinus mrigala* than *Catla catla*	Hussain et al. (2016)
Pb, Cu, Ni and Cd	*Wallago attu*, *Sperata sarwari* and *Labeo rohita*	The presence of heavy metals caused a DNA fragmentation in liver, gills, kidney, and muscles in the exposed fishes	Sultana et al. (2020)
Pb, Hg and Cd	*Pelophylax ridibundus*, *Xenopus laevis* and *Xenopus tropicalis*	HM mediated stress responses were observed in all three frog species	Kaczor-Kamińska et al. (2020)
Cu, Cr, Cd, Pb and Zn	*Barbus barbus*, *Squalius cephalus* and *Chondrostoma nasus*	Presence of HMs caused histopathological alterations in the studied fish species. The impacts were more prolonged in kidney	Onita (Mladin) et al. (2021)

caused a substantial increase in abnormal D-larvae percentage and DNA strand gets broken.

A study on genotoxic effects of HMs (Cd, Cu, Mn. Zn, Pb, Cr, Sn, and Hg) in two edible fish species *Cirrhinus mrigala* and *Catla catla* were studied by Hussain et al. (2016). They found that HMs caused DNA damage in both the fish species. Sultana et al. (2020) reported significant DNA fragmentation in fish species (*Wallago attu*, *Sperata sarwari* and *Labeo rohita*) exposed to Pb, Cu, Ni and Cd contaminated water. In a recent study done by Kaczor-Kamińska et al. (2020), it was reported that the availability of Pb, Hg, and Cd in the water bodies produced stress responses in three species of frogs, that is, *Pelophylax ridibundus*, *Xenopus laevis* and *Xenopus tropicalis*. They found alterations in gene expression and activities of cystathionine γ-lyase, 3-mercaptopyruvate sulfurtransferase and rhodanese in heart, liver, skeletal muscles, brain, kidney, and testes in frogs. Chen et al. (2000) assessed accumulation of Zn, As, Hg, Cd, and Pb zooplankton and fish sampled from 20 lakes in the north-eastern United States.

4.5 Remediation strategies for heavy metal contamination in aquatic ecosystem

Several physical, chemical, and biological treatment methods have been developed to remove HMs from the contaminated water bodies (Sharma et al., 2021). Ion exchange, membrane filtration, precipitation, coagulation and electrocoagulation, phytoremediation, phycoremediation, absorption, adsorption, ultrafiltration, reverse osmosis, nanofiltration, etc., are some of the commonly used technologies for the removal of HMs (Meunier et al., 2006; Gautam et al., 2014; Peng and Guo, 2020; Sharma et al., 2021). Separation of HMs using physicochemical techniques become challenging because of good solubility of majority of HM salts in solution.

In many studies, bioremediation has been found significantly effective for both the detoxification and removal of HMs from the contaminated water bodies (Gautam et al., 2014; Ayangbenro and Babalola, 2017; Yasar et al., 2018; Sharma et al., 2021). Detoxification method using different microorganisms is suitable for the elements having different ionic forms. For instance, methyl mercury can be converted into a lesser toxic form, that is, Hg(II) and a nontoxic form, that is, Hg(0) and Cr(VI) to Cr(III) which is also a less mobile and less toxic form of chromium (Wu et al., 2010). Rhizofiltration has been also found to be an efficient technique for the extraction of HMs. Several species of algae and macrophytes have been extensively studied and found to have potential removal of HMs from the contaminated water bodies (Table 4.3).

Macrophytes like *E. crassipes*, *Cyperus alopecuroides*, *P. australis*, *Ranunculus sceleratus*, and *T. domingensis*, *Trapa natans*, etc., have been reported as potential plant species that accumulate a variety of HMs in their roots and shoots (Saha et al., 2017; Eid et al., 2019, 2020a,b; Abdelaal et al., 2021; Huynh et al., 2021). These plants are habituated to survive in harsh conditions such as in the presence of high concentration of metals, water having high total dissolved solids. Aquatic macrophytes have some other merits to be used for the removal of toxic metals like easy to cultivate, high biomass, fast growing nature, etc.

Many algal species have been reported to accumulate significant amount of HMs. Algal species such as *Chlorella* sps., *Scenedesmus* sps., *Neochloris minuta*, *Chlamydomonas reinhardtii*,

TABLE 4.3 Heavy metal removal potential of algal and plant species.

Plant species (algae/macrophytes)	HMMs	Remarks	Reference
Eichhornia crassipes	Cd	The removal of Cd was found from 82% to 92%	Prakash et al. (1987)
Lemna minor and *Eichornia crassipes*	As	The removal rate of *Lemna minor* for As was found 140 mg As/ha day and of Water Hyacinth was 600 mg As/ha day	Alvarado et al. (2008)
Eichhornia crassipes	Zn and Cr	Removal of Zn was 95% and of Cr was 84%	Mishra and Tripathi (2008)
Eichornia crassipes, Typha latifolia and *Monochoria hastata*	Cu, Ni, Pb, Fe, Mn, and Zn	*Monochoria hastata* was found to have the maximum bioconcentration factor (BCF) for root as 4.32 and shoot as 2.70 for Mn. For *Typha latifolia*, BCF was maximum for roots (163.5) and respective shoots 86.46 for Fe which was followed by 7.3 and 5.8 for root and shoot in case of the Mn respectively	Hazra et al. (2015)
Eichhornia crassipes	Ag, Ba, Cd, Mo, and Pb	Accumulate Mo, Pb, and Ba with BCF values 24,360, 18,800 and 10,040 respectively	Romanova et al. (2016)
Trapa natans	Na^{+}, K^{+}, Ca^{2+}, Mg^{2+}, Cd, Cr, Cu, Fe, Mn, Pb, Zn	The contents of metals Cd, Cu, Fe, Mn and Zn were translocated in the of *Trapa natans* leaves and of Cr and Pb were accumulated in the roots of *Trapa natans* in significant amount	Kumar and Chopra (2017)
Pistia stratiotes	Fluoride (F)	Maximum removal efficiency was found 5 mg/L F concentration was found to be 38.89% after 8 days of exposure	Karmakar et al. (2018)
Nostoc muscorum and *Trichormus variabilis*	Cd	*Nostoc muscorum* was found more efficient than *Trichormus variabilis* for removing Cd. *Nostoc muscorum* achieved Cd removal efficiency of 93.4% and *Trichormus variabilis* removed up to 89.13%	El-Hameed et al. (2021)

Cyanidioschizon merolae, and *Neochloris alveolaris* are efficient in the removal of HMs (Hanikenne et al., 2005; Gómez-Jacinto et al., 2015; Asiandu and Wahyudi, 2021; Danouche et al., 2021; Giarikos et al., 2021; Spain et al., 2021). Majority of algae have been reported to grow smoothly in the contaminated water bodies because they utilize the ionic compounds as a nutrient. Presence of low concentration of several HMs like As, Pb, Cd, Zn, Fe, Cu, in the growing medium enhance the growth of the algal species viz. *Phormidium ambiguum, Pseudochlorococcum typicum* and *Scenedesmus quadricauda, D. tertiolecta, Nostoc minutum, Chlorella* sp. and *Monoraphidium minutum* (Knauer and Hemond, 2000; El-Sheekh et al., 2003; Nishikawa et al., 2003; Sacan et al., 2007; Karadjova et al., 2008; Shanab et al., 2012; Ferrari et al., 2013). Potential aquatic plants can be used for the removal of HMMs from the contaminated water bodies.

4.6 Conclusion

HM and metalloids are potentially toxic to all forms of life. Their presence deteriorates the physicochemical and biological characteristics of the ecosystem, especially aquatic ecosystems. Scholarly literature has well documented that HMMs like Cd, Cu, Ni, Zn, Hg, As, Cr and others adversely affect the life of aquatic plants and animals. Aquatic plants, including microalgae, macroalgae, macrophytes, etc. are common plant groups that are sensitive towards HMMs. Aquatic plants species like *E. crassipes*, *Ipomoea aquatica*, *P. minimum*, *D. tertiolecta*, *T. weissflogii*, *Synechococcus* sp., *C. braunii*, *C. vulgaris*, *S. suiae*, *P. autumnale*, *N. palea* and *U. confervicolum*, etc., are common plants that express the toxic effects in the presence of HMMs. If the productivity of these plant species is hampered, it might result in the reduction of ecosystem productivity. The accumulated HMs also gets transferred to the animals and get biomagnified which enhance the level of HMs at higher trophic levels. There should be proper monitoring of industrial wastewater before their discharge into the nearby surface water bodies. Although, there are several technologies viz. ion exchange, adsorption, etc. that have been developed for the remediation of contaminated water, rhizofiltration has received attention due to its environment-friendly nature and cost-effectiveness. The application of aquatic macrophytes like *E. crassipes*, *I. aquatic*, *T. natans* and some species of algae viz. *C. braunii*, *C. vulgaris* have been found efficient for the removal of HMs from the contaminated water bodies.

References

Abdelaal, M., Mashaly, I.A., Srour, D.S., et al., 2021. Phytoremediation perspectives of seven aquatic macrophytes for removal of heavy metals from polluted drains in the Nile delta of Egypt. O Biologico 10 (6), 560. Available from: https://doi.org/10.3390/biology10060560.

Abedi, Z., Khalesi, M.K., Kohestan, E.S., 2013. Biochemical and hematological profiles of common carp (*Cyprinus carpio*) under sublethal effects of trivalent chromium. Iranian Journal of Toxicology 7, 782–792.

Ahmad, I., Maria, V.L., Oliveira, M., Pacheco, M., Santos, M.A., 2006. Oxidative stress and genotoxic effects in gill and kidney of *Anguilla anguilla* L. exposed to chromium with or without pre-exposure to β-naphthoflavone. Mutation Research - Genetic Toxicology and Environmental Mutagenesis 608 (1), 16–28. Available from: https://doi.org/10.1016/j.mrgentox.2006.04.020.

Al-Abbawy, D.A., Al-Thahaibawi, B.M.H., Al-Mayaly, I.K., Younis, K.H., 2021. Assessment of some heavy metals in various aquatic plants of Al-Hawizeh Marsh, southern of Iraq. Biodiversitas Journal of Biological Diversity 22 (1). Available from: https://doi.org/10.13057/biodiv/d220141.

Albu, P., Herman, H., Balta, C., et al., 2021. Correlation between heavy metal-induced histopathological changes and trophic interactions between different fish species. Applied Sciences 11 (9), 3760. Available from: https://doi.org/10.3390/app11093760.

Alvarado, S., Guédez, M., Lué-Merú, M.P., et al., 2008. Arsenic removal from waters by bioremediation with the aquatic plants Water Hyacinth (*Eichhornia crassipes*) and Lesser Duckweed (*Lemna minor*). Bioresource Technology 99, 8436–8440. Available from: https://doi.org/10.1016/j.biortech.2008.02.051.

Angulo-Bejarano, P.I., Puente-Rivera, J., Cruz-Ortega, R., 2021. Metal and metalloid toxicity in plants: an overview on molecular aspects. Plants 10 (4), 635. Available from: https://doi.org/10.3390/plants10040635.

Asiandu, A.P., Wahyudi, A., 2021. Phycoremediation: heavy metals green-removal by microalgae and its application in biofuel production. Journal of Environmental Treatment Techniques 9, 647–656. Available from: https://doi.org/10.47277/JETT/9(3)647.

Ates, A., Türkmen, M., Tepe, Y., 2015. Assessment of heavy metals in fourteen marine fish species of four Turkish seas. Indian Journal of Marine Sciences 44 (1), 49–55.

Authman, M.M., Ibrahim, S.A., El-Kasheif, M.A., Gaber, H.S., 2013. Heavy metals pollution and their effects on gills and liver of the Nile catfish *Clarias gariepinus* inhabiting El-Rahawy drain. Egypt Global Veterinary 10 (2), 103–115.

Ayangbenro, A.S., Babalola, O.O., 2017. A new strategy for heavy metal polluted environments: a review of microbial biosorbents. International Journal of Environmental Research and Public Health 14 (1), 94. Available from: https://doi.org/10.3390/ijerph14010094.

Bai, L., Liu, X.L., Hu, J., et al., 2018. Heavy metal accumulation in common aquatic plants in rivers and lakes in the Taihu Basin. International Journal of Environmental Research and Public Health 15, 2857. Available from: https://doi.org/10.3390/ijerph15122857.

Bakulski, K.M., Seo, Y.A., Hickman, R.C., Brandt, D., Vadari, H.S., Hu, H., et al., 2020. Heavy metals exposure and Alzheimer's disease and related dementias. Journal of Alzheimer's Disease: JAD 76 (4), 1215–1242. Available from: https://doi.org/10.3233/JAD-200282.

Balali-Mood, M., Naseri, K., Tahergorabi, Z., Khazdair, M.R., Sadeghi, M., 2021. Toxic mechanisms of five heavy metals: mercury, lead, chromium, cadmium, and arsenic. Frontiers in Pharmacology 12, 643972. Available from: https://doi.org/10.3389/fphar.2021.643972.

Boldrocchi, G., Monticelli, D., Omar, Y.M., Bettinetti, R., 2019. Trace elements and POPs in two commercial shark species from Djibouti: implications for human exposure. The Science of the Total Environment 669, 637–648. Available from: https://doi.org/10.1016/j.scitotenv.2019.03.122.

Branković, S., Pavlović-Muratspahić, D., Topuzović, M., Glišić, R., Milivojević, J., Đekić, V., 2012. Metals concentration and accumulation in several aquatic macrophytes. Biotechnology & Biotechnological Equipment 26 (1), 2731–2736. Available from: https://doi.org/10.5504/BBEQ.2011.0086.

Carfagna, S., Lanza, N., Salbitani, G., et al., 2013. Physiological and morphological responses of Lead or Cadmium exposed *Chlorella sorokiniana* 211-8K (Chlorophyceae). Springer Plus 2, 147. Available from: https://doi.org/10.1186/2193-1801-2-147.

Chen, C.Y., Stemberger, R.S., Klaue, B., Blum, J.D., Pickhardt, P.C., Folt, C.L., 2000. Accumulation of heavy metals in food web components across a gradient of lakes. Limnology and Oceanography 45 (7), 1525–1536. Available from: https://doi.org/10.4319/lo.2000.45.7.1525.

Clément, B., Lamonica, D., 2018. Fate, toxicity and bioconcentration of cadmium on *Pseudokirchneriella subcapitata* and *Lemna minor* in mid-term single tests. Ecotoxicology 27 (2), 132–143. Available from: https://doi.org/10.1007/s10646-017-1879-z.

Danouche, M., Ghachtouli, N.E., Arroussi, H.E., 2021. Phycoremediation mechanisms of heavy metals using living green microalgae: physicochemical and molecular approaches for enhancing selectivity and removal capacity. Heliyon 7, 07609. Available from: https://doi.org/10.1016/j.heliyon.2021.e07609.

Echeveste, P., Silva, J.C., Lombardi, A.T., 2017. Cu and Cd affect distinctly the physiology of a cosmopolitan tropical freshwater phytoplankton. Ecotoxicology and Environmental Safety 143, 228–235. Available from: https://doi.org/10.1016/j.ecoenv.2017.05.030.

Eid, E.M., Shaltout, K.H., Moghanm, F.S., Youssef, M.S.G., El-Mohsnawy, E., Haroun, S.A., 2019. Bioaccumulation and translocation of nine heavy metals by *Eichhornia crassipes* in Nile Delta, Egypt: perspectives for phytoremediation. International Journal of Phytoremediation 21, 821–830. Available from: https://doi.org/10.1080/15226514.2019.1566885.

Eid, E.M., Galal, T.M., Sewelam, N.A., Talha, N.I., Abdallah, S.M., 2020a. Phytoremediation of heavy metals by four aquatic macrophytes and their potential use as contamination indicators: a comparative assessment. Environmental Science and Pollution Research 27, 12138–12151. Available from: https://doi.org/10.1007/s11356-020-07839-9.

Eid, E.M., Galal, T.M., Shaltout, K.H., et al., 2020b. Biomonitoring potential of the native aquatic plant *T. domingensis* by predicting trace metals accumulation in the Egyptian Lake Burullus. Science of the Total Environment 714, 136603. Available from: https://doi.org/10.1016/j.scitotenv.2020.136603.

Elbeshti, R.T., Elderwish, N.M., Abdelali, K.M., Tastan, Y., 2018. Effects of heavy metals on fish. Menba Journal of Fisheries Faculty 4 (1), 36–47.

El-Hameed, M.M.A., Abuarab, M.E., Al-Ansari, N., et al., 2021. Phycoremediation of contaminated water by cadmium (Cd) using two cyanobacterial strains (*Trichormus variabilis* and *Nostoc muscorum*). Environmental Sciences Europe 33, 135. Available from: https://doi.org/10.1186/s12302-021-00573-0.

El-Moselhy, K.M., Othman, A., El-Azem, H.A., El-Metwally, M., 2014. Bioaccumulation of heavy metals in some tissues of fish in the Red Sea, Egypt. Egyptian Journal of Basic and Applied Sciences 1, 97–105. Available from: https://doi.org/10.1016/j.ejbas.2014.06.001.

El-Sheekh, M.M., el-Naggar, A.H., Osman, M.E.H., el-Mazaly, E., 2003. Effect of cobalt on growth, pigments and the photosynthetic electron transport in *Monoraphidium minutum* and *Nitzchia perminuta*. Brazilian Journal of Plant Physiology 15, 159–166. Available from: https://doi.org/10.1590/S1677-04202003000300005.

Engin, M.S., Uyanik, A., Kutbay, H.G., 2015. Accumulation of heavy metals in water, sediments and wetland plants of Kizilirmak Delta (Samsun, Turkey). International Journal of Phytoremediation 17 (1), 66–75. Available from: https://doi.org/10.1080/15226514.2013.828019.

Ferrari, S.G., Silva, P.G., Gonzalez, D.M., Navoni, J.A., Silva, H.J., 2013. Arsenic tolerance of cyanobacterial strains with potential use in biotechnology. Revista Argentina de Microbiologia 45, 174–179. Available from: https://doi.org/10.1016/S0325-7541(13)70021-X.

Franklin, N.M., Stauber, J.L., Lim, R.P., Petocz, P., 2002. Toxicity of metal mixtures to a tropical freshwater alga (*Chlorella* sp): the effect of interactions between copper, cadmium, and zinc on metal cell binding and uptake. Environmental Toxicology and Chemistry / SETAC 21 (11), 2412–2422. Available from: https://doi.org/10.1002/etc.5620211121.

Gautam, G.J., Chaube, R., 2018. Differential effects of heavy metals (cadmium, cobalt, lead and mercury) on oocyte maturation and ovulation of the catfish *Heteropneustes fossilis*: an in vitro study. Turkish Journal of Fisheries and Aquatic Sciences 18, 1205–1214. Available from: https://doi.org/10.4194/1303-2712-v18_10_07.

Gautam, R.K., Sharma, S.K., Mahiya, S., Chattopadhyaya, M.C., 2014. Contamination of heavy metals in aquatic media: transport, toxicity and technologies for remediation. Water: Presence, Removal and Safety 1–24. Available from: https://doi.org/10.1039/9781782620174-00001.

Giarikos, D.G., Brown, J., Razeghifard, R., Vo, D., Castillo, A., Nagabandi, N., 2021. Effects of nitrogen depletion on the biosorption capacities *of Neochloris minuta* and *Neochloris alveolaris* for five heavy metals. Applied Water Science 11, 1–15. Available from: https://doi.org/10.1007/s13201-021-01363-y.

Gómez-Jacinto, V., García-Barrera, T., Gómez-Ariza, J.L., Garbayo-Nores, I., Vílchez-Lobato, C., 2015. Elucidation of the defence mechanism in microalgae *Chlorella sorokiniana* under mercury exposure. Identification of Hg–phytochelatins. Chemico-Biological Interactions 238, 82–90. Available from: https://doi.org/10.1016/j.cbi.2015.06.013.

Göthberg, A., Greger, M., Holm, K., Bengtsson, B.E., 2004. Influence of nutrient levels on uptake and effects of mercury, cadmium, and lead in water spinach. Journal of Environmental Quality 33 (4), 1247–1255. Available from: https://doi.org/10.2134/jeq2004.1247.

Govind, P., Madhuri, S., 2014. Heavy metals causing toxicity in animals and fishes. Research Journal of Animal, Veterinary and Fishery Sciences 2 (2), 17–23.

Han, T.W., Tseng, C.C., Cai, M., Chen, K., Cheng, S.Y., Wang, J., 2020. Effects of cadmium on bioaccumulation, bioabsorption, and photosynthesis in *Sarcodia suiae*. International Journal of Environmental Research and Public Health 17 (4), 1294. Available from: https://doi.org/10.3390/ijerph17041294.

Hanikenne, M., Krämer, U., Demoulin, V., Baurain, D., 2005. A comparative inventory of metal transporters in the green alga *Chlamydomonas reinhardtii* and the red alga *Cyanidioschizon merolae*. Plant Physiology 137 (2), 428–446. Available from: https://doi.org/10.1104/pp.104.054189.

Hassanzadeh, M., Zarkami, R., Sadeghi, R., 2021. Uptake and accumulation of heavy metals by water body and *Azolla filiculoides* in the Anzali wetland. Applied Water Science 11, 91. Available from: https://doi.org/10.1007/s13201-021-01428-y.

Hazlina, A.Z., Devanthiran, L., Fatimah, H., 2019. Morphological changes and DNA damage in *Chlorella vulgaris* (UMT-M1) induced by Hg^{2+}. Malaysian Applied Biology 48, 27–33.

Hazra, M., Avishek, K., Pathak, G., 2015. Phytoremedial potential of *Typha latifolia*, *Eichornia crassipes* and *Monochoria hastata* found in contaminated water bodies across Ranchi city (India). International Journal of Phytoremediation 17 (9), 835–840. Available from: https://doi.org/10.1080/15226514.2014.964847.

He, Z.L., Yang, X.E., Stoffella, P.J., 2005. Trace elements in agroecosystems and impacts on the environment. Journal of Trace Elements in Medicine and Biology: Organ of the Society for Minerals and Trace Elements (GMS) 19 (2–3), 125–140. Available from: https://doi.org/10.1016/j.jtemb.2005.02.010.

Hindarti, D., Larasati, A.W., 2019. Copper (Cu) and Cadmium (Cd) toxicity on growth, chlorophyll-a and carotenoid content of phytoplankton *Nitzschia* sp. IOP Conference Series: Environmental Earth Sciences 236 (1), 012053. Available from: https://doi.org/10.1088/1755-1315/236/1/012053.

Hong, Y.J., Liao, W., Yan, Z.F., Bai, Y.C., Feng, C.L., Xu, Z.X., et al., 2020. Progress in the research of the toxicity effect mechanisms of heavy metals on freshwater organisms and their water quality criteria in China. Journal of Chemistry. Available from: https://doi.org/10.1155/2020/9010348. Article ID 9010348.

Huang, M., Men, Q., Meng, X., Fang, X., Tao, M., 2017. Chronic toxic effect of lead on male testis tissue in adult *Pelophylax nigromaculata*. Nature Environment & Polution Technology 16, 213–218.

HuiLing, O., XiangZhen, K., Wei, H.E., Ning, Q., QiShuang, H.E., Yan, W., et al., 2012. Effects of five heavy metals at sub-lethal concentrations on the growth and photosynthesis of *Chlorella vulgaris*. Chinese Science Bulletin 57, 3363–3370. Available from: https://doi.org/10.1007/s11434-012-5366-x.

Hussain, B., Sultana, T., Sultana, S., Mahboob, S., Al-Ghanim, K.A., Nadeem, S., 2016. Variation in genotoxic susceptibility and biomarker responses in *Cirrhinus mrigala* and *Catla catla* from different ecological niches of the Chenab River. Environmental Science and Pollution Research 23 (14), 14589–14599. Available from: https://doi.org/10.1007/s11356-016-6645-x.

Hussain, F., Eom, H., Toor, U.A., Lee, C.S., Oh, S., 2021. Rapid assessment of heavy metal-induced toxicity in water using micro-algal bioassay based on photosynthetic oxygen evolution. Environmental Engineering Research 26 (6), 2021. Available from: https://doi.org/10.4491/eer.2020.391.

Huynh, A.T., Chen, Y.C., Tran, B.N.T., 2021. A small-scale study on removal of heavy metals from contaminated water using water hyacinth. Processes 9, 1802. Available from: https://doi.org/10.3390/pr9101802.

Intamat, S., Phoonaploy, U., Sriuttha, M., Tengjaroenkul, B., Neeratanaphan, L., 2016. Heavy metal accumulation in aquatic animals around the gold mine area of Loei province, Thailand. Human and Ecological Risk Assessment: An International Journal 22 (6), 1418–1432. Available from: https://doi.org/10.1080/10807039.2016.1187062.

Jayawardena, U.A., Angunawela, P., Wickramasinghe, D.D., Ratnasooriya, W.D., Udagama, P.V., 2017. Heavy metal–induced toxicity in the Indian green frog: biochemical and histopathological alterations. Environmental Toxicology and Chemistry / SETAC 36 (10), 2855–2867. Available from: https://doi.org/10.1002/etc.3848.

Joshi, P.S., 2011. Studies on the effects of zinc sulphate toxicity on the detoxifying organs of fresh water fish *Clarias batrachus* (Linn.). Golden Research Thoughts 1, 1–4.

Kaczor-Kamińska, M., Sura, P., Wróbel, M., 2020. Multidirectional changes in parameters related to sulfur metabolism in frog tissues exposed to heavy metal-related stress. Biomolecules 10 (4), 574. Available from: https://doi.org/10.3390/biom10040574.

Kamal, M., Ghaly, A.E., Mahmoud, N., Cote, R., 2004. Phytoaccumulation of heavy metals by aquatic plants. Environment International 29 (8), 1029–1039. Available from: https://doi.org/10.1016/S0160-4120(03)00091-6.

Kapkov, V.I., Belenikina, O.A., Fedorov, V.D., 2011. Effect of heavy metals on marine phytoplankton. Moscow University Biological Sciences Bulletin 66, 32–36. Available from: https://doi.org/10.3103/S0096392511010056.

Karadjova, I.B., Slaveykova, V.I., Tsalev, D.L., 2008. The biouptake and toxicity of arsenic species on the green microalga *Chlorella salina* in seawater. Aquatic Toxicology (Amsterdam, Netherlands) 87, 264–271. Available from: https://doi.org/10.1016/j.aquatox.2008.02.006.

Karmakar, S., Mukherjee, J., Mukherjee, S., 2018. Biosorption of fluoride by water lettuce (*Pistia stratiotes*) from contaminated water. International Journal of Environmental Science and Technology 15, 801–810. Available from: https://doi.org/10.1007/s13762-017-1439-3.

Knauer, K., Hemond, H., 2000. Accumulation and reduction of arsenate by the freshwater green alga *Chlorella* sp. (Chlorophyta). Journal of Phycology 36, 506–509. Available from: https://doi.org/10.1046/j.1529-8817.2000.99056.x.

Kubrak, O.I., Husak, V.V., Rovenko, B.M., Poigner, H., et al., 2012. Tissue specificity in nickel uptake and induction of oxidative stress in kidney and spleen of goldfish *Carassius auratus*, exposed to waterborne nickel. Aquatic Toxicology (Amsterdam, Netherlands) 118-119, 88–96. Available from: https://doi.org/10.1016/j.aquatox.2012.03.016.

Kumar, V., Chopra, A.K., 2017. Phytoremediation potential of water caltrop (*Trapa natans* L.) using municipal wastewater of the activated sludge process-based municipal wastewater treatment plant. Environmental Technology 39 (1), 12–23. Available from: https://doi.org/10.1080/09593330.2017.1293165.

Kumar, P., Palanivel, S., Sarasu, M.R., 2011. Sublethal effects of chromium on some biochemical profiles of the fresh water teleost, *Cyprinus carpio*. International Journal of Applied Biology and Pharmaceutical Technology 2, 295–300.

Kundu, D., Mondal, S., Dutta, D., Haque, S., Ghosh, A.R., 2016. Accumulation and contamination of lead in different trophic levels of food chain in sewage-fed East Kolkata Wetland, West Bengal, India. International Journal of Environmental Science and Technology 2, 61–68.

Lee, H.J., Park, M.K., Seo, Y.R., 2018. Pathogenic mechanisms of heavy metal induced-Alzheimer's disease. Toxicology and Environmental Health Sciences 10, 1–10. Available from: https://doi.org/10.1007/s13530-018-0340-x.

Loustau, E., Ferriol, J., Koteiche, S., et al., 2019. Physiological responses of three mono-species phototrophic biofilms exposed to copper and zinc. Environmental Science and Pollution Research 26 (34), 35107–35120. Available from: https://doi.org/10.1007/s11356-019-06560-6.

Ma, L., Azad, M.G., Dharmasivam, M., Richardson, V., Quinn, R.J., Feng, Y., et al., 2021a. Parkinson's disease: alterations in iron and redox biology as a key to unlock therapeutic strategies. Redox Biology 41, 101896. Available from: https://doi.org/10.1016/j.redox.2021.101896.

Ma, Y., Wang, G., Wang, Y., Dai, W., Luan, Y., 2021b. Mercury uptake and transport by plants in aquatic environments: a meta-analysis. Applied Sciences 11, 8829. Available from: https://doi.org/10.3390/app11198829.

Mahjoub, M., Fadlaoui, S., El Maadoudi, M., Smiri, Y., 2021. Mercury, lead, and cadmium in the muscles of five fish species from the Mechraâ-Hammadi Dam in Morocco and health risks for their consumers. Journal of Toxicology. Available from: https://doi.org/10.1155/2021/8865869. ID 8865869.

Mai, H., Cachot, J., Brune, J., Geffard, O., Belles, A., Budzinski, H., et al., 2012. Embryotoxic and genotoxic effects of heavy metals and pesticides on early life stages of Pacific oyster (*Crassostrea gigas*). Marine Pollution Bulletin 64 (12), 2663–2670. Available from: https://doi.org/10.1016/j.marpolbul.2012.10.009.

Malar, S., Vikram, S.S., Favas, P.J., Perumal, V., 2016. Lead heavy metal toxicity induced changes on growth and antioxidative enzymes level in water hyacinths [*Eichhornia crassipes* (Mart.)]. Bot Studio 55 (1), 54. Available from: https://doi.org/10.1186/s40529-014-0054-6.

Marinova, G., Ivanova, J., Pilarski, P., Chernev, G., Chaneva, G., 2018. Effect of heavy metals on the green alga scenedesmus incrassatulus. Oxidation Communications 41 (2), 318–328.

Mehana, E.S.E., Khafaga, A.F., Elblehi, S.S., et al., 2020. Biomonitoring of heavy metal pollution using acanthocephalans parasite in ecosystem: an updated overview. Animals 10 (5), 811. Available from: https://doi.org/10.3390/ani10050811.

Melignani, E., Faggi, A.M., de Cabo, L.I., 2019. Growth, accumulation and uptake of *Eichhornia crassipes* exposed to high cadmium concentrations. Environmental Science and Pollution Research 26 (22), 22826–22834. Available from: https://doi.org/10.1007/s11356-019-05461-y.

Meunier, N., Drogui, P., Montané, C., Hausler, R., Mercier, G., Blais, J.F., 2006. Comparison between electrocoagulation and chemical precipitation for metals removal from acidic soil leachate. Journal of Hazardous Materials 137 (1), 581–590. Available from: https://doi.org/10.1016/j.jhazmat.2006.02.050.

Miao, A.J., Wang, W.X., Juneau, P., 2005. Comparison of Cd, Cu, and Zn toxic effects on four marine phytoplankton by pulse-amplitude-modulated fluorometry. Environmental Toxicology and Chemistry 24 (10), 2603–2611. Available from: https://doi.org/10.1897/05-009r.1.

Mishra, K., Tripathi, B.D., 2008. Concurrent removal and accumulation of heavy metals by the three aquatic macrophytes. Bioresource Technology 99, 7091–7097. Available from: https://doi.org/10.1016/j.biortech.2008.01.002.

Mishra, K.K., Rai, U.N., Prakash, O., 2007. Bioconcentration and phytotoxicity of Cd in *Eichhornia crassipes*. Environmental Monitoring and Assessment 130 (1-3), 237–243. Available from: https://doi.org/10.1007/s10661-006-9392-5.

Naeem, S., Ashraf, M., Babar, M.E., Zahoor, S., Ali, S., 2021. The effects of some heavy metals on some fish species. Environmental Science and Pollution Research International 28 (20), 25566–25578. Available from: https://doi.org/10.1007/s11356-021-12385-z.

Naggar, A.Y., Khalil, M.S., Ghorab, M.A., 2018. Environmental pollution by heavy metals in the aquatic ecosystems of Egypt. Open Access Journal of Toxicology 3, 555603. Available from: https://doi.org/10.19080/OAJT.2018.03.555603.

Naorbe, M.C., Serrano Jr, A.E., 2018. Effects of heavy metals on cell density, size, specific growth rate and chlorophyll a of *Tetraselmis tetrathele* under controlled laboratory conditions. Aquaculture, Aquarium, Conservation & Legislation 11 (3), 589–597.

Nasir, M., Ansari, T.M., Javed, H., Yasin, G., Khan, A.A., Shoaib, M., 2020. Analytical quantification of copper in frogs (*Rana tigrina*) found from various aquatic habitats. African Journal of Biotechnology 19 (2), 121–128. Available from: https://doi.org/10.5897/AJB2017.16013.

Nishikawa, K., Yamakoshi, Y., Uemura, I., Tominaga, N., 2003. Ultrastructural changes in Chlamydomonas acidophila (Chlorophyta) induced by heavy metals and polyphosphate metabolism. FEMS Microbiology Ecology 44 (2), 253–259. Available from: https://doi.org/10.1016/S0168-6496(03)00049-7.

Nwabunike, M.O., 2016. The effects of bioaccumulation of heavy metals on fish fin over two years. Journal of Fisheries and Livestock Production 4, 2. Available from: https://doi.org/10.4172/2332-2608.1000170.

Odubanjo, G.O., Oyetibo, G.O., Ilori, M.O., 2021. Ecological risks of heavy metals and microbiome taxonomic profile of a freshwater stream receiving wastewater of textile industry. Frontiers in Environmental Science 9, 554490. Available from: https://doi.org/10.3389/fenvs.2021.554490.

Onita (Mladin), B., Albu, P., Herman, H., Balta, C., Lazar, V., Fulop, A., et al., 2021. Correlation between heavy metal-induced histopathological changes and trophic interactions between different fish species. Applied Sciences 11, 3760. Available from: https://doi.org/10.3390/app11093760.

Pandiyan, J., Mahboob, S., Govindarajan, M., et al., 2021. An assessment of level of heavy metals pollution in the water, sediment and aquatic organisms: a perspective of tackling environmental threats for food security. Saudi Journal of Biological Sciences 28 (2), 1218–1225. Available from: https://doi.org/10.1016/j.sjbs.2020.11.072.

Peng, H., Guo, J., 2020. Removal of chromium from wastewater by membrane filtration, chemical precipitation, ion exchange, adsorption electrocoagulation, electrochemical reduction, electrodialysis, electrodeionization, photocatalysis and nanotechnology: a review. Environmental Chemistry Letters 18, 2055–2068. Available from: https://doi.org/10.1007/s10311-020-01058-x.

Perumal, K., Antony, J., Muthuramalingam, S., 2021. Heavy metal pollutants and their spatial distribution in surface sediments from Thondi coast, Palk Bay, South India. Environmental Sciences Europe 33, 63. Available from: https://doi.org/10.1186/s12302-021-00501-2.

Prakash, O., Mehrotra, I., Kumar, P., 1987. Removal of cadmium from water by water hyacinth. Journal of Environmental Engineering 113, 352–365.

Qian, H., Li, J., Sun, L., Chen, W., Sheng, G.D., Liu, W., et al., 2009. Combined effect of copper and cadmium on *Chlorella vulgaris* growth and photosynthesis-related gene transcription. Aquatic Toxicology (Amsterdam, Netherlands) 94 (1), 56–61. Available from: https://doi.org/10.1016/j.aquatox.2009.05.014.

Rajeshkumar, S., Li, X., 2018. Bioaccumulation of heavy metals in fish species from the Meiliang Bay, Taihu Lake, China. Toxicology Reports 5, 288–295. Available from: https://doi.org/10.1016/j.toxrep.2018.01.007.

Rakib, M.R.J., Jolly, Y.N., Enyoh, C.E., et al., 2021. Levels and health risk assessment of heavy metals in dried fish consumed in Bangladesh. Scientific Reports 11, 14642. Available from: https://doi.org/10.1038/s41598-021-93989-w.

Romanova, T.E., Shuvaeva, O.V., Belchenko, L.A., 2016. Phytoextraction of trace elements by water hyacinth in contaminated area of gold mine tailing. International Journal of Phytoremediation 18, 190–194. Available from: https://doi.org/10.1080/15226514.2015.1073674.

Sacan, M.T., Oztay, F., Bolkent, S., 2007. Exposure of *Dunaliella tertiolecta* to lead and aluminum: toxicity and effects on ultrastructure. Biological Trace Element Research 120, 264–272. Available from: https://doi.org/10.1007/s12011-007-8016-4.

Saha, P., Shinde, O., Sarkar, S., 2017. Phytoremediation of industrial mines wastewater using water hyacinth. International Journal of Phytoremediation 19, 87–96. Available from: https://doi.org/10.1080/15226514.2016.1216078.

Sfakianakis, D.G., Renieri, E., Kentouri, M., Tsatsakis, A.M., 2015. Effect of heavy metals on fish larvae deformities: a review. Environmental Research 137, 246–255. Available from: https://doi.org/10.1016/j.envres.2014.12.014.

Shaheen, T., Akhtar, T., 2012. Assessment of chromium toxicity in *Cyprinus carpio* through hematological and biochemical blood markers. Turkish Journal of Zoology 36 (5), 682–690. Available from: https://doi.org/10.3906/zoo-1102-21.

Shanab, S., Essa, A., Shalaby, E., 2012. Bioremoval capacity of three heavy metals by some microalgae species (*Egyptian Isolates*). Plant Signaling & Behavior 7 (3), 392–399. Available from: https://doi.org/10.4161/psb.19173.

Sharma, R., Agrawal, P.R., Kumar, R., Gupta, G., Ittishree, 2021. Current scenario of heavy metal contamination in water. In: Ahamad, A., Siddiqui, S.I., Singh, P. (Eds.), Contamination of Water. Health Risk Assessment and Treatment Strategies. Elsevier, pp. 49–64. Available from: https://doi.org/10.1016/B978-0-12-824058-8.00010-4.

Singh, J., Kalamdhad, A.S., 2011. Effects of heavy metals on soil, plants, human health and aquatic life. International Journal of Research in Chemistry and Environment 1 (2), 15–21.

Spain, O., Plöhn, M., Funk, C., 2021. The cell wall of green microalgae and its role in heavy metal removal. Physiologia Plantarum 1–10. Available from: https://doi.org/10.1111/ppl.13405.

Srivastav, A.K., Srivastav, S., Suzuki, N., 2016. Acute toxicity of a heavy metal cadmium to an anuran, the Indian skipper frog *Rana cyanophlyctis*. Iranian Journal of Toxicology 10 (5), 39–43. Available from: https://doi.org/10.29252/arakmu.10.5.39.

Sultana, S., Jabeen, F., Sultana, T., AL-Ghanim, K.A., Al-Misned, F., Mahboob, S., 2020. Assessment of heavy metals and its impact on DNA fragmentation in different fish species. Brazilian Journal of Biology = Revista Brasleira de Biologia 80 (4). Available from: https://doi.org/10.1590/1519-6984.221849.

Tamele, I.J., Loureiro, P.V., 2020. Lead, mercury and cadmium in fish and shellfish from the Indian Ocean and Red Sea (African Countries): public health challenges. Journal of Marine Science and Engineering 8, 344. Available from: https://doi.org/10.3390/jmse8050344.

Thanomsangad, P., Tengjaroenkul, B., Sriuttha, M., Neeratanaphan, L., 2020. Heavy metal accumulation in frogs surrounding an e-waste dump site and human health risk assessment. Human and Ecological Risk Assessment: An International Journal 26, 1313–1328. Available from: https://doi.org/10.1080/10807039.2019.1575181.

Wu, G., Kang, H., Zhang, X., Shao, H., Chu, L., Ruan, C., 2010. A critical review on the bio-removal of hazardous heavy metals from contaminated soils: issues, progress, eco-environmental concerns and opportunities. Journal of Hazardous Materials 174, 1–8. Available from: https://doi.org/10.1016/j.jhazmat.2009.09.113.

Yasar, A., Zaheer, A., Tabinda, A.B., Khan, M., Mahfooz, Y., Rani, S., et al., 2018. Comparison of reed and water lettuce in constructed wetlands for wastewater treatment. Water Environment Research: A Research Publication of the Water Environment Federation 90, 129–135. Available from: https://doi.org/10.2175/106143017X14902968254728.

Yousif, R., Choudhary, M.I., Ahmed, S., Ahmed, Q., 2021. Bioaccumulation of heavy metals in fish and other aquatic organisms from Karachi Coast, Pakistan. Nusantara Bioscience 13, 73–84. Available from: https://doi.org/10.13057/nusbiosci/n130111.

Yozukmaz, A., Yabanli, M., Sel, F., 2018. Heavy metal bioaccumulation in *Enteromorpha intestinalis* (L.) Nees, a macrophytic algae: the example of Kadin Creek (Western Anatolia). Brazilian Archives of Biology and Technology 61, 18160777. Available from: https://doi.org/10.1590/1678-4324-2018160777.

Zeng, J., Yang, L., Wang, W.X., 2009. Cadmium and zinc uptake and toxicity in two strains of *Microcystis aeruginosa* predicted by metal free ion activity and intracellular concentration. Aquatic Toxicology (Amsterdam, Netherlands) 91 (3), 212–220. Available from: https://doi.org/10.1016/j.aquatox.2008.11.004.

CHAPTER

5

Metal in water: an assessment of toxicity with its biogeochemistry

Bipradeep Mondal and Nirmali Bordoloi

Department of Environmental Sciences, Central University of Jharkhand, Ranchi, Jharkhand, India

5.1 Introduction

Among 118 elements present in the periodic table, 95 are metallic in nature. The term "metal" indicates an element that is a good conductor of heat and electricity having high density with electrical resistance directly proportional to absolute temperature (Pandey et al., 2011). Elements like metals are intrinsic properties of Earth. Therefore the presence of metals in natural ecosystems like water and soil is imperative. Contamination in environment by metals are a result of both natural processes and anthropogenic activities (Fig. 5.1) (Kumar et al., 2019). Metal contamination in water can result from atmospheric precipitation, erosion of surface deposits of minerals, volcanic activity, etc. Simultaneously, rapid unplanned urbanization interfering Earth system processes have accelerated metal pollution through discharge of agricultural, municipal, domestic, and industrial waste products (Salomons and Forstner, 2012a,b). Several literatures have articulated that combustion of fossil fuels, extraction of metals from ores with increasing variety have aggravated metal pollution to a serious dimension (Boman and Wagner, 2003; Pandey et al., 2011; Li et al., 2018; Wang et al., 2019; Chen et al., 2022). Trace elements such as arsenic (As), cadmium (Cd), lead (Pb), uranium (U), beryllium (Be), radon (Rn), nickel (Ni), mercury (Hg), fluorine (F), manganese (Mn), selenium (Se), chlorine (Cl), thorium (Th), cobalt (Co) and many more are present in coal fly ash (Kumar et al., 2016; Leelarungroj et al., 2018; Park et al., 2021). These elements get released during combustion of coal based fuels, with size less than 0.1 μm and greater than 100 μm. Interacting with natural waters, metals initiate complex reactions and accumulate in sediments and organisms to generate toxic effects. Heavy metals have a significant importance in ecotoxicology as they persist for a longer period of time causing adverse effects to organisms in water as well as human beings. Metals have the potential of bioaccumulation and biomagnification in food chain (Yin et al., 2019).

Metals in Water
DOI: https://doi.org/10.1016/B978-0-323-95919-3.00017-3

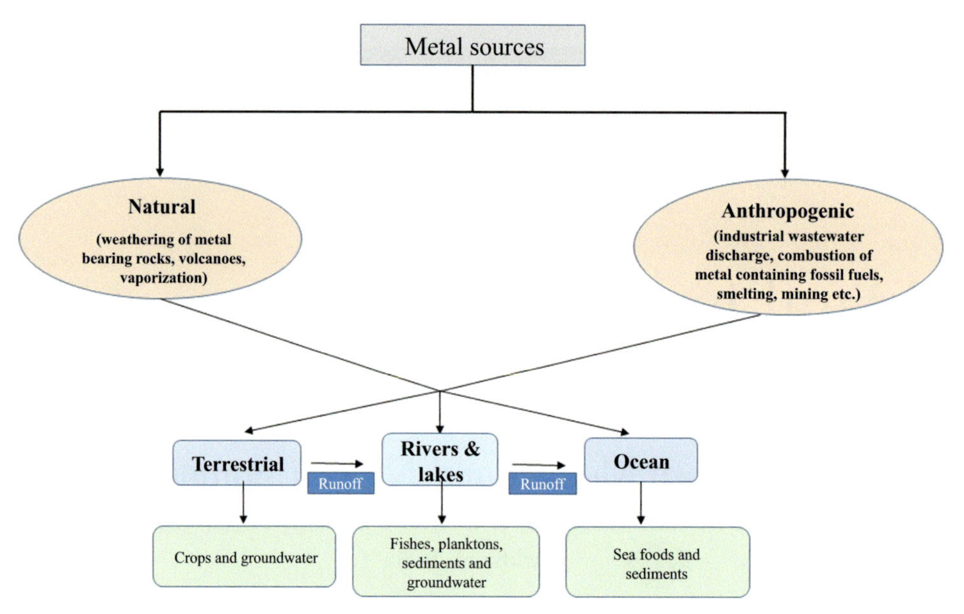

FIGURE 5.1 Dispersion of metals in environment.

Significantly, a larger part of heavy metals except lead, which gets released into the environment are transferred by different types of aquatic ecosystems providing different biogeochemistry (Wojciechowska et al., 2019; Miranda et al., 2021). Although, Earth's crust contains < 0.1% of toxic heavy metals but increasing anthropogenic activities like releasing industrial effluents, pesticides, herbicides, fertilizers, detergents, use of metallic compounds in war weapons, smelting and mining of ores, etc., have magnified the concentration of heavy metals and their salts in the environment (Fig. 5.1) (Lee and Von Lehmden, 1973; Vareda et al., 2019). Generally toxicity of metal rises with atomic number and electropositivity viz. toxicity of metal salts increases from Nitrates < Chlorides < Bromides < Acetates < Iodides < Perchlorates < Sulfates < Phosphates < Carbonates < Fluorides < Hydroxides < Oxides (Pandey et al., 2011). Nevertheless, many metals can contradict with the general determination of toxicity due to pertaining same electropositivity. In such cases density and atomic number of metals are the determinants of toxicity (Ali and Khan, 2018). With the increase in density and atomic number of metals, there is a simultaneous increase of toxicity level (Mehrandish et al., 2019). In fact many literatures have indicated that a number of metals and metalloids such as Arsenic, Chromium, and Nickel happen to be carcinogenic in nature (Sunderman, 1981). Urbanization and industrialization are the main backbone behind current metal pollution. Countries like China and India are mostly affected with metal contamination on a global stage. In China, surface waters of Zhuzhou, Yunnan Province, Shaoguan, etc., are found to be highly polluted with heavy metals, with high concentrations of Pb (30.1 μg/L), Cu (96.8 μg/L), Hg (46.1 μg/L), Cr (7.7 μg/L), and Zn (98.3 μg/L) in Wen-Rui Tang River (ZongTang et al., 2010; Qu et al., 2018). The Central Pollution Control Board of India has identified 43 industrial groups responsible for contaminating waterbodies (Saha and Paul, 2019). In the eastern part of India, cities like Ramgarh, Ranchi, etc., in Jharkhand (Tirkey et al., 2016; Kumar et al., 2020), Chakdah,

Ranaghat, Baruipur, Durgapur, Alipurduar, etc., in West Bengal (Bhattacharya et al., 2010a,b; Adimalla, 2019) are found to be most affected with metal pollution and toxicity as compared to other parts of India (Kumar et al., 2019).

In order to safeguard the water ecosystems with efficient protection and management plans, potential contamination sites should be neutralized with the involvement of both scientific community and local decision makers. Regular monitoring of water contamination is a prerequisite with proper environment quality indicators and contamination indices and water sustainability indices for decision makers to analyze and process data before developing a management plan. Reduction of metal wastes from point sources like industries, treatment of wastewater using microbes, chemical precipitation, and adsorption, etc., are the most viable techniques to minimize metal pollution in global waters. Water, being an essential natural resource, must be recycled and reused after proper treatment in developing countries like India, Bangladesh, as these countries experiences water scarcity and metal pollution makes it more difficult to replenish that scarcity. As compared to developed countries, the developing countries persist a higher rate of population increment, therefore more socioeconomic disparity, unequal distribution of natural resources and huge production of wastewater exists (Byrnes and Bumb, 2017; Gallego-Schmid and Tarpani, 2019; Kuznets, 2019; Biswas et al., 2021). These scenarios have led to issues with natural resources like water as there remains a gap between freshwater supply and demand.

5.2 Metal pollution in aquatic environment and impact on flora and fauna

Water is one of the critical natural resource that is irreplaceable and essential for survival for all life forms, it is also an important asset for socioeconomic growth and sustainable development of a country (Bytyci et al., 2018). However, indiscriminate and rapid industrial growth and other human activities have put this vital entity under serious threat with poor management and water pollution (Pobi et al., 2019; Proshad et al., 2020). Prevalence of inadequate and incompetent laws have allowed majority of industries to discharge their wastewater effluents in rivers, lakes or near surface waterbodies. Consequently, existence of these industrial endpoints near rivers and lakes, have deteriorated the quality of water and contaminated these surface waterbodies with heavy metals like Cd, As, Hg, Pb, Cr, Zn, etc. These metals possess a specific density greater than 5 g/cm^3 (Jaishankar et al., 2014). Heavy metals are a great concern to aquatic flora and fauna, due to their long residence time and bioaccumulation. Higher concentration of heavy metals in aquatic environment can show toxic effects and bring imbalance to ecosystem functions (Proshad et al., 2020). The biota and water sediments are the main accumulators of metal in aquatic environment (Gbaruko and Friday, 2007). The amount of metal accumulation changes with seasonal fluctuations. The chemical effects of metal on aquatic organisms are highly dependent on its binding capability with lipid and composition of biological tissue (Bower et al., 1978). Generally fishes are the determinant of metal pollution as they accumulate a large amount of certain metals like Pb, Fe, Zn, Cd, Cu, and Mn, etc. (Baby et al., 2011). So, as the trophic level rise in food chain, biomagnification occurs with increasing concentration of metal. Although a trace amount of some metals are beneficial for all life forms but excessive amount can show toxic effect resulting into detrimental. Fishes like Zebrafish, medaka, Shellfish, juvenile loach show a great response to metal toxicity even in low doses. As for example, while assessing acute sublethal rate of cadmium toxicity for 24 and 48 h, juvenile loach showed a response with LC_{50} 1.22 and 0.85 mg/L respectively (Zhou et al., 2008). During food ingestion, the fishes eventually intake heavy metals through their gills, viscera, and mucus, etc. (Coombs, 1980; Lall and Kaushik, 2021). Consequently, fishes experiences disruption in fecundity, low reproductive rate and metabolism (Zhou et al., 2008). Among other aquatic organisms, plankton shows consistent kinetics on metal pollution. They uptake the metal through passive diffusion method, where the ligands bind with cell membranes and exchange ions following diffusion to enter the internal components of the organism (Coombs, 1980). Zooplankton is more sensitive than phytoplankton. In summer under direct toxicity of metals like cadmium, copper, and mercury, etc., there occurs disruption in zooplankton predation and nutrient regeneration (Kerrison et al., 1988). Reports suggest that zooplanktons like copepods are most tolerant to heavy metal toxicity. With increased concentration of metals in water, the zooplanktons like *Bosmina longirostris* and *Daphnia cucullata* experiences decline in population density and diversity (Kerrison et al., 1988). Similarly, several researchers found a high concentration of mercury that can reduce the population size of rotifers (like *Keratella cochlearis* and *Asplanchna priodonta*) (Kerrison et al., 1988). Apart from that, aquatic plants accumulate metals in roots and stems and leaves. Aquatic plants have a large potential to uptake high amount of heavy metals from water/sediments through active and passive absorption (Bai et al., 2018). The common toxic effects of heavy metals witnessed in aquatic plants are typically chlorosis, inhibition of growth and photosynthesis, low accumulation of biomass and ultimately senescence (Singh et al., 2016). Generally these toxic effects are exerted through common mechanisms like—(1) completion in absorption of similar nutrient cations by root such As and Cd compete with P and Zn; (2) disruption in protein synthesis as metals have affinity toward binding with −SH (sulphydryl) group of functional proteins (Li et al., 2006; Gill et al., 2013; Singh et al., 2016); (3) damaging macromolecules through initiation of reactive oxygen species (Sharma and Dietz, 2009; DalCorso et al., 2013; Singh et al., 2016). Therefore, repercussions of metal pollution in aquatic environment and its flora and fauna are widely complex, with toxic effects proportional to concentration and accumulation rate of metals in water.

5.3 Metal contamination in groundwater

The unique chemical properties of water due to its hydrogen bonds and polarization helps metal salts to get adsorb, dissolve, and absorb in natural waters (Mendie, 2005). These metals are stubborn, recalcitrant, and toxic in nature causing harm to human beings and other organisms entering the food chain (Marcovecchio et al., 2007). Apart from surface waterbodies, freshwater aquifers are highly affected due to metal pollution. Naturally, Earth's crust consists ample amount of metals like Nickel (Ni), Manganese (Mn), Aluminum (Al), Arsenic (As), Lead (Pb), etc., but it's the anthropogenic activities that amplify the trace amount into a significantly higher concentration in our environment. As similar to surface waterbodies, groundwater also get highly affected metal pollution. Landfill leaching, sewage leakage, industrial discharge are the major common activities that aggravate the metal pollution in groundwater (Sankhla and Kumar, 2019). Related to these activities along with natural processes, the major sources of groundwater pollution are as follows:

1. Natural sources:
   As mentioned earlier, earth have adequate amount of trace elements like As, Pb, Cd, Ni, Al, Zn, etc. The natural processes like weathering of metal-bearing rocks, volcanic eruptions and vaporization unleash the metals present in earth's crust and contribute groundwater pollution (Bagul et al., 2015; Ali et al., 2019).
2. Anthropogenic activities:
   In today's world, the evolution of new epoch "Anthropocene" the human beings are overexploiting the natural resources for the sake of urbanization, industrialization and so-called socioeconomic growth. Consequently, increased mining, smelting of metal ores, burning excessive metal containing fossil fuels, discharging industrial wastes have contaminated the groundwater with heavy metals. Application of chemical fertilizers in agricultural field containing metals have also contributed in groundwater pollution (Ali et al., 2019). Generally, phosphate fertilizers are produced from phosphate rock by acidulation as a result the final product contains all the metal that existed in that rock (Dissanayake and Chandrajith, 2009). As depicted below in Fig. 5.2, these metals enters soil as well as a large part of it gets leached into the aquifers contaminating the groundwater.
   Evidently, a large amount of global water demand is satisfied by groundwater. With increasing metal contamination in aquifers, there are some constant major problems that are needed to be addressed, such as (1) drinking metal contaminated groundwater can initiate harmful effects on human being targeting the organs like kidney, brain and liver (Sankhla and Kumar, 2019); (2) application of groundwater containing metal in agricultural field can cause biomagnification with rising trophic levels through food chain (Ali et al., 2019; Khan et al., 2021; Kristanti et al., 2021); (3) daily usage of contaminated water in countries like Bangladesh, India with heavy metal pollution such as arsenic can cause irritation leading to skin cancer, etc. (Bhattacharya et al., 2010a,b; Biswas et al., 2020; Sengupta et al., 2021) Therefore, it is necessary to restrict metals from the sources getting leached in the groundwater as the natural aquifers play a significant role in survival of fairly large community around the world.

Application of Phosphate/other chemical fertilizers (metal source) in agricultural fields

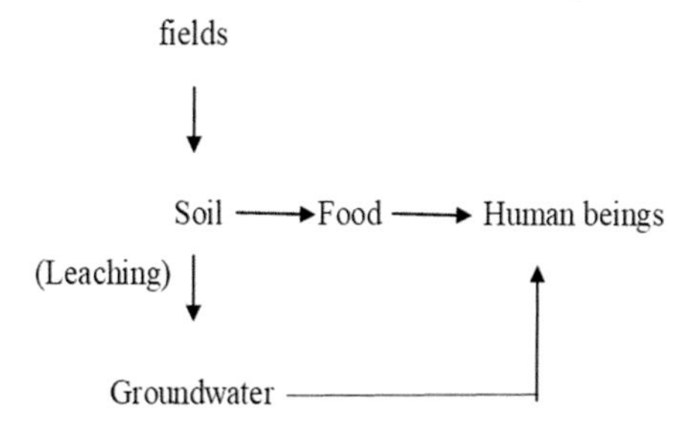

FIGURE 5.2 Process of groundwater contamination with metals via chemical fertilization in agricultural fields.

5.4 Metal contamination in marine environment

In nature, metals persists as both forms- biologically essential and nonbiologically essential. For instance iron (Fe) is an essential part of hemoglobin, calcium is the backbone behind the formation bones similarly Zinc (Zn) and Copper (Cu) are also biologically essential metals (Kennish, 2017). On the other hand metals like cadmium (Cd), mercury (Hg), lead (Pb) and tin (Sn) fall under the category of nonbiologically essential metals. These elements are quite tolerable at low amount but the rising concentration in natural ecosystems of marine, freshwater, soil and atmosphere can very much deleterious. The main sources behind metal pollution in marine environment are the same as any other aquatic environment such as the combustion of metal containing fossil fuels, weathering of rocks, use of chemical fertilizers and industrial wastewater discharge, etc. (Nammalwar, 1983). Although the pathway is different, ocean accumulates secondary metal pollutants as all the primary metals stored in river, lakes sediments gets dumped into the oceans and seas through riverine influx (Nammalwar, 1983). The threshold concentration of all heavy metals depends upon the chemical forms of metal, physiological status of organisms, presence of other metals and physicochemical properties of marine environmental such as pH, salinity, etc. (Manahan, 2017). Eventually exposure with high concentration of metals result to be fatal for marine organisms. Metals generally damage tissues and DNA, inhibits growth and make the organisms unable to regenerate damaged tissues (Furness and Rainbow, 2018). Although most of the marine organisms are capable of detoxify metal from their bodies via excretion, and metallothioneins (Schreiber and Burger, 2001). Studies showed that metal toxicity in marine ecosystems doesn't kill much organisms like fishes but have great potential to damage these organism to a significant level (Amara et al., 2018; López-Berenguer et al., 2020; Rand et al., 2020). It is impossible to prevent complete metal pollution from the humongous marine environment, but can be minimized through legalization and adequate management of such huge water resource.

5.5 Biogeochemistry of metal complexes in natural water

In order to achieve highest stability form, metals get bonded or coordinated with other species present in natural water or wastewater. Generally numerical form of metal ions in water are depicted by M^{n+}. These metals ions binds with water molecules or any other species (electron donor) to form hydrated metal cations $M(H_2O)_X{}^{n+}$(Manahan, 2017). These metal cations in aqueous solution are in constant search of stability through chemical reaction involving acid—base. As for example (Manahan, 2017)

$$Fe(H_2O)_6^{3+} FeOH \rightleftarrows (H_2O)_5^{2+} + H^+ \text{ (precipitation)} \tag{5.1}$$

$$Fe(H_2O)_6^{3+} \rightleftarrows Fe(OH)_3(s) + 3H_2O + 3H^+ \tag{5.2}$$

$$Fe(H_2O)_6^{3+} \rightleftarrows Fe(OH)_3(s) + 3H_2O + e^- + 3H^+ \text{ (oxidation and reduction)} \tag{5.3}$$

The above reactions justify the reason behind the less availability hydrated form of Iron in water, that is, $Fe(H_2O)_6{}^{3+}$ (Manahan, 2017). The behavior of metal in water is highly complex as it depends upon various external factors like pH, oxidation, and reductions in natural and wastewaters (Salomons and Forstner, 2012a,b). The metal tends to precipitate in aquatic environment causing interactions with different water molecules, solid particles present in water, and loss and addition of different gases. Among various types precipitates of metal ions in water, some major types exist such as oxidation type (precipitation occurs due to oxidation of swamp waters), reducing type (metals like Ag, Cu, etc., with lower-valency getting reduced by interaction with organic matters), and alkalinity type (precipitation occur due to increase in pH of water), etc. (Rose and Elliott, 2000). Simultaneously, the degree at which metals will dissolve in waters highly depend on the nature of metal species. Metals binds with inorganic or organic anions to form metal complexes and this phenomenon is known as complexation. The anions binding with one metal ion are known as ligands whereas ligands getting attached with two or more metal ions is known chelation (Manahan, 2017).

Metals/metalloids (Hg, As, Sb, Se, etc.) in water often binds with carbon atoms to form organometallic/organometalloid compounds which are often found to be highly toxic in nature. The formation of organometallic and organometalloid compounds in water sediments executed by aquatic microbes is widely known as *methylation*. Among all the metals mercury (Hg) is methylated to the highest extent (Clarkson, 1990; Mason, 2012). These bonds are covalent in nature with the intention to fill the d and f orbitals of metals and metalloids (Craig and Jenkins, 2004). Sulfate reducing bacteria (SRB) are the major aquatic organisms that mediate methylation of Hg into CH_3Hg (methyl mercury) (Compeau and Bartha, 1985; Gilmour et al., 1992). Eventually, in various aquatic ecosystems mercury is found to interact with sulfide forming neutrally charged Hg-S species (HOHgSH and $Hg(SH)_2$) (Benoit et al., 2001a,b). Consequently, the increase in bioavailability of such compounds (Hg-S) makes it easier for the SRB's to execute methylation of mercury (Benoit et al., 2003; Hammerschmidt and Fitzgerald, 2004; Sunderland et al., 2006; Heyes et al., 2006). Later studies have also found that iron (Fe) reducing bacteria of genera *Geobacter* have similar potentials like SRB to methylate Hg into its organometallic form (Kerin et al., 2006; Fleming et al., 2006). Organic compound of mercury (CH_3Hg) is much more toxic than elemental mercury, as identified in Japan during 1960s. Cases with methyl mercury being accumulated in fish and biomagnified into humans causing the great *minamata* disease were the major concern during the 1960s in Japan (Clarkson, 1990, 1994). Perhaps in case of arsenic (As) the organic form is less toxic than is inorganic form (Hare et al., 2020). In terms of oxyanions, the bioavailability of inorganic forms of arsenic such as arsenic acid (H_3AsO_4) and arsenous acid (H_3AsO_3) in water are higher as compared to nutrients of phosphates (Sun, 2010; Mason, 2012). Simultaneously, the pK_a values of arsenic acid ($pK_{a1} = 2.22$, $pK_{a2} = 6.98$ and $pK_{a3} = 11.53$) are similar to pK_a values of phosphate nutrients ($pK_{a1} = 2.1$, $pK_{a2} = 7.2$, $pK_{a3} = 12.4$) (Mason, 2012), as a result it is suitable for most of the bacteria, yeasts, fungi and phytoplankton, etc., to degrade their cell materials and perform methylation of As (Bentley and Chasteen, 2002; Edmonds and Francesconi, 2003; Plant et al., 2005). Apart from these cases there are other cases involving different metal(loid)s undergoing the process of methylation, such as Cd forming $(CH_3)_2Cd$, although organic form of Cd is unstable in water (Pongratz and Heumann, 1999; Feldman, 2003). On the other hand Zn do not form any kind of stable methylated compounds in aquatic systems. Therefore there exists a decent number of studies concentrating on methylation of metals (Wang et al., 2016; Ebenebe et al., 2017; D'Itri, 2020), with further researches going on.

TABLE 5.1 Methods to assess presence of metals in water samples.

Methods	Concentration (g)
Gravitimetry	$<10^{-9}$
Titrimetry	10^{-9}–10^{-10}
Fluorescence reactions	10^{-10}–10^{-11}
Kinetic reactions	10^{-11}–10^{-13}
Atomic absorption spectroscopy (AAS)	10^{-10}–10^{-12}
Gas chromatography	10^{-11}–10^{-13}
Mass spectrometry	10^{-12}–10^{-15}

Methods as cited by Salomons, W. and Forstner, U., 2012a. Metal Pollution in the Aquatic Environment. Springer Science and Business Media B.V. Available at: https://books.google.co.in/books?hl = en&lr = &id = 9PLuCAAAQBAJ&oi = fnd&pg = PA1&dq = Metal + pollution + in + aquatic + environment + &ots = 4STc4Qg0-I&sig = 24SYQsur9t8rQRv-G1E9wUoNxXM&redir_esc = y#v = onepage&q = Metal. pollution.in.aquatic.environment&f = false (Accessed: 9 October 2021); Salomons, W. and Forstner, U., 2012b. Metals in the Hydrocycle. Springer-Verlag Berlin Heidelberg. https://doi.org/10.1007/978-3-642-69325-0.

The presence of metals in aquatic environment are generally assessed by various techniques as mentioned in Table 5.1, via collecting water samples from desired aquatic medium.

5.6 Impacts of heavy metals toxicity on human beings

As addressed by international legislative and health bodies, the major threats to human health are heavy metals like arsenic (As), cadmium (Cd), mercury (Hg), and lead (Pb), etc. Large doses and long duration of these metal accumulation can result in skin, lung, and liver cancer. Depending upon low to high exposure, heavy metals show acute and chronic symptoms both in adults and children (Table 5.2).

The acute symptoms are generally seen for short period of time. Although humans unknowingly develop chronic symptoms with accumulation heavy metals in low doses for a longer period of time (Baby et al., 2011). Unaccessibility of fresh drinking water, consumption of metal accumulated fishes and crops like rice wheat can gradually accumulate as well as biomagnify metal concentration in human beings. As a consequence there develops chronic symptoms which gradually lead towards experiencing detrimental effects like cancer and arsenicosis, *ItaiItai*, Minamata, etc. Although a large portion of heavy metals is excreted by human on daily basis but a fair amount is stored in kidney which can lead to kidney dysfunction in future.

In recent times, many scientists have speculated that low level exposures to heavy metals (Cd, As, Hg, Pb, etc.) in humans especially males have significant effects on various *reproductive functions* (Wirth and Mijal, 2010; Selvaraju et al., 2014). Literatures of male population with abnormality in reproductive functions are still less in number as compared to other symptoms around the world. But there are ample evident of males experiencing

TABLE 5.2 Acute and chronic symptoms of heavy metal toxicity in human (Adults and Children).

Heavy metal	Acute symptoms	Chronic symptoms
Arsenic	Fever, hepatomegaly, cardiac arrhythmia, sensory loss in peripheral nervous system, etc.	Loss in appetite and weight, diarrhea, neurotoxicity in central and peripheral nervous system, arsenicosis.
Lead	Interference in heme synthesis leading to hematological damage, etc.	Loss of teeth, constipation, kidney dysfunction (*Fanconi syndrome*), increased blood pressure, tumors in respiratory systems, etc.
Mercury	Corrosive bronchitis, interstitial pneumonitis, acrodynia (in children), etc.	Loss of memory, severe depression, ataxia, minamata, loss of vision and hearing ability, etc.
Cadmium	Vomiting, abdominal cramps, headaches.	*Itailtai* (decalcification of bones), kidney damage (Aminoaciduria, Proteinuria, etc.), osteoporosis, liver dysfunction, etc.

Courtesy (Pandey et al., 2011; Manahan, 2017). Pandey, K., Shukla, J.P., Trivedi, S.P., 2011. Fundamentals of Toxicology, fourth rev ed. New Central Book Agency; Manahan, S., 2017. Environmental Chemistry, tenth ed. CRC Press. doi:10.1201/9781315160474.

changes in hormones and reproductive functions like *low sperm count, less semen volume,* and *low sperm motility,* etc., due to low level exposure of heavy metals on a regular basis(Sun et al., 2017; Badr and El-Habit, 2018). Some of the major effects on male reproductive functions found due to heavy metal exposures are mentioned categorically

1. Cadmium (Cd):

   Cadmium (Cd) is naturally found as an associate in ores of zinc, copper and lead, but increased anthropogenic activities have made this heavy metal abundant to natural world getting exposed frequently (ATSDR, 2008). The major routes of cadmium exposure to humans are consumptions of contaminated shellfish, rice and smoking (Jayasinghe et al., 2018). Generally, Cadmium is stored in kidney and blood streams of human body with a biological half-life of 7–26 years in kidney and 3–4 months in blood. There is a possibility of male reproductive hormones and sperm quality getting affected due to infusion of cadmium in blood. Xu et al. (1993) stated that a potent number of Asian men containing cadmium in blood stream (mean 1.35 mg/L) were positive correlation was found with defects in sperm midpiece ($r = 0.42$, $P < .05$) and immature ($r = 0.45$, $P < .05$). Similar cases were seen in Nigeria, Croatia, and infertility clinics of Michigan (USA) (Jurasovic et al., 2004; Akinloye et al., 2006). These studies are limited at this stage but fairly exposes the susceptibility of male reproductive hormones and functions with cadmium toxicity.

2. Lead (Pb):

   Humans get exposed to lead (Pb) in two ways, mainly occupational and nonoccupational (Gennart et al., 1992). Occupational exposure comprises inhalation of lead containing dusts and fumes, whereas, nonoccupational exposure of lead involves consuming food and drinking water contaminated with lead (Gittleman et al., 1994). After absorption of lead, human body stores majority of lead in blood and bones with half-life of 35 days and 20–30 years respectively (Wirth and Mijal, 2010). Consequently stored lead in blood can manipulate the sperm quality such as reducing its motility, count with increasing abnormal morphology (Alexander et al., 1998). Such cases were witnessed in Croatian men with smoking habits(Jurasovic et al., 2004), where their blood stream containing lead had positive correlation with abnormal sperm head morphology ($r = 0.209$, $P < .01$) and negative correlation with sperm motility ($r = 0.179$, $P < .05$) (Telisman et al., 2000). In addition to these men with smoking habits were also found with prostate cancer due to lead infusion in blood streams (Pant et al., 2003). Therefore, there is a potential correlation between lead and male reproductive functions.

3. Mercury (Hg):
   Organic mercury is the most potent toxic form of mercury disrupting health of humans and animals. Inorganic mercury are eventually used to preserve vaccines, skin creams, extraction of gold, etc. (Clarkson and Magos, 2008). Simultaneously water and its sediments are the biggest medium that contains organic mercury in the form of methyl mercury that gets bioaccumulated and even biomagnified with increasing trophic level (Rignell-Hydbom et al., 2007). Such an example was found in Singapore where men consuming mercury contaminated fish were having higher level of mercury in blood stream than men with low fish consumptions (Chia et al., 1992). Similarly, on Hong Kong, subfertile men were likely to have 40% higher mercury stored in hair than fertile men in comparison (Dickman et al., 1998). Further studies revealed that blood stream containing mercury above 8 $\mu g/L$ have higher risk of lowering sperm quality and other reproductive parameters in male (Wirth and Mijal, 2010).

4. Arsenic (As):
   The major pathways of arsenic (As) absorption in human being are inhalation of air containing arsenic, saw dusts, burning of hazardous waste containing arsenic and groundwater contaminated with arsenic (Bhattacharyya et al., 2021; Sengupta et al., 2021). Arsenic is naturally present in rocks, but due to overexploitation of natural resources like groundwater, As has entered in our food chain. Countries like India (mainly eastern part) and Bangladesh are highly affected to arsenic as their groundwaters are contaminated with arsenic naturally (Bhattacharyya and Sengupta, 2020). Although, humans exposed to arsenic experiences skin irritation, and arsenicosis, etc., but, several studies have found As having high correlation with male reproductive functions. Studies in Michigan, USA infertility labs found few cases where men containing arsenic in blood stream have higher risk of low sperm motility (Meeker et al., 2008). Though, more and more researches are a prerequisite before coming to any conclusion in reference to linkage between arsenic and male reproductive functions.

5.7 Impacts of metals on wildlife

The assessment of metal toxicity in wildlife includes rigorous processes with variety of dimensions. The dynamic nature of aquatic and terrestrial organisms makes wildlife toxicity an interdisciplinary study which changes with venues (research labs, vivarium, mesocosm, field, etc.) and risk—exposure relationship (Rattner, 2009). In the early 20th century metal toxicity on wildlife was only known through waterfowls injected with lead shots

(Bowles, 1908). Later to protect quail and song bird's thallium and strychnine were shot to rodent in order to have predator control (Peterle, 1991). Studies have found that the overuse of nitrogen and phosphate fertilizers are responsible for increase in concentration of cd, Hg, and Pb, etc., in soil and plants (Rattner, 2009; Alves et al., 2016). These compounds get accumulated in food chain and widely affect the wildlife community. With modernization in agricultural practices, in late 1950s there arrived various pesticides and insecticides like calcium arsenate, methyl mercury, and lead arsenate, etc. (Peterle, 1991). Consequently the organic and inorganic forms of these metals are highly abundant in natural waters and food chain. Terrestrial and aquatic life forms were at risk as they are getting exposed to compounds like mercury and methyl mercury through agricultural drainwater, and industrial effluents, etc. Studies have found that on regular exposure to mercury or its organic compounds birds experiences developmental abnormalities, decreased embryo weights, micromelia, deformities in skeletal structures, etc. (Hoffman and Moore, 1979; Wolfe et al., 1998). Similar scenarios are found with wild organisms mainly birds getting exposed to inorganic compounds of Arsenic (As). Symptoms like slowness, jerkiness, falling, hyperactivity, fluffed feathers, drooped eyelids, huddled position, unkempt appearance, loss of righting reflex, immobility are much more visible in birds, poisoned with inorganic trivalent arsenite (Hudson and Haegele, 1984; Eisler, 1994). As water sediments are the major source of mercury, arsenic, lead and other metal containing pesticides fishes like shellfish, *Cyprinus carpio*, and *Grass carp*, etc., are highly vulnerable to metal toxicity (Huseen and Mohammed, 2019). Generally lead shows severe toxicity in aquatic fishes as compared to other metals. The metals get accumulated in muscle, liver and gills as a result, symptoms like reduction in liver size, immaturity of gills are common in nature (Eisler, 1994). Consequently consumption of these species lead to biomagnification of metals with increasing trophic level. Therefore like any other organism wildlife are highly susceptible to metal toxicity and with rapid urbanization these metals enter into the biogeochemical cycle exerting these different degrees toxicity in the natural world.

5.8 Importance of climate change behind metal concentrations and distributions

The world has entered a new geological epoch "Anthropocene," witnessing the climate change and global warming (Biswas et al., 2021). Since industrial revolution the degree of heavy metal contamination in soil, air and water has aggravated. Activities like intensive agricultural practices, mining, and combustion of fossil fuels, smelting ore, and unsafe waste disposal have incorporated heavy metals in aquatic and terrestrial environments. Previously, IPCC's fifth report (2014) hypothesized climate change to be an influence to hydrology of catchments. By the latest report of IPCC (2021) it is already claimed that there is an increase in the rates of precipitation and evapotranspiration, thus climate change affecting hydrology of different waterbodies. These changes affect flow of surface waters and encourages higher leaching of metals to groundwater. Hence, climate change serves as a cue to larger distribution of heavy metals in aquatic environments (Fig. 5.3) (Jericevic et al., 2012). A research study based on a model prediction revealed that Cd and Zn loads were supposed to get increased by large margins in the River Dommel of Belgium, as a resultant of increasing climate change in future. Another incident that is prevalent due to climate change is the atmospheric precipitation of heavy metals. Such cases are evident for lead (Pb). During mid-20th century lead

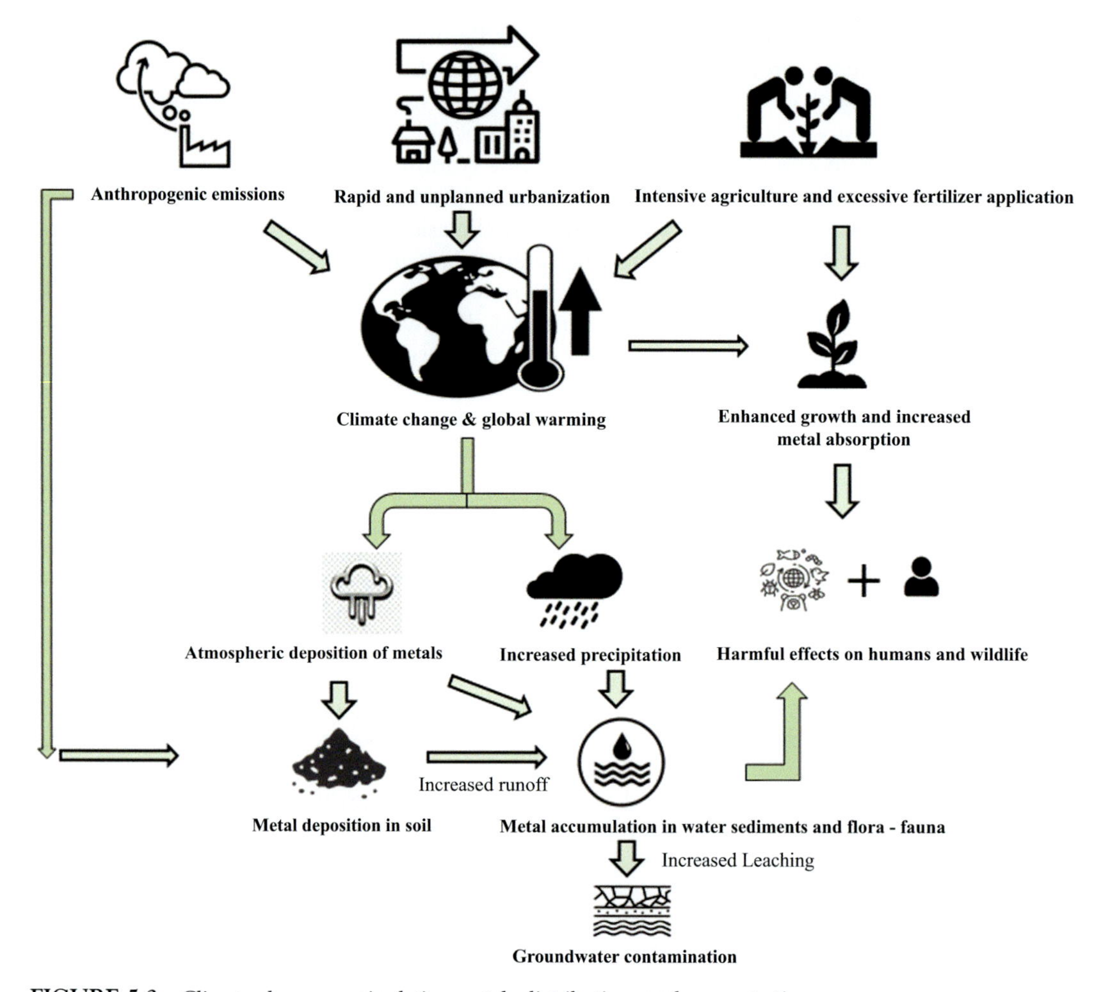

FIGURE 5.3 Climate change manipulating metals distributions and concentrations.

was heavily used as an antiknocking agent in gasoline (Prüss-Üstün et al., 2016; Laveskog, 2020). As a result most of the vehicles using Pb containing gasoline released a fair amount of Pb in the atmosphere which then got transferred and deposited into waterbodies through precipitation and surface runoff (Masters and Ela, 1974; Masters, 1997; Prüss-Üstün et al., 2016). Similarly, the introduction of green revolution with modern technologies and methods have also contributed towards metal contamination in the biosphere. The emergence of pesticides *viz.* lead arsenates, and mercurial fungicides, etc., are the major sources of arsenic, lead, and mercury, etc., in water bodies and soil (Evenson and Gollin, 2003). Moreover, the distribution of these heavy metals have been amplified, as result of increased surface runoff and leaching, induced by climate change and global warming. Concurrently, water sediments are a significant hub for metal storage (Islam et al., 2015). Nevertheless, a majority of metals are present in rocks. Anthropogenic activities acclimated with climate change have increased the bioavailability of metals like As, Pb, Hg, Cd, Cu, Zn, Fe and Cl, etc. (Rothwell et al., 2007; Monteiro et al., 2011; Visser et al., 2012; Schroth et al., 2015; Zhang et al., 2018) (Fig. 5.3). Studies on lakes of Chenghai and Qionghai (China) have found positive correlation of Cd and Pb distribution with annual average air temperatures, where the anthropogenic sources were phosphate fertilizers and combustion of fossil fuels (Zhang et al., 2018). Several studies have reviewed the potential of climate change in altering the quality of surface waters and groundwater as climate change have the control on precipitation patterns and quantities in reference to metal contamination (Aulenbach, 2013; García-Ordiales et al., 2016; Frogner-Kockum et al., 2020).

The surge of CO_2 concentration has indirectly accelerated chemical weathering (due to both pH level and temperature), perhaps this is one of the importance factor that have encouraged metal exposure to soil and marine environments (Prathumratana et al., 2008; Benitez-Gilabert et al., 2010; Frogner-Kockum et al., 2020). Metal like mercury stored in water sediments undergoes methylation by algae turning into a potential toxic agent. This contamination-transformation of metals to it toxic in waterbodies as well as other ecosystems have proven to be detrimental for all life—forms. Metals in the form of organic and inorganic can enter the food chain getting magnified with increasing trophic level. Such cases were evident in Japan (1950s) when methyl mercury got accumulated in human through consumption of fishes containing organic form of methyl mercury, resulting into a deleterious disease *mina-mata* (Rignell-Hydbom et al., 2007; Pandey et al., 2011). Adjacently, elevated CO_2 levels and temperature enhance growth of aquatic plants which helps these plants to absorb metals at a better rate (Duan et al., 2014). This phenomenon influences binding capacity of metals like Zn, Cl, As, Hg, etc., with organic matters (Zhang et al., 2018). Climate change is thus responsible for distributions, speciation, and bioavailability of metals across soil-aquatic environments by manipulating the patterns of hydrological cycle in a dynamic nature.

5.9 Conclusions

Metals are ubiquitous in aquatic environment both in particulate and dissolved forms. They are naturally present in rocks but increased anthropogenic activities have excavated these stored metals leading to contamination of soil and aquatic systems. This chapter has been dedicated towards the biogeochemistry of metals (As, Cd, Hg, Zn, Cl, Pb, etc.) in aquatic environment. A deep discussion has been depicted on different aspects of metal contamination in aquatic environment and its relativity toward entering the food chain. Heavy metals are found to be have deleterious effects on aquatic communities as well as human beings. Bioaccumulation of heavy metal in aquatic animals is a common phenomenon. Under changing global climate, retention of heavy metals in plant and its organic matter is also a matter concern. It is quite evident that consumption of such plants and animals on a regular basis acts as a potential source of health issues in human beings. Water is an essential compound for sustainability of all life forms. It is a component of superlative class defined by its chemical properties that helps metals to get easily combined with. Increasing CO_2 level, chemical weathering, and climate change have assisted metals in combining with water more easily. Transformation of inorganic metals to their organic forms are realized as more toxic than the inorganic ones. Usage of such water are found to be the reason behind serious health issues like diarrhea, bronchitis, vomiting, *mina-mata, itai-itai* and even cancer, etc. With increasing population, it is quite evident that human world is still progressing towards urbanization with parallel generation of wastewater, contamination of aquatic environments and overexploitation of natural resources. Eventually metal toxicity poses a crucial challenge for governments and scientific community.

Nevertheless, the gradual awareness of metal toxicity have led the international organizations and legislatives to constrain the usage of heavy metals. As for example according to Clean Air Act, usage of lead as antiknocking agent in gasoline has been banned. Similarly usage of methyl mercury in batteries and fluorescent lights are prohibited. Simultaneously the government are now focusing toward wastewater treatment as well as reduction metal from the sources. The UN-sustainable development goals (specifically SDG 6) are one of the major initiatives towards achieving sustainability of natural resources like water and protecting it from pollutants like metals. Development of wastewater treatment plants with maximum efficiency are the now the priority of governments all over the world. Wastewater containing metals can be purified by techniques involving reverse osmosis, chemical precipitation, lime coagulation, ion exchange, etc. Similarly phytoremediation involving water hyacinth and aquaculture seedlings of Indian mustard, etc., can be used in aquatic environment to extract and absorb metals from water. Now a day's treatment with constructed wetlands are fruitful strategy to minimize metals concentration from water resources. It is known to us that wetland are the kidney of Earth, it has the humongous potential to absorb and metals and other trace nutrients. As a result this constructed wetland is a viable choice to remediate metal contaminated water along with existing conventional treatment methods. Apart from these landfills containing heavy metals should reside of concrete beds as to avoid leaching toward groundwater. Strict regulations are still needed for some industries who discharge their wastewater in river or lakes streams. Hence there is a need for synergistic participation of government, decision makers and common people to mitigate the problem of metal toxicity in natural water.

References

Adimalla, N., 2019. Heavy metals pollution assessment and its associated human health risk evaluation of urban soils from Indian cities: a review. Environmental Geochemistry and Health 42 (1), 173–190. Available from: https://doi.org/10.1007/S10653-019-00324-4.

Akinloye, O., Arowojolu, A.O., Shittu, O.B., Anetor, J.I., 2006. Cadmium toxicity: a possible cause of male infertility in Nigeria. Reproductive Biology 6 (1), 17–30.

Alexander, B.H., Checkoway, H., Faustman, E.M., Netten, C.V., 1998. Contrasting associations of blood and semen lead concentrations with semen quality among lead smelter workers. American Journal of Industrial Medicine 34 (4), 464–469. Available from: https://doi.org/10.1002/(sici)1097-0274(199811)34:4<464::aid-ajim20>3.0.co;2-z.

Ali, H., Khan, E., 2018. What are heavy metals? Long-standing controversy over the scientific use of the term 'heavy metals'-proposal of a comprehensive definition. Toxicological & Environmental Chemistry 100 (1), 6–19. Available from: https://doi.org/10.1080/02772248.2017.1413652.

Ali, H., Khan, E., Ilahi, I., 2019. Environmental chemistry and ecotoxicology of hazardous heavy metals: environmental persistence, toxicity, and bioaccumulation. Journal of Chemistry. Available from: https://doi.org/10.1155/2019/6730305.

Alves, L.R., Reis, A.R.dos, Gratão, P.L., 2016. Heavy metals in agricultural soils: from plants to our daily life. Científica 44 (3), 346–361. Available from: https://doi.org/10.15361/1984-5529.2016V44N3P346-361.

Amara, I., Miled, W., Slama, R.B., Ladhari, N., 2018. Antifouling processes and toxicity effects of antifouling paints on marine environment. A review. Environmental Toxicology and Pharmacology 57, 115–130. Available from: https://doi.org/10.1016/J.ETAP.2017.12.001.

ATSDR, 2008. Toxicological profile for Cadmium (Draft for Public Comment). <https://scholar.google.com/scholar?hl=en&as_sdt=0%2C5&scioq=ATSDR+%281999%29+Toxicological+Profile+for+Mercury+%5BUpdate%5D.+(Ed.)+Agency+for+Toxic+Substances+and+Disease+Registry.+US+Department+of+Health+and+Human+Services.+Public+Health+Service%2C+Atlanta> (accessed 09.11.21).

Aulenbach, B.T., 2013. Improving regression-model-based streamwater constituent load estimates derived from serially correlated data. Journal of Hydrology 503, 55–66. Available from: https://doi.org/10.1016/J.JHYDROL.2013.09.001.

Baby, J., Raj, J.S., Biby, E.T., Sankarganesh, P., Jeevitha, M.V., Ajisha, S.U., et al., 2011. Toxic effect of heavy metals on aquatic environment. International Journal of Biological and Chemical Sciences 4 (4), 939–952. Available from: https://doi.org/10.4314/ijbcs.v4i4.62976.

Badr, F.M., El-Habit, O., 2018. Heavy metal toxicity affecting fertility and reproduction of males. Bioenvironmental Issues Affecting Men's Reproductive and Sexual Health 293–304. Available from: https://doi.org/10.1016/B978-0-12-801299-4.00018-9.

Bagul, V.R., Shinde, D.N., Chavan, R.P., Patil, C.L., Pawar, R.K., 2015. New perspective on heavy metal pollution of water. Journal of Chemical and Pharmaceutical Research 7 (12), 700–705.

Bai, L., Liu, X.L., Hu, J., Li, J., Wang, Z.L., Han, G., et al., 2018. Heavy metal accumulation in common aquatic plants in rivers and lakes in the Taihu Basin. International Journal of Environmental Research and Public Health 15 (12), 2857. Available from: https://doi.org/10.3390/IJERPH15122857.

Benitez-Gilabert, M., Alvarez-Cobelas, M., Angeler, D.G., 2010. Effects of climatic change on stream water quality in Spain. Climatic Change 103, 339–352.

Benoit, J.M., Gilmour, C.C., Mason, R.P., 2001a. The influence of sulfide on solid-phase mercury bioavailability for methylation by pure cultures of *Desulfobulbus propionicus* (1pr3). Environmental Science & Technology 35 (1), 127–132.

Benoit, J.M., Gilmour, C.C., Mason, R., 2001b. Aspects of bioavailability of mercury for methylation in pure cultures of *Desulfobulbus propionicus* (1pr3). Applied and Environmental Microbiology 67 (1), 51–58.

Benoit, J.M., Gilmour, C.C., Heyes, A., Mason, R.P., Miller, C., 2003. Geochemical and biological controls over methylmercury production and degradation in aquatic ecosystems. Biogeochemistry of Environmentally Important Trace Elements 262–296.

Bentley, R., Chasteen, T.G., 2002. Microbial methylation of metalloids: arsenic, antimony, and bismuth. Microbiology and Molecular Biology Reviews 66 (2), 250–271.

Bhattacharyya, K., Sengupta, S., 2020. Arsenic management options in soil-plant-food chain, pp. 1721. bausabour.ac.in.

Bhattacharya, P., Chandra, A.C., Majumdar, J., Santra, S.C., 2010a. Uptake of arsenic in rice plant varieties cultivated with arsenic rich groundwater. Environment Asia 3 (2), 34–37. Available from: http://www.tshe.org/EA.

Bhattacharya, P., Jacks, G., Nath, B., Chatterjee, D., Biswas, A., Halder, D., et al., 2010b. Natural arsenic in coastal groundwaters in the bengal delta region in West Bengal, India. Management and Sustainable Development of Coastal Zone Environments 146–160. Available from: https://doi.org/10.1007/978-90-481-3068-9_10.

Bhattacharyya, K., Sengupta, S., Pari, A., Halder, S., Bhattacharya, P., Pandian, B., et al., 2021. Characterization and risk assessment of arsenic contamination in soil–plant (vegetable) system and its mitigation through water harvesting and organic amendment. Environmental Geochemistry and Health 43, 1–4.

Biswas, J.K., Warke, M., Datta, R., Sarkar, D., 2020. Is arsenic in rice a major human health concern? Current Pollution Reports 6 (2), 37–42. Available from: https://doi.org/10.1007/S40726-020-00148-2.

Biswas, J.K., Mondal, B., Priyadarshini, P., Abhilash, P.C., Biswas, S., Bhatnagar, A., 2021. Formulation of Water Sustainability Index for India as a performance gauge for realizing the United Nations Sustainable Development Goal 6. Ambio 1–19. Available from: https://doi.org/10.1007/S13280-021-01680-1.

Boman, J., Wagner, A., 2003. Biomonitoring of trace elements in muscle and liver tissue of freshwater fish. Related papers. Spectrochimica Acta. Part B: Atomic Spectroscopy 58 (12), 2215–2226. Available from: https://doi.org/10.1016/j.sab.2003.05.003.

Bower, P.M., Simpson, H.J., Williams, S.C., Li, Y.H., 1978. Heavy metals in the sediments of foundry cove, cold spring, New York. Environmental Science and Technology 12 (6), 683–687.

Bowles, J.H., 1908. Lead Poisoning in Ducks, 25. Auk, pp. 312–313. Available from: http://doi.org/10.2307/4070529.

Byrnes, B.H., Bumb, B.L., 2017. Population growth, food production and nutrient requirements. Nutrient Use in Crop Production 1–27. Available from: https://doi.org/10.1201/9780203745281-1.

Bytyci, P., Fetoshi, O., Durmishi, B., Etemi, F.Z., Cadraku, H., Ismaili, M., et al., 2018. Status assessment of heavy metals in water of the Lepenci River Basin, Kosova. Journal of Ecological Engineering 19 (5), 19–32. Available from: https://doi.org/10.12911/22998993/91273.

Chen, L.-C., Maciejczyk, P., Thurston, G., 2022. Metals and air pollution. Handbook on the Toxicology of Metals. Academic Press, pp. 137–182. Available from: https://doi.org/10.1016/B978-0-12-823292-7.00004-8.

Chia, S.E., Ong, C.N., Lee, S.T., Tsakok, F.H., 1992. Blood concentrations of lead, cadmium, mercury, zinc, and copper and human semen parameters. Systems Biology in Reproductive Medicine 29 (2), 177–183.

Clarkson, T.W., 1990. Human health risks from methylmercury in fish. Environmental Toxicology and Chemistry: An International Journal 9 (7), 957–961.

Clarkson, T.W., 1994. The toxicology of mercury and its compounds. Mercury Pollution Integration and Synthesis. pp. 631–641.

Clarkson, T.W., Magos, L., 2008. The toxicology of mercury and its chemical compounds. Critical Reviews in Toxicology 36 (8), 609–662. Available from: https://doi.org/10.1080/10408440600845619.

Compeau, G.C., Bartha, R., 1985. Sulfate-reducing bacteria: principal methylators of mercury in anoxic estuarine sediment. Applied and Environmental Microbiology 50 (2), 498–502.

Coombs, T.L., 1980. Heavy metal pollutants in the aquatic environment, animals and environmental fitness: physiological and biochemical aspects of adaptation and ecology, pp. 283–302. doi:10.1016/B978-0-08-024938-4.50021-5.

Craig, P.J., Jenkins, R.O., 2004. Organometallic compounds in the environment: an overview. Organic Metal and Metalloid Species in the Environment 1–15.

D'Itri, F.M., 2020. The biomethylation and cycling of selected metals and metalloids in aquatic sediments. Environmental & Ecological Toxicology 163–214. Available from: https://doi.org/10.1201/9780367810894-6.

DalCorso, G., Manara, A., Furini, A., 2013. An overview of heavy metal challenge in plants: from roots to shoots. Metallomics 5 (9), 1117–1132. Available from: https://doi.org/10.1039/C3MT00038A.

Dickman, M.D., Leung, C.K.M., Leong, M.K.H., 1998. Hong Kong male subfertility links to mercury in human hair and fish. Science of the Total Environment 214 (1–3), 165–174. Available from: https://doi.org/10.1016/S0048-9697(98)00062-X.

Dissanayake, C.B., Chandrajith, R., 2009. Phosphate mineral fertilizers, trace metals and human health. Journal of the National Science Foundation of Sri Lanka 37 (3), 153–165. Available from: https://doi.org/10.4038/jnsfsr.v37i3.1219.

Duan, D., Ran, Y., Cheng, H., Chen, J., Wan, G., 2014. Contamination trends of trace metals and coupling with algal productivity in sediment cores in Pearl River Delta, South China. Chemosphere 103, 35–43. Available from: https://doi.org/10.1016/j.chemosphere.2013.11.011.

Ebenebe, P.C., Shale, K., Sedibe, M., Tikilili, P., Achilonu, M.C., 2017. South African mine effluents: heavy metal pollution and impact on the ecosystem. International Journal of Chemical Science 15 (4), 198. Available from: http://ir.cut.ac.za/handle/11462/1628.

Edmonds, J.S., Francesconi, K.A., 2003. Organoarsenic compounds in the marine environment. Organometallic Compounds in the Environment 195–222.

Eisler, R., 1994. A review of Arsenic hazards to plants and animals with emphasis on fishery and wildlife resources. Arsenic in the Environment, Part II: Human Health and Ecosystem Effects 185–259.

Evenson, R.E., Gollin, D., 2003. Assessing the impact of the green revolution, 1960 to 2000. Science (New York, N.Y.) 300 (5620), 758–762. Available from: https://doi.org/10.1126/SCIENCE.1078710.

Feldman, J., 2003. Volatilization of metals from a landfill site. In: Cai, Y., Braids, O.C. (Eds.), Biogeochemistry of Environmentally Important Trace Elements. *ACS Series, Washington, DC*, pp. 128–140.

Fleming, E.J., Mack, E.E., Green, P.G., Nelson, D.C., 2006. Mercury methylation from unexpected sources: molybdate-inhibited freshwater sediments and an iron-reducing bacterium. Applied and Environmental Microbiology 72 (1), 457–464.

Frogner-Kockum, P., Goransson, G., Haeger-Eugensson, M., 2020. Impact of climate change on metal and suspended sediment concentrations in urban waters. Frontiers in Environmental Science 8, 269. Available from: https://doi.org/10.3389/FENVS.2020.588335/BIBTEX.

Furness, R.W., Rainbow, P.S., 2018. Heavy metals in the marine environment. Heavy Metals in the Marine Environment. Available from: https://doi.org/10.1201/9781351073158.

Gallego-Schmid, A., Tarpani, R.R.Z., 2019. Life cycle assessment of wastewater treatment in developing countries: a review. Water Research 153, 63–79. Available from: https://doi.org/10.1016/J.WATRES.2019.01.010.

García-Ordiales, E., Esbrí, J.M., Covelli, S., López-Berdonces, M.A., Higueras, P.L., Loredo, J., 2016. Heavy metal contamination in sediments of an artificial reservoir impacted by long-term mining activity in the Almadén mercury district (Spain). Environmental Science and Pollution Research 23 (7), 6024–6038. Available from: https://doi.org/10.1007/S11356-015-4770-6/FIGURES/7.

Gbaruko, B.C., Friday, O.U., 2007. Bioaccumulation of heavy metals in some fauna and flora. International Journal of Environmental Science and Technology 4 (2), 197–202. Available from: https://doi.org/10.1007/BF03326274.

Gennart, J.P., Roels, H., Ghyselen, P., Ceulemans, E., Lauwerys, R., 1992. Fertility of male workers exposed to cadmium, lead, or manganese. American Journal of Epidemiology 135 (11), 1208–1219. Available from: https://doi.org/10.1093/OXFORDJOURNALS.AJE.A116227.

Gill, S.S., Hasanuzzaman, M., Nahar, K., Macovei, K., Tuteja, N., 2013. Importance of nitric oxide in cadmium stress tolerance in crop plants. Plant Physiology and Biochemistry 63, 254–261. Available from: https://doi.org/10.1016/J.PLAPHY.2012.12.001.

Gilmour, C.C., Henry, E.A., Mitchell, R., 1992. Sulfate stimulation of mercury methylation in freshwater sediments. Environmental Science & Technology 26 (11), 2281–2287.

Gittleman, J.L., Engelgau, M.M., Shaw, J., Wille, K.K., Seligman, P.J., 1994. Lead poisoning among battery reclamation workers in Alabama. Journal of Occupational Medicine 36 (5), 526–532.

Hammerschmidt, C.R., Fitzgerald, W.F., 2004. Geochemical controls on the production and distribution of methylmercury in near-shore marine sediments. Environmental Science & Technology 38 (5), 1487–1495.

Hare, V., Chowdhary, P., Singh, A.K., 2020. Arsenic toxicity: adverse effect and recent advance in microbes mediated bioremediation. Microorganisms for Sustainable Environment and Health 53–80.

Heyes, A., Mason, R.P., Kim, E.H., Sunderland, E., 2006. Mercury methylation in estuaries: insights from using measuring rates using stable mercury isotopes. Marine Chemistry 102 (1-2), 134–147.

Hoffman, D.J., Moore, J.M., 1979. Teratogenic effects of external egg applications of methyl mercury in the mallard, Anas platyrhynchos. Teratology 20 (3), 453–461. Available from: https://doi.org/10.1002/TERA.1420200315.

Hudson, R., Haegele, M., 1984. Handbook of toxicity of pesticides to wildlife. US Department of the Interior, Fish and Wildlife Service. <https://books.google.com/books?hl = en&lr = &id = 7u0UAQAAIAAJ&oi = fnd&pg = PP7&dq = Hudson, + R. + H. + Tucker. + R. + K., + and + Haegele, + M. + A. + (1984). + Handbook + of + t + ox + icily + of + pesticides + to + wildlife. + U.S. + Fish + Witdl. + Sen. + Resour. + Publ. + 153, + 1-90. &ots = QNG6rsG97O&sig = Py0I_0z> (accessed 10.11.21).

Huseen, H.M., Mohammed, A.J., 2019. Heavy metals causing toxicity in fishes. Journal of Physics: Conference Series 1294 (6). Available from: https://doi.org/10.1088/1742-6596/1294/6/062028.

IPCC, 2014. Chapter Climate Change 2014 Synthesis Report Summary for Policymakers Summary for Policymakers. https://www.ipcc.ch/site/assets/uploads/2018/06/AR5_SYR_FINAL_SPM.pdf.

IPCC, 2021. Summary for policymakers. In: Climate Change 2021. The Physical Science Basis. Contribution of Working Group I to the Sixth Assessment Report of the Intergovernmental Panel on Climate Change [Masson Delmotte, V., P. Zhai, A. Pirani, S.L. Connors, C. Péan, S. Berger, N. Caud, Y. Chen, L. Goldfarb, M.I. Gomis, M. Huang, K. Leitzell, E. Lonnoy, J.B.R. Matthews, T.K. Maycock, T. Waterfield, O. Yelekçi, R. Yu, and B. Zhou (eds.)]. Cambridge University Press.

Islam, M.S., Ahmed, M.K., Raknuzzaman, M., Habibullah-Al-Mamun, M., Islam, M.K., 2015. Heavy metal pollution in surface water and sediment: a preliminary assessment of an urban river in a developing country. Ecological Indicators 48, 282–291. Available from: https://doi.org/10.1016/J.ECOLIND.2014.08.016.

Jaishankar, M., Tseten, T., Anbalagan, N., Mathew, B.B., Beeregowda, K.N., 2014. Toxicity, mechanism and health effects of some heavy metals. Interdisciplinary Toxicology 7 (2), 60. Available from: https://doi.org/10.2478/INTOX-2014-0009.

Jayasinghe, T.M., Sandaruwan, K.P.G.L., De Silva, D.W.L.U., Jinadasa, K., 2018. Total Diet Study approach in estimating mercury and cadmium levels using selected fish: a case study from Sri LankaWomen in Fisheries View project Fatty Acid Composition of Three Different Marine Fish Under Different Culinary Process View projectCeylon Journal of Science 47 (3), 275–279. Available from: https://doi.org/10.4038/cjs.v47i3.7534.

Jericevic, A., Ilyin, I., Vidic, S., 2012. Modelling of heavy metals: study of impacts due to climate change. National Security and Human Health Implications of Climate Change 125, 175–189. Available from: https://doi.org/10.1007/978-94-007-2430-3_15.

Jurasovic, J., Cvitkovi, P., Pizent, A., Colak, B., Telisman, S., 2004. Semen quality and reproductive endocrine function with regard to blood cadmium in Croatian male subjects. Biometals: An International Journal on the Role of Metal Ions in Biology, Biochemistry, and Medicine 17 (6), 735–743. Available from: https://doi.org/10.1007/S10534-004-1689-7.

Kennish, M.J., 2017. Practical Handbook of Estuarine and Marine Pollution. CRC Press.

Kerin, E.J., Gilmour, C.C., Roden, E., Suzuki, M.T., Coates, J.D., Mason, R., 2006. Mercury methylation by dissimilatory iron-reducing bacteria. Applied and Environmental Microbiology 72 (12), 7919–7921.

Kerrison, P.H., Annoni, D., Zarini, S., Ravera, O., Moss, B., 1988. Effects of low concentrations of heavy metals on plankton community dynamics in a small, shallow, fertile lake. Journal of Plankton Research 10 (4), 779–812. Available from: https://doi.org/10.1093/PLANKT/10.4.779.

Khan, R., Saxena, A., Shukla, S., Sekar, S., Senapathi, V., Wu, J., 2021. Environmental contamination by heavy metals and associated human health risk assessment: a case study of surface water in Gomti River Basin, India. Environmental Science and Pollution Research 28 (40), 56105–56116. Available from: https://doi.org/10.1007/S11356-021-14592-0/FIGURES/8.

Kristanti, R.A., Jie Ngu, W., Yuniarto, A., Hadibarata, T., 2021. Rhizofiltration for removal of inorganic and organic pollutants in groundwater: a review. Biointerface Research in Applied Chemistry 11, 12326–12347. Available from: https://doi.org/10.33263/BRIAC114.1232612347.

Kumar, A., Samadder, S.R., Elumalai, S.P., 2016. Recovery of trace and heavy metals from coal combustion residues for reuse and safe disposal: a review. JOM 68 (9), 2413–2417. Available from: https://doi.org/10.1007/S11837-016-1981-3.

Kumar, V., Parihar, R.D., Sharma, A., Bakshi, P., Sidhu, G.P.S., Bali, A.S., et al., 2019. Global evaluation of heavy metal content in surface water bodies: a meta-analysis using heavy metal pollution indices and multivariate statistical analyses. Chemosphere 236, 124364. Available from: https://doi.org/10.1016/j.chemosphere.2019.124364.

Kumar, S., Toppo, S., Kumar, A., Tewari, G., Beck, A., Bachan, V., et al., 2020. Assessment of heavy metal pollution in groundwater of an industrial area: a case study from Ramgarh, Jharkhand, India. International Journal of Environmental Analytical Chemistry . Available from: https://doi.org/10.1080/03067319.2020.1828391.

Kuznets, S., 2019. Economic growth and income inequality. The Gap Between Rich and Poor: Contending Perspectives on the Political Economy of Development 25–37. Available from: https://doi.org/10.4324/9780429311208-4/ECONOMIC-GROWTH-INCOME-INEQUALITY-SIMON-KUZNETS.

Lall, S.P., Kaushik, S.J., 2021. Nutrition and metabolism of minerals in fish. Animals 11 (9), 2711. Available from: https://doi.org/10.3390/ANI11092711.

Laveskog, A., 2020. Gasoline additives: past, present, and future. Biological Effects of Organolead Compounds 5–12. Available from: https://doi.org/10.1201/9780429282980-2.

Lee, R.E., Von Lehmden, D.J., 1973. Trace metal pollution in the environment. Journal of the Air Pollution Control Association 23 (10), 853–857. Available from: https://doi.org/10.1080/00022470.1973.10469854.

Leelarungroj, K., Likitlersuang, S., Chompoorat, T., Janjaroen, D., 2018. Leaching mechanisms of heavy metals from fly ash stabilised soils. Waste Management and Research 36 (7), 616–623. Available from: https://doi.org/10.1177/0734242X18775494.

Li, W.X., Chen, T.B., Huang, Z.C., Lei, M., Liao, X.Y., 2006. Effect of arsenic on chloroplast ultrastructure and calcium distribution in arsenic hyperaccumulator *Pteris vittata* L. Chemosphere 62 (5), 803–809. Available from: https://doi.org/10.1016/j.chemosphere.2005.04.055.

Li, J., Zheng, B., He, Y., Zhou, Y., Chen, X., Ruan, S., et al., 2018. Antimony contamination, consequences and removal techniques: a review. Ecotoxicology and Environmental Safety 156, 125–134. Available from: https://doi.org/10.1016/j.ecoenv.2018.03.024.

López-Berenguer, G., Peñalver, J., Martínez-López, E., 2020. A critical review about neurotoxic effects in marine mammals of mercury and other trace elements. Chemosphere 246, 125688. Available from: https://doi.org/10.1016/J.CHEMOSPHERE.2019.125688.

Manahan, S., 2017. Environmental Chemistry, tenth ed. CRC Press.

Marcovecchio, J., Botte, S.E., Freije, H., 2007. Heavy metals, major metals, trace elements, Handbook of Water Analysis, second ed. CRC Press, London, pp. 275–311. Available from: https://www.researchgate.net/publication/284026619.

Mason, R.P., 2012. The methylation of metals and metalloids in aquatic systems. Methylation—From DNA, RNA and Histones to Diseases and Treatment 71–301.

Masters, G.M., 1997. Introduction to Environmental Science and Engineering. Prentice-Hall, Upper Saddle River, NJ. Available from: https://catsr.vse.gmu.edu/SYST460/Aviation_Environmet_Water_Workbook.pdf.

Masters, G. M., Ela, W. P., 1974. Introduction to Environmental Science & Technology–3rd. <https://catsr.vse.gmu.edu/SYST460/Aviation_Environmet_Air_Workbook.pdf> (accessed 10.11.21).

Meeker, J.D., Rossano, M.G., Protas, B., Diamond, M.P., Puscheck, E., Daly, D., et al., 2008. Cadmium, lead, and other metals in relation to semen quality: human evidence for molybdenum as a male reproductive toxicant. Environmental Health Perspectives 116 (11), 1473–1479. Available from: https://doi.org/10.1289/EHP.11490.

Mehrandish, R., Rahimian, A., Shahriary, A., 2019. Heavy metals detoxification: a review of herbal compounds for chelation therapy in heavy metals toxicity. Journal of Herbmed Pharmacology 8 (2), 69–77. Available from: https://doi.org/10.15171/jhp.2019.12.

Mendie, U., 2005. The nature of water. In: The Theory and Practice of Clean Water Production for Domestic and Industrial Use. Lagos: Lacto-Medals Publishers, 1, p. 21.

Miranda, L.S., Wijesiri, B., Ayoko, G.A., Egodawatta, P., Goonetilleke, A., 2021. Water-sediment interactions and mobility of heavy metals in aquatic environments. Water Research 202, 117386. Available from: https://doi.org/10.1016/J.WATRES.2021.117386.

Monteiro, F.F., Cordeiro, R.C., Santelli, R.E., Machado, W., Evangelista, H., Villar, L.S., et al., 2011. Sedimentary geochemical record of historical anthropogenic activities affecting Guanabara Bay (Brazil) environmental quality. Environmental Earth Sciences 65 (6), 1661–1669. Available from: https://doi.org/10.1007/S12665-011-1143-4.

Nammalwar, P., 1983. Heavy metals pollution in the marine environment. Science Report 20 (3), 158–160.

Pandey, K., Shukla, J.P., Trivedi, S.P., 2011. Fundamentals of Toxicology, fourth rev ed. New Central Book Agency.

Pant, N., Banerjee, A.K., Pandey, S., Mathur, N., Saxena, D.K., Srivastava, S.P., 2003. Correlation of lead and cadmium in human seminal plasma with seminal vesicle and prostatic markers. Human and Experimental Toxicology 22 (3), 125–128. Available from: https://doi.org/10.1191/0960327103ht336oa.

Park, H., Wang, L., Yun, J.H., 2021. Coal beneficiation technology to reduce hazardous heavy metals in fly ash. Journal of Hazardous Materials 416, 125853. Available from: https://doi.org/10.1016/J.JHAZMAT.2021.125853.

Peterle, T., 1991. Wildlife toxicology. <https://www.cabdirect.org/cabdirect/abstract/19930513431> (accessed 10.11.21).

Plant, J.A., Kinniburgh, D.G., Smedley, P.L., Fordyce, F.M., Klinck, B.A., 2005. Arsenic and selenium. Environmental Geochemistry 9, 17–66.

Pobi, K.K., Satpati, S., Dutta, S., Nayek, S., Saha, R.N., Gupta, S., 2019. Sources evaluation and ecological risk assessment of heavy metals accumulated within a natural stream of Durgapur industrial zone, India, by using multivariate analysis and pollution indices. Applied Water Science 9 (3), 1–16. Available from: https://doi.org/10.1007/S13201-019-0946-4.

Pongratz, R., Heumann, K.G., 1999. Production of methylated mercury, lead, and cadmium by marine bacteria as a significant natural source for atmospheric heavy metals in polar regions. Chemosphere 39 (1), 89–102.

Prathumratana, L., Sthiannopkao, S., Kim, K.W., 2008. The relationship of climatic and hydrological parameters to surface water quality in the lower Mekong River. Environment International 34 (6), 860–866. Available from: https://doi.org/10.1016/j.envint.2007.10.011.

Proshad, R., Islam, S., Tusher, T.R., Zhang, D., Khadka, S., Gao, J., et al., 2020. Appraisal of heavy metal toxicity in surface water with human health risk by a novel approach: a study on an urban river in vicinity to industrial areas of Bangladesh. Toxin Reviews 0 (0), 1–17. Available from: https://doi.org/10.1080/15569543.2020.1780615.

Prüss-Üstün, A., Wolf, J., Corvalán, C., Bos, R., Neira, M., 2016. Preventing Disease Through Healthy Environments: A Global Assessment of the Burden of Disease From Environmental Risks. World Health Organization. Available from: https://books.google.com/books?hl = en&lr = &id = HQ8LDgAAQBAJ&oi = fnd&pg = PP1&dq = PREVENTING + DISEASE + THROUGH + HEALTHY + ENVIRONMENTS&ots = QGWCiwlOcj&sig = J6StmoJidp8_1KQNdiVjWeHE2Fs.

Qu, L., Huang, H., Xia, F., Liu, Y., Dahlgren, R.A., Zhang, M., et al., 2018. Risk analysis of heavy metal concentration in surface waters across the rural-urban interface of the Wen-Rui Tang River, China. Environmental Pollution 237, 639–649. Available from: https://doi.org/10.1016/j.envpol.2018.02.020.

Rand, G.M., Wells, P.G., McCarty, L.S., 2020. Introduction to aquatic toxicology. Fundamentals of Aquatic Toxicology 3–67. Available from: https://doi.org/10.1201/9781003075363-2.

Rattner, B.A., 2009. History of wildlife toxicology. Ecotoxicology (London, England) 18 (7), 773–783. Available from: https://doi.org/10.1007/s10646-009-0354-x.

Rignell-Hydbom, A., Axmon, A., Lundh, T., Jönsson, B.A., Tiido, T., Spano, M., 2007. Dietary exposure to methyl mercury and PCB and the associations with semen parameters among Swedish fishermen. Environmental Health: A Global Access Science Source 6 (1), 1–10. Available from: https://doi.org/10.1186/1476-069X-6-14/TABLES/3.

Rose, S., Elliott, W.C., 2000. The effects of pH regulation upon the release of sulfate from ferric precipitates formed in acid mine drainage. Applied Geochemistry 15 (1), 27–34. Available from: https://doi.org/10.1016/S0883-2927(99)00015-3.

Rothwell, J.J., Evans, M.G., Daniels, S.M., Allott, T.E.H., 2007. Baseflow and stormflow metal concentrations in streams draining contaminated peat moorlands in the Peak District National Park (UK). Journal of Hydrology 341 (1–2), 90–104. Available from: https://doi.org/10.1016/J.JHYDROL.2007.05.004.

Saha, P., Paul, B., 2019. Assessment of heavy metal toxicity related with human health risk in the surface water of an industrialized area by a novel technique. Human and Ecological Risk Assessment: An International Journal 25 (4), 966–987. Available from: https://doi.org/10.1080/10807039.2018.1458595.

Salomons, W., Forstner, U., 2012a. Metal Pollution in the Aquatic Environment. Springer Science and Business Media B.V. <https://books.google.co.in/books?hl = en&lr = &id = 9PLuCAAAQBAJ&oi = fnd&pg = PA1&dq = Metal + pollution + in + aquatic + environment + &ots = 4STc4Qg0-I&sig = 24SYQsur9t8rQRv-G1E9wUoNxXM&redir_esc = y#v = onepage&q = Metal.pollution.in.aquatic.environment&f = false> (accessed 09.10.21).

Salomons, W., Forstner, U., 2012b. Metals in the Hydrocycle. Springer-Verlag Berlin Heidelberg. https://doi.org/10.1007/978-3-642-69325-0.

Sankhla, M.S., Kumar, R., 2019. Contaminant of heavy metals in groundwater & its toxic effects on human health & environment. International Journal of Environmental Sciences & Natural Resources 18 (5), 1–5. Available from: https://doi.org/10.19080/ijesnr.2019.18.555996.

Schreiber, E., Burger, J., 2001. Biology of Marine Birds. CRC Press.

Schroth, A.W., Giles, C.D., Isles, P.D.F., Xu, Y., Perzan, Z., Druschel, G.K., 2015. Dynamic coupling of iron, manganese, and phosphorus behavior in water and sediment of shallow ice-covered eutrophic lakes. Environmental Science and Technology 49 (16), 9758–9767. Available from: https://doi.org/10.1021/ACS.EST.5B02057/SUPPL_FILE/ES5B02057_SI_001.PDF.

Selvaraju, S., Jodar, M., Krawetz, S.A., 2014. The influence of environmental contaminants and lifestyle on testicular damage and male fertility, pp. 185–203. doi:10.1007/7653_2014_13.

Sengupta, S., Bhattacharyya, K., Mandal, J., Bhattacharya, P., Halder, S., Pari, A., 2021. Deficit irrigation and organic amendments can reduce dietary arsenic risk from rice: introducing machine learning-based prediction models from field data. Agriculture, Ecosystems & Environment 319, 107516. Available from: https://doi.org/10.1016/J.AGEE.2021.107516.

Sharma, S.S., Dietz, K.J., 2009. The relationship between metal toxicity and cellular redox imbalance. Trends in Plant Science 14 (1), 43–50. Available from: https://doi.org/10.1016/J.TPLANTS.2008.10.007.

Singh, S., Parihar, P., Singh, R., Singh, V.P., Prasad, S.M., 2016. Heavy metal tolerance in plants: role of transcriptomics, proteomics, metabolomics, and ionomics. Frontiers in Plant Science 6, 1143. Available from: https://doi.org/10.3389/fpls.2015.01143.

Sun, H. (Ed.), 2010. Biological Chemistry of Arsenic, Antimony and Bismuth. John Wiley & Sons.

Sun, J., Yu, G., Zhang, Y., Liu, X., Du, C., Wang, L., et al., 2017. Heavy metal level in human semen with different fertility: a meta-analysis. Biological Trace Element Research 176 (1), 27–36. Available from: https://doi.org/10.1007/S12011-016-0804-2.

Sunderland, E.M., Gobas, F.A., Branfireun, B.A., Heyes, A., 2006. Environmental controls on the speciation and distribution of mercury in coastal sediments. Marine Chemistry 102 (1-2), 111–123.

Sunderman, F. W., 1981. Nickel, disorders of mineral metabolism, pp. 201–232. Available from: https://doi.org/10.1016/B978-0-12-135301-8.50011-X.

Telisman, S., Cvitkovic, P., Jurasovic, J., Pizent, A., Gavella, M., Rocic, B., 2000. Semen quality and reproductive endocrine function in relation to biomarkers of lead, cadmium, zinc, and copper in men. Environmental Health Perspectives 108 (1), 45–53. Available from: https://doi.org/10.1289/EHP.0010845.

Tirkey, P., Bhattacharya, T., Chakraborty, S., 2016. Arsenic and other metals in the groundwater samples of Ranchi city, Jharkhand, India. Current Science 76–80. Available from: https://www.jstor.org/stable/24906614.

Vareda, J.P., Valente, A.J.M., Duraes, L., 2019. Assessment of heavy metal pollution from anthropogenic activities and remediation strategies: a review. Journal of Environmental Management 246, 101–118. Available from: https://doi.org/10.1016/j.jenvman.2019.05.126.

Visser, A., Kroes, J., Van Vliet, M.T.H., Blenkinsop, S., Fowler, H.J., Broers, H.P., 2012. Climate change impacts on the leaching of a heavy metal contamination in a small lowland catchment. Journal of Contaminant Hydrology 127 (1–4), 47–64. Available from: https://doi.org/10.1016/J.JCONHYD.2011.04.007.

Wang, J., Wu, M., Lu, G., Si, Y., 2016. Biotransformation and biomethylation of arsenic by *Shewanella oneidensis* MR-1. Chemosphere 145, 329–335. Available from: https://doi.org/10.1016/J.CHEMOSPHERE.2015.11.107.

Wang, Y.Y., Chai, L.Y., Yang, W.C., 2019. Arsenic distribution and pollution characteristics. Arsenic Pollution Control in Nonferrous Metallurgy. Springer, Singapore, pp. 1–15. Available from: https://doi.org/10.1007/978-981-13-6721-2_1.

Wirth, J.J., Mijal, R.S., 2010. Adverse effects of low level heavy metal exposure on male reproductive function. Systems Biology in Reproductive Medicine 56 (2), 147–167. Available from: https://doi.org/10.3109/19396360903582216.

Wojciechowska, E., Nawrot, N., Walkusz-Miotk, J., Matej-Łukowicz, K., Pazdro, K., 2019. Heavy metals in sediments of urban streams: contamination and health risk assessment of influencing factors. Sustainability 11 (3), 563. Available from: https://doi.org/10.3390/SU11030563.

Wolfe, M.F., Schwarzbach, S., Sulaiman, R.A., 1998. Effects of mercury on wildlife: a comprehensive review. Environmental Toxicology and Chemistry 17 (2), 146–160. doi: 10.1897/1551-5028(1998)017<0146:EOMOWA>2.3.CO;2.

Xu, B., Chia, S.E., Tsakok, M., Ong, C.N., 1993. Trace elements in blood and seminal plasma and their relationship to sperm quality. Reproductive Toxicology 7 (6), 613–618. Available from: https://doi.org/10.1016/0890-6238(93)90038-9.

Yin, K., Yin, K., Wang, Q., Lv, M., Chen, L., 2019. Microorganism remediation strategies towards heavy metals. Chemical Engineering Journal 360, 1553–1563. Available from: https://doi.org/10.1016/J.CEJ.2018.10.226.

Zhang, H., Huo, S., Yeager, K.M., Xi, B., Zhang, J., He, Z., et al., 2018. Accumulation of arsenic, mercury and heavy metals in lacustrine sediment in relation to eutrophication: impacts of sources and climate change. Ecological Indicators 93, 771–780. Available from: https://www.sciencedirect.com/science/article/pii/S1470160X1830400X.

Zhou, Q., Zhang, J., Fu, J., Shi, J., Jiang, G., 2008. Biomonitoring: an appealing tool for assessment of metal pollution in the aquatic ecosystem. Analytica Chimica Acta 606 (2), 125–150. Available from: https://doi.org/10.1016/j.aca.2007.11.018.

ZongTang, L., ChunHai, L., GangYa, Z., 2010. Application of principal component analysis to the distributions of heavy metals in the water of lakes and reservoirs in Yunnan Province. Research of Environmental Sciences 23 (4), 459–466. Available from: https://www.cabdirect.org/cabdirect/abstract/20103149191.

CHAPTER

6

Assessment and impact of metal toxicity on wildlife and human health

Nitin Verma[1], *Mahesh Rachamalla*[2], *P. Sravan Kumar*[2] *and Kamal Dua*[3]

[1]Chitkara University School of Pharmacy, Chitkara University, Solan, Himachal Pradesh, India [2]Department of Biology, University of Saskatchewan, Saskatoon, SK, Canada [3]Discipline of Pharmacy, Graduate School of Health, University of Technology Sydney, Broadway, NSW, Australia

6.1 Introduction

Over the past few decades production and production of heavy metals (HMs) increased many folds worldwide due to increased economic activities. Due to increased anthropological activities levels of HMs in environmental compartments like water, soil, sediment, and air were increased by many folds due to use in various industrial applications. Due to raising levels of these metal pollutants in the environment which is further impact health and quality of various species on earth especially aquatic species due to release of large quantities of industrial effluents (Frantz et al., 2012). Increasing levels of these persisting HMs in environment will posses' serious risk to quality and sustainability of ecosystems. Over the last few years from various scientific publications, it is evident that exposure to HMs can cause serious health impacts in wide range of species including aquatic and terrestrial species. Some of the most noted impacts are endocrine dysfunction, impaired nervous system, genotoxicity, physiological and behavioral abnormalities (Burger and Gochfeld, 2000; Dauwe et al., 2004). Currently there are so many HMs were detected in various environmental compartments like Lead (Pb), Arsenic (As), Cobalt (Co), Cadmium (Cd), Nickel (Ni), Copper (Cu), Zinc (Zn), Iron (Fe), Manganese (Mn), and Chromium (Cr) there mounting evidence of literature available on each of these metals toxicity potentials in various organisms. Despite of various impact of HMs and huge literature available, currently in this chapter we have emphasized majorly an overview of HMs distribution,

DOI: https://doi.org/10.1016/B978-0-323-95919-3.00002-1

environmental fate, biomagnification, bioaccumulation potentials, and on toxic effects an bioremediations strategy.

6.2 Metals and its function in life

6.2.1 Heavy metals

HMs are generally referred to as the metals that have an atomic number more than 20 and have more than 5 g/cm^3 elemental density. Examples of the most abundant and common HM in environmental perspective are Mercury (Hg), Chromium (Cr), Zinc (Zn), Arsenic (As), and Cadmium (Cd). Some of the less common HMs include Cobalt (Co), Manganese (Mn), Iron (Fe), Nickel (Ni). HMs are among the most serious environmental contaminants which can cause severe ecological, environmental, and economic complications. HMs exposure is also a key concern because of their toxic, persistence, and bioaccumulation nature (Ayari et al., 2010; Briffa et al., 2020).

6.2.2 Essential and nonessential heavy metals

Metals are very important in physiological and biological function for all species, which are one of key elements for survival. However, in nature there are so many metals available in which some of them all essential for physiological function and survival of organism whereas some of them are not required for normal function of organism. These metals are categorized as (1) essential metals and (2) nonessential metals. Essential metals are absolutely necessary for an organism's many physiological and biochemical functions. On the other side nonessential metals do not have any physiological role in organism function (Fraga, 2005). However, the category of essential and nonessential metal including HMs will vary with various species and organisms according to their habitat and physiological need (Ali et al., 2019; Ali and Khan, 2019). The most common HMs which are essential for human body function are Fe, Cu, Mn, and Zn that are needed in very less quantities, at higher levels these metals can cause toxic effects. In ecotoxicological perspective essential HMs possess a very narrow index between beneficial and toxic levels. For aquatic species Zn and Cu are the most hazardous metals; long term exposure to these metals can cause detrimental impacts on various organisms including humans. Among all nonessential HMs Pb, Cd, Hg, and As are considered as the most notoriously toxic metalloids and metals. Being nonessential HMs, even at exposure at very low concentrations can lead to severe toxic impacts (Kim et al., 2019; Thakur et al., 2021).

6.3 Sources of heavy metals within environment

Natural processes like volcanic eruptions and weathering of metal-bearing rock formations are primary sources of contamination emitted into the atmosphere and anthropological activities like mining, industrial wastewater dumping and sewage sludge (Spiegel, 2002). Another major source also included the combustion of fossil fuel which causes release of metals into environment. Production of phosphate containing fertilizers are also key contributors of release of HMs, which is majorly due to sourcing of phosphate from phosphate rock which naturally contains various HMs like As, Cd, Pb, Uranium (U), and Cr (Dissanayake and Chandrajith, 2009). Cd are among the most toxic metals that is possibly concentrated in sedimentary rocks and is widely disseminated in the environment. Phosphate-based fertilizers are the key contributors of HM distribution in environmental compartments like water, air, soil, and sediment. Application of these fertilizers will further potentiate the distribution of HMs leading to transfer into food chain and leads exposure to humans directly or indirectly (Grant, 2011). Apart from fertilizers, other anthropological activities also cause leaching of HMs significantly to environment such as textile and tannery industry effluents, severe mining for rare metals, plastic industry, pigment manufacturing, plating, arsenic containing pesticides, marble industry and various household and industrial wastes (Ali et al., 2013; Grant and Sheppard, 2008; Hashem et al., 2017; Mulk et al., 2015; Sabiha-Javied et al., 2009; Madhav et al., 2018).

6.4 Contamination of natural waters, sediments, and soils by heavy metals

Water and soil contamination are the most two hazardous dangers that plague the majority of the world's countries. A specific water body's contamination can always be traced back to an industrial, sewage, or agricultural drainage. Toxic HMs are transported by wastewater and enter the aquatic and soil system via a various method, one being irrigation. Agricultural drainage water including pesticides and fertilisers, as well as industrial effluents and runoffs, along with sewage effluents, contribute massive amounts of HMs and inorganic anions to water bodies or sediment. Industrial, petroleum pollution, and sewage discharge are the major anthropogenic sources of metals (Mohamed Hassan et al., 2013; Mishra et al., 2022). Research was done in peri-urban Bangalore, India, where city effluent from four water bodies was utilized to grow vegetable crops. Analyses found significant amounts of Cd and Cr in all of the tank's fluids, much above the permitted values. The HM content of vegetables followed a similar pattern, aside from the fact that concentrations of HMs in soil have been below permitted levels during research according to the regulations. Higher levels of Cr in vegetables might be due to effluents released into the tanks from nearby industrial facilities, as well as long-term usage of the tank waters for vegetable growth. The Cr content in vegetables near this tank was approximately five times greater than in an uncontaminated area. Leafy vegetables like amaranthus and spinach had the highest mean concentration, followed by root vegetables like radish and carrot (Varalakshmi and Ganeshamurthy, 2010). Metal contamination of soil is largely caused by metal dust deposition. Cd and Zn pollution in soils near heavy industry, ore mines, and untreated wastewater irrigation regions should be given special attention (Ahamad et al., 2021). Contamination of HMs in soils and plants was studied in polluted areas, even those with heavy industry, smelting, and metal mining, untreated wastewater irrigation districts. Polygonum hydropiper has collected 1061 mg/kg of Zn within the shoots while developing in a sewage pond on disturbed soils. Rumex acetosa L., particularly flourished nearby a smelter, had roots and shoots that accumulated over 900 mg/kg of Zn (Wang et al., 2011a). This clearly shows that HMs are directly absorbed from contaminated soils and stored by plants. The amount of HM contaminants in the environment is increasing as a result of human activities. As a consequence of burning and smelting of coal, waste, and oil, industrial pollution would be emitted in to sky. The variable amounts of HMs (Cr and As) within coal had affected HM emissions into the environment in coal-burning areas (Adriano, 2001). Contamination by HMs is a severe ecological concern in developing countries, especially in burgeoning moderate cities, as a result of unregulated levels of pollution created by potential causes like industrial growth and an increase in traffic using petroleum fuels. And after processing at wastewater treatment plants, HMs are rarely eliminated from sewage, providing a risk of build-up of HMs of soil and, as a consequence, food web (Fytianos et al., 2001). Aerial metal resources include, air released from chimneys or ducts, vapor streams or gas, and short discharges such as dust from storage areas and trash dumps. Metals originating from the air are typically released as particles in the gas stream. High-temperature activities can also volatilize certain metals, such as Cd, As, and Pb. Until a reducing air is maintained, these metals will convert to oxides and shrink as tiny particles (Evanko and Dzombak, 1997; Ahamad et al., 2021). Major cause of soil contamination is the floating release of Pb from the combustion of fuel containing tetraethyl lead; this significantly increases the amount of Pb in urban and roadside soils (Ahamad et al., 2021). Tyres and lubricating fluids are two sources of zinc and cadmium that may be applied to soils near roadways (Duzgoren-Aydin, 2007). Other key issue which leads to soil pollution is electronic waste (e-waste), which is now the main source of municipal waste over last century as a result of rapid expansion of information systems as well as the constant updating of electronic devices (Schmidt, 2002). Due to cheaper labor cost and far less severe environmental rules, most industrialized nations choose to transfer e-waste to developing nations for disposal and recycling (for example, Bangladesh, China, India, Vietnam and Pakistan) (Ni and Zeng, 2009). Given the economic advantages, e-waste had also caused considerable pollution (e.g., water and soil contamination) since it is commonly illicitly handled in workshops and backyards utilizing basic methods like strong acid digestion, open burning of dismantled elements to retrieve metals, and dropping of irrecoverable components (Tang et al., 2010).

The distribution and extraction of mineral compounds from its natural sources has been accompanied by increasing industrialization across the world. Many types of hazardous materials enter the environment during mining and smelting processes, producing significant environmental concerns (Angelovičová and Fazekašová, 2014). HM pollution of groundwater is frequently linked to mining and the following ores processing. HMs can arrive in groundwater in soluble form as well as in pairing with products cleaned off from the ground, where they would travel long distances (Frankowski et al., 2009). Huge quantities of HMs have been observed in and near old metalliferous mine as a result of mine dumping waste and dissemination into neighboring cultivated fields, stream systems, and food sources (Jung, 2001). Contamination of the surface environment comes from mining and milling activities such as grinding ore and concentrating ores, with mill waste water and mine or the dispersion of metals for abandoned mine materials. Because semi-arid soils are generally sparsely vegetated, climatic variables such as large rainfall events have a significant impact on metal dispersion (Navarro et al., 2008). Depending upon geochemical features and grade of mineralization of the tailings, the amount and degree of HM pollution surrounding mines varies (Johnson et al., 2000). Some metals dissolve and enter solution when rainfall reaches the soil, whereas precipitated or adsorbed and migrate with sand particles. At the same time precipitation, coprecipitation, and sorption processes in and near mines reduce the metals produced by sulfide oxidation (Kim et al., 2002; Berger et al., 2000). Soil pollution with toxic metals, that is followed by rising urbanization and industrialization, is indeed a major worry due to the possible impact on human and animals (Li et al., 2019). Corrosion, air deposition, ion erosion, leaching, sediment resurfacing, and the Metal evaporation from such a contaminated source of water into water and soil too can lead to pollution (Cai et al., 2019). Natural activities like volcanic outbreaks and weathering also contribute to the pollution of the environment (Wu et al., 2018). Electroplating industrial effluent cause pollution soil, water and air and it is a significant cause of pollution since it releases hazardous compounds and HMs by water, solid waste, air emissions within environment with HM contamination identified by different industrial businesses (Akshaya et al., n.d.; Baby et al., 2011).

6.5 Human exposure to heavy metals

HMs are poisonous, bioaccumulative, bioavailable, and non-biodegradable by nature. Humans, plants, and a variety of living species use polluted water supplies, which has serious health consequences (Yu et al., 2008). HMs like As, Cd, Hg and Pb are soluble in fat and can penetrate biological barriers. Because of their biological amplification throughout the food web, HMs are detrimental to both human health and the nearby ecology. HMs may enter the body through three routes air, water and soil. Food consumption, dust/soil ingestion, inhalation, and dermal exposure are also an evident route of exposure of HMs. Humans may directly interact with more hazardous substances and associated contaminants through interaction with nearby dust, soil, air, and water, and through sources of food, in addition to the direct occupational (informal or formal) contact (Ji et al., 2013).

HMs are prone to contaminate drinking water sources (e.g., surface water, groundwater, and seawater). The ingestion of HMs by drinking water is practically inevitable. Small and rural communities, as well as individuals, are particularly susceptible. Because of the considerable dangers to human health, exposure to a few HMs (e.g., As, Cd, and Pb) is an urgent problem (Kaur et al., 2020; Chowdhury et al., 2016). Because of the possible health dangers to nearby residents, metal pollution from mining activities is a major worry. The quantities of Cd, Zn, Cu, As, Pb, in environmental models and foods cultivated near the mines in Korea, Goseong, were studied, and possible health hazards amid local inhabitants were assessed. Cu, As, and Zn pollution levels in the soil around the mines surpassed the soil quality standards. Cu, Cd, Zn, and Pb concentrations in crop sample obtained from research region were considerably greater than that of crop samples collected from reference area. The maximum allowed amount of 0.2 mg Cd/kg was surpassed in certain rice samples obtained from the research region (Ji et al., 2013). Fish is a rich source of protein; however, some anthropogenic activities can pollute their environment with high quantities of HMs. Cadmium, for example, is a persistent metal that can build up in muscles, crustaceans, and fish, resulting in human seafood poisoning. Daily, most individuals, particularly women, utilize cosmetics and their components. Toxic exposure causes cancers of the bones, heart, kidneys, liver, and other organs. Because of the pervasive nature of these elements, HM contaminants in cosmetic goods are inevitable, although they should be eliminated whenever technically possible. Cosmetics create local (eye, skin) interactions that are used on face, lips, eyes, and mucosa (Mishra et al., 2022). Given the enormous and mostly uncontrollable human exposure to cosmetics and its components, such products should be rigorously assessed for safety before being marketed (Human Exposure to Heavy Metals from Cosmetics: Oriental Journal of Chemistry, n.d.).

In addition to natural human activity different metals are exposed in different ways. Natural degassing of crust of the earth, for example, discharges mercury (Hg) into atmosphere (Nordberg et al., 2007). It appears as mercury vapor with in environment, and is the principal mechanism of worldwide dissemination for the contaminant. Returns to a surface of the earth and transforms to a water-soluble condition. This is converted back to or mono-methyl mercury components or mercury vapor by microorganisms, typically bacteria. Mono-methyl mercury has the potential to infiltrate the marine food chain via fish and planktons, among many other marine plants. As a consequence, people are commonly subjected to mercury with ingesting such species (Habiba et al., 2017; Sagiv et al., 2012). Consumption of (Pb) lead-contaminated meals, as well as absorption of lead particulates produced by combustion of lead-containing products such as household paint and gasoline are the two main pathways of lead exposure (Roy et al., 2009). Chromium (Cr) enters the body by absorption through the skin, inhalation, and ingestion. Occupational exposure to chromium is mostly through skin contact and inhalation. Ingestion of contaminated water and food, on the other hand, accounts for a significant percentage of general population exposure (Zhitkovich, 2005). Ingestion of contaminated foods like absorption of tobacco smoking, shellfish, metal handling improperly, and consuming cadmium-contaminated water all are ways for cadmium (Cd) to enter the human body. Notably, cadmium exposure occurs largely through plants. A plant's ability to absorb cadmium depends on its ability to absorb it from the fertilizers, irrigation water and airborne cadmium fallout (Olympio et al., 2018). For example, Rice (*Oryza Sativa*) consumption has been found to be the most significant source of Cd in humans (Olympio et al., 2018). Rice is now a crucial human meal, particularly in Asia, it is the backbone of diet in regions like China and South Asia. When rice fields are contaminated with hazardous HMs, the plant bioaccumulates the toxins. Public health is concerned about the transmission of HMs from rice plant roots to the stem, grains and leaves of rice (Rice, health, and toxic metals—Rice Today, n.d.). Rice contamination with carcinogenic HMs is a particular health risk for the underdeveloped nations (Besante et al., 2011). Another important HM barium (Ba) exposure occurs mostly through breath of polluted air and the consumption of contaminated drink and food. The barium compounds solubility within the circulation determines their absorption. Insoluble chemicals (barium carbonate and barium sulfate) are more easily absorbed into circulation than highly soluble ones (barium nitrate and barium chloride) (Medical Management Guidelines—Letter A, Toxic Substance Portal, ATSDR, n.d.).

The impact of HMs on children's health has already been proven being more severe than on adults. A child's lifetime exposure to chemicals might begin in the pregnancy through placental transfer at an early age. Nutrition is the main source of exposure after childhood. Breast milk, on the other hand, may include a variety of exogenous hazardous substances (dietary xenobiotics), such as HMs, which can harm the health of breastfed infants (Al Osman et al., 2019). A study was conducted in an Iranian city to determine the amounts of arsenic (As), and lead (Pb), chromium (Cr) with in breast milk of 100 nursing mothers. About 70% of examined breast samples collected surpassed the WHO-recommended metal levels. The number of maternal samples collected which exceeded the acceptable values for components is particularly concerning (Samiee et al., 2019) (Table 6.1).

Smoking cigarettes is indeed a key cause of Cd as well as other hazardous HMs contained in tobacco leaves (Arunakumara et al., 2013). Because polluted tobacco leaves are being used to produce cigarette, toxic HM accumulation in tobacco is indeed a risk to humans. Tobacco is grown with commercial inorganic fertilizers, notably phosphates, which are rich in HMs and hence dangerous. HMs are taken up by tobacco roots in large quantities during growth and then translocated from the soil to the leaves (Regassa and Chandravanshi, 2016). HMs inhaled when smoking is quickly absorbed into the body via the lungs and into the circulation, in which they can move to those other parts of the body. Cigarette smokers' blood carries more dangerous HMs than nonsmokers' bloodstream (del Piano et al., 2008; Regassa and Chandravanshi, 2016).

6.6 Biomagnification and bioaccumulation of heavy metals in the human food chains

Biomagnification of HMs in environmental compartments contributes key role in distribution of HMs at various tropic levels. Biomagnification is referred as rise in the levels of pollutants in along with food chain and leading to increased accumulation in successive tropic level. In environment soil to plant allocation of HMs is the most important routes of entry to organisms via food (Cai et al., 2015; Singh et al., 2010). Water supplied to agricultural fields through irrigation consists of wastewater causes significant accumulation of metals in soil and sediment which further lead to accumulation in crops, vegetables also in milk. Recent research studies reveled that wheat which is most cultivated crop in Asia has more tendency to accumulate metals than corn. Also, some studies on rice indicated that usage of water contaminated with HMs is very high concern because rice consumes lot of water also crop get exposed to metals since the germination stage (Arunakumara

TABLE 6.1 WHO's permissible limit of some heavy metals.

Metals	WHO's permissible limit (mg/l)
As	0.01
Mn	0.02
Cu	0.02
Fe	0.30
Zn	3.00
Pb	0.01
Cr	0.0003
Cd	0.05
Se	0.02

et al., 2013). Several studies published recently also point points out increasing accumulation of HMs in vegetables also emerging risk to human population. Recent finding suggest levels of metals are reported very high in vegetable which are grown in soils which are irrigated with contaminated water when compared to control soil (Khan et al., 2015; Singh et al., 2011; Zia et al., 2016). For further understanding of dietary exposure of metal impacts are very important to minimize the danger for human population (Singh et al., 2011). The biomagnification potential of metalloids and HMs in various tropic levels is measured through tropic magnification factors (TMFs) and biomagnification factors (BMFs).

Generally, BMF is the ratio of concentration of HMs in organism-to-organism diet which can be calculated by using the below equation.

$$\text{BMF} = \frac{\text{Metal conc. in the organism}}{\text{Metal conc. in the organism's diet or prey}}$$

Whereas TMFs, can be calculated from the slope of log transformed concentrations of organism against tropic levels of the organism in food web. TMFs will provide holistic quantification of metal potential of biomagnification in food chain (Yarsan and Yipel, 2013).

Bioaccumulation of metals is very important parameter which will help to understand accumulation potential of metals in organism and its movement in food chain. Bioaccumulation is a process which requires active metabolic system in an organism which leads to uptake into intracellular spaces with the help of importer complexes which enables the translocation pathway through lipid layers. After entering intracellular spaces HMs sequestered by proteins and ligands (Diep et al., 2018). Various organisms such as aquatic and terrestrial animals get exposed to HMs from environment and their food sources leading to bioaccumulation. Various studies have demonstrated that HMs accumulation potential in invertebrates. Detected levels of HMs in invertebrates can vary differently based on their habitant and species. In comparison to omnivorous and herbivorous insects it was detected that Carnivorous and predatory insects' levels of HMs are high. In a study conducted in industrial area of Gujrat, Pakistan revealed that accumulation of HMs in three different species as follows dragonfly *(Crocothemis Servilia)* > grasshopper *(Oxya hyla hyla)* > butterfly *(Danaus chrysippus)* (Azam et al., 2015). On contrary, accumulation of HMs were observed more in omnivorous fish compared to carnivorous fish. In fish species HMs enter body in multiple ways like (1) gills, (2) skin and (3) alimentary tract, among which soluble fraction will go through gills and particulate matter will enter through alimentary tract (Štrbac et al., 2014). Bioaccumulation of HMs in avian species also very well studied, regardless of species and feeding habits, studies suggest that accumulation of HMs in different feeding regimes will be like following carnivorous > omnivorous and insectivorous > grainivores (Abbasi et al., 2015). Bioaccumulation of HMs in mammalian species is considered to be very less in comparison to the species above discussed, however there are plenty of literature is suggesting causing various toxicities due to exposure of HMs like Arsenic in general population (Kumar et al., 2021).

6.7 Assessment of toxicity of heavy metals

Blood, urine, hair, and nails are all biomarkers of HM exposure. A HM blood test is a collection of assays that assess the presence of potentially hazardous metals in the bloodstream. Lead, mercury, arsenic, and cadmium are the most often examined metals and copper, zinc, aluminum, and thallium are some of the less often examined metals in blood.

The kidneys discharge urine, that contains metabolic by-products like toxins, salts, and water which end up within the bloodstream (Marchiset-Ferlay et al., 2012). Urine reveals the dangers of consuming HMs from polluted water. In addition, several studies utilized urine as a biomarker for Cd, Cr, Mn, Ni, Phosphorous (P), As and Hg daily excretion (Gil et al., 2011; Li et al., 2011). Furthermore, in a study, urine levels were utilized as a typical biomarker of dosages of Cd, Silver (Ag), Cu, Barium (Ba), Cobalt (Co), Aluminum (Al), As, Beryllium (Be), Cr, Fe, Cesium (Cs), Ni, Mn, Selenium (Se), Pb, Titanium (Ti), Uranium (U), Strontium (Sr), Zn and Vanadium (V) whereby a comprehensive HM analysis was performed could demonstrate the influence from recent daily drinking water (Ivanenko et al., 2013). As a result of their short residence periods of 2–3 h and 3–4 days, HMs in blood and urine may be used to precisely measure acute exposure to HMs in the body. Hair and nails contain HMs that have been detached from the metabolic process, and as a result, they may be used as a reliable indicator of long-term exposure (Hughes, 2006). Hair sample analysis is said to be less invasive, easier to preserve and carry, and less risky to use. It has become more crucial in identifying the level of exposure to HMs in humans. Furthermore, because the observed contaminant levels might represent exposure over a lengthy period of time, a hair sample can be a helpful evaluation tool. Quantitative examination of untreated hair samples has proven to be very effective in studying environments polluted by hazardous substances. The concentration of sulfur declines progressively throughout irradiation, but the concentration of other elements, such as arsenic and mercury, does not change significantly in hairs collected from different regions of the body. These findings show that the content of mercury and arsenic in hairs may be used to estimate human exposure to these hazardous elements. The absence of connection between trace element concentrations in hair and other target organs (e.g., liver, kidney) or bodily fluids (e.g., blood, urine) is the most significant limitation of hair analysis (39, 40). Hair samples from residents and workers at such an e-waste recycling plant were analyzed and processed for trace elements and HMs in research. Cu, Pb, and Cd had elevated levels, and the levels of all elements were discovered in the following order: Pb $>$ Cu $>>$ Mn $>$ Ba $>$ Cr $>$ Ni $>$ Cd $>$ As $>$ V. According to this study, human scalp hair could be a useful biomarker for detecting the amount of exposure to HMs in residents and workers in areas where exposure to HMs is widespread (Wang et al., 2009).

The use of a nail as a biomarker in the assessment of HM environmental exposure was also suggested. Toxic metal concentrations in nail tissue have been observed to be orders of magnitude greater than in bodily fluids and other accessible tissues. It is not only convenient for sample collection, storage, and preparation for analysis; it is also a cost-effective approach that is resistant to infections and contaminations (Were et al., 2008; Sera et al., 2011). Furthermore, because toenails grow slower than fingernails, there is more time for metal to accumulate (the former can represent 2–12 months of exposure). Many studies have found that the amount of arsenic in a toenail correlates with the amount of arsenic in their drinking water.

Sediment is a vital and active component of the river basin due to the diversity of ecosystem and habitat. It is believed that HMs enter rivers via the water-soil contact and the water-atmosphere interface, where they are rapidly dilute, transported over hundreds of kilometers, and ultimately deposited in sediment (Ali et al., 2016; Resongles et al., 2014). Over 90% of the total HM burden in aquatic environments is linked to suspended particulate matter and sediments. On top of that, due to the wide range of external redox circumstances and pH levels, HMs in sediments are discharged into the surrounding water, posing a toxicity risk to species (Wang et al., 2012). Thus, sediment quality is an excellent indication of contamination in the water column, where HMs tend to accumulate. When calculating metal bioavailability in sediment, pH is a significant consideration. Whenever the pH lowers, metal ions and H + fight for binding sites in sediment, causing metal complexes to disintegrate and metal ions to be released into to the overlying water (Decena et al., 2018).

HM enrichment in soils has been observed to be significant. It is believed that the parent material (lithogenic source) and anthropogenic source are accountable for the discharge of metalloids and HMs into soils (Alloway, 2013). The composition of parent rock, the degree of weathering, and chemical, biological, and physical, properties of soil and climate conditions, all have an impact on the occurrence and delivery of HMs to the soil (Arunakumara et al., 2013). The bioavailability of HMs in soils is crucial for its outcome with in ecosystem and uptake by plants. Diverse HMs have varying levels of bioavailability in soil due to metal speciation and soil physicochemical characteristics (Ali et al., 2019). To determine Investigators had been using a range of methodologies to investigate the environmental interaction and possible environmental hazard of HMs leaching in soils, such as mineralogical assessment, a three-stage breakpoint cluster region (BCR) sequential extraction process, a dynamic leaching test, as well as the Hakanson Potential Ecological Risk Index Method (67). Cu, Pb, and Zn, Cd levels in soils around a lead–zinc mine in Shangyu, Zhejiang Province, China, had been evaluated in research, and its toxicity then assessed using toxicity characteristic leaching procedure (TCLP) as per United States Environmental Protection Agency (USEPA) (Min et al., 2013). HM contamination of fish has emerged as a major global concern, posing a hazard not only to fish but also to its consumers (Rahman et al., 2012). The increased accumulation of HMs in metabolically active tissues of fish such as gills, liver and kidneys, is typically explained by presence of metal-binding proteins known as metallothioneins (MTs) in these tissues following HM exposure (Mansouri et al., 2012). In a research, fish samples (*Oreochromis niloticus*) were taken at random from different locations in the lake over eight months and tested for the levels of five main HMs (Mn, Cd, Zn, Pb and Cu). HMs were found in significant quantities in the meat of fish samples. The estimated Pb value was up to 38 times higher than the maximum permitted limit. Lake fish may now be used as a biomarker to assess HM toxicity, according to the findings. This confirms that lake fish can be used as biomarker to evaluate HM toxicity (Berndtsson and Hassan, n.d.). Several marine species of animals were also recommended as bioindicators of HM pollution. For example, the date mussel (Lithophaga lithophaga) has now been recommended as a good bioindicator of marine pollution (Miedico et al., 2015).

6.8 Effects of toxic heavy metals on human health

HM toxicity has been found to be a serious risk, with a variety of health issues associated with this. Despite the fact that such metals serve no biological function, its harmful effects remain in a way which is hazardous to human health and its appropriate function.

These may appear like they're a body part occasionally, and so they can potentially interact with metabolic pathways. HMs become hazardous if they're not absorbed by the body and concentrate in soft tissues.

Lead: The damage caused by Pb to humans is mediated by two separate mechanisms: the first is the direct production of reactive oxygen species (ROS) (O_2, H_2O_2, $ONOO^-$) and the second is the depletion of antioxidants by inhibiting enzymes like glutathione reductase (GR) and delta-aminolevulinic acid dehydratase (ALAD) (Ahamed et al., 2005; Nuran Ercal et al., 2005). By attaching to sulfhydryl protein, lead inhibits enzyme function, reduces trace mineral uptake, and impairs structural protein production. This also lowers the quantity and delivery of sulfhydryl antioxidant reserves in the body (Patrick, 2006). High amounts of lead exposure can have devastating effects on the brain and kidneys, and can result in death. Lead has the potential to have an impact on every organ and system in the body. Male reproductive organs can be damaged by long-term exposure at high levels. Lead exposure during pregnancy has been linked to miscarriage in some women (Patrick, 2006). Pb is a carcinogen that can harm mitochondria, induce apoptosis, and deplete glutathione intracellularly. This has also been demonstrated to influence expression of genes by replacing Zn for all other components in proteins, hence lowering the proteins' interaction to DNA (Koh, et al., 2014; Sabath and Robles-Osorio, 2012).

Arsenic: Organic and inorganic arsenic molecules are both extremely harmful to humans. The inorganic form is more harmful and accumulates in exposed organisms. As in its trivalent form is the most poisonous and generally interacts with protein thiol groups, whereas because this is less poisonous in its pentavalent type, but it does have an unrivaled oxidative phosphorylation ability (Ahmed et al., 2013; He et al., 2014). Inorganic arsenic is a well-documented carcinogen that has been linked to several cancers, including lung, skin, bladder and liver cancers. Minimal exposure may cause vomiting and nausea, as well as decreased white and red blood cell production, irregular heartbeats, and damage to blood vessels. Lengthy reduced exposure causes skin darkening with pricking sensation which create tiny warts on the body parts such as palms and soles (Chen et al., 2009). Lung illness, neurological difficulties, hypertension, peripheral vascular disease, cardiovascular disease, diabetes mellitus, internal malignancies, all can be caused by long-term exposure high-level exposure (Smith et al., 2000). Arsenic has been shown to cause epigenetic alterations in cells such as histone alteration, microRNA control and DNA methylation (Wang et al., 2011b) (Fig. 6.1).

Cadmium: Cadmium poisoning occurs from the biological system's interaction with the HM. Cd is a carcinogenic element. Regular smokers have been subjected to far more cadmium as nonsmokers. High cadmium exposure causes significant lung damage. Excessive consumption irritates the gut, leading in diarrhea and vomiting. Prolonged exposure generates an accumulation with in kidneys, that causes kidneys disease, brittle bones and lung problems (Sobha et al., 1970). Cadmium disrupts oxidative phosphorylation processes by interfering with sulfhydryl enzymes as well as other cell ligands (Jomova and Valko, 2011). When Cd is exposed to the liver, it causes metallothionein production, which leads to hepatocellular injury in humans. Lipid peroxidation, DNA damage, and protein carbonylation are all facilitated by oxidative stress and hepatic marker dysfunction (Kang et al., 2013; Rashid et al., 2013) (Fig. 6.2).

Mercury: Microorganisms in soil and water convert mercury to methylmercury (MeHg), a bioaccumulating poison. Nausea, lung damage, diarrhea, vomiting, amplified

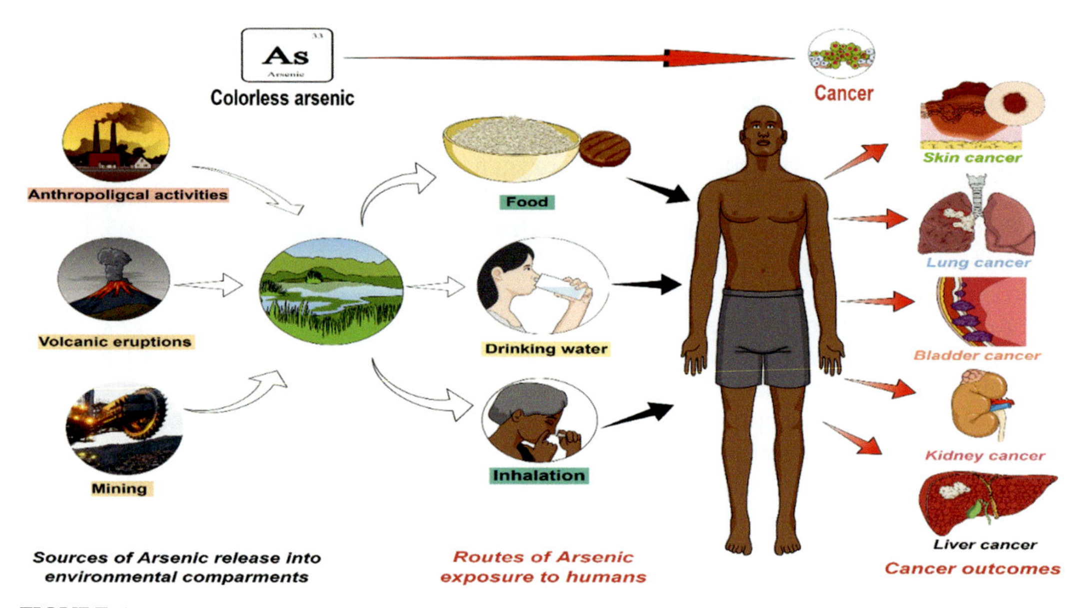

FIGURE 6.1 Arsenic life cycle from origins from sources to human exposure and possible cancerous outcomes inn humans upon exposure through various routes.

blood pressure or heart rate, eye irritation and skin rashes may occur after short-range exposure to higher amounts of metallic mercury vapors (Weldon et al., 2000). Because of its electrophilic nature, mercury, particularly MeHg, is hazardous owing to its tendency to interact with nucleophiles such as sulfhydryl (–SH) or selenohydryl (–SeH) (Farina et al., 2011). Mercury, in any form, is very toxic to the neurological system. Low-dose Hg has the capacity to influence a growing fetus' neural systems, preventing the growth of behavioral domains such as cognitive, motor, sensory, processes (Kovacic and Somanathan, 2014). Low-level Hg exposure causes cell cytotoxicity, and increased β-amyloid, oxidative stress that is linked to neurological illnesses such as Alzheimer's and Parkinson's in adults (Carocci et al., 2014; Goldman, 2014) (Fig. 6.3).

Chromium: Chromium (VI) compounds are recognized carcinogens, but Chromium (III) is a necessary nutrient for human health (Chen et al., 2020; Zhitkovich, 2005). Breathing excessive amounts can cause nasal lining irritation, runny nose, ulcers, and breathing problems including cough, asthma, wheezing or shortness of breath. As a consequence of dermal contact, skin ulcers can form. Allergic responses such as significant skin redness and edema have been reported (Chen et al., 2020). When exposed for an extended period of time, the liver, kidneys, circulatory system, and nerves might be harmed, and the skin can become irritated (Chu et al., 2021; Madhav et al., 2021).

Nickel: Nickel has a wide range of toxicological effects and is well-known for its carcinogenicity when exposed to humans. Hair Loss is the most noticeable side effect of nickel poisoning. Allergic dermatitis can be caused by skin contact with metallic nickel, or by soluble derivatives of nickel. Nickel intake results in substantial weight loss

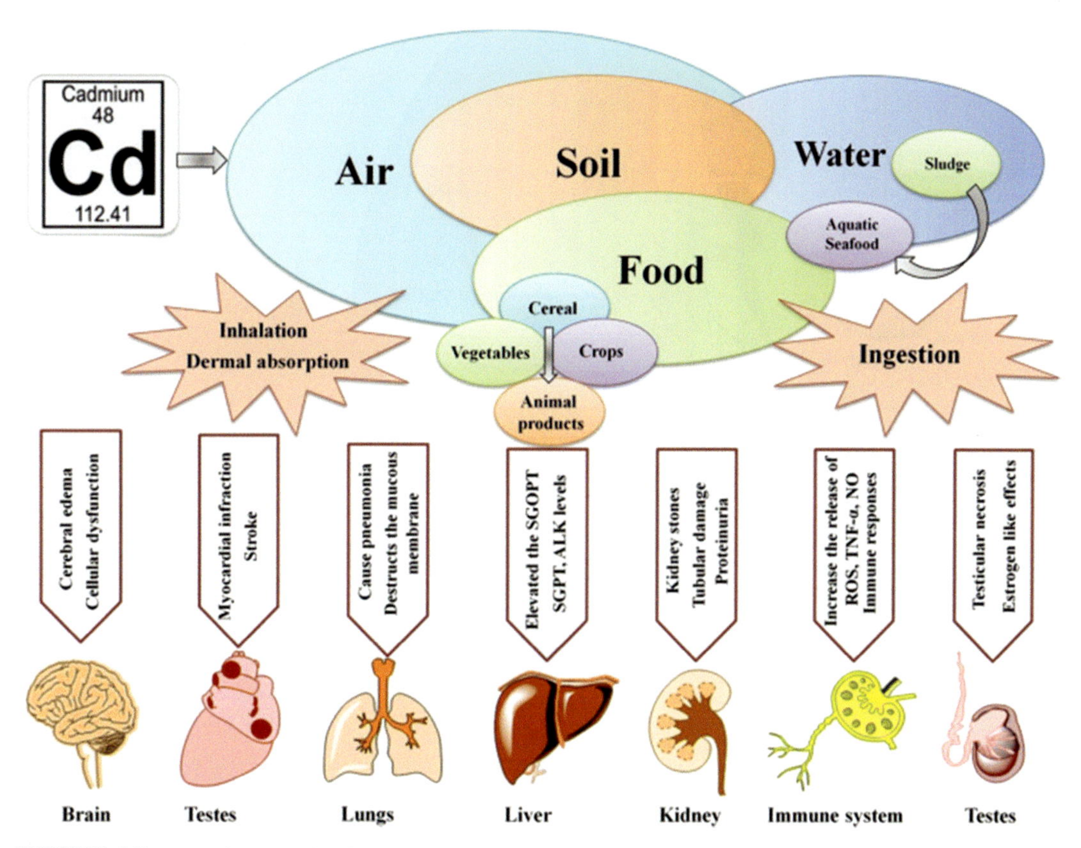

FIGURE 6.2 Bioindicators of cadmium toxicity.

(Duda-Chodak and Blaszczyk, 2008). People that inhaled dust carrying particular nickel constituents suffered most serious health consequences from exposure of nickel, including chronic bronchitis, impaired lung function, nasal sinus and lung cancer (Grimsrud and Andersen, 2010).

Zinc: Hypertension, nausea, and stomach damage are just a few of the side effects that might occur zinc exposure. Zinc increases the risk of cardiovascular disease. It may potentially have neurotoxic effects on human health. When zinc is used in excess, it might lead to psychological problems. Zinc intake causes a variety of physiological and neurological effects (Duruibe et al., 2007; Karikari et al., 2020).

Copper: Acute copper poisoning is commonly linked to unintentional consumption, although certain people may be more vulnerable to the harmful consequences of excessive Cu intake owing to a genetic predisposition condition. It is possible that a high intake of Cu by humans might cause significant mucosal irritation and corrosion, as well as extensive capillary damage, kidney failure and central nervous system damage, all of which can lead to depression. There may be severe gastrointestinal discomfort as well as kidney and liver necrosis (Argun et al., 2007; Stern et al., 2007).

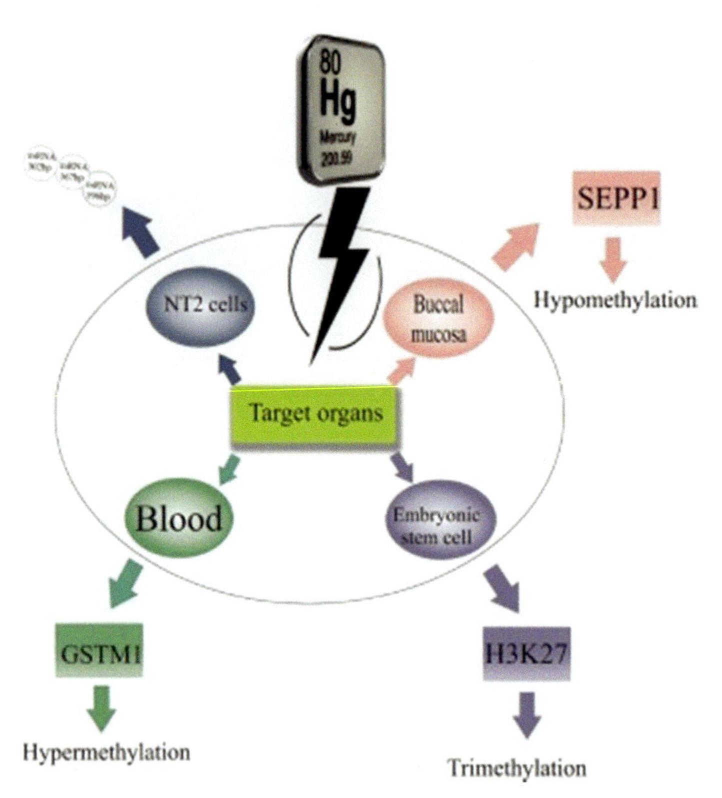

FIGURE 6.3 Mercury exposure and toxicity.

6.9 Monitoring and analysis of heavy metals in the environment

Presence of HMs in atmosphere posses' huge risk to ecosystem, surprisingly even at low levels of presence in environment HMs can cause significant impact on performance. It is very essential also important to monitor the levels of HMs in various environmental compartments, however tracing and detecting HMs is very challenging due to requirement of sophisticated methodology. Over the last few decades science and technology advancements made tremendously, currently there are so many techniques are available to detect levels of HMs in various samples, in many useful techniques currently available the most used are AAS (atomic absorption spectroscopy), ICP-Ms (inductively coupled plasma Mass spectroscopy), LIBS (laser-induced breakdown spectroscopy) and AFS (atomic fluorescence spectrometry) (Jin et al., 2020).

6.10 Conclusion

HMs distribution in environment has increased significantly over the last few years at an alarming rate, which is going to impact all wildlife and other species as well. Mounting evidence suggested that exposure to HMs can happen on various routes including through water, air, and diet. Despite available evidence on impact, there is a large area of research that needs to be done to improve the efficiency and accuracy of detection techniques; also much more emphasis is needed to be done on making economic options of detections methods. Also, there is a need to promote regulatory push from government side to minimize and mitigate the exposure of HMs in environment. There is also significant need to improve the infrastructure and wastewater treatments capacity especially in Asian nations such as Pakistan, Indian, and China where most of the people are affected with HMs pollution from industrial waste.

References

Abbasi, N.A., Jaspers, V.L.B., Chaudhry, M.J.I., Ali, S., Malik, R.N., 2015. Chemosphere 120, 527–537.

Adriano, D.C., 2001. Trace Elements in Terrestrial Environments. Springer, pp. 1–27.

Ahamad, A., Raju, N.J., Madhav, S., Gossel, W., Ram, P., Wycisk, P., 2021. Potentially toxic elements in soil and road dust around Sonbhadra industrial region, Uttar Pradesh, India: source apportionment and health risk assessment. Environmental Research 202, 111685.

Ahamed, M., Verma, S., Kumar, A., Siddiqui, M.K.J., 2005. Science of the Total Environment 346, 48–55.

Ahmed, M.K., Habibullah-Al-Mamun, M., Parvin, E., Akter, M.S., Khan, M.S., 2013. Experimental and Toxicologic Pathology 65, 903–909.

Akshaya, N., Kumari, S.B., Kumar, M.M., Drishya, M.K., Sujila, T., Gopinathan, S., n.d. IOSR Journal of Pharmacy and Biological Sciences (IOSR-JPBS) 12, 37–41.

Al Osman, M., Yang, F., Massey, I.Y., 2019. Biometals: An International Journal on the Role of Metal Ions in Biology, Biochemistry, and Medicine 32, 4–32, 563–573.

Ali, H., Khan, E., 2019. Human and Ecological Risk Assessment 25, 1353–1376.

Ali, Z., Malik, R.N., Qadir, A., 2013. 29, 676–692. Available from: https://doi.org/10.1080/02757540.2013.810728

Ali, M.M., Ali, M.L., Islam, M.S., Rahman, M.Z., 2016. Environmental Nanotechnology, Monitoring & Management 5, 27–35.

Ali, H., Khan, E., Ilahi, I., 2019. Journal of Chemistry 2019.

Alloway, B.J., 2013. Sources of heavy metals and metalloids in soils. In: Alloway, B.J. (Ed.), Heavy Metals in Soils. Springer, Dordrecht, pp. 11–50.

Angelovičová, L., Fazekašová, D., 2014. Soil and Water Research 9, 18–24.

Argun, M.E., Dursun, S., Ozdemir, C., Karatas, M., 2007. Journal of Hazardous Materials 141, 77–85.

Arunakumara, K.K.I.U., Walpola, B.C., Yoon, M.-H., 2013. Reviews in Environmental Science and Bio/Technology 12, 4–12, 355–377.

Ayari, F., Hamdi, H., Jedidi, N., Gharbi, N., Kossai, R., 2010. International Journal of Environmental Science & Technology 7, 3–7, 465–472.

Azam, I., Afsheen, S., Zia, A., Javed, M., Saeed, R., Sarwar, M.K., et al., 2015. BioMed Research International 2015.

Baby, J., Raj, J., Biby, E., Sankarganesh, P., Jeevitha, M., Ajisha, S., et al., 2011. International Journal of Biological and Chemical Sciences 4.

Berger, A.C., Bethke, C.M., Krumhansl, J.L., 2000. Applied Geochemistry 15, 655–666.

Berndtsson, R., Hassan, M., n.d. Assessment of Heavy Metals Pollution and Microbial Contamination in Water, Sediments and Fish of Lake Manzala, Egypt.

Besante, J., Niforatos, J., Mousavi, A., 2011. 12, 121–123. https://doi.org/10.1080/15275922.2011.577521

Briffa, J., Sinagra, E., Blundell, R., 2020. Heliyon 6, e04691.

Burger, J., Gochfeld, M., 2000. Science of the Total Environment 257, 37–52.

Cai, L.M., Xu, Z.C., Qi, J.Y., Feng, Z.Z., Xiang, T.S., 2015. Chemosphere 127, 127–135.

Cai, L.-M., Wang, Q.-S., Luo, J., Chen, L.-G., Zhu, R.-L., Wang, S., et al., 2019. Science of the Total Environment 650, 725–733.

Carocci, A., Rovito, N., Sinicropi, M.S., Genchi, G., 2014. Reviews of Environmental Contamination and Toxicology 229, 1–18.

Chen, Y., Parvez, F., Gamble, M., Islam, T., Ahmed, A., Argos, M., et al., 2009. Toxicology and Applied Pharmacology 239, 184–192.
Chen, Y., Yu, W., Zheng, R., Li, J.Y., Zhang, L., Wang, Q., et al., 2020. Science of the Total Environment 736, 139185.
Chowdhury, S., Mazumder, M.A.J., Al-Attas, O., Husain, T., 2016. Science of the Total Environment 569–570, 476–488.
Chu, J., Hu, X., Kong, L., Wang, N., Zhang, S., He, M., Ouyang, W., Liu, X., Lin, C., 2021. Science of the Total Environment 771. Available from: https://doi.org/10.1016/j.scitotenv.2020.144643.
Dauwe, T., Janssens, E., Kempenaers, B., Eens, M., 2004. Environmental Pollution 129, 125–129.
Decena, S.C.P., Arguilles, M.S., Robel, L.L., 2018. Polish Journal of Environmental Studies 27, 1983–1995.
del Piano, L., Abet, M., Sorrentino, C., Barbato, L., Sicignano, M., Cozzolino, E., et al., 2008. Uptake and distribution of lead in tobacco (Nicotiana tabacum L.). Journal of Applied Botany and Food Quality .
Diep, P., Mahadevan, R., Yakunin, A.F., 2018. Frontiers in Bioengineering and Biotechnology 29, 157.
Dissanayake, C.B., Chandrajith, R., 2009. Phosphate mineral fertilizers, trace metals and human health. Journal of the National Science Foundation of Sri Lanka .
Duda-Chodak, A., Blaszczyk, U., 2008. Journal of Elementology 13, 685–693.
Duruibe, J.O., Ogwuegbu, M.O.C., Egwurugwu, 2007. International Journal of Physical Sciences 2, 112–118.
Duzgoren-Aydin, N.S., 2007. Science of the Total Environment 385, 182–195.
Evanko, C.R., Dzombak, D.A., 1997. Remediation of Metals-Contaminated Soils and Groundwater. Technology Evaluation Report, 61Paper Publishing.
Farina, M., Rocha, J.B.T., Aschner, M., 2011. Life Sciences 89, 555–563.
Fraga, C.G., 2005. Molecular Aspects of Medicine 26, 235–244.
Frankowski, M., Sojka, M., Zioła-Frankowska, A., Siepak, M., Murat-Błałejewska, S., 2009. Oceanological and Hydrobiological Studies 38, 51–61.
Frantz, A., Pottier, M.A., Karimi, B., Corbel, H., Aubry, E., Haussy, C., et al., 2012. Environmental Pollution 168, 23–28.
Fytianos, K., Katsianis, G., Triantafyllou, P., Zachariadis, G., 2001. Bulletin of Environmental Contamination and Toxicology 67, 3–67, 0423–0430.
Gil, F., Hernández, A.F., Márquez, C., Femia, P., Olmedo, P., López-Guarnido, O., et al., 2011. Science of the Total Environment 409, 1172–1180.
Goldman, S.M., 2014. 54, 141–164. Available from: https://doi.org/10.1146/annurev-pharmtox-011613-135937.
Grant, C.A., 2011. Pedologist 54, 143–155.
Grant, C.A., Sheppard, S.C., 2008. 14, 210–228. Available from: https://doi.org/10.1080/10807030801934895.
Grimsrud, T.K., Andersen, A., 2010. Journal of Occupational Medicine and Toxicology 5, 1–6.
Habiba, G., Abebe, G., Bravo, A.G., Ermias, D., Staffan, Å., Bishop, K., 2017. Biological Trace Element Research 175, 237–243.
Hashem, M.A., Nur-A-Tomal, M.S., Mondal, N.R., Rahman, M.A., 2017. Environmental Chemistry Letters 15, 3–15, 501–506.
He, J., Wang, M., Jiang, Y., Chen, Q., Xu, S., Xu, Q., et al., 2014. Environmental Health Perspectives 122, 255–261.
Hughes, M.F., 2006. Environmental Health Perspectives 114, 1790–1796.
Human Exposure to Heavy Metals from Cosmetics: Oriental Journal of Chemistry [WWW Document], n.d. http://www.orientjchem.org/vol27no1/human-exposure-to-heavy-metals-from-cosmetics/ (accessed 10.9.21).
Ivanenko, N.B., Ivanenko, A.A., Solovyev, N.D., Zeimal', A.E., Navolotskii, D.V., Drobyshev, E.J., 2013. Talanta 116, 764–769.
Ji, K., Kim, J., Lee, M., Park, S., Kwon, H.-J., Cheong, H.-K., et al., 2013. Environmental Pollution (Barking, Essex: 1987) 178, 322–328.
Jin, M., Yuan, H., Liu, B., Peng, J., Xu, L., Yang, D., 2020. Analytical Methods 12, 5747–5766.
Johnson, R.H., Blowes, D.W., Robertson, W.D., Jambor, J.L., 2000. Journal of Contaminant Hydrology 41, 49–80.
Jomova, K., Valko, M., 2011. Toxicology 283, 65–87.
Jung, M.C., 2001. Applied Geochemistry 16, 1369–1375.
Kang, M.-Y., Cho, S.-H., Lim, Y.-H., Seo, J.-C., Hong, Y.-C., 2013. Occupational and Environmental Medicine 70, 268–273.

Karikari, A.Y., Asmah, R., Anku, W.W., Amisah, S., Agbo, N.W., Telfer, T.C., et al., 2020. Aquaculture Research 51, 2041–2051.

Kaur, M., Kumar, A., Mehra, R., Kaur, I., 2020. Environmental Geochemistry and Health 42, 77–94.

Khan, A., Khan, S., Khan, M.A., Qamar, Z., Waqas, M., 2015. Environmental Science and Pollution Research 22, 18–22, 13772–13799.

Kim, J.J., Kim, Y.S., Kumar, V., 2019. Journal of Trace Elements in Medicine and Biology 54, 226–231.

Kim, M.-J., Ahn, K.-H., Jung, Y., 2002. Chemosphere 49, 307–312.

Koh, D.H., Bhatti, P., Coble, J.B., Stewart, P.A., Lu, W., Shu, X.O., et al., 2014. Journal of Exposure Science & Environmental Epidemiology 24 (1), 9–16.

Kovacic, P., Somanathan, R., 2014. Systems Biology of Free Radicals and Antioxidants 567–580. 9783642300189.

Kumar, A., Rahman, M.S., Ali, M., Kumar, R., Niraj, P.K., Akhouri, V., et al., 2021. Toxicology and Environmental Health Sciences 13, 287–297.

Li, P., Feng, X., Qiu, G., Wan, Q., 2011. Science of the Total Environment 409, 4484–4488.

Li, C., Zhou, K., Qin, W., Tian, C., Qi, M., Yan, X., et al., 2019. 28, 380–394. Available from: https://doi.org/10.1080/15320383.2019.1592108.

Madhav, S., Ahamad, A., Singh, P., Mishra, P.K., 2018. A review of textile industry: wet processing, environmental impacts, and effluent treatment methods. Environmental Quality Management 27 (3), 31–41.

Madhav, S., Raju, N.J., Ahamad, A., Singh, A.K., Ram, P., Gossel, W., 2021. Hydrogeochemical assessment of groundwater quality and associated potential human health risk in Bhadohi environs, India. Environmental Earth Sciences 80 (17), 1–14.

Mansouri, B., Ebrahimpour, M., Babaei, H., 2012. Toxicology and Industrial Health 28, 361–368.

Marchiset-Ferlay, N., Savanovitch, C., Sauvant-Rochat, M.-P., 2012. Environment International 39, 150–171.

Medical Management Guidelines—Letter A | Toxic Substance Portal | ATSDR [WWW Document], n.d. Available from: https://wwwn.cdc.gov/TSP/MMG/MMGLanding.aspx (accessed 10.11.21).

Miedico, O., Ferrara, A., Tarallo, M., Pompa, C., Bisceglia, D., Chiaravalle, A.E., 2015. 98, 877–885. https://doi.org/10.1080/02772248.2015.1128434.

Min, X.B., Xie, Xd, Chai, L.Y., Liang, Y.J., Li, M., Ke, Y., 2013. Transactions of Nonferrous Metals Society of China 23, 208–218.

Mishra, R., Kumar, A., Singh, E., Kumar, S., Tripathi, V.K., Jha, S.K., et al., 2022. Current status of available techniques for removal of heavy metal contamination in the river ecosystem. Ecological Significance of River Ecosystems. Elsevier, pp. 217–234.

Mohamed Hassan, E., Berndtsson, R., Hassan, M., Abd El-Azim, H., et al., 2013. Life Science Journal 10, 1097–8135.

Mulk, S., Azizullah, A., Korai, A.L., Khattak, M.N.K., 2015. Environmental Monitoring and Assessment 187, 2–187, 1–23.

Navarro, M.C., Pérez-Sirvent, C., Martínez-Sánchez, M.J., Vidal, J., Tovar, P.J., Bech, J., 2008. Journal of Geochemical Exploration 96, 183–193.

Ni, H.-G., Zeng, E.Y., 2009. Environmental Science and Technology 43, 3991–3994.

Nordberg, G.F., Fowler, B.A., Nordberg, M., Friberg, L.T., 2007. Handbook on the Toxicology of Metals.

Nuran Ercal, B.S.P., Hande Gurer-Orhan, B.S.P., Nukhet Aykin-Burns, B.S.P., 2005. Current Topics in Medicinal Chemistry 1, 529–539.

Olympio, K.P.K., Silva, J.P.D.R., Silva, A.S.D., Souza, V.C.O., Buzalaf, M.A.R., Barbosa Jr, F., et al., 2018. Environmental Pollution (Barking, Essex: 1987) 240, 831–838.

Patrick, L., 2006. Lead toxicity, a review of the literature. Part 1: Exposure, evaluation, and treatment.Alternative. Medicine Review 11 (1), 2–22.

Rahman, M.S., Molla, A.H., Saha, N., Rahman, A., 2012. Food Chemistry 134, 1847–1854.

Rashid, K., Sinha, K., Sil, P.C., 2013. Food and Chemical Toxicology 62, 584–600.

Regassa, G., Chandravanshi, B.S., 2016. SpringerPlus 5, 1–5, 1–9.

Resongles, E., Casiot, C., Freydier, R., Dezileau, L., Viers, J.Ô., Elbaz-Poulichet, F., 2014. Science of the Total Environment 481, 509–521.

Rice, health, and toxic metals—Rice Today [WWW Document], n.d. https://ricetoday.irri.org/rice-health-and-toxic-metals/ (accessed 10.9.21).

Roy, A., Bellinger, D., Hu, H., Schwartz, J., Ettinger, A.S., Wright, R.O., et al., 2009. Environmental Health Perspectives 117, 1607–1611.

Sabath, E., Robles-Osorio, M.L., 2012. Nefrología (English Edition) 32, 279–286.
Sabiha-Javied, Mehmood, T., Chaudhry, M.M., Tufail, M., Irfan, N., 2009. Microchemical Journal 91, 94–99.
Sagiv, S.K., Thurston, S.W., Bellinger, D.C., Amarasiriwardena, C., Korrick, S.A., 2012. Archives of Pediatrics & Adolescent Medicine 166, 1123–1131.
Samiee, F., Vahidinia, A., Javad, M.T., Leili, M., 2019. Science of the Total Environment 650, 3075–3083.
Schmidt, C.W., 2002. Environmental Health Perspectives 110, A188.
Sera, K., Futatsugawa, S., Murao, S., Clemente, E., 2011. 12, 125–136. https://doi.org/10.1142/S0129083502000226.
Singh, A., Sharma, R.K., Agrawal, M., Marshall, F.M., 2010. Food and Chemical Toxicology 48, 611–619.
Singh, R., Gautam, N., Mishra, A., Gupta, R., 2011. Indian Journal of Pharmacology 43, 246.
Smith, A.H., Lingas, E.O., Rahman, M., 2000. Bulletin of the World Health Organization 78, 1093–1103.
Sobha, K., Poornima, A., Harini, P., Veeraiah, K., 1970. Kathmandu University Journal of Science, Engineering and Technology 3, 1–11.
Spiegel, H., 2002. Turkish Journal of Chemistry 26, 815–824.
Stern, B.R., Solioz, M., Krewski, D., Aggett, P., Aw, T.-C., Baker, S., et al., 2007. Journal of Toxicology and Environmental Health Part B, Critical Reviews 10, 157–222.
Štrbac, S., Šajnović, A., Budakov, L., Vasić, N., Kašanin-Grubin, M., Simonović, P., et al., 2014. 30, 169–186. https://doi.org/10.1080/02757540.2013.841893.
Tang, X., Shen, C., Shi, D., Cheema, S.A., Khan, M.I., Zhang, C., et al., 2010. Journal of Hazardous Materials 173, 653–660.
Thakur, M., Rachamalla, M., Niyogi, S., Datusalia, A.K., Flora, S.J.S., 2021. International Journal of Molecular Sciences 22, 10077.
Varalakshmi, L.R., Ganeshamurthy, A.N., 2010. Heavy metal contamination of water bodies, soils and vegetables in peri urban areas of Bangalore city of India.
Wang, T., Fu, J., Wang, Y., Liao, C., Tao, Y., Jiang, G., 2009. Environmental Pollution (Barking, Essex: 1987) 157, 2445–2451.
Wang, Q.-R., Cui, Y.-S., Liu, X.-M., Dong, Y.-T., Christie, P., 2011a. 38, 823–838. Available from: https://doi.org/10.1081/ESE-120018594.
Wang, Z., Zhao, Y., Smith, E., Goodall, G.J., Drew, P.A., Brabletz, T., et al., 2011b. Toxicological Sciences: An Official Journal of the Society of Toxicology 121, 110–122.
Wang, S., Xing, D., Jia, Y., Li, B., Wang, K., 2012. Applied Geochemistry 27, 343–351.
Weldon, M.M., Smolinski, M.S., Maroufi, A., Hasty, B.W., Gilliss, D.L., Boulanger, L.L., et al., 2000. Western Journal of Medicine 173, 15.
Were, F.H., Njue, W., Murungi, J., Wanjau, R., 2008. Science of the Total Environment 393, 376–384.
Wu, W., Wu, P., Yang, F., Sun, D.l, Zhang, D.X., Zhou, Y.K., 2018. Science of the Total Environment 630, 53–61.
Yarsan, E., Yipel, M., 2013. Article in Journal of Molecular Biomarkers & Diagnosis.
Yu, R., Yuan, X., Zhao, Y., Hu, G., Tu, X., 2008. Journal of Environmental Sciences 20, 664–669.
Zhitkovich, A., 2005. Importance of Chromium-DNA Adducts in Mutagenicity and Toxicity of Chromium(VI). Chemical Research in Toxicology 18, 3–11. Available from: http://dx.doi.org/10.1021/tx049774+.
Zia, M.H., Watts, M.J., Niaz, A., Middleton, D.R.S., Kim, A.W., 2016. Environmental Geochemistry and Health 39, 4–39, 707–728.

CHAPTER

7

Metal contamination in water resources due to various anthropogenic activities

Amrita Daripa[1], *Lal Chand Malav*[2], *Dinesh K. Yadav*[3] *and Sudipta Chattaraj*[1]

[1]ICAR-National Bureau of Soil Survey and Land Use Planning, Nagpur, Maharashtra, India [2]ICAR-National Bureau of Soil Survey and Land Use Planning, Regional Center, Udaipur, Rajasthan, India [3]ICAR-Indian Institute of Soil Science, Bhopal, Madhya Pradesh, India

7.1 Introduction

Water quality is an immediate concern for humans. Therefore, water quality is very crucial for a healthy and sustainable ecosystem. Over the last few decades, water pollution is considered as a severe global threat due to its effects on the ecosystem, environment, and human health. Urbanization, climate change, and industrialization release a diversity of pollutants, posing a major threat to water quality (Zamora-Ledezma et al., 2021). Water bodies are being contaminated by both inorganic and organic pollutants. Presently, heavy metals contamination is one of the most urgent environmental issues globally (Gupta et al., 2013). Heavy metals are ever-present in the environment, and it has been overshadowed due to both natural/native and anthropological sources (human-induced) and processes (Kumar et al., 2017). Heavy metals refers to a huge set of elements with a density more than 5 g/cm^3 and is the most generally used and recognized term. Heavy metal pollution of soil, water and plants (crops, vegetables and trees) is one of the prevailing critical environmental predicaments mostly in developing countries including India because of their toxicity, biomagnification capacity and their nonbiodegradability (Cervantes-Ramírez et al., 2018). Heavy metals can enter the food chain after being absorbed by plants in soil solution, and so humans can be exposed to them. The contamination of these toxic or nonbiodegradable metals poses potential menaces to the ecosystem (Rani et al., 2022). It can

Metals in Water
DOI: https://doi.org/10.1016/B978-0-323-95919-3.00022-7

impair the health of living beings through several absorption pathways such as direct ingestion, cutaneous contact, inhalation, and oral intake (Hou et al., 2018). Heavy metals are abundant in the environment and are regarded major chemical food pollutants. Heavy metals are most commonly inorganic pollutants which are tremendously persistent in the environment, ubiquitous, bioaccumulate, and also have the consequences on the environment and aquatic biota, including humans (Arianmehr and Nouti-Zahi, 2016). However, anthropogenic activities lead to increase in heavy metals such as Ni, Cu, Co, Pb, Cd, Hg, Zn, and As, etc., in water systems. Heavy metals enter water bodies through agricultural runoff (pesticides/fertilizers), industrial discharge, sewage sludge, wet and dry deposition, and other sources (Gautam et al., 2015).

7.1.1 Point and nonpoint sources of heavy metals in water resources

Point source: The direct localized identifiable sources of contaminant are called point source. Direct pipe outlets of some of the industries to the nearby water body contribute in this kind of pollution. Such as thermal power plants, ore refineries, textile factories, sewage treatment plants, leather industry, etc. (Schweitzer and Noblet, 2018). Other examples include an oil spill from a water tanker. Storm sewer discharge are an add on that affects frequently the area near it. *Nonpoint source* when diverse nonidentifiable diffuse sources of pollutants enter groundwater or surface water environment it is called nonpoint sources (NPSs). The examples of NPSs include urban waste runoff, agricultural fields' runoff and other diffused runoff to water bodies. In case of NPS the pollutants enters the environment in one place and cause an adverse effect at hundreds or even thousands of miles away, therefore it is sometimes also referred as trans-boundary pollution. Examples of NPS of pollution include radioactive waste from nuclear reprocessing plants that travels miles through the oceans and contaminate nearby countries with the pollutants. Other examples include storm flows from urban land use that include heavy metals generated through road traffic, atmospheric deposition, etc. (Herngren et al., 2006; Lau and Stenstrom, 2005). These heavy metal constituents are in turn activated by storm water and are dynamically partitioned either as dissolved-dissociated in water or adsorbed onto fine sediments and particulate organic matter (Pitt et al., 1995). These dissolved heavy metals when enter into water system have the potential to cause acute and long term toxicity to water inhabitants (Hatje et al., 2003; Marsalek et al., 1999).

7.2 Metal contamination in water resources through agriculture activity

Agriculture plays a major role in contaminating water. A huge amount of agrochemicals, drug residues, organic matter, saline drainage and sediments from farms are discharged into water bodies causing their pollution. These pollutants pose threat to aquatic ecosystems, human health and other productive activities of the water bodies. The three main divisions of agriculture system-cropping, livestock and aquaculture have intensified and expanded to meet the food demand of the growing population. This is causing an elevated agricultural pressure on water quality (FAO, 2014, 2016). Globally the intensive use of inputs such as chemical fertilizers and pesticides are the prime drivers to increase crop production. The growth and intensification of livestock sector is faster than crop production. The associated waste generated through it mainly manure, veterinary medicines and other growth hormones has critical inferences for water quality (FAO, 2006). Aquaculture sector such as marine, brackish water and fresh water environments has also documented a tremendous growth in recent years worldwide (FAO, 2016). Greater use of antibiotics, fungicides, fish excreta and feed residues has led to deterioration of water quality. The metal contamination in water bodies through crop production activities occur mainly through runoff of soil sediments from metal contaminated agricultural fields to the water. The agricultural runoff mainly contributes in nonpoint source. To understand the phenomenon of water body contamination through agriculture activity the heavy metal contamination of soil has to be discussed first. The sources of metal contamination of water contamination through agricultural activity are discussed in detail in following points.

Fertilizers: To complete the lifecycle and growth of a plant it requires macro and micronutrients. These micronutrients are heavy metal group. When soils are deficient in nutrients, fertilizers are added into the soil to meet up the plant demand for its growth and development. In intensive farming systems huge amount of fertilizers are regularly added to soils to provide adequate N, P, K and other micronutrients to the crop. However heavy metals (e.g., Cd and Pb) are present as impurities with the compounds that supply nutrient elements to the soil. On continued fertilizer application these impurities are continuously added to the soil thus significant increase in their content in the soil is recorded. Some phosphatic fertilizers has traces of Cd and other potentially toxic elements (Wuana and Okieimen, 2011; Kelepertzis, 2014; Tóth et al., 2016).

Pesticides: Pesticides that are used in agriculture also contain trace metals that are harmful in nature. These pesticides have heavy metals like Cu, Hg, Mn, Pb, or Zn as impurity in their compounds, such as Bordeaux mixture a Cu-containing copper oxychloride and fungicidal sprays (copper sulfate). To control some parasitic insects in fruit orchards lead arsenate was used for many years that contains harmful Pb and As in its composition. Use of these formulations containing harmful metals in it greatly exceed background concentrations in soil as reported from many sites; in case of extensive runoff and leaching of heavy metals from these sites to water such contamination has the potential to cause water pollution (Syafrudin et al., 2021).

Manures and other biosolids: Different forms of biosolids are applied in agricultural soil to increase its fertility and productivity; but application of these biosolids comes with a cost by buildup of toxic heavy metals in the soil. Although most manure are seen as valuable fertilizers to soil but are also a supplier of heavy metals. Concentration of heavy metals such as As, Cu, and Zn in manure and biosolids are aggravated by the health products used for the health improvement of cattle and poultry industry (Basta et al., 2005).

7.3 Metal contamination in water resources through industrial activity

The worldwide inclination of industrialization and urbanization for socioeconomic development has led to a rise in the anthropogenic heavy metals contamination in the environment. The various industries participating in heavy metal contamination of water bodies are as follows:

Thermal power plants: The emission of heavy metals and other wide range of gaseous pollutants occurs through the fossil fuel combustion such as coal in thermal power plants, which reach to the surrounding environment such as soil and water once it is released into the air. The ash generated from power plants contains arsenic (As), boron (B), selenium (Se), vanadium (V), molybdenum (Mo), cadmium (Cd)and aluminum (Al) at extremely toxic concentration (Gupta et al., 2002). Experiments carried out at different locations of the world have also reported the elevated heavy metal concentrations in the ash (Mandal and Sengupta, 2006; Fiket et al., 2016). An elevated value of heavy metals has been reported from soil near thermal power plants (Keegan et al., 2006; Ćujić et al., 2016). As a solid waste material fly ash is generated through coal combustion. Currently in India coal-based thermal power plants generates about 131 mt fly ash per annum (Singh et al., 2014). Most of the ash is left to settle in the settling ponds. The effluent outlets from these ponds enter directly into local waterways and contaminate it. Underlying water table also gets affected due to the large volume of the leachate from the ash that percolates through the unlined ash pond (Praharaj et al., 2002). The leaching of soluble ions present in fly ash causes alteration of water quality around the thermal power plants. The leaching characteristics of fly ash containing soluble ions has the potential to contaminate groundwater aquifers (Singh et al., 2014).

Phosphatic fertilizer industry: Phosphorus fertilizers are used in agriculture to increase the crop output. Heavy metal impurities are present in superphosphate fertilizers in addition to nutrient elements (Oyedele et al., 2006). There are many reported cases of Pb poisoning across the world (Galadima and Garba, 2012; Agwaramgbo et al., 2014). Recent studies also reported toxicity of As and other elements like Cd and Pb, impacting human health by causing chronic kidney diseases at Srilanka. Cd can accumulate in plant system in excess amount without showing any phytotoxic symptoms (Moustakas et al., 2001; Kirkham, 2006; Al-faiyz et al., 2007).

Leather industry: The leather industry serves platform to recycle the by-products of meat industry but during the process of preparing leather generates a large amount of pollutants (Joesph and Nithya, 2009; Sathish et al., 2016). The leather production process involves14–15 operational steps in which four steps are main with subsequent processing steps (Durai and Rajasimman, 2011; Saravanbhavan et al., 2006). Each operational step generates harmful pollutants that induce air, soil, and water pollution. Trivalent chromium (Cr^{3+}) and hexavalent chromium (Cr^{6+}) are the predominant chemical compounds of concern from this industry. "Severe allergic contact dermatitis" arise due to the formation of Cr^{6+} from $Cr^{3+}$and this also leach into soils polluting groundwater (Shi et al., 2016; Saikia et al., 2017). Presently in leather industry, trivalent chromium is heavily used to process leather due to its time and economic advantages (Zuriaga-Agustí et al., 2015). Pollution in water bodies from leather industries are caused when untreated chemically infused effluent water is being discharged from tanneries to surface water bodies (Chen et al., 2017). Globally, 145 billion gallons of wastewater is generated by the tanning industry annually (Sathish et al., 2016). The industry is responsible for almost 40% of global Cr pollution and is considered to be "one of the most polluting industries in the world" (Bien et al., 2017).

Textile industry: This industry is the largest generator of wastewater rich in organic and inorganic pollutants harmful for the environment during cloth making processes (Dung et al., 2013; Bilinska et al., 2016). 17%–20% of total wastewater is generated from the operations like textile dyeing and finishing (Holkar et al., 2016). In this wastewater the heavy metals (HMs) and metalloids along with recalcitrant biodegradables are mostly present and exert a toxicological effect (Dung et al., 2013). Cr, As, Hg, Pb, Cd, Fe, Zn and Cu and other formulations of these HMs toxicants are the most common as reported in textile effluents (Kaur et al., 2021).

7.4 Metal contamination in water resources through mining activity

Globally, contamination of water environments due to heavy metal from mining industry has gained considerable attention by researchers and policy makers (Förstner and Wittmann, 1983; Meng et al., 2016). Metals originate into the environment through a number of processes that include natural processes and anthropogenic activities. Volcanic activity, bedrock weathering and erosion are among the natural processes whereas various industries such as, agriculture: fertilizer and pesticides; mining: metal smelting and refining, etc., are anthropogenic activities (Shi et al., 2013; Kumar et al., 2017). Studies documented that in river basins pollution the heavy metals from mining activity is the most critical contributor (Zhou et al., 2007). The mining activity leads to alterations in landscapes, habitats destruction, soil and water contamination and land resources degradation thus it is also called most dominant anthropogenic activities (Acosta et al., 2011; Moreno-Jiménez et al., 2011). Mine operations often contribute in removal of large quantities of overburden waste materials and eventually cause metal release from stable form into the environment (Geelani et al., 2013). The tailings from mine operations are frequently dispersed to the surrounding environment like rivers, arable land and ground waters through erosion, wind action, etc., contributing pollutants mostly in the form of metals in air, soil, vegetation, surface and ground waters eventually causing pollution in aquatic habitats, agricultural lands (Ochieng et al., 2010; Matthews et al., 2012; Liu et al., 2015). Leaching of tailing and overburden drainage bring in significant amount of metals into the adjacent land, air and water resources and cause severe adverse effects (Abdul-Wahab and Marikar, 2012).

7.4.1 Copper mine

In a Cu mine catchment at Uganda ore processing center tailing sites and leachate from that was observed to be the main source of trace as well as heavy metals pollution in river Nyamwamba. An elevated concentrations of Ni, Pb, As, Cu, and Coin river catchment water especially the Co concentrations was reported. Nyamwamba river water was 53% of contaminated both along the mine zone and downstream. Also, metals like Fe, Mn and Co exceeded drinking water standards (Abraham and Susan, 2017). These HM accumulated in the atmospheric dust later settle down in nearby land mass and water body causing its pollution. On critical evaluation of an ore processing centers, leachate of abandoned mines, operational tailings dams and other auxiliary activities are the most common sources of heavy metal pollution. The wastewater discharged from tailing dam into the river also pollutes it with metals. Cu mining contributes to elevated concentrations of heavy metals (Gabrielyan et al., 2018a,b). Studies on Cu mine across the world has reported elevated concentration of heavy metals like Cu, Ni, Co, Cr, Mn, Pb, As, Cd, Hg and Zn in soil (Giri et al., 2017; Demková et al., 2016; Angelovičová, and Fazekašová, 2014).

7.4.2 Coal mine

Coal Production is of utmost important for energy generation while being crucial for the economic welfare of the nation. The mining of this precious natural resource results in a hazardous consequences due to the accumulation of heavy metals that are nonbiodegradable and damage the environment. The coal mining areas having high water table has a more tendency to undergo heavy metals accumulation and destruction of the ecological environment occurs through coal mining subsidence ponds. As and Pb content in the water bodies has shown slightly exceeded limits near coal mine area (Tan et al., 2020). The mining operation has led to the contamination of the water bodies due to leaching of heavy metals. As compared to the opencast mines samples from underground mines reported slightly lower metal contamination. In conventional coal mining methods a practicable and effective technology needs to be integrated in order to contain and prevent the leaching of metals causing contamination of the environment. The HM concentration in different land use types surrounding mine area is reported by many researchers. They reported high and toxic ranges of heavy metals in the soil. However they reported higher concentration cropland as compared to forest which is rarely disturbed by human induced activities (Zheng et al., 2016). Pandey et al. (2017) studied three sites at Jharia coalfield area, good plantation area—residential area with vehicular movement and transport of coal and near to mining activity, coal handling plant, vehicular movement, industrial activity. Highest value for Pb and Cd was observed at mining activity site and other metals such as Cu, Cr and Ni was observed in residential area with transport of Coal. The site containing good plantation has lowest HM content in soil. In a lignite coal mine tailings in Gujrat, India all the HM Pb, Cd, Cu, Cr, Co, Ni are in toxic range and the concentration of Cu was highest at 199 mg/kg (Ladwani et al., 2012). Other studies done at Sonepur Bazari, Raniganj, India and Shandong coal mine, China the concentration of the HM are in toxic range (Masto et al., 2015; Manna and Maiti, 2018; Li et al., 2017).

7.4.3 Iron mine

Pollution of the environment occurs through mining activities combined with ore processing (Mitchell, 2009; Acheampong et al., 2013). An iron-manufacturing plant discharges contain heavy metals enriched substantial amount of wastewater (Hu et al., 2014). This heavy metal loaded discharges are left untreated and added into nearby water body results in an elevated concentration of heavy metal in streams, rivers, and ground waters. Priority heavy metals are both toxic and lethal to human, these metals are found in the soils in near proximity to iron ore mine at toxic concentration (Schneegurt et al., 2001; Pena et al., 2004). Studies conducted globally to understand heavy metal accumulation near iron ore mine has reported high level of metals and metalloids (Stephen and Oladele, 2012).

7.5 Heavy metals contamination in water resources through transport activity

7.5.1 Pollution from road traffic

Water plays a very important role in planet Earth. It is one of the basis on which life cycle depend and global biodiversity survives (Bagul et al., 2015). This life supporting resource of the earth is exposed to harmful anthropogenic pollutants like heavy metals. Heavy metals that are released into the environment is leading to the deterioration of the environmental health because of its long time persistence in nature. The effects arising out of pollution is a global issue that has taken pace during last few decades, in addition to undermining economic growth it also jeopardizes the health of innumerable people (Mateo-Sagasta et al., 2017). However, anthropogenic activities play a major role in generating heavy metals in sediment and water polluting the aquatic environment (Yi et al., 2011). The vehicular traffic in urban area is reported to be the prime source of heavy metals (Adamiec et al., 2016; Men et al., 2018). Usually on roads and junctions during dry phase substantial amount of heavy metals from various vehicular activities build up. These deposited metals on roads are consequently washed-away by storm water from precipitation and downpour in the form of runoff gets settled down in reservoirs. Seasonal variation, content of vegetation on urban roads often controls the pollutant load that is flowing down to reservoirs with surface runoff and storm water drainage system (Sojka et al., 2018). Heavy metal pollution from transport sector mainly arises due to exhaust emitted from car and other wear and tear (Yuen et al., 2012). In transport sector pollution due to road surface abrasion and road dust contamination from tire wear has remained a well-recognized source of pollution. Many researchers have confirmed the elevated amounts of heavy metals present in the dust by conducting detailed analyses of tire wear dust (Adachia and Tainoshob, 2004; Schauer et al., 2006; Hjortenkrans et al., 2007).

The road abrasions due to vehicular operations are responsible for heavy metals in road dust. Concrete motorways are responsible for accelerated tire abrasion (Duong and Lee, 2011). Vehicle speed also strongly determines heavy metal content in road dust. Higher speed result in more tire wear and increased fuel combustion for example the dust from roads with average speed range of 80 to 90 km/h has higher heavy metal concentration as compared to the roads with 70–80 km/h (Duong and Lee, 2011). Motorways have twice road dust metal concentration as compared to the roundabouts (Duong and Lee, 2011). Brake dust contains Fe with significant amounts of heavy metals including Pb and Cd. Moreover, ground or over headlines used by trolley buses, trams and trains adds to the street dusts heavy metals (Shi et al., 2011; Zgłobicki et al., 2019). Apparently, areas with high volumes of traffic presence of heavy metals are common as brake wear, tire wear and motor oil, emits Zn, Cr, Ni, Cd (Ferreira et al., 2016). Lubricant combustion, engine corrosion, particulate filters automotive coatings, elements corrosion, catalytic converters are some of the various exhaust-related sources that are the potent cause of metals emission (Sýkorová et al., 2012). Vehicle exhaust on urban road surfaces is also a potential main contributor to Pb pollution (Hong et al., 2018) where use of leaded petrol is still in practice.

7.5.2 Human toxicity

Heavy metal toxicity induces a number of health issues in humans. Heavy metal toxicity may show acute or chronic effects based on the concentration and rate of exposure. The symptoms include alteration in the functioning of major organs in humans. Heavy metal toxicity may lead to damage of these systems. Long term heavy metal exposure of the human body may lead to diseases like Parkinson's disease, Alzheimer's disease, muscular dystrophy, multiple sclerosis. Moreover, chronic long term toxic heavy metal exposure may lead to cancer (Jarup, 2003; Rani et al., 2022). Heavy metal cycling in different strata is illustrated in Fig. 7.1. Health risks of some toxic heavy metals are illustrated below.

Arsenic: Acute As poisoning may cause destruction of gastrointestinal tissue, blood vessels, heart and brain. Chronic As toxicity usually affects skin manifestations such as pigmentation and keratosis that is also called arsenicosis (Martin and Griswold, 2009).

Lead: Central nervous system and gastrointestinal tract in children and adults are mostly affected in lead poisoning. Acute exposure can cause headache, vertigo, arthritis, etc., while chronic exposure result in weight loss, dyslexia mental retardation, kidney damage, muscular weakness, brain damage, autism, psychosis, hyperactivity coma and apparently may lead to death (Martin and Griswold, 2009). The Pb poisoning can impact major organs of the body; though it is preventable but remains a dangerous metal on exposure.

Mercury: Hg can easily form inorganic and organic compounds by combining with other elements in which methyl-Hg compound is highly carcinogenic. Elevated levels of metallic, inorganic and organic Hg compound upon exposure can damage the developing fetus and major organs like kidney and brain (Alina et al., 2012). Organic Hg like methyl-Hg is lipophilic in nature and harm human system by easily penetrating cell membranes. Nervous

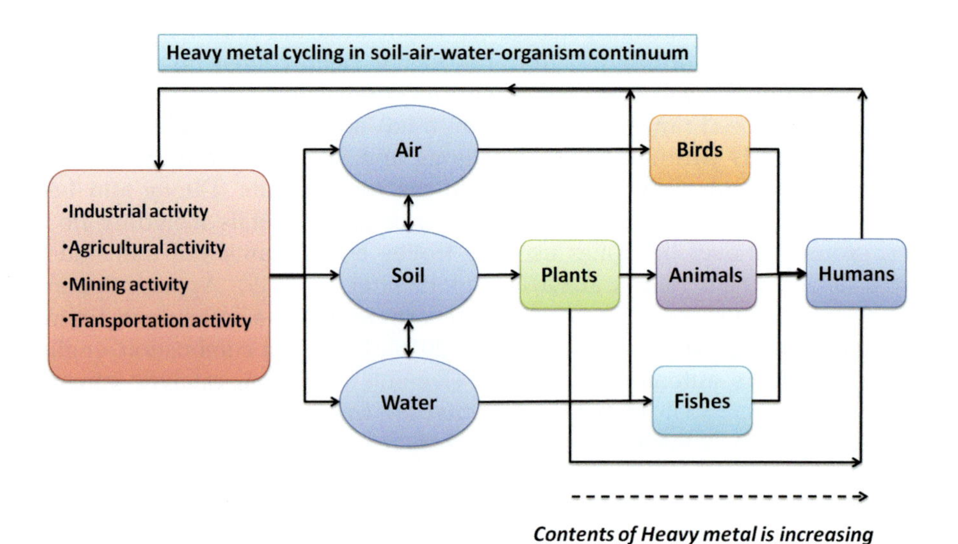

FIGURE 7.1 Heavy metal cycling in different strata.

system is mainly affected by mercury and its compounds; upon short duration exposure lead to diarrhea, vomiting, nausea, high blood pressure, etc. (Martin and Griswold, 2009).

Cadmium: Cd causes several deleterious health effects in humans. As reabsorption of Cd by the kidney limits its excretion human body cant excrete Cd this causes the health effects of Cd exposure are worsened. Inhalation of Cd upon short-term exposure can cause respiratory irritation and severe harm to the lungs. Long-term exposure to Cd and its ingestion in higher dose leads to its deposition in lungs and bones. Thus Cd can cause lung and bone damage (Bernard, 2008).

Chromium: The hexavalent (Cr IV) form of Cr is the most toxic species among the all the species of Cr reported. Major symptoms shown are anemia, impacting male reproductive system by damaging sperm, ulcers in the stomach and small intestine. Chromium (VI) compounds are lethal in extremely high doses (Shekhawat et al., 2015). Cr compounds induce DNA damage in many different ways. Cr has a potential to induce carcinogenic tumors in stomach (Matsumoto et al., 2006).

7.6 Mode of action of heavy metals in the environment

Accumulation of heavy metals even if present in traces accumulates in the food chain and eventually causes life threatening diseases. Heavy metals transport from water body to human through food chain is depicted in Fig. 7.2. Degeneration of health condition has also been reported such as senility and neuro-degeneration (Jan et al., 2015). However some heavy metals such as Cu, Ni, Se, Zn, Mn, etc., are essential elements required in trace amount for appropriate metabolism and development, but are equally toxic and deleterious when present in elevated concentrations in living organisms (Stephen and Oladele, 2012). Toxicity and carcinogenicity of metals such as As, Cd, Cr, Pb, and Hg (Tchounwou et al., 2014) to the environment is mainly executed by production of reactive oxygen species (ROS) and consequently oxidative stress. These five heavy metals are systemic toxicants and even at lower points of exposure are known to induce risks to the human system (Jan et al., 2015).

Heavy metals solubility, partitioning, and mobility are influenced by a variety of physicochemical parameters such as salinity, pH, redox potential (Eh), dissolved oxygen (DO), concentration and type of ligands and chelating agents. Heavy metals enter into water bodies through different natural and anthropogenic sources, then pass through sediments, biota and fish to enter into the human body (Fig. 7.2). Even at relatively low levels, HMs cause histo-pathological problems in fish. According to several experts, the eco-environmental consequences of heavy metals in aquatic ecosystems were determined by their bioavailable percentage rather than their total concentrations. Heavy metal bioavailability in water bodies has a significant impact on water quality. Because sediments are a major sink for most pollutants, accumulating heavy metals in sediments could pose a long-term threat to water bodies if remobilization/redox processes occur. The chemical dynamics of heavy metals also depended on processes like as adsorption-desorption, sedimentation-resuspension in water. Several factors determine the toxicity of heavy metals in living organism, namely, dose, route of exposure, and chemical species, as well as the genetics, age, gender, and nutritional status of exposed individuals. Heavy metal

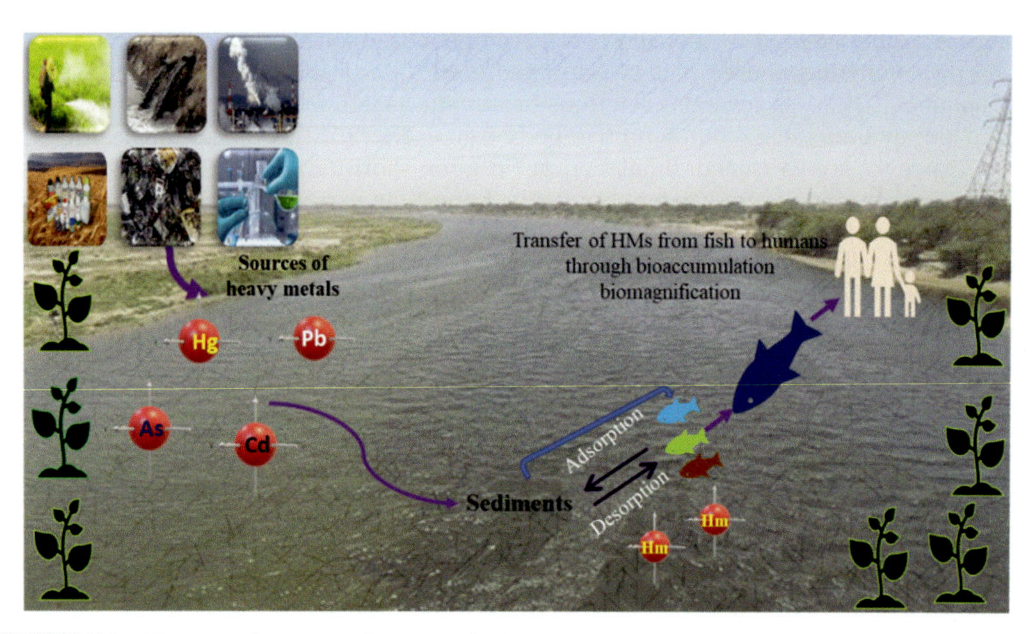

FIGURE 7.2 Heavy metals transport from water body to human through food chain.

may occur naturally in soil based on the origin of soil (Zhou and Guo, 2015). However, various anthropogenic activities play a major role in causing heavy metal contamination of the soil ecosystem as recognized by several reports (Al-Khashman and Shawabkeh, 2006; Banat et al., 2005; Kasassi et al., 2008).

7.7 Conclusion

All the anthropogenic activities discussed in this chapter are contributing largely to the pollution of water body with heavy metals. Mostly the industrial activities like textile, leather, phosphatic fertilizer, etc., are the point source of water pollution whereas pollution due to agriculture, mining, and transport activity are contributing in NPS of pollution as they are diffused sources. Pollution of water body with heavy metals leads to deterioration in water quality. Although some heavy metals are required in traces for cellular, metabolic, and hormonal functioning in humans, but if the limit exceeds that may cause severe health hazards. As, Pb, Cd, and Cr are the heavy metals of the highest concern as they do not participate in any physiological activities in humans, animals and plants rather they show toxic symptoms. All the four metals have carcinogenic properties that induce cancer in humans. Therefore, it has to be the paramount interest of policy makers to protect and improve quality of water bodies. Several approaches and statistical methods are in use to track the heavy metal sources in the ecosystem viz. positive matrix factorization (PMF), principal component analysis (PCA), and hierarchical cluster analysis (HCA). These statistical approaches suggest the contamination sources of heavy metals. To assess the degree of heavy metal contamination and risk posed in sediments of water body various methods are used viz. sediment quality guidelines (SQGs), geoaccumulation indices (I_{geo}), enrichment factors (EFs), contamination factors (CFs), and pollution load indices (PLIs). To control water pollution through point source, the wastewater generated in industrial facilities has to be treated up to WHO recommended limits before its discharge into water bodies. To control NPS pollution installation of suitable engineering structures can be undertaken in contaminated area to reduce land runoff. In addition to that phytoremediation with suitable plant species which can uptake heavy metals from water system can be done. Regulation of heavy metal pollution of the environment with enforcing strict policies is to be undertaken. Pollution of water body can be regulated with the following strategies:

- Afforestation with plants, shrub, and grasses to prevent soil loss through runoff.
- Soil test to be conducted for optimization of fertilizer use.
- Proper disposal of heavy metal contaminated products.
- Replace use of permeable surfaces with impermeable surface like concrete and wood.
- Effective conservation practices like construction of terraces on slopes, grass waterways planting and creating wetlands to slow down pollutants in areas vicinity to waterways.
- Use of proper filtration and treatment of point source of pollution.
- Tertiary treatment of wastewater along with phytoremediation technology to be done before its discharge into water bodies.

References

Abdul-Wahab, S.A., Marikar, F.A., 2012. The environmental impact of gold mines, pollution by heavy metals. Central European Journal of Engineering 2 (2), 304–313.

Abraham, M.R., Susan, T.B., 2017. Water contamination with heavy metals and trace elements from Kilembe copper mine and tailing sites in Western Uganda; implications for domestic water quality. Chemosphere 169, 281–287.

Acheampong, M.A., Paksirajan, K., Lens, P.N.L., 2013. Assessment of the effluent quality from a gold mining industry in Ghana. Environmental Science and Pollution Research 20 (6), 3799–3811.

Acosta, J., Faz, A., Martinez-Martinez, S., Zornoza, R., Carmona, D., Kabas, S., 2011. Multivariant statistical and GIS-based approach to evaluate heavy metals behaviour in mine sites for future reclamation. Journal of Geochemical Exploration 109 (1–3), 8–17.

Adachia, K., Tainoshob, Y., 2004. Characterization of heavy metal particles embedded in tire dust. Environment International 30, 1009–1017.

Adamiec, E., Jarosz-Krzemińska, E., Wieszała, R., 2016. Heavy metals from non-exhaust vehicle emissions in urban and motorway road dusts. Environmental Monitoring Assessment 188 (6). Available from: https://doi.org/10.1007/s10661-016-5377-1.

Agwaramgbo, L., Iwuagwu, A., Alinnor, J., 2014. Lead removal from contaminated water by corn and palm nut husks. British Journal of Applied Science and Technology 4, 4992–4999.

Al-faiyz, Y.S., El-Garaway, M.M., Assubaie, F.N., Al-Eed, M.A., 2007. Impact of phosphate fertilizer on cadmium accumulation in soil and vegetable crops. Bulletin of Environmental Contamination Toxicology 78, 358–362.

Alina, M., Azrina, A., Mohd Yunus, A.S., Mohd Zakiuddin, S., Mohd Izuan Effendi, H., Muhammad Rizal, R., 2012. Heavy metals (mercury, arsenic, cadmium, plumbum) in selected marine fish and shellfish along the straits of Malacca. International Food Research Journal 19 (1), 135–140.

Al-Khashman, O.A., Shawabkeh, R.A., 2006. Metals distribution in soils around the cement factory in Southern Jordan. Environmental Pollution 140, 387–394.

Angelovičová, L., Fazekašová, D., 2014. Contamination of the soil and water environment by heavy metals in the former mining area of Rudňany (Slovakia). Soil and Water Research 9 (1), 18–24.

Arianmehr, S., Nouti-Zahi, G., 2016. Effects of agricultural pollution on water and subterranean soil of bahokalat river of Iran with reference to heavy metals. Bioscience Biotechnology Research Communications . Available from: https://doi.org/10.21786/bbrc/9.1/11.

Bagul, V.R., Shinde, D.N., Chavan, R.P., Patil, C.L., Pawar, R.K., 2015. New perspective on heavy metal pollution of water. Journal of Chemical and Pharmaceutical Research 7 (12), 700–705.

Banat, K., Howari, M.F.M., Al-Hamad, A.A., 2005. Heavymetals in urban soils of central Jordan. Should we worry about their environmental risks? Environmental Research 97 (3), 258–273.

Basta, N.T., Ryan, J.A., Chaney, R.L., 2005. Trace element chemistry in residual-treated soil: key concepts and metal bioavailability. Journal of Environmental Quality 34 (1), 49–63.

Bernard, A., 2008. Cadmium & its adverse effects on human health. Indian Journal Medical Research 128 (4), 557–564.

Bien, J., Celary, P., Wystalska, K., 2017. The problems in achieving sustainable development in the tannery industry in regard to sewage sludge management. Journal of Ecological Engineering 13–20.

Bilinska, L., Gmurek, M., Ledakowicz, S., 2016. Comparison between industrial and simulated textile wastewater treatment by AOPs – biodegradability, toxicity and cost assessment. Chemical Engineering Journal 306, 550–559. Available from: https://doi.org/10.1016/j.cej.2016.07.1.

Cervantes-Ramírez, L.T., Ramírez-López, M., Mussali-Galante, P., et al., 2018. Heavy metal biomagnification and genotoxic damage in two trophic levels exposed to mine tailings: a network theory approach. Revista Chilena de Historia Natural 91, 6. Available from: https://doi.org/10.1186/s40693-018-0076-7.

Chen, L., Wang, L., Wu, X., Ding, X., 2017. A process-level water conservation and pollution control performance evaluation tool of cleaner production technology in little industry. Journal of Cleaner Production 1137–1143.

Ćujić, M., Dragović, S., Đorđević, M., Dragović, R., Gajić, B., 2016. Environmental assessment of heavy metals around the largest coal fired power plant in Serbia. Catena 139, 44–52.

Demková, L., Jezný, T., Bobuľská, L., 2016. Assessment of soil heavy metal pollution in a former mining area – before and after the end of mining activities. Soil & Water Research 17221/107.

Dung, T.T.T., Cappuyns, V., Swennen, R., Phung, N.K., 2013. From geochemical background determination to pollution assessment of heavy metals in sediments and soils. Reviews. Environmental Science Biotechnology 12, 335–353. Available from: https://doi.org/10.1007/s11157-013-9315-1.

Duong, T., Lee, B.K., 2011. Determining contamination level of heavy metals in road dust from busy traffic areas with different characteristics. Journal of Environmental Management 92 (3), 554–562.

Durai, G., Rajasimman, M., 2011. Biological treatment of tannery wastewater. A review. Journal of Environmental Science and Technology 1–17.

FAO, 2006. Livestock's Long Shadow. Food and Agriculture Organization of the United Nations, Rome.

FAO, 2014. Area Equipped for Irrigation. Infographic. AQUASTAT: FAO's Information System on Water and Agriculture. Food and Agriculture Organization of the United Nations (FAO), Rome. Available at. Available from: http://www.fao.org/nr/water/aquastat/infographics/Irrigation_eng.pdf.

FAO, 2016. The State of World Fisheries and Aquaculture: Contributing to Food Security and Nutrition for All. Food and Agriculture Organization of the United Nations (FAO), Rome.

Ferreira, A.J.D., Soares, D., Serrano, L.M.V., Walsh, R.P.D., Dias-Ferreira, C., Ferreira, C.S.S., 2016. Roads as sources of heavy metals in urban areas. The Covões catchment experiment, Coimbra, Portugal. Journal of Soils and Sediments 16 (11), 2622–2639. Available from: https://doi.org/10.1007/s11368-016-1492-4.

Fiket, Z., Meduni, C., Kniewald, G.G., 2016. Rare earth elements distribution in soil nearby thermal power plant. Environmental Earth Science 75, 598 [CrossRef].

Förstner, U., Wittmann, G., 1983. Metal Pollution in the Aquatic Environment. Springer, Berlin, Germany, ISBN 9783540128564.

Gabrielyan, A.V., Shahnazaryan, G.A., Minasyan, S.H., 2018a. Distribution and identification of sources of heavy metals in the Voghji river basin impacted by mining activities (Armenia). Environmental Biogeochemistry of Elements and Emerging Contaminants Special issue, *journal of chemistry*.

Gabrielyan, A.V., Shahnazaryan, G.A., Minasyan, S.H., 2018b. Distribution and identification of sources of heavy metals in the Voghji River basin impacted by mining activities (Armenia). Journal of Chemistry 2018, 9. Available from: https://doi.org/10.1155/2018/7172426. Article ID 7172426.

Galadima, A., Garba, Z.N., 2012. Heavy metals pollution in Nigeria. causes and consequences. Elixir Pollution 45, 7917–7922.

Gautam, R.K., Sharma, S.K., Mahiya, S., Chattopadhyay, M., 2015. Contamination of Heavy Metals in Aquatic Media: Transport, Toxicity and Technologies for Remediation. Published by the Royal Society of Chemistry. Available from: http://www.rsc.org, http://doi.org/10.1039/9781782620174-00001.

Geelani, S.M., Bhat, S.J.A., Geelani, S.H., Haq, S.S., Mir, N.A., Qazi, G., et al., 2013. Mining and its impacts on environment with special reference to Indian. International Journal of Current Research 5 (12), 3586–3590.

Giri, S., Singh, A.K., Mahato, M.K., 2017. Metal contamination of agricultural soils in the copper mining areas of Singhbhum shear zone in Indian. Journal of Earth System Science 126, 49.

Gupta, D., Rai, U., Tripathi, R., Inouhe, M., 2002. Impacts of fly-ash on soil and plant responses. Journal of Plant Research 115, 401–409.

Gupta, D.K., Huang, H.G., Corpas, F.J., 2013. Lead tolerance in plants: strategies for phytoremediation. Environmental Science and Pollution Research 20, 2150–2161. Available from: https://doi.org/10.1007/s11356-013-1485-4.

Hatje, V., Apte, S.C., Hales, L.T., Birch, G.F., 2003. Dissolved trace metal distributions in Port Jackson estuary (Sydney Harbour), Australia. Marine Pollution Bulletin 46 (6), 719–730.

Herngren, L., Goonetilleke, A., Ayoko, G.A., 2006. Analysis of heavy metals in road-deposited sediments. Analytica Chimica Acta 571 (2), 270–278.

Hjortenkrans, D.S.T., Bergbäck, B.G., Häggerud, A.V., 2007. Metal emissions from brake linings and tires: case studies of Stockholm, Sweden 1995/1998 and 2005. Environmental Science & Technology 41, 5224–5230.

Holkar, C.R., Jadhav, A.J., Pinjari, D.V., Mahamuni, N.M., Pandit, A.B., 2016. A critical review on textile wastewater treatments: possible approaches. Journal of Environmental Management 182, 351–366. Available from: https://doi.org/10.1016/j.jenvman.2016.07.090.

Hong, N., Zhu, P., Liu, A., Zhao, X., Guan, Y., 2018. Using an innovative flag element ratio approach to tracking potential sources of heavy metals on urban road surfaces. Environmental Pollution 243, 410–417. Available from: https://doi.org/10.1016/j.envpol.2018.08.098.

Hou, S., Zheng, N., Tang, L., Ji, X., Li, Y., Hua, X., 2018. Pollution characteristics, sources, and health risk assessment of human exposure to Cu, Zn, Cd and Pb pollution in urban street dust across China between 2009 and 2018. Environment International 128, 430–437. Available from: https://doi.org/10.1016/j.envint.2019.04.046. ISSN 0160-4120.

Hu, X.F., Jiang, Y., Shu, Y., Hu, X., Liu, L., Luo, F., 2014. Effects of mining wastewater discharges on heavy metal pollution and soil enzyme activity of the paddy fields. Journal of Geochemical Exploration 147 (PB), 139–150.

Jan, A.T., Azam, M., Siddiqui, K., Ali, A., Choi, I., Haq, Q.M.R., 2015. Heavy metals and human health. Mechanistic insight into toxicity and counter defense system of antioxidants. International Journal of Molecular Sciences 16 (12), 29592–29630.

Jarup, L., 2003. Hazards of heavy metal contamination. British Medical Bulletin 68 (1), 167–182.

Joesph, K., Nithya, N., 2009. Material flows in the life cycle of leather (Keep). Journal of Cleaner Production 676–682.

Kasassi, A., Rakimbei, P., Karagiannidis, A., Zabaniotou, A., Tsiouvaras, K., Nastis, A., et al., 2008. Soil contamination by heavy metals measurements from a closed unlined land fill. Bio-Resource Technology 99 (18), 8578–8584.

Kaur, J., Bhatti, S.S., Bhat, S.A., Nagpal, A.K., Kaur, V., Katnoria, J.K., 2021. Evaluating potential ecological risks of heavy metals of textile effluents and soil samples in vicinity of textile industries. Soil Systems 5, 63. Available from: https://doi.org/10.3390/soilsystems5040063.

Keegan, T.J., Farago, M.E., Thornton, I., Hong, B., Colvile, R.N., Pesch, B., et al., 2006. Dispersion of as and selected heavy metals around a coal-burning power station in Central Slovakia. Science of the Total Environment 358, 61–71.

Kelepertzis, E., 2014. Accumulation of heavy metals in agricultural soils of Mediterranean insights from Argolida basin, Peloponnese, Greece. Geoderma 221, 82–90. Available from: https://doi.org/10.1016/j.geoderma.2014.01.007.

Kirkham, M.B., 2006. Cadmium in plants on polluted soils effects of soil factors, hyperaccumulation, and amendments. Geoderma 137, 19–32.

Kumar, M., Ramanatahn, A.L., Tripathi, R., Farswan, S., Kumar, D., Bhattacharya, P.A., 2017. Study of trace element contamination using multivariate statistical techniques and health risk assessment in groundwater of Chhaprola Industrial Area, Gautam Buddha Nagar, Uttar Pradesh, India. Chemosphere 166, 135–145.

Ladwani, K.D., Ladwani, K.D., Manik, V.S., Ramteke, D.S., 2012. Assessment of heavy metal contaminated soil near coal mining area in Gujarat by toxicity characteristics leaching procedure. International Journal of Life Sciences Biotechnology and Pharma Research 1 (4), 73–80.

Lau, S.L., Stenstrom, M.K., 2005. Metals and PAHs adsorbed to street particles. Water Research 39 (17), 4083–4092.

Li, F., Qiu, Z.Z., Zhang, J.D., 2017. Investigation, pollution mapping and simulative leakage health risk assessment for heavy metals and metalloids in groundwater from a typical brownfield, middle China. International Journal of Environmental Research and Public Health 14 (7), 768. Available from: https://doi.org/10.3390/ijerph14070768.

Liu, R., Liu, J., Zhang, Z., Borthwick, A., Zhang, K., 2015. Accidental water pollution risk analysis of mine tailings ponds in guanting reservoir watershed, Zhangjiakou City, China. International Journal of Environmental Research and Public Health 12 (12), 15269–15284.

Mandal, A., Sengupta, D., 2006. An assessment of soil contamination due to heavy metals around a coal-fired thermal power plant in India. Environmental Geology 51, 409–420.

Manna, A., Maiti, R., 2018. Geochemical contamination in the mine affected soil of Raniganj Coalfield – a river basin scale assessment. Geoscience Frontiers 9 (5), 1577–1590.

Marsalek, J., Rochfort, Q., Brownlee, B., Mayer, T., Servos, M., 1999. An exploratory study of urban runoff toxicity. Water Science and Technology 39 (12), 33–39.

Martin, S., Griswold, W., 2009. Human health effects of heavy metals. Environmental Science and Technology 15, 1–6. Briefs for Citizens.

Masto, R.E., Sheik, S., Nehru, G., Selvi, V.A., George, J., Ram, L.C., 2015. Assessment of environment soil quality around Sonepur Bazar mine of Raniganj coalfield, India. Solid Earth 6, 811–821.

Mateo-Sagasta, J., Zadeh, S.M., Turral, H., Burke, J., 2017. Water pollution from agriculture: a global review. Executive summary.

Matsumoto, S.T., Mantovani, M.S., Malaguttii, M.I.A., Dias, A.L., Fonseca, I.C., Marin-Morales, M.A., 2006. Genotoxicity and mutagenicity of water contaminated with tannery effluents, as evaluated by the micronucleus test and comet assay using the fish *Oreochromis niloticus* and chromosome aberrations in onion root-tips. Genetics Molecular Biology 29 (1), 148–158.

Matthews, A., Omono, C., Kakulu, S., 2012. Impact of mining and agriculture on heavy metal levels in environmental samples in Okehi Local Government Area of Kogi State. International Journal of Pure and Applied Sciences and Technology 12 (2), 66–77.

Men, C., Liu, R., Wang, Q., Guo, L., Shen, Z., 2018. The impact of seasonal varied human activity on characteristics and sources of heavy metals in metropolitan road dusts. Science Total Environment 637–638, 844–854. Available from: https://doi.org/10.1016/j.scitotenv.2018.05.059.

Meng, Q., Zhang, J., Zhang, Z., Wu, T., 2016. Geochemistry of dissolved trace elements and heavy metals in the Dan River Drainage (China): distribution, sources, and water quality assessment. Environmental Science and Pollution Research 23, 8091–8103.

Mitchell, J.W., 2009. An assessment of leadmine pollution using macroinvertebrates at Greenside Mines, Glenridding. Earth & Environmental Science 4, 27–57.

Moreno-Jimenez, E., Beesley, L., Lepp, N.W., Dickinson, N.M., Hartley, W., Clemente, R., 2011. Field sampling of soil pore water to evaluate trace element mobility and associated environmental risk. Environmental Pollution 159, 3078–3085.

Moustakas, N.K., Akoumianakis, K.A., Passam, H.C., 2001. Cadmium accumulation and its effects on yield of lettuce, radish, and cucumber. Communication. Soil Science. Plant Analysis 32, 1793–1802.

Ochieng, G.M., Seanego, E.S., Nkwonta, O.I., 2010. Impacts of mining on water resources in South Africa. A review. Scientific Research and Essays 5 (22), 3351–3357.

Oyedele, D.J., Asonugho, C., Awotoye, O.O., 2006. Heavy metals in soil and accumulation by edible vegetables after phosphate fertilizer application. Electronic Journal of Environmental, Agricultural and Food Chemistry 5 (4), 1446–1453.

Pandey, B., Mukherjee, A., Agrawal, M., Singh, S., 2017. Assessment of seasonal and site specific variations in soil physical, chemical and biological properties around opencast coal mines. Pedospheres 10, 60431–60434.

Pena, E., Suárez, J., Sánchez-Tembleque, F., Jácome, A., Puertas, J., 2004. Characterization of polluted runoff in a granite mine, Galicia, Spain. In: Jarvis, A.P., Dudgeon, B.A., Younger, P.L. (Eds.), Proceedings of the International Mine Water Association Symposium, 1. University of Newcastle, Newcastle upon Tyne, pp. 185–194.

Pitt, R., Field, R., Lalor, M., Brown, M., 1995. Urban stormwater toxicpollutants: assessment, sources, and treatability. Water Environmental Research 67 (3), 260–275.

Praharaj, T., Powell, M.A., Hart, B.R., Tripathy, S., 2002. Leachability of elements from subituminous coal fly ash from India. Environment International 27, 609–615. Available from: https://doi.org/10.1016/S0160-4120(01)00118-0.

Rani, L., Srivastav, A.L., Kaushal, J., Grewal, A.S., Madhav, S., 2022. Heavy metal contamination in the river ecosystem. Ecological Significance of River Ecosystems. Elsevier, pp. 37–50.

Saikia, P., Goswami, T., Dutta, D., Dutta, N.K., Sengupta, P., Neog, D., 2017. Development of a flexible composite from leather industry waste and evaluation of their physico-chemical properties. Clean Technology Environmental Policy 2171–2178.

Saravanbhavan, S., Thanikaivelan, P., Rao, J.R., Ramasami, T., 2006. Reversing the conventional. Leather processing sequence for cleaner leather production. Environmental Science and Technology 1069–1075.

Sathish, M., Madhan, B., Sreeram, K.J., Rao, J.R., Nair, B.U., 2016. Alternative carrier medium for sustainable leather manufacturing - a review and perspective. Journal of Cleaner Production 49–58.

Schauer, J.J., Lough, G.C., Shafer, M.M., Christensen, W.F., Arndt, M.F., DeMinter, J.T., et al., 2006. Characterization of metals emitted from motor vehicles. Research Report. Health Effects Institute 133:1–76. Discussion 77–88.

Schneegurt, M.A., Jain, J.C., Menicucci, J.A., Brown, S.A., Kemner, K.M., Garofalo, D.F., 2001. Biomass byproducts for the remediation of wastewaters contaminated with toxic metals. Environmental Science and Technology 35 (18), 3786–3791.

Schweitzer, L., Noblet, J., 2018. Water Contamination and Pollution. Green Chemistry.

Shekhawat, K., Chatterjee, S., Joshi, B., 2015. Chromium toxicity and its health hazards. International Journal of Advanced Research 7 (3), 167–172.

Shi, G., Chen, Z., Bi, C., Wang, L., Teng, J., Li, Y., et al., 2011. A comparative study of health risk of potentially toxic metals in urban and suburban road dust in the most populated city of China. Atmosphere Environment 45 (3), 764–771. Available from: https://doi.org/10.1016/j.atmosenv.2010.08.039.

Shi, X., Chen, L., Wang, J., 2013. Multivariate analysis of heavy metal pollution in street dusts of Xianyang city, NW China. Environmental Earth Sciences 69, 1973–1979.

Shi, J., Puig, R., Sang, J., Lin, W., 2016. A comprehensive evaluation of physical and environmental performances for wet-white leather manufacture. Journal of Cleaner Production 1512–1519.

Singh, R.K., Gupta, N.C., Guha, B.K., 2014. PH dependence leaching characteristics of selected metals from coal fly ash and its impact on groundwater quality. International Journal of Chemical and Environmental Engineering 5, 218–222.

Sojka, M., Jaskula, J., Siepak, M., 2018. Heavy metals in bottom sediments of reservoirs in the lowland area of western Poland concentrations, distribution, sources and ecological risk. Water (Switzerland) 11, 1–20. Available from: https://doi.org/10.3390/w11010056.

Stephen, O.O., Oladele, O., 2012. Baseline studies of some heavy metals in top soils around the iron - ore mining field Itakpe North Central Nigeria. International Journal of Mining Engineering and Mineral Processing 1 (3), 107–114.

Syafrudin, M., Kristanti, R.A., Yuniarto, A., Hadibarata, T., Rhee, J., Al-onazi, W.A., et al., 2021. Pesticides in drinking water—a review. International Journal of Environmental Research and Public Health 18, 468. Available from: https://doi.org/10.3390/ijerph18020468.

Sýkorová, I., Havelcová, M., Trejtnarová, H., Kotlík, B., 2012. Toxicologically important trace elements and organic compounds investigated in size-fractionated urban particulate matter collected near the Prague highway. Science of the Total Environment 437, 127–136. Available from: https://doi.org/10.1016/j.scitotenv.2012.07.030.

Tan, M., Wang, K., Xu, Z., Li, H., Qu, J., 2020. Study on heavy metal contamination in high water table coal mining subsidence ponds that use different resource reutilization methods. Water 12, 3348. Available from: https://doi.org/10.3390/w12123348.

Tchounwou, P.B., Yedjou, C.G., Patlolla, A.K., Sutton, D.J., 2014. Heavy metals toxicity and the environment. National Institute of Health Public Access 101, 133–164.

Tóth, G., Hermann, T., Da Silva, M.R., Montanarella, L., 2016. Heavy metals in agricultural soils of the European Union with implications for food safety. Environmental Pollution 88, 299–309. Available from: http://doi.org/10.1016/j.envint.2015.12.017.

Wuana, R.A., Okieimen, F.E., 2011. Heavy metals in contaminated soils. A review of sources, chemistry, risks and best available strategies for remediation. International Scholarly Research Network ISRN Ecology 2011, 20. Article ID 402647.

Yi, Y., Yang, Z., Zhang, S., 2011. Ecological risk assessment of heavy metals in sediment and human health risk assessment of heavy metals in fishes in the middle and lower reaches of the Yangtze River basin. Environmental Pollution 159 (10), 2575–2585.

Yuen, J.Q., Olin, P.H., Lim, H.S., Benner, S.G., Sutherland, R.A., Ziegler, A.D., 2012. Accumulation of potentially toxic elements in road deposited sediments in residential and light industrial neighborhoods of Singapore. Journal Environmental Management 101, 151–163. Available from: https://doi.org/10.1016/j.jenvman.2011.11.017.

Zamora-Ledezma, C., Negrete-Bolagay, D., Figueroa, F., Zamora-Ledezma, E., Ni, M., Alexis, F., et al., 2021. Heavy metal water pollution: a fresh look about hazards, novel and conventional remediation methods. Environmental Technology & Innovation 22, 101504. Available from: https://doi.org/10.1016/j.eti.2021.101504. ISSN 2352-1864.

Zgłobicki, W., Telecka, M., Skupiński, S., 2019. Assessment of short-term changes in street dust pollution with heavy metals in Lublin (E Poland)—levels, sources and risks. Environmental Science Pollution Research 26, 35049–35060. Available from: https://doi.org/10.1007/s11356-019-06496-x.

Zheng, R., Ma, C., Zhao, J., Wang, L., Jiang, G., 2016. Land use effects on the distribution and speciation of heavy metals and arsenic in coastal soils on changing island in the Yangtze River Estuary, China. Pedosphere 26 (1), P74–P84.

Zhou, H., Guo, X., 2015. Soil heavy metal pollution evaluation around mine area with traditional and ecological assessment methods. Journal of Geoscience and Environment Protection 3, 28–33.

Zhou, J.M., Dang, Z., Cai, M.F., Liu, C.Q., 2007. Soil heavy metal pollution around the dabaoshan mine, Guangdong Province, China. Pedosphere 17, 588–594.

Zuriaga-Agustí, E., Galiana-Aleixandre, M.V., Bes-Pia, A., Mendoza-Roca, J.A., Risueno-Puchades, V., Segarra, V., 2015. Pollution reduction in an ecofriendly chrome-free tanning and evaluation of the biodegradation by composting of the tanned leather wastes. Journal of Cleaner Production 874–881.

Further reading

Ahamad, A., Raju, N.J., Madhav, S., Gossel, W., Wycisk, P., 2019. Impact of non-engineered Bhalswa landfill on groundwater from Quaternary alluvium in Yamuna flood plain and potential human health risk, New Delhi, India. Quaternary International 507, 352–369.

Ahamad, A., Raju, N.J., Madhav, S., Khan, A.H., 2020. Trace elements contamination in groundwater and associated human health risk in the industrial region of southern Sonbhadra, Uttar Pradesh, India. Environmental Geochemistry and Health 42 (10), 3373–3391.

Al Khade, AMF, 2015. The impact of phosphorus fertilizers on heavy metals content of soils and vegetables grown on selected farms in Jordan. Agro Technology 5, 1. Available from: https://doi.org/10.4172/2168-9881.1000137.

Awasthi, S.K., 2000. Prevention of Food Adulteration Act No 37 of 1954. Central and State Rules as Amended for 1999, third ed. Ashoka Law House, New Delhi.

Baekken, T., 1993. Environmental effects of asphalt and tyrewear by road traffic. Nordic seminar or Arbejdsrapporter 1992:628, Copenhagen, Denmark.

Bradl, H. (Ed.), 2002. Heavy Metals in the Environment: Origin, Interaction and Remediation, vol. 6. Academic Press, London.

Clemens, S., 2006. Toxic metal accumulation, responses to exposure and mechanisms of tolerance in plants. Biochemical 88, 1707–1719. Available from: https://doi.org/10.1016/j.biochi.2006.07.003.

Duffus, J.H., 2002. Heavy metals—a meaningless term? Pure and Applied Chemistry 74 (5), 793–807.

European Union, 2002. Heavy metals in Wastes, European Commission on Environment. <http://ec.europa.eu/environment/waste/studies/pdf/heavymetalsreport.pdf> (accessed 21.07.14).

FAO, 2017. Water pollution from agriculture a global review.

Hansen, E., Lassen, C., Stuer-Lauridsen, F., Kjølholt, J., 2002. Heavy metals in waste, European Commission DG ENV. E3, Brussels.

Kabata-Pendias, A., Pendias, H., 2001. Trace Metals in Soils and Plants, second ed. CRC Press, Boca Raton, FL.

Lee, P.K., Touray, J.C., Baillif, P., Ildefonse, J.P., 1997. Heavy metal contamination of settling particles in a retention pond along the A-71 motorway in Sologne, France. The Science of the Total Environment 201, 1–15.

Legret, M., Pagotto, C., 1999. Evaluation of pollutant loadings in the runoff waters from a major rural highway. The Science of the Total Environment 235, 143–150.

Liang, G., Gong, W., Li, B., Zuo, J., Pan, L., Liu, X., 2019. Analysis of heavy metals in foodstuffs and an assessment of the health risks to the general public via consumption in Beijing, China. International Journal of Environmental Research and Public Health 16, 909.

Madhav, S., Raju, N.J., Ahamad, A., Singh, A.K., Ram, P., Gossel, W., 2021. Hydrogeochemical assessment of groundwater quality and associated potential human health risk in Bhadohi environs, India. Environmental Earth Sciences 80 (17), 1–14.

MDH, 1999. Minnesota Department of Health Screening Evaluation of Arsenic, Cadmium, and Lead Levels in Minnesota Fertilizer Products.

Pan, Y., Li, H., 2016. Investigating heavy metal pollution in mining brownfield and its policy implications. A case study of the Bayan Obo rare earth mine, Inner Mongolia. Environmental Management in China 57 (4), 879–893.

Pandey, B., Agrawal, M., Singh, S., 2016. Effects of coal mining activities on soil properties with special reference to heavy metals. In: Raju, N.J. (Ed.), Geostatistical and Geospatial Approaches for the Characterization of Natural Resources in the Environment. Capital Publishing Company, Chapter 56.

Pierzynski, G.M., Sims, J.T., Vance, G.F., 2000. Soils and Environmental Quality, second ed. CRC Press, London, UK.

WHO/FAO/IAEA. World Health Organization. Switzerland: Geneva, 1996. Trace Elements in Human Nutrition and Health.

CHAPTER

8

Impact assessment of heavy metal pollution in surface water bodies

Soumya Pandey and Neeta Kumari

Department of Civil and Environmental Engineering, Birla Institute of Technology, Mesra, Ranchi, Jharkhand, India

8.1 Introduction

Water is the most crucial resource that has made life possible on Earth. Although the freshwater resources are globally present, their distribution is uneven. Currently, the water quality of the resources has deteriorated globally. This is believed to be the result of decades of water mis-management (Ranjan et al., 2017). Water pollution has become a global challenge (Fusco et al., 2020) for both developing and developed countries, yet the drive for development in developing countries have worsened the situation (Ouyang et al., 2018). According to the United States Environmental Protection Agency (EPA), around 1/3rd of the globe is suffering from water pollution (Tirgar et al., 2020). Around 1 billion population does not have access to clean drinking water (Gikas et al., 2020), while more than 2.2 million people residing in developing countries. They are facing death due to lack of clean water (Raveesh et al., 2015). As per world research agencies like WHO (World Health Organization) and UNICEF (United Nations Children's Fund), around 60% of infants are under life-threatening situations as the contaminated water can cause communicable, bacterial and parasitic diseases in them (Nadikatla et al., 2020). Among all the other pollutants, heavy metals are the most notorious pollutant as they can result in ecotoxicity along with being fatal for human lives by causing chronic health issues, like cancer, neurological and reproductive disorders (Zhang et al., 2018). Heavy metals are present in abundance and are naturally found in soil and water resources because of weathering and erosion processes. They are naturally present in the earth's crust, in the form of carbonates, sulfates, silicates, aluminates, oxides or even in their elemental form (Wu et al., 2020). Anthropogenic activities like industrialization, urbanization, smoke emissions from industries, mine drains, waste disposal, use of fertilizers, effluent and sludges disposal etc (Bhardwaj et al., 2017; Li et al., 2019; Nandan et al., 2017; Sahoo et al., 2018) have led to

DOI: https://doi.org/10.1016/B978-0-323-95919-3.00004-5

increase of toxic gases and aerosols, chemicals and hazardous waste into the environment which are rich in heavy metals (Doostmohammadi et al., 2017; Varol, 2020; de Nobile et al., 2021; Chakraborty et al., 2021; Ren et al., 2021; Gayathri et al., 2021).

The hazardous impact of these contaminants are clearly visible in the water resources, as water acts as a universal solvent and dissolves contaminants and pollutants readily, making it quite vulnerable to pollution (Nazneen et al., 2019; Ahamad et al., 2020). They leach and transport to the water bodies and sediment beds from soils with precipitation and runoff. As per the Sustainable Development Goal (SDG)'s formed by the United Nations in 2015, SDG 6 was specifically designated as "Life below water." One major step to reach the goal is to improve the quality of water resources by controlling the discharge of sewage and effluents as these have been found to be major source of heavy metal influx in the water bodies (Mateo-Sagasta et al., 2017). The major threat has been that heavy metals are non-biodegradable, tend to bioaccumulate in the environment and magnify throughout the food chain. Although, some of these metals support the functioning of life but most of them tend to be toxic. Toxic heavy metals can potentially be carcinogenic, mutagenic, neurodegenerative, and can also cause various skin disorders, organ failure, digestive system damage (Ahamad et al., 2018; Lipy et al., 2021). In higher concentrations, they can be fatal (Zhitkovich, 2011). Other than humans even the aquatic and terrestrial plants, animals and other resource qualities are equally damaged. Many methods have been adopted to analyze and remove heavy metals from water, sediments, human, and aquatic organism. Some commonly used detection techniques are electrochemical, spectroscopic technique, optical technique etc. whereas some analysis are done using geospatial and indexing methods (Gallo et al., 2018; Bibak et al., 2020). Compared to other methods, Indexing methods are becoming extremely popular in analysis of water quality contaminated with heavy metal as they are easy, and give the definite status of water in the result (Pandey et al., 2021). In this chapter, the anthropogenic sources of heavy metal and their impact on the aquatic ecosystem are presented.

8.2 Analysis of heavy metal contamination in water bodies

Every reported pollutant compound or element present in the aquatic environment is defined by world regulation and country-level agencies. Many times, countries will adopt the international standard of framework like USEPA environmental regulation, Organization for Economic Cooperation and Development (OECD) and the Asia Water Council (AWC) are followed (OECD-FAO, 2012; Xueman et al., 2020; Zhou et al., 2020). These standards help in monitoring and analysis of physical, chemical and biological parameters along with biomarkers of the region. The technique selected for analysis is based upon the cost-effectiveness, feasibility, precise and sensitivity to required metal traces (El-Feky et al., 2018; Mohanty et al., 2018). To determine the sources, spatial distribution and relation between the metals, research are conducted globally using physicochemical and biological parameters of water, sediment, and aquatic microbes (Kumar et al., 2019; Ren et al., 2020). Many advanced detection techniques have been developed such as geospatial techniques like geographic information system (GIS), Remote sensing, Spectroscopic technique, Electrochemical technique, Optical technique, Geostatistics technique, and Machine learning (Chanapathi and Thatikonda, 2020; Penman et al., 2020; Ren et al., 2020).

Spectroscopic techniques such as atomic absorption (AAS), X-ray fluorescence (XRF), inductively coupled plasma mass (ICP-Ms), inductively coupled plasma optical emission spectroscopy (ICP-OES), inductively coupled plasma atomic emission (ICP-AES), ion chromatography ultraviolet-visible (IC-UV Vis) be sensitive, with femtomolar range, multiple detections, and modifiable detection range (Zamora-Ledezma et al., 2021). Although these techniques are quite efficient and accurate, the equipment required for analysis are expensive with high maintenance cost and need skilled labor for sample preparation and accurate analysis. This adds to their disadvantage. Similarly, the electrochemical process uses potentiometric, voltametric, colorimetric equipment, biosensors, etc. (Liao et al., 2020; Payal and Tomer, 2021). The optical techniques use optical sensors for detection. Although the optical sensors are believed to be an easy technique yet their range and sensitivity are not remarkable (Salvo et al., 2018).

More than 30 Water Quality Index (WQI) models have been developed over the years since 1960 till date. They are usually based on the availability of data and ecological significance of the water quality factors. Most WQI use around 4–13 parameters depending upon availability and feasibility (Singh et al., 2020). Some of these Indexing methods are the Canadian Water Quality Index (CCME WQI), National Sanitation Foundation Water Quality Index (NSSFWQI), Horton WQI, Oregon WQI, Weighted WQI, British Council WQI, etc. (Yulistia et al., 2018; Prathap and Chakraborty, 2019; Mokarram et al., 2020; Gayathri et al., 2021). These index methods help in bringing forward a cumulative single unit index through which the water quality can be labeled between good to bad. Specific analysis of the detection data of the heavy metal pollution is analyzed using indexes like geo-accumulation index, contamination index, heavy metal pollution index (HPI), enrichment factors, entropy weighted heavy metal contamination index (EHCI) etc. (Yang et al., 2018; Zhang et al., 2018; Elsagh et al., 2021; Gayathri et al., 2021; Ravanipour et al., 2021). These indexing methods help in bringing forward a cumulative single unit index through which the water quality can be labeled between good to bad. Depending upon the detection and status of the water resource a feasible, cost-effective, eco-friendly remediation technique is applied to resolve the pollution issues.

8.3 Presence of heavy metals in the water bodies of Asian countries

8.3.1 India

The development of the country is dependent on the proper use of its water resources. India has about 17.7% of the global population, but only around 4% of the world's water resources are available in India. This has put immense pressure on water management. Some geographically crucial rivers that flow in India are Yamuna (Sharma and Kansal, 2011), Ganga (Datta and Litt, 2020), Mahanadi (Jin et al., 2018; Hussain et al., 2020), Subarnarekha (Chaturvedi et al., 2018; Mishra et al., 2019), Krishna, Godavari (Hussain et al., 2017; Ranjith et al., 2019; Kharake and Raut, 2021), Indus, Damodar (Chakraborty et al., 2021), Narmada (Dipti et al., 2018), etc. But most of these rivers and their tributaries are heavily contaminated by urban and industrial sludges and sewage effluents (Singh et al., 2020). The river Yamuna has no ecological water flow due to anthropogenic activities in its catchment area that covers multiple states of India. The river has been acting as a sink for industrial, domestic, and urban wastes (Laumonier and Nasi, 2018; Bowes et al., 2020). River Ganga also flows through many states and all the surface runoff from agricultural farms, industrial effluents and sewage, and religious waste like flowers, incense stick etc., all get dumped in the river (Panditharathne et al., 2019; Boral et al., 2020; Mishra and Kumar, 2021). This has enriched the water with heavy metal and organic waste. Similar is the situation with most of the surface waters in the urban and metro cities whereas, groundwater contamination has been found in Assam, West Bengal, Odisha, Chhattisgarh, Karnataka, and many small water bodies of Punjab, Maharashtra, Madhya Pradesh Jharkhand, Tamil Nādu etc. (Fusco et al., 2020; Imtiaz et al., 2018).

In Jharkhand, the mineral capital of India, the major rivers are suffering from metal toxicity, while depending upon the source and non-point source of pollution the level of toxicity vary in the river bodies. River Subarnarekha, Koel, Damodar etc., flow through the cradle of industries (Singh, 2016). Damodar River is many times more polluted from heavy metals than the permissible limits set by WHO and IS standards (Singh and Mandhyan, 2017). The metals like cadmium were found to be six times higher than permissible limits prescribed by the WHO (Mohanta et al., 2019). The mine water collected from coalfields of the steel city of Bokaro, Jharkhand showed that the water have been heavily polluted due to mining, geogenic and metals leaching from other sources (Mahato et al., 2017; Tiwari et al., 2017; Verma et al., 2020). In its stretch of Jharkhand, the river Subarnarekha comes across many different land use and land cover. The untreated domestic sewage, industrial effluent, mine drains and agricultural runoff have been disposed in the river. The Subarnarekha watershed includes the capital city of Ranchi and its suburbs. Its tributaries like Jumar, Harmu, etc., also add more pollutants in the river (Kumar et al., 2016). Subarnarekha is found to be rich in heavy metal mostly due to anthropogenic sources (Chaturvedi et al., 2018; Singh and Giri, 2018).

8.3.2 Bangladesh

Bangladesh is one of the most densely populated countries and has seen massive enhancement in terms of industrialization. The country has a plenty of water resources (230 large and small water body) to meet its daily population demand, but the quality of water is uncertain (Hasan et al., 2016; Chowdhury et al., 2020). This is mostly due to increase in urbanization and industrialization, much of these water bodies have become contaminated with heavy metal, coliforms, organic and inorganic metals, especially with arsenic and chromium (Rahman and Zhang, 2018; Safiur Rahman et al., 2021). Most of the groundwater has been polluted with arsenic leading to many health hazards reported in Bangladesh. Although the status of pollution in river bodies have been reported by multiple researchers, but the health condition of the population has not been reported adequately. Uddin and Jeong (2021) have analyzed the peripheral rivers and study the pollution levels in last 40 years. The research showed that the concentration of metals was in order of water $<$ fish $<$ sediment. Even the crops were found to have heavy metal contamination as they were irrigated with polluted water. The status of pollution in river bodies of Bangladesh is quite concerning, as it may affect the health of the people and economy of the country with time (Adeloju et al., 2021).

8.3.3 Pakistan

Pakistan being the sixth most populated country, faces immense water quality and quantity challenges (Imtiaz et al., 2018; Khalid et al., 2020). Water shortage and pollution, especially due to heavy metals, has become a huge problem as reported by many research (Rasool et al., 2016; Gharahi and Zamani-Ahmadmahmoodi, 2020). In 2004, Arsenic contamination of drinking water led to death of 40 people in the city of Hyderabad, Pakistan. Furthermore, areas of Sindh, Punjab, Lahore, have been found to affected by arsenic (Imtiaz et al., 2018; Imran et al., 2019; Iqbal et al., 2021). Also, in 2005 Pakistan council research in water resources has declared six cities of Punjab as contaminated with varying level of heavy metal toxicity (Ullah et al., 2018; Imran et al., 2019). Like other developing countries, Pakistan also has a poor implementation of environmental laws, especially in terms of industrial effluent discharge being dump directly into rivers. World bank reported in 2016, that heavy metal contamination of drinking water has led to increase in non-communicable and heart diseases leading to around 50% diseases and 40% death of the population, especially residing in the downstream or low land areas (Imran et al., 2019; Imran and Mehmood, 2020). Many studies have suggested immediate need for monitoring and implementation of stricter laws for controlling untreated effluent discharge in the water bodies (Imtiaz et al., 2018; Mannan et al., 2019; Raza et al., 2021).

8.3.4 China

Among the Asian countries, China is the largest global producer and consumer of heavy metals like antimony, manganese, iron, tin, tungsten, zinc etc. (Yu et al., 2017). But increased use of metals in agriculture in form of fertilizers, insecticides, pesticides etc., urban, commercial, domestic, mining sectors have influx the water bodies with metal contaminants. For example, the city of Shanghai depends on the Yangtze River for its basic purposes like drinking cleaning, industrial, domestic and commercial use (Ouyang et al., 2018; Ahmad, 2020). But Yangtze River has been considered as one of the most deteriorated rivers of the world by WWF (World Wild Fund for Nature). Whereas in South China around 200 million tons of industrial effluents and sewage waste are disposed of annually into the Pearl River (Sun et al., 2018). The inshore of the Bohai river are also contaminated with mercury, cadmium, lead and chromium (Zhou et al., 2020). the sources of this contamination are found to be anthropogenic activities like electronic industries, mining, agriculture etc. These rivers flow through the major cities and hence there is a huge accumulation of metals in their estuaries (Ouyang et al., 2017). In cities like Guangdong, Guangzhou, Shenzhen, Dongguan, Foshan, Jiangmen, Zhao Qing etc. the practice of electronic manufacturing is prominent hence leading to heavy metal contamination of soil, air, and water. The prominent metals found in these regions are chromium, zinc, copper, lead, nickel, cadmium. Even the agricultural lands of Dongguan and Huizhou are contaminated with mercury and lead as lands are being irrigated with wastewater (Yue et al., 2019). Moreover, in the cities like Qingyuan and Shantou, the recycling of electronic waste has again polluted the land and water resources. Other than these anthropogenic activities like mining, smelting, agriculture have also equally polluted the farms and urban regions of many cities in China. This threatens the sustainability of human lives, especially children and infants, crops, plants, and aquatic ecosystems of the regions (Xu et al., 2020; Zhou et al., 2020). Also, along with surface waters, around 60% of the country's groundwater has become contaminated. This possesses more problems for even common household activities.

8.3.5 Indonesia

Indonesia has been suffering from water stress and metal contamination for a long period (Ali, 2019). The water scarcity results due to poor water management throughout the country. For instance, the capital city of Jakarta is facing heavy challenges in meeting basic water demands for its population (Goh, 2019; Pingping et al., 2019; Cristable et al., 2020). Around 30% of population has no access to water supply network and is heavily dependent on the groundwater for their basic need. Unfortunately, 50% of available groundwater resources are polluted by sewage and 10% are polluted by heavy metals like Iron and Manganese (Amalo et al., 2018). Moreover, uncontrolled extraction of water is causing the city to sink faster. It has been estimated that around 95% of northern regions of the capital city will drown by 2050. This is quite alarming as the residing inhabitants may either have to relocate or suffer huge economic losses. Furthermore, around 24 million Indonesian population do not have proper sewage and sanitation facilities (Goh, 2019). Most sewages are directly discharged in water bodies hence around 13 rivers are immensely polluted. Also, the septic tanks have poor construction and often leak into the groundwater making them contaminated (Cristable et al., 2020). These not only lead to wastage of water resources but also risk the outbreak of water-borne diseases like typhoid, cholera etc., and the situation has further worsened in the COVID-19 pandemic.

8.3.6 Iran

In the Country of Iran, many major rivers have been reported to be contaminated with heavy metals. The Zayandeh-Rud River in the central plateau of Iran has been subjected to effluent discharge from varying point and non-point sources (Karimian et al., 2020). The amount of Arsenic found in the samples of Algae collected from downstream of the river has been in higher concentration than any other river reported (Karimian et al., 2020). In the Bandar Abbas of the Persian Gulf, the analysis of sediments showed the presence of mercury contamination, posing deadly challenges to aquatic lives (Elsagh et al., 2021). In the Iranian side of the Caspian Sea, nutrients like the stable isotope of nitrogen and heavy metal pollutants like Chromium, Nickel, Lead, and Zinc have been prominently found (Costantini et al., 2021). The major reason attributed to this pollution is the discharge of wastewaters from industries and domestic regions, along with surface runoff from urban, rural, and mining areas into the sea (Elsagh et al., 2021). In the Bazman watershed, South-eastern Iran the groundwater quality is low to moderately polluted although the level of contamination was not dangerous (Rezaei et al., 2019). The research on water quality and pollution indices of Kor River in Iran was done. The study showed that the water has been significantly polluted due to anthropogenic activities like industrial effluents, agricultural runoffs, and the influx of domestic sewage in water bodies (Sheykhi and Samani, 2020).

8.4 Anthropogenic sources of heavy metal in aquatic environment

The significant way of depositing harmful metals into the ecosystem occurs via a series of pathways and activities which can be both natural and anthropogenic. The sources of heavy metals in the aquatic ecosystem are geogenic and anthropogenic (Fig. 8.1). The anthropogenic sources can be divided into agricultural fields, industrial regions, domestic and commercial sources, atmospheric sources, pharmaceutical sources, sewage discharge and effluents, leachates, mining.

Among the various anthropogenic sources, one crucial factor has been the industries. Since the beginning of industrialization, the amount of heavy metal influx into the ecosystem has been tremendous. Industries like chemical making, manufacturing, smelting, electronic, and electroplating, weaponry, production, rubber, automobile etc., are majorly responsible for causing pollution. These industries emit toxic smoke that causes atmospheric pollution and a huge deterioration in air quality. Furthermore, the effluent discharges, which are mostly unfiltered or raw, are released directly into the water bodies like rivers, lakes, ponds etc (Rai et al., 2012; Shekhar et al., 2017).

8.4.1 Industrial effluents

These industrial effluents and waste are rich in heavy metals which do not dissolve easily but settle down in sediments bed. This not only pollutes the river or the respective

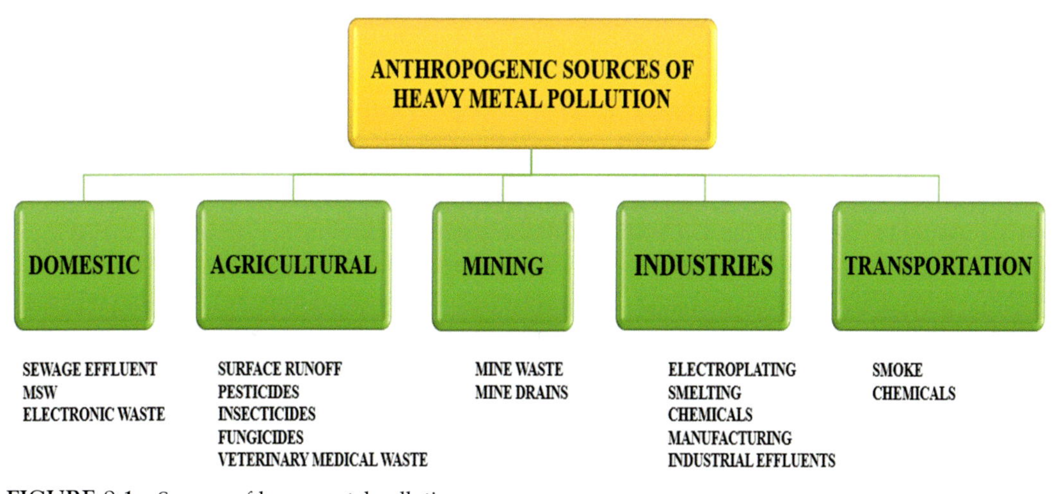

FIGURE 8.1 Sources of heavy metal pollution.

water bodies but create danger for the aquatic organism (Angela et al., 2015). The septic zone in which the effluents are discharged have extremely high levels of toxicity. The fishes are absent from these zones but instead sludge worms, bacteria, fungi, and other microbes are present. The Dissolved oxygen (Chakroborty et al., 2017) curve of this zone usually has a severe dip and are lower than 2 ppm, whereas clean water has around 8 ppm of dissolved oxygen. The decomposition zone and recovery zones also have little aquatic organisms like fish, carp etc. present. But the recovery is subjected to time provided to the rivers to clean themselves or decompose the foreign compounds.

Unfortunately, the metals do not easily dissolve or disintegrate in the environment and hence say there for a longer period causing pollution (Zhao et al., 2018). Some heavy metals are specific to certain industries, like lead in paints and ceramics (Wang et al., 2017; Saleh et al., 2019), Tin and tungsten in electronic industries, chromium in leather industries (Das et al., 2020), Arsenic, cadmium, iron etc., in the coal mining and industries (Pandey et al., 2016; Prathap and Chakraborty, 2019). In countries like Iran, China, the USA, India, mass death of fish has been reported due to heavy metal toxicity in the aquatic surroundings. In Iran, canned fish samples were found to have metal toxicity. In India, Paveda- Talaab, Nalagarh river, Ropar, Dhana lake, Indrayani river, Dhanushkodi, Kamini river, Periyar river have suffered from mass fish-killing due to aquatic toxicity caused by heavy metals (Sigdel, 2017; Laxmi Mohanta et al., 2020).

8.4.2 Sewage effluents

Sewage effluents are mostly generated from domestic and commercial regions. But Domestic sewage effluents do not usually contain hazardous metals in larger quantities. They mostly comprise human fecal, animal waste, wastes from the slaughterhouse, greywater, black water, brown water etc. When the sewage is disposed of in water bodies not only does it become contaminated but also enriched with nutrients and metals. This causes nutrient and metal pollution in the water bodies. The excess nutrients allow an immense growth of algae in the water bodies. This initially causes eutrophication and then causes massive algal blooms. Algal blooms and eutrophication create surface shade that does not allow the sunlight to pass in the water bodies (Caruana et al., 2020; Irandoost et al., 2021). They also inhibit the availability of oxygen in the water bodies as they consume it themselves. Hence the aquatic plants do not get sunlight or oxygen and eventually die. Even the aquatic animals and organisms which are dependent on these plants for food or sunlight and oxygen also die (Xia et al., 2020). As the oxygen levels lower and the water body becomes dead the water starts to stink with the decomposition of animals and plants in water bodies (Pastor and Hernández, 2012). In the USA and China, nutrient pollution has become a major environmental challenge. The solid waste generated from general households consists of glass, paper, wood, plastic, metal caps, batteries, organic waste etc. These wastes are deposited either in landfills or in open dump yards (Asibor et al., 2016; US EPA, 2020).

8.4.3 Leachate

Leachate can penetrate the ground with time and contaminate the surface and groundwater resources. The real problem with leachate is that its chemical composition is hard to predict as it changes with the composition of the waste present (Essienubong et al., 2019). This requires the constant monitoring of the leachate produced. Also, the nearby soil and water bodies need to be monitored closely to avoid outbreak of any disease. Many countries lack proper landfill facilities. Hence the waste is thrown in water bodies like sea or ocean or on open land, polluting them. When the precipitation falls on these wastes, they form leachate that flows to nearby water bodies with runoff polluting them even more.

8.4.4 Mining

Mining activities and mine drain from abandoned mines are majorly responsible for heavy metal pollution in both soils and water bodies. They are considered as the conventional anthropogenic sources of metal pollution. In mining processes, the deeper soil layers are excavated at a mass scale that is rich in heavy metal particles. Once the metals are extracted the remaining smaller fines are left in the soil as mine waste (Ahirwal and Maiti, 2016; Doostmohammadi et al., 2017). With precipitation, this mine waste forms mine drains that flow to waterbodies and seep into the soil. This not only contaminates soil but also the groundwater and surface water (Elmayel et al., 2019). The acid mine drains are rich in compounds of parent mineral, have low pH, making them acidic and high content of essential and nonessential heavy metal, along with other anions.

8.4.5 Agriculture

Another major factor comes from agricultural fields. The farm waste comprising of poultry waste, cans and bottles of chemicals, fertilizers, pesticides, insecticides, veterinary medicines, syringes etc. are rich in heavy metals (Hossain et al., 2017). The use of pesticides insecticides, fungicides, herbicides in the crop production processes leads to an increased amount of heavy metal in the soil and nearby water resources (Shahradnia et al., 2021). These chemicals are rich in heavy metals and bioaccumulate in the soils and crops. They biologically magnify throughout the food chain (Lipy et al., 2021). The heavy metals also destroy the microbes present in the soil or challenge their sustainability in the soil (Singhal et al., 2021). While spraying these chemicals on-farm, they are dispersed into the atmosphere as aerosols and again cause health issues if inhaled. These aerosols also get spread on larger areas near the farmlands with airflow contaminating the surrounding (Büntgen et al., 2020). In agricultural farms along with rainfall, irrigation water also forms runoff and wash away the sediments and heavy metals into the larger water bodies (Matta et al., 2020). Also, the practices of using recycled wastewater for irrigation add more heavy metals to the soil and water reserves.

8.4.6 Acid rain

Many areas of the USA and China have been suffering from acid rains. These rains have some content of metal compounds that add more toxicity to the ecosystem (Foster et al., 2019). The presence of heavy metals in aquatic systems can be determined by analyzing the physiochemical and biological parameters of water quality. The water quality parameters are e in finding the sources of contaminants but monitoring the quality of water bodies is crucial for maintaining the water ecosystem (Ling et al., 2010). The contributor of heavy metals is assessed by considering the spatial and temporal differences in the quality parameters.

8.4.7 Transportation

Another anthropogenic source of heavy metal pollution co be urban traffic (Nawrot et al., 2020). Urban roads are sort of sink for the heavy metals that accumulate on the surfaces of road and get deposited on the side roads or sediments (Rastegari Mehr et al., 2017). Heavy metal mostly accumulates from abrasion of tires, fuel residues and fluid leakage (oil, grease, and coolant) pavement degradation. The dust exposed to air due to the movement of vehicles tends to mix with the metal deposit deposited on the road. And these processes are much more visible during the summer or dry weather (Madhav et al., 2021). Whereas, on rainy days, accumulated heavy tend to either remain as roadside deposits or get removed by stormwater into larger bodies (Men et al., 2020; Ahamad et al., 2021). The amount of chemicals released from the vehicle is found to be rich in heavy metals. The most widespread heavy metals found are zinc, cadmium, copper, lead, nickel, chromium, cobalt and mercury (Acar and Özkul, 2020). In recent years it is nearly impossible to identify the chemical characteristics of particle used in the automobile industry, as in numerous new materials are introduced every year (Nazneen et al., 2018; Nawrot et al., 2020). The components of the braking system have been known to emit zinc, nickel, cadmium, copper, and chromium, mostly among other heavy metal (Nawrot et al., 2020).

8.5 Metal toxicity and its impact on aquatic environment

Contamination of heavy metals in the aquatic biomes has turned out to be a persistent and chronic issue for the preservation of ecosystem bio-diversities. Sources of heavy metal contamination could be from both geogenic sources and anthropogenic sources (Elango and Jagadeshan, 2018). There have been numerous cases reported globally on the toxicological effects of heavy metals on the aquatic organism, especially in surface water with major influence from anthropogenic sources (Ewaid, 2017). These toxicological effects are dependent on the level of exposure, the form of compounds of metals, their utility (trace or nonessential metal) and their amount (Li et al., 2019). A heavy metal that acts as major pollutants is Cadmium, Copper, Cobalt, Mercury, Lead, Chromium, Nickel, Zinc, Iron (Hussain et al., 2017). These metals neither biodegrade nor metabolize (Ayilara et al., 2020). Hence, they become a threat to the aquatic ecosystem as they can easily penetrate the food chain, accumulate, and enhance biologically (Lipy et al., 2021).

8.5.1 Sediment bed

The heavy metal can be detected in the aquatic ecosystem via analysis of sediment, water, and organism (Chakrabarti, 2019). It has been found that the levels of heavy metal present in water are less compared to sediment and organisms. Heavy metals present in soil resources do not pose harm usually (Das et al., 2020). The sediments collect the heavy metals through chemical reactions, percolation and transmission through pores, and other physical processes (Mishra et al., 2017; Kumar et al., 2019). The heavy metal is deposited on the sediments through adsorption, rainfall, runoff, diffusion, chemical reaction, biological processes (Pandiyan et al., 2020). From these sediments, the heavy metals are leached into water bodies and absorbed by aquatic organisms and plants (Meraj et al., 2018).

8.5.2 Macro flora and fauna

Plants and other organisms like fish, shrimps, prawns etc., easily ingest heavy metals as a part of their food (Ahmed and Akter, 2017; Ewaid and Abed, 2017). The impact of heavy metals is not very widely researched on microbes like bacteria, fungi, protozoa or ciliates, polychaetes, crustaceans, worms, etc. (Guldhe et al., 2017; Andrade et al., 2020). Some research shows that some species of protozoa can have a major impact from heavy metals, like variation in feeding capacity, growth, stresses in oxidation and respiration. These changes occur due to alterations in microstructural and membrane (Andrade et al., 2020). Furthermore, these organisms are the primary members of the food chain and can transfer the heavy metal to higher levels in the food chain.

Fish has been the main food for a major population, especially residing in the coastal cities globally, as they are rich in protein and omega-3 fatty acids. In research conducted on Baiyangdian lake of China, it was found that omnivorous fish accumulate more metal than carnivorous Fish, except for Mercury (Zerizghi et al., 2020). In the research done on the Gulf of Gela, the fishes showed a high risk of dietary exposure to heavy metals like Cadmium and lead. The inorganic elemental load on the water bodily was found to be majorly due to industrial pollution (La Torre et al., 2020; Salvo et al., 2018). Similar is the case of Fundao dam, Brazil, where the release of mine waste from tailing dams resulted in severe pollution of the water bodies nearby. Fishes are in a state of oxidative stress and suffering from heavy metal contamination (Alberto et al., 2020; Almeida et al., 2020). Liu et al. (2020a,b) found that fish gills and crab gills along with hepatopancreas had elevated concentrations of heavy metals. The fish gills accumulate the heavy metal via ion exchange from water. The presence of heavy metal in the tissues are found to be influenced by the location and diet of the organism it is studied in (Maurya and Malik, 2019). Seafood is found to be a significant source of arsenic in human diets, causing cancers in humans especially when present in its inorganic form (He et al., 2021). Some other biomarkers used to assess the contamination of heavy metals in water are Metallothionein, Heat shock protein-70, cytochrome P450−1A (Savassi et al., 2020).

8.5.3 Micro flora and fauna

The ciliates or other protozoa species are not considered widely as the indicator for heavy metal assessment as they do not sustain in the laboratory conditions well (Andrade et al., 2020). Moreover, there is a lack of standardized experimental procedures for these species. But the species of algae, fishes and other benthic organisms like prawns, shrimp etc. act as crucial indicators. Furthermore, Fishes and other benthic organisms are at top of the aquatic food chain (Danovaro et al., 2016; Irandoost et al., 2021). The concentration of heavy metals in the organisms can be much higher than in the water around them. Hence, they can pose more toxicity when consumed. Fishes accumulate heavy metals easily. They can also act as a pathway for the transmission of heavy metal pollutants in other consumers and can cause serious health issues for humans when consumed (Liu et al., 2020a,b).

Algae can be considered as the easiest biological indicator in water bodies. They tend to show rapid change with any alteration in the physical, chemical, and biological changes in the water quality of their residing water bodies. Studying the presence in the algae specimen can give a better idea of heavy metal contamination as there is no bioaccumulation in them (Bibak et al., 2020, 2021). Bibak et al. (2020) found in their research that differently species of algae can be used to determine the concentration of heavy metals. Brown algae can be used to find the concentration of heavy metals like manganese, iron, aluminum, lead, nickel, cobalt, cadmium, whereas the red algae show other metals like Magnesium, Zinc, Copper, Chromium, Arsenic. Green algae did not show much absorption of heavy metal in the experiment analysis. The toxicity of heavy metals in the microorganism can lead to more contamination in the higher-level organism of the food chain (Bibak et al., 2020).

8.6 Impact of metal toxicity on human health

Heavy metals are elements having at least the density of $5\,g/cm^3$ (Hait and Tare, 2012). Some heavy metals like Iron, copper, Zinc, Nickel, Selenium etc. (Luo et al., 2016; Singh and Kamal, 2017; Çelik et al., 2018; Kaur et al., 2018) are important for human health but within permissible limits. Humans can easily get exposed to these heavy metals by inhalation or by oral consumption. According to WHO 2017 report, out of 35 metals that have been toxic to humans, 23 are heavy metals like Chromium, Gold, Silver, Iron, Copper, Manganese, Zinc, and Nickel etc. Metals become toxic for human health and the ecosystem only when they cross the permissible limits established by world research organizations like WHO, UNICEF, etc. Also, every country has established their standards like the Bureau of Indian Standard (BIS) in India, etc. Some of these metals like Copper, Nickel, Zinc, and Iron etc. are all crucial for the functioning of human health. But some other heavy metals like Arsenic, Mercury, Cadmium, and Lead are non-essentials for life. Hence these have been categorized as carcinogens of varying classes depending upon the threat it poses. Some commonly reported cases are of Minamata disease that is mercury poisoning in infants and adults and arsenic poisoning (Hasan et al., 2019). But in most cases heavy metal-induced disease has similar symptoms to other disease hence it is hard to differentiate the symptoms. Many deaths have been reported in Pakistan, India, and Bangladesh due to arsenic and mercury poisoning (Sharma et al., 2019).

In general, pollution can be considered as a factor that alters or harms the physical, biological, and chemical properties of water (Duru et al., 2019; Al-Hamdany et al., 2020). Unlike organic metal compounds, heavy metals cannot be biodegraded to safer forms. Hence, they stay in the ecosystem for a long period. Metals are elements without which life is not possible. They constitute 2.5% of the human body. They are required in less quantity yet their proper distribution in organs and cells ensure the proper functioning of the human body (Bing et al., 2016). Mainly human bodies need trace metals like Copper, Zinc, Iron, etc. for intracellular and DNA-binding processes. Most of these metals are toxic even in traceable quantities (lead, cadmium, chromium) and can be carcinogenic (Prasad et al., 2020; Lipy et al., 2021). Heavy metal-induced carcinogenesis inhibits DNA repair and its cross-linking with proteins because of the formation of reactive oxygen species, leading to oxidative stress in the cells and tissues. The heavy metals impact on human health is mentioned in Table 8.1.

TABLE 8.1 Heavy metals: sources and human health impact.

Heavy metals	Sources	Effect	References
Chromium	Mines, mineral resources	Corrosion, allergies, irritation of the nose, nose ulcers, swelling and redness of the skin, cardiovascular diseases, respiratory diseases, hematological diseases, renal, hepatic, neurological effects, death, damage to DNA, stomach diseases	Zhitkovich (2011), Bourgin et al. (2017), Kaur et al. (2018), ALabdeh et al. (2020), Kumar et al. (2020); Pramanik et al. (2020)
Manganese	Welding, fuel addition, production industries	Neurological disease, tremors, gait disorder, postural instability, and cognitive disorder, gait disturbances, bradykinesia, micrographia, memory and cognitive dysfunction, and mood disorder, manganism	Ghosh and Banerjee (2012), Kaur et al. (2018), Chen et al. (2020)
Iron	Asbestos, geogenic, mining, effluents	Gastrointestinal effects, vomiting, hypotension lethargy, hepatic necrosis, tachycardia, Ulcers, metabolic acidosis, death, cancer, Damage to DNA	Sánchez-Navarro et al. (2015), Gilbreath et al. (2019), Liao et al. (2020), Valenzuela-Diaz et al. (2020)
Copper	Mining, pesticides, chemical industries, metal industries, piping etc.	Liver disease, hepatitis, lymph node cancer, leukemia, brain, liver and breast cancer, diabetes	Sánchez-Navarro et al. (2015), Wan et al. (2018), Askari et al. (2020)
Zinc	Manufacturing industries, refineries, Plating, plumbing sectors, ores, excessive zinc supplement, paints, lead, industrial chemicals, cleaners, solvents, fumes, varnish rubber etc.	Abdominal pain, headache, irritability, lethargy, anemia, dizziness, gastric irritation, vomiting	Luo et al. (2016), Askari et al. (2020), Chen et al. (2020), Kumar et al. (2020), Selvam et al. (2020)
Arsenic	Pesticides, smelting industries, fungicides etc.	Poisoning, nausea, keratosis, leukocytes, damaged blood vessels, heart damage, pulmonary disease, skin lesions, hypertension, diabetes, death, bladder cancer, skin cancer, colon cancer	Tirkey et al. (2017), Wan et al. (2018), Hasan et al. (2019), Kumar et al. (2020)

(Continued)

TABLE 8.1 (Continued)

Heavy metals	Sources	Effect	References
Cadmium	Smelting, electroplating, fertilizers, pesticides, nuclear plants, Volcanic eruption, ores, cement production, weathering, forest fires	Vomiting, diarrhea, lung damage, respiratory issues, bone mineralization, bone damage, osteoporosis, renal dysfunction, kidney damage, kidney stones, prostate cancer	Luo et al. (2016), Milivojević et al. (2016), Wan et al. (2018), Yalçın et al. (2018), ALabdeh et al. (2020), Tahmasebi et al. (2020)
Mercury	Pesticides, electronic wastes, batteries, Paper industries, latex gloves, gold mining, cement production, combustion of fossil fuels	Kidney damage, brain damage, fetus damage, tremor, irritability, memory issues, skin rashes, diarrhea, headache, fatigue, hair loss, insomnia, depression, impaired speech, malformation, Minamata disease	Reza and Singh (2010), Milivojević et al. (2016), Wan et al. (2018), Li et al. (2019), ALabdeh et al. (2020)
Lead	Paints, smoke, automobiles, coal burning, mining, welding of piping network	Headache, loss of appetite, abdominal pain, fatigue, sleeplessness, hallucinations, vertigo, renal dysfunction, hypertension, arthritis, birth defects, mental retardation, autism, psychosis, allergies, paralysis, weight loss, dyslexia, hyperactivity, muscular weakness, kidney damage, brain damage, coma	Hermosilla et al. (2015), Wan et al. (2018), Maurya et al. (2019)

The toxicity of heavy metals is dependent on their state of oxidation. This determines the physiology of biological toxicity of the metals in humans and other forms of life. When these metals enter the Human body or any living organism, they combine with the enzymes, proteins and alter the DNA molecules (Zhitkovich, 2011; Maurya et al., 2019). This leads to improper functioning of the organism or the human body causing carcinogenic, genotoxic, mutagenic etc., diseases. In terms of aquatic bodies, surface water bodies are more vulnerable to heavy metal pollution as they have more exposure compared to groundwater bodies (Begum et al., 2021). But contamination of groundwater is more dangerous as it is hard to detect and cannot be easily cleaned or renewed (Zahra, 2015). Overall, any form of contaminated water body can be dangerous to both human health and the ecosystem.

8.7 Summary

The influx of heavy metals in water resources has increased due to various human activities like urbanization, agriculture, mining, transportation, industrialization, etc. The use of harmful chemicals enriched with heavy metals, in these activities has caused an increase in the influx of toxic metals into the aquatic environment. These sources of chemicals are in the form of pesticides, herbicides, insecticides, fertilizers, animal vaccines, and medicines from the agricultural sector. Leachate from waste disposal, untreated industrial effluents, sewage discharge and mine drain flow through runoff and are discharged directly in waterbodies. The increased toxicity results in various adverse effects like loss of microbial and aquatic biomes, chlorosis, hinder germination, growth, cause low biomass formation, affect the process of photosynthesis, decrease nutrient absorption, generate free radicals that harm the membranes and weaken the cell structure in aquatic plants. Toxic heavy metals can potentially be carcinogenic, mutagenic, neurodegenerative, etc., they can also cause various skin disorder, organ failure, digestive system damage. In higher concentrations, it can be fatal. Worldwide metal toxicity in water bodies has caused major economic losses, the closing of the aquaculture sector on a small and large scale, food poisoning in humans and animals, destruction of crops, infertility of associated soil resources etc. many developing countries have found to be in more critical condition than the developed countries in terms of pollution management and remediation. Detection and monitoring of metal are crucial in aquatic bodies, hence methods like electrochemical, spectroscopy, optical sensors etc. are frequently used. The collected data are more commonly analyzed using indexing techniques like WQI, HPI, accumulation index etc. These indexing techniques help in providing a broader view of the status of the aquatic body. Metal recycling from the wastewater and polluted water bodies can be both economic and ecological gain. There is an urgent need to control the metal toxicity in water resources and apply advanced techniques for remediation.

References

Acar, R.U., Özkul, C., 2020. Investigation of heavy metal pollution in roadside soils and road dusts along the Kütahya–Eskişehir Highway. Arabian Journal of Geosciences 135. Available from: https://doi.org/10.1007/s12517-020-5206-2.

Adeloju, S.B., Khan, S., Patti, A.F., 2021. Arsenic contamination of groundwater and its implications for drinking water quality and human health in under-developed countries and remote communities—a review. Applied Sciences Switzerland 114, 1–25. Available from: https://doi.org/10.3390/app11041926.

Ahamad, A., Madhav, S., Singh, P., Pandey, J., 2018. Assessment of groundwater quality with special emphasis on nitrate contamination in parts of Varanasi City, Uttar Pradesh, India. Applied Water Science 84, 1–13. Available from: https://doi.org/10.1007/s13201-018-0759-x.

Ahamad, A., Raju, N.J., Madhav, S., 2020. Trace elements contamination in groundwater and associated human health risk in the industrial region of southern Sonbhadra, Uttar Pradesh, India. Environmental Geochemistry and Health 4210, 3373–3391. Available from: https://doi.org/10.1007/s10653-020-00582-7.

Ahamad, A., Raju, N.J., Madhav, S., Gossel, W., Ram, P., Wycisk, P., 2021. Potentially toxic elements in soil and road dust around Sonbhadra industrial region. Source Apportionment and Health Risk Assessment', Environmental Research, Uttar Pradesh, India, p. 111685. Available from: http://doi.org/10.1016/j.envres.2021.111685.

Ahirwal, J., Maiti, S.K., 2016. Assessment of soil properties of different land uses generated due to surface coal mining activities in tropical Sal Shorea robusta forest, India. Catena 140, 155–163. Available from: https://doi.org/10.1016/j.catena.2016.01.028.

Ahmad, E., 2020. Multilevel responses to risks, shocks and pandemics: lessons from the evolving Chinese governance model. Journal of Chinese Governance 00, 1–29. Available from: https://doi.org/10.1080/23812346.2020.1813395.

Ahmed, K.R., Akter, S., 2017. Analysis of landcover change in southwest Bengal delta due to floods by NDVI, NDWI and K-means cluster with landsat multi-spectral surface reflectance satellite data. Remote Sensing Applications: Society and Environment 168–181. Available from: https://doi.org/10.1016/j.rsase.2017.08.010.

Al-Hamdany, N.A.S., Al-Shaker, Y.M.S., Al-Saffawi, A.Y.T., 2020. Water quality assessment using the nsfwqi model for drinking and domestic purposes: a case study of groundwater on the left side of mosul city, iraq. Plant Archives 201, 3079–3085.

ALabdeh, D., Karbassi, A.R., Omidvar, B., Sarang, A., 2020. Speciation of metals and metalloids in Anzali Wetland, Iran. International Journal of Environmental Science and Technology 173, 1411–1424. Available from: https://doi.org/10.1007/s13762-019-02471-8.

Alberto, A., Ferreira, C., Souza, F., De, Magno, R., Melo, C., et al., 2020. Effects of metal contamination on liver in two fi sh species from a highly impacted Neotropical river: a case study of the Fundão dam, Brazil. Ecotoxicology and Environmental Safety 190. Available from: https://doi.org/10.1016/j.ecoenv.2020.110165.

Ali, M.I., 2019. Status of water quality in watersheds due to mining industry activities. International Journal of Scientific and Technology Research 812, 1799–1805.

Almeida, D., Nascimento, L.P., Abreu, A.T., De Arias, H., Júnior, N., 2020. Geochemical evaluation of bottom sediments affected by historic mining and the rupture of the Fundão dam, Brazil. Environmental Science and Pollution Research International 4365–4375.

Amalo, L.F., Ma'Rufah, U., Permatasari, P.A., 2018. Monitoring 2015 drought in West Java using Normalized Difference Water Index (NDWI). IOP Conference Series: Earth and Environmental Science 149 (1), 6–13. Available from: https://doi.org/10.1088/1755-1315/149/1/012007.

Andrade, J., Boas, V., Jaqueline, S., Vinicius, M., Senra, X., Rico, A., et al., 2020. Ciliates as model organisms for the ecotoxicological risk assessment of heavy metals: a meta – analysis. Ecotoxicology and Environmental Safety 199, 110669. Available from: https://doi.org/10.1016/j.ecoenv.2020.110669.

Angela, C.B., Javier, C.J., Teresa, G.M., Marisa, M.H., 2015. Hydrological evaluation of a peri-urban stream and its impact on ecosystem services potential. Global Ecology and Conservation 3, 628–644. Available from: https://doi.org/10.1016/j.gecco.2015.02.008.

Asibor, G., Edjere, O., Ebighe, D., 2016. Leachate characterization and assessment of surface and groundwater water qualities near municipal solid waste dump site at Okuvo, Delta State, Nigeria. Ethiopian Journal of Environmental Studies and Management 94, 523. Available from: https://doi.org/10.4314/ejesm.v9i4.11.

Askari, M.S., Alamdari, P., Chahardoli, S., Afshari, A., 2020. Quantification of heavy metal pollution for environmental assessment of soil condition. Environmental Monitoring and Assessment 1923. Available from: https://doi.org/10.1007/s10661-020-8116-6.

Ayilara, M.S., Olanrewaju, O.S., Babalola, O.O., Odeyemi, O., 2020. Waste management through composting: challenges and potentials. Sustainability Switzerland 1211, 1–23. Available from: https://doi.org/10.3390/su12114456.

Begum, S., Yuhana, N.Y., Md Saleh, N., Kamarudin, N.H.N., Sulong, A.B., 2021. Review of chitosan composite as a heavy metal adsorbent: material preparation and properties. Carbohydrate Polymers 259, 117613. Available from: https://doi.org/10.1016/j.carbpol.2021.117613.

Bhardwaj, R., Gupta, A., Garg, J.K., 2017. Evaluation of heavy metal contamination using environmetrics and indexing approach for River Yamuna, Delhi stretch, India. Water Science 311, 52–66. Available from: https://doi.org/10.1016/j.wsj.2017.02.002.

Bibak, M., Sattari, M., Tahmasebi, S., Agharokh, A., Namin, J.I., 2020. Marine macro-algae as a bio-indicator of heavy metal pollution in the marine environments, Persian Gulf, 49. pp. 357–363.

Bibak, M., Sattari, M., Tahmasebi, S., Kafaei, R., Sorial, G.A., Ramavandi, B., 2021. Trace and major elements concentration in fish and associated sediment – seawater. Northern Shores of the Persian Gulf 6, 2717–2729.

Bing, H., Zhou, J., Wu, Y., Wang, X., Sun, H., Li, R., 2016. Current state, sources, and potential risk of heavy metals in sediments of Three Gorges Reservoir, China. Environmental Pollution 214, 485–496. Available from: https://doi.org/10.1016/j.envpol.2016.04.062.

Boral, S., Sen, I.S., Tripathi, A., Sharma, B., Dhar, S., 2020. Tracking dissolved trace and heavy metals in the Ganga river from source to sink: a baseline to judge future changes. Geochemistry, Geophysics, Geosystems 2110. Available from: https://doi.org/10.1029/2020GC009203.

Bourgin, M., Borowska, E., Helbing, J., Hollender, J., Kaiser, H.P., Kienle, C., et al., 2017. Effect of operational and water quality parameters on conventional ozonation and the advanced oxidation process O_3/H_2O_2: kinetics of micropollutant abatement, transformation product and bromate formation in a surface water. Water Research 122, 234–245. Available from: https://doi.org/10.1016/j.watres.2017.05.018.

Bowes, M.J., Read, D.S., Joshi, H., Sinha, R., Ansari, A., Hazra, M., et al., 2020. Nutrient and microbial water quality of the upper Ganga River, India: identification of pollution sources. Environmental Monitoring and Assessment 1928. Available from: https://doi.org/10.1007/s10661-020-08456-2.

Büntgen, U., González-Rouco, J.F., Luterbacher, J., Stenseth, N.C., Johnson, D.M., 2020. Extending the climatological concept of "Detection and Attribution" to global change ecology in the Anthropocene. Functional Ecology 3411, 2270–2282. Available from: https://doi.org/10.1111/1365-2435.13647.

Caruana, A.M.N., Le Gac, M., Hervé, F., Rovillon, G.A., Geffroy, S., Malo, F., et al., 2020. Alexandrium pacificum and Alexandrium minutum: harmful or environmentally friendly? Marine Environmental Research 160. Available from: https://doi.org/10.1016/j.marenvres.2020.105014.

Çelik, A., Yaman, H., Turan, S., Kara, A., Kara, F., Zhu, B., et al., 2018. Composite water management index. Journal of Materials Processing Technology 11, 1–8. Available from: https://doi.org/10.1016/j.ijfatigue.2019.02.006%0Ahttp.

Chakrabarti, T., 2019. Assessment of seasonal variations in physico-chemical parameters in Panchet reservoir, Dhanbad district, Jharkhand. Annals of Plant and Soil Research 214, 390–394.

Chakroborty, M., Kumar Bundela, A., Agrawal, R., Sabu, B., Kumar, M., Raipat, B.S., et al., 2017. Studies on physico chemical parameters of 5 water bodies of Ranchi Jharkhand. Balneo Research Journal 82, 51–57. Available from: https://doi.org/10.12680/balneo.2017.142.

Chakraborty, B., Roy, S., Bera, A., Adhikary, P.P., Bera, B., Sengupta, D., et al., 2021. Eco-restoration of river water quality during COVID-19 lockdown in the industrial belt of eastern India. Environmental Science and Pollution Research 2820, 25514–25528. Available from: https://doi.org/10.1007/s11356-021-12461-4.

Chanapathi, T., Thatikonda, S., 2020. Evaluation of sustainability of river Krishna under present and future climate scenarios. Science of the Total Environment 738, 140322. Available from: https://doi.org/10.1016/j.scitotenv.2020.140322.

Chaturvedi, A., Bhattacharjee, S., Singh, A.K., Kumar, V., 2018. A new approach for indexing groundwater heavy metal pollution. Ecological Indicators 87, 323–331. Available from: https://doi.org/10.1016/j.ecolind.2017.12.052. June 2017.

Chen, C., Chaudhary, A., Mathys, A., 2020. Nutritional and environmental losses embedded in global food waste. Resources, Conservation and Recycling 160, 104912. Available from: https://doi.org/10.1016/j.resconrec.2020.104912.

Chowdhury, M., Hasan, M.E., Abdullah-Al-Mamun, M.M., 2020. Land use/land cover change assessment of Halda watershed using remote sensing and GIS. Egyptian Journal of Remote Sensing and Space Science 231, 63–75. Available from: https://doi.org/10.1016/j.ejrs.2018.11.003.

Costantini, M.L., Agah, H., Fiorentino, F., Irandoost, F., Trujillo, F.J.L., Careddu, G., et al., 2021. Nitrogen and metal pollution in the southern Caspian Sea: a multiple approach to bioassessment. Environmental Science and Pollution Research 288, 9898–9912. Available from: https://doi.org/10.1007/s11356-020-11243-8.

Cristable, R.M., Nurdin, E., Wardhana, W., 2020. Water quality analysis of Saluran Tarum Barat, West Java, based on National Sanitation Foundation-Water Quality Index NSF-WQI. IOP Conference Series: Earth and Environmental Science 4811. Available from: https://doi.org/10.1088/1755-1315/481/1/012068.

Danovaro, R., Molari, M., Corinaldesi, C., Delleanno, A., 2016. Macroecological drivers of archaea and bacteria in benthic deep-sea ecosystems. Science Advances 2.

Das, P.K., Das, B.P., Dash, P., 2020. Chromite mining pollution, environmental impact, toxicity and phytoremediation: a review. Environmental Chemistry Letters vi. Available from: https://doi.org/10.1007/s10311-020-01102-w.

Datta, R., Litt, T.B.D., 2020. Water quality of the Ganges in West-Bengal during covid-19 lockdown and unlock period shows the betterment of bod level: An analytical study Water quality of the Ganges in West-Bengal during covid-19 lockdown and unlock period shows the betterment of bod. International Journal of Environmental and Ecology Research 2 (1), 29–34.

Dipti, T., Tiwari, H.L., Rohtash, S., 2018. Hydrological modelling in Narmada basin using remote sensing and GIS with SWAT model and runoff prediction in Patan Watershed. International Journal of Advance Research 42, 344–352. Available from: http://www.IJARIIT.com.

Doostmohammadi, R., Olfati, M., Roodsari, F.G., 2017. Mining pollution control using biogrouting. Journal of Mining Science 532, 367–376. Available from: https://doi.org/10.1134/S1062739117022248.

Duru, C.E., Enedoh, M.C., Duru, I.A., 2019. Physicochemical assessment of borehole water in a reclaimed section of Nekede Mechanic Village, Imo State, Nigeria. Chemistry Africa 24, 689–698. Available from: https://doi.org/10.1007/s42250-019-00077-8.

El-Feky, M.M.M., Alprol, A.E., Heneash, A.M.M., Abo-Taleb, H.A., Omer, M.Y., 2018. Evaluation of water quality and plankton for mahmoudia canal in Northern west of Egypt. Egyptian Journal of Aquatic Biology and Fisheries 225, 461–474. Available from: https://doi.org/10.21608/ejabf.2019.26384.

Elango, L., Jagadeshan, G., 2018. Clean and Sustainable Groundwater in India, Springer Hydrogeology . Available from: http://link.springer.com/10.1007/978-981-10-4552-3.

Elmayel, I., Higueras, P.L., Bouzid, J., Noguero, E.M.G., Elouaer, Z., 2019. Recent Advances in Geo-Environmental Engineering, Geomechanics and Geotechnics, and Geohazards. Springer International Publishing. Available from: http://doi.org/10.1007/978-3-030-01665-4.

Elsagh, A., Jalilian, H., Ghaderi Aslshabestari, M., 2021. Evaluation of heavy metal pollution in coastal sediments of Bandar Abbas, the Persian Gulf, Iran: mercury pollution and environmental geochemical indices. Marine Pollution Bulletin 167, 112314. Available from: https://doi.org/10.1016/j.marpolbul.2021.112314.

Essienubong, I.A., Okechukwu, E.P., Ejuvwedia, S.G., 2019. Effects of waste dumpsites on geotechnical properties of the underlying soils in wet season. Environmental Engineering Research 242, 289–297. Available from: https://doi.org/10.4491/EER.2018.162.

Ewaid, S.H., 2017. Water quality evaluation of Al-Gharraf river by two water quality indices. Applied Water Science 77, 3759–3765. Available from: https://doi.org/10.1007/s13201-016-0523-z.

Ewaid, S.H., Abed, S.A., 2017. Water quality index for Al-Gharraf River, southern Iraq. Egyptian Journal of Aquatic Research 432, 117–122. Available from: https://doi.org/10.1016/j.ejar.2017.03.001.

Foster, K.R., Davidson, C., Tanna, R.N., Spink, D., 2019. Introduction to the virtual special issue monitoring ecological responses to air quality and atmospheric deposition in the Athabasca Oil Sands region the wood Buffalo environmental Association's Forest health monitoring program. Science of the Total Environment 686, 345–359. Available from: https://doi.org/10.1016/j.scitotenv.2019.05.353.

Fusco, F., Allocca, V., Coda, S., Cusano, D., Tufano, R., De Vita, P., 2020. Quantitative assessment of specific vulnerability to nitrate pollution of shallow alluvial aquifers by process-based and empirical approaches. Water 12 (1), 269.

Gallo, A., Zannoni, D., Valotto, G., Nadimi-Goki, M., Bini, C., 2018. Concentrations of potentially toxic elements and soil environmental quality evaluation of a typical Prosecco vineyard of the Veneto region NE Italy. Journal of Soils and Sediments 1811, 3280–3289. Available from: https://doi.org/10.1007/s11368-018-1999-y.

Gayathri, S., Krishnan, K.A., Krishnakumar, A., Maya, T.M.V., Dev, V.V., Antony, S., et al., 2021. Monitoring of heavy metal contamination in Netravati river basin: overview of pollution indices and risk assessment. Sustainable Water Resources Management 72, 1–15. Available from: https://doi.org/10.1007/s40899-021-00502-2.

Gharahi, N., Zamani-Ahmadmahmoodi, R., 2020. Evaluation of groundwater quality for drinking purposes: a case study from the Beheshtabad Basin, Chaharmahal and Bakhtiari Province, Iran. Environmental Earth Sciences 794, 1–12. Available from: https://doi.org/10.1007/s12665-020-8816-9.

Ghosh, A.R., Banerjee, R., 2012. Qualitative evaluation of the Damodar River water flowing over the coal mines and industrial area. International Journal of Scientific and Research Publications 210, 1–6. Available from: http://www.ijsrp.org.

Gikas, G.D., et al., 2020. Comparative evaluation of river chemical status based on WFD methodology and CCME water quality index. Science of the Total Environment 745, 140849. Available from: https://doi.org/10.1016/j.scitotenv.2020.140849.

Gilbreath, A., McKee, L., Shimabuku, I., Lin, D., Werbowski, L.M., Zhu, X., et al., 2019. Multiyear water quality performance and mass accumulation of PCBs, mercury, methylmercury, copper, and microplastics in a bioretention rain garden. Journal of Sustainable Water in the Built Environment 54, 04019004. Available from: https://doi.org/10.1061/jswbay.0000883.

Goh, K., 2019. Urban waterscapes: the hydro-politics of flooding in a Sinking City. International Journal of Urban and Regional Research 432, 250–272. Available from: https://doi.org/10.1111/1468-2427.12756.

Guldhe, A., Singh, B., Renuka, N., Singh, P., Misra, R., Bux, F., 2017. Bioenergy: a sustainable approach for cleaner environment. Phytoremediation Potential of Bioenergy Plants. Springer Singapore, pp. 47–62. Available from: http://doi.org/10.1007/978-981-10-3084-0_2.

Hait, S., Tare, V., 2012. Transformation and availability of nutrients and heavy metals during integrated composting-vermicomposting of sewage sludges. Ecotoxicology and Environmental Safety 79, 214–224. Available from: https://doi.org/10.1016/j.ecoenv.2012.01.004.

Hasan, M.R., Khan, M.Z.H., Khan, M., Aktar, S., Rahman, M., Hossain, F., et al., 2016. Heavy metals distribution and contamination in surface water of the Bay of Bengal coast. Cogent Environmental Science 21. Available from: https://doi.org/10.1080/23311843.2016.1140001.

Hasan, M.K., Shahriar, A., Jim, K.U., 2019. Water pollution in Bangladesh and its impact on public health. Heliyon 58, e02145. Available from: https://doi.org/10.1016/j.heliyon.2019.e02145.

He, X., Chen, C., Lu, C., Wang, L., Chen, H., 2021. Spatial-temporal distribution of red tide in coastal China. IOP Conference Series: Earth and Environmental Science 7831. Available from: https://doi.org/10.1088/1755-1315/783/1/012141.

Hermosilla, T., Wulder, M.A., White, J.C., Coops, N.C., Hobart, G.W., 2015. Regional detection, characterization, and attribution of annual forest change from 1984 to 2012 using Landsat-derived time-series metrics. Remote Sensing of Environment 170, 121–132. Available from: https://doi.org/10.1016/j.rse.2015.09.004.

Hossain, M.Z., Fragstein, P.V., Niemsdorff, P.V., Heß, J., 2017. Effect of different organic wastes on soil propertie s and plant growth and yield: a review. Scientia Agriculturae Bohemica 484, 224–237. Available from: https://doi.org/10.1515/sab-2017-0030.

Hussain, J., Husain, I., Arif, M., Gupta, N., 2017. Studies on heavy metal contamination in Godavari river basin. Applied Water Science 78, 4539–4548. Available from: https://doi.org/10.1007/s13201-017-0607-4.

Hussain, J., Dubey, A., Hussain, I., Arif, M., Shankar, A., 2020. Surface water quality assessment with reference to trace metals in River Mahanadi and its tributaries, India. Applied Water Science 108, 1–12. Available from: https://doi.org/10.1007/s13201-020-01277-1.

Imran, M., Mehmood, A., 2020. Analysis and mapping of present and future drivers of local urban climate using remote sensing: a case of Lahore, Pakistan. Arabian Journal of Geosciences 136. Available from: https://doi.org/10.1007/s12517-020-5214-2.

Imran, U., Ullah, A., Shaikh, K., Mehmood, R., Saeed, M., 2019. Health risk assessment of the exposure of heavy metal contamination in surface water of lower Sindh, Pakistan. SN Applied Sciences 16, 1–10. Available from: https://doi.org/10.1007/s42452-019-0594-1.

Imtiaz, F., Nisa, Z., Ahmed, S., Ijaz, S., Tabasum, M., 2018. Physicochemical assessment of soil and drinking water quality; a case study of urban area of Green Town, Lahore. Pakistan Journal of Science 701, 40–47.

Iqbal, Z., Abbas, F., Mahmood, A., Ibrahim, M., Gul, M., Yamin, M., et al., 2021. Human health risk of heavy metal contamination in groundwater and source apportionment. International Journal of Environmental Science and Technology 0123456789. Available from: https://doi.org/10.1007/s13762-021-03611-9.

Irandoost, F., Agah, H., Rossi, L., Calizza, E., Careddu, G., Costantini, M.L., 2021. Stable isotope ratios δ13C and δ15N and heavy metal levels in macroalgae, sediment, and benthos from the northern parts of Persian Gulf and the Gulf of Oman. Marine Pollution Bulletin 163, 111909. Available from: https://doi.org/10.1016/j.marpolbul.2020.111909. July 2020.

Jin, L., Whitehead, P.G., Rodda, H., Macadam, I., Sarkar, S., 2018. Simulating climate change and socio-economic change impacts on flows and water quality in the Mahanadi River system, India. Science of the Total Environment 637–638, 907–917. Available from: https://doi.org/10.1016/j.scitotenv.2018.04.349.

Karimian, S., Chamani, A., Shams, M., 2020. Evaluation of heavy metal pollution in the Zayandeh-Rud River as the only permanent river in the central plateau of Iran. Environmental Monitoring and Assessment 1925. Available from: https://doi.org/10.1007/s10661-020-8183-8.

Kaur, M., Kumar, A., Mehra, R., Mishra, R., 2018. Human health risk assessment from exposure of heavy metals in soil samples of Jammu district of Jammu and Kashmir, India. Arabian Journal of Geosciences 1115. Available from: https://doi.org/10.1007/s12517-018-3746-5.

Khalid, S., Shahid, M., Natasha, Shah, A.H., Saeed, F., Ali, M., et al., 2020. Heavy metal contamination and exposure risk assessment via drinking groundwater in Vehari, Pakistan. Environmental Science and Pollution Research 2732, 39852–39864. Available from: https://doi.org/10.1007/s11356-020-10106-6.

Kharake, A.C., Raut, V.S., 2021. An assessment of water quality index of Godavari river water in Nashik city, Maharashtra. Applied Water Science 116, 1–11. Available from: https://doi.org/10.1007/s13201-021-01432-2.

Kumar, A., Rai, A.K., Pandey, A.C., 2016. Geoinformatics based site suitability modeling for future urban development using MCDM- analytic hierarchy process techniques. Remote Sensing for Natural Resources Management and Monitoring 378–399. February 2019.

Kumar, A., Sinha, D.K., Mishra, B.J., 2019. Physio-chemical parameters of surface water of hatia region at Ranchi, Jharkhand. IOSR Journal of Environmental Science, Toxicology and Food Technology 1311, 33–38. Available from: https://doi.org/10.9790/2402-1311013338.

Kumar, V., Sharma, A., Pandita, S., Bhardwaj, R., Thukral, A.K., Cerda, A., 2020. A review of ecological risk assessment and associated health risks with heavy metals in sediment from India. International Journal of Sediment Research 355, 516–526. Available from: https://doi.org/10.1016/j.ijsrc.2020.03.012.

La Torre, G.L., Cicero, N., Bartolomeo, G., Rando, R., Vadalà, R., Santini, A., Salvo, A., 2020. Assessment and monitoring of fish quality from a coastal ecosystem under high anthropic pressure: A case study in southern Italy. International Journal of Environmental Research and Public Health 17 (9), 3285.

Laumonier, Y., Nasi, R., 2018. The last natural seasonal forests of Indonesia: implications for forest management and conservation. Applied Vegetation Science 213, 461–476. Available from: https://doi.org/10.1111/avsc.12377.

Laxmi Mohanta, V., Naz, A., Kumar Mishra, B., 2020. Distribution of heavy metals in the water, sediments, and fishes from Damodar river basin at steel city, India: a probabilistic risk assessment. Human and Ecological Risk Assessment 262, 406–429. Available from: https://doi.org/10.1080/10807039.2018.1511968.

Li, C., Zhou, K., Qin, W., Tian, C., Qi, M., Yan, X., et al., 2019. A review on heavy metals contamination in soil: effects, sources, and remediation techniques. Soil and Sediment Contamination 284, 380–394. Available from: https://doi.org/10.1080/15320383.2019.1592108.

Li, P., Tian, R., Liu, R., 2019. Solute geochemistry and multivariate analysis of water quality in the Guohua Phosphorite Mine, Guizhou Province, China. Exposure and Health 112, 81–94. Available from: https://doi.org/10.1007/s12403-018-0277-y.

Liao, J., Chang, F., Han, X., Ge, C., Lin, S., 2020. Wireless water quality monitoring and spatial mapping with disposable whole-copper electrochemical sensors and a smartphone. Sensors and Actuators, B: Chemical 306, 127557. Available from: https://doi.org/10.1016/j.snb.2019.127557.

Ling, D.J., Huang, Q.C., Ouyang, Y., 2010. Identification of most susceptible soil parameters for Latosol under simulated acid rain stress using principal component analysis. Journal of Soils and Sediments 107, 1211–1218. Available from: https://doi.org/10.1007/s11368-010-0231-5.

Lipy, E.P., Hakim, M., Mohanta, L.C., Islam, D., Lyzu, C., Roy, D.C., et al., 2021. Assessment of heavy metal concentration in water, sediment and common fish species of Dhaleshwari River in Bangladesh and their health implications. Biological Trace Element Research 4295–4307.

Liu, Q., Liao, Y., Xu, X., Shi, X., Zeng, J., Chen, Q., Shou, L., 2020a. Heavy metal concentrations in tissues of marine fish and crab collected from the middle coast of Zhejiang Province, China. Environmental monitoring and assessment 192 (5), 1–12.

Liu, W., Bailey, R.T., Andersen, H.E., Jeppesen, E., Nielsen, A., Peng, K., et al., 2020b. Quantifying the effects of climate change on hydrological regime and stream biota in a groundwater-dominated catchment: a modelling approach combining SWAT-MODFLOW with flow-biota empirical models. Science of the Total Environment 745, 140933. Available from: https://doi.org/10.1016/j.scitotenv.2020.140933.

Luo, Z., Bin, He, J., Polle, A., Rennenberg, H., 2016. Heavy metal accumulation and signal transduction in herbaceous and woody plants: paving the way for enhancing phytoremediation efficiency. Biotechnology Advances 1131–1148. Available from: https://doi.org/10.1016/j.biotechadv.2016.07.003. Elsevier Inc.

Madhav, S., Raju, N.J., Ahamad, A., 2021. A study of hydrogeochemical processes using integrated geochemical and multivariate statistical methods and health risk assessment of groundwater in Trans-Varuna region, Uttar PradeshSpringer NetherlandsEnvironment, Development and Sustainability. Available from: http://doi.org/10.1007/s10668-020-00928-2.

Mahato, M.K., Singh, G., Singh, P.K., Singh, A.K., Tiwari, A.K., 2017. Assessment of mine water quality using heavy metal pollution index in a coal mining area of Damodar River Basin, India. Bulletin of Environmental Contamination and Toxicology 991, 54–61. Available from: https://doi.org/10.1007/s00128-017-2097-3.

Mannan, A., Liu, J., Zhongke, F., Khan, T.U., Saeed, S., Mukete, B., et al., 2019. Application of land-use/land cover changes in monitoring and projecting forest biomass carbon loss in Pakistan. Global Ecology and Conservation 1735, e00535. Available from: https://doi.org/10.1016/j.gecco.2019.e00535.

Mateo-Sagasta, J., Zadeh, S.M., Turral, H., 2017. Water pollution from agriculture: a global review. Executive Summary 35. Available from: http://www.fao.org/3/a-i7754e.pdf.

Matta, G., Nayak, A., Kumar, A., Kumar, P., 2020. Water quality assessment using NSFWQI, OIP and multivariate techniques of Ganga River system, Uttarakhand, India. Applied Water Science 109, 1–12. Available from: https://doi.org/10.1007/s13201-020-01288-y.

Maurya, P.K., Malik, D.S., 2019. Bioaccumulation of heavy metals in tissues of selected fish species from Ganga river, India, and risk assessment for human health. Human and Ecological Risk Assessment 254, 905–923. Available from: https://doi.org/10.1080/10807039.2018.1456897.

Maurya, S., Rashk-E-Eram, Naik, S.K., Choudhary, J.S., Kumar, S., 2019. Heavy metals scavenging potential of Trichoderma asperellum and Hypocrea nigricans isolated from acid soil of Jharkhand. Indian Journal of Microbiology 591, 27–38. Available from: https://doi.org/10.1007/s12088-018-0756-7.

Men, C., Liu, R., Xu, L., Wang, Q., Guo, L., Miao, Y., et al., 2020. Source-specific ecological risk analysis and critical source identification of heavy metals in road dust in Beijing, China. Journal of Hazardous Materials 388, 121763. Available from: https://doi.org/10.1016/j.jhazmat.2019.121763.

Meraj, G., Romshoo, S.A., Ayoub, S., Altaf, S., 2018. Geoinformatics based approach for estimating the sediment yield of the mountainous watersheds in Kashmir Himalaya, India. Geocarto International 3310, 1114–1138. Available from: https://doi.org/10.1080/10106049.2017.1333536.

Milivojević, J., Krstić, D., Šmit, B., Djekić, V., 2016. Assessment of heavy metal contamination and calculation of its pollution index for Uglješnica River, Serbia. Bulletin of Environmental Contamination and Toxicology 975, 737–742. Available from: https://doi.org/10.1007/s00128-016-1918-0.

Mishra, S., Kumar, A., 2021. Estimation of physicochemical characteristics and associated metal contamination risk in the Narmada river, India. Environmental Engineering Research 261, 1–11. Available from: https://doi.org/10.4491/eer.2019.521.

Mishra, H., Denis, D.M., Suryavanshi, S., Kumar, M., Srivastava, S.K., Denis, A.F., et al., 2017. Hydrological simulation of a small ungauged agricultural watershed Semrakalwana of Northern India. Applied Water Science 76, 2803–2815. Available from: https://doi.org/10.1007/s13201-017-0531-7.

Mishra, S., Singh, R., Kumar, B., 2019. The estimation of heavy metals in Subarnarekha river at Mau Bhandar and Galudih Barrage, Jharkhand. International Journal of Engineering Applied Sciences and Technology 0404, 84–86. Available from: https://doi.org/10.33564/ijeast.2019.v04i04.014.

Mohanta, L.C., Niloy, M.N.H., Chowdhury, G.W., Islam, D., Lipy, E.P., 2019. Heavy metals in water, sediment and three fish species of Dhaleshwari river, Savar, Bangladesh. Journal of Zoology 472, 263–272. Available from: https://doi.org/10.3329/bjz.v47i2.44337.

Mohanty, A.K., Lingaswamy, M., Rao, V.G., Sankaran, S., 2018. Impact of acid mine drainage and hydrogeochemical studies in a part of Rajrappa coal mining area of Ramgarh District, Jharkhand State of India. Groundwater for Sustainable Development 7, 164–175. Available from: https://doi.org/10.1016/j.gsd.2018.05.005.

Mokarram, M., Saber, A., Sheykhi, V., 2020. Effects of heavy metal contamination on river water quality due to release of industrial effluents. Journal of Cleaner Production 277, 123380. Available from: https://doi.org/10.1016/j.jclepro.2020.123380.

Nadikatla, S.K., Mushini, V.S., Mudumba, P.S.M.K., 2020. Water quality index method in assessing groundwater quality of Palakonda mandal in Srikakulam district, Andhra Pradesh, India. Applied Water Science 101, 1–14. Available from: https://doi.org/10.1007/s13201-019-1110-x.

Nandan, A., Yadav, B.P., Baksi, S., Bose, D., 2017. Recent scenario of solid waste management in India. World Scientific News 66, 56–74. Available from: http://www.worldscientificnews.com.

Nawrot, N., Wojciechowska, E., Rezania, S., Walkusz-miotk, J., Pazdro, K., 2020. The effects of urban vehicle traffic on heavy metal contamination in road sweeping waste and bottom sediments of retention tanks. Science of the Total Environment 749, 141511. Available from: https://doi.org/10.1016/j.scitotenv.2020.141511.

Nazneen, S., Singh, S., Raju, N.J., 2018. Heavy metal fractionation in core sediments and potential biological risk assessment from Chilika lagoon, Odisha state, India. Quaternary International. pp. 0–1. Available from: http://doi.org/10.1016/j.quaint.2018.05.011.

Nazneen, S., Raju, N.J., Madhav, S., Ahamad, A., 2019. Spatial and temporal dynamics of dissolved nutrients and factors affecting water quality of Chilika lagoon. Arabian Journal of Geosciences 12 (7), 1–23.

de Nobile, F.O., Calero Hurtado, A., Prado, R.D.M., de Souza, H.A., Anunciação, M.G., Palaretti, L.F., et al., 2021. A novel technology for processing urban waste compost as a fast-releasing nitrogen source to improve soil properties and broccoli and lettuce production. Waste and Biomass Valorization. Available from: https://doi.org/10.1007/s12649-021-01415-z.

OECD-FAO, 2012. Agricultural Outlook 2011–2020, Organisation for Economic Co-Operation and Development. Available from: https://doi.org/10.1787/agr_outlook-2011-en.

Ouyang, W., Cai, G., Tysklind, M., Yang, W., Hao, F., Liu, H., 2017. Temporal-spatial patterns of three types of pesticide loadings in a middle-high latitude agricultural watershed. Water Research 122, 377–386. Available from: https://doi.org/10.1016/j.watres.2017.06.023.

Ouyang, W., Wang, Y., Lin, C., He, M., Hao, F., Liu, H., et al., 2018. Heavy metal loss from agricultural watershed to aquatic system: a scientometrics review. Science of the Total Environment 637–638, 208–220. Available from: https://doi.org/10.1016/j.scitotenv.2018.04.434.

Pandey, B., Agrawal, M., Singh, S., 2016. Ecological risk assessment of soil contamination by trace elements around coal mining area. Journal of Soils and Sediments 161, 159–168. Available from: https://doi.org/10.1007/s11368-015-1173-8.

Pandey, S., Kumari, N., Priya, S., 2021. Soil quality and pollution assessment around Jumar watershed of Jharkhand, India. Arabian Journal of Geosciences. Available from: https://doi.org/10.1007/s12517-021-09091-y.

Panditharathne, D.L.D., Abeysingha, N.S., Nirmanee, K.G.S., Mallawatantri, A., 2019. Application of revised universal soil loss equation Rusle model to assess soil erosion in "kalu Ganga" River Basin in Sri Lanka. Applied and Environmental Soil Science 2019. Available from: https://doi.org/10.1155/2019/4037379.

Pandiyan, J., Mahboob, S., Govindarajan, M., Al-ghanim, K.A., 2020. An assessment of level of heavy metals pollution in the water, sediment and aquatic organisms: a perspective of tackling environmental threats for food security. Saudi Journal of Biological Sciences. Available from: https://doi.org/10.1016/j.sjbs.2020.11.072.

Pastor, J., Hernández, A.J., 2012. Heavy metals, salts and organic residues in old solid urban waste landfills and surface waters in their discharge areas: determinants for restoring their impact. Journal of Environmental Management 95 (Suppl). Available from: https://doi.org/10.1016/j.jenvman.2011.06.048.

Payal, R., Tomer, T., 2021. Determination of heavy metal ions from water. Applied Water Science 1 (1), 255–272. Available from: https://doi.org/10.1002/9781119725237.ch9.

Penman, T.D., Cirulis, B., Marcot, B.G., 2020. Bayesian decision network modeling for environmental risk management: a wildfire case study. Journal of Environmental Management 270, 110735. Available from: https://doi.org/10.1016/j.jenvman.2020.110735.

Pingping, L., Kang, S., Zhou, M., Lyu, J., Aisyah, S., Binaya, M., et al., 2019. Water quality trend assessment in Jakarta: a rapidly growing Asian megacity. PLoS One 1–17.

Pramanik, A.K., Majumdar, D., Chatterjee, A., 2020. Factors affecting lean, wet-season water quality of Tilaiya reservoir in Koderma District, India during 2013–2017. Water Science 341, 85–97. Available from: https://doi.org/10.1080/11104929.2020.1765451.

Prasad, S., Saluja, R., Joshi, V., Garg, J.K., 2020. Heavy metal pollution in surface water of the Upper Ganga River, India: human health risk assessment. Environmental Monitoring and Assessment 19211. Available from: https://doi.org/10.1007/s10661-020-08701-8.

Prathap, A., Chakraborty, S., 2019. Hydro chemical characterization and suitability analysis of groundwater for domestic and irrigation uses in open cast coal mining areas of Charhi and Kuju, Jharkhand, India. Groundwater for Sustainable Development 9, 100244. Available from: https://doi.org/10.1016/j.gsd.2019.100244.

Rahman, K.M.A., Zhang, D., 2018. Effects of fertilizer broadcasting on the excessive use of inorganic fertilizers and environmental sustainability. Sustainability Switzerland 103. Available from: https://doi.org/10.3390/su10030759.

Rai, A.K., Paul, B., Kishor, N., 2012. A study on the sewage disposal on water quality of Harmu River in Ranchi City Jharkhand, India. International Journal of Plant, Animal and Environmental Sciences 21, 102–106.

Ranjan, A.K., Prakash, S., Anand, A., Verma, S.K., Murmu, L., Kumar, P.B.S., 2017. Groundwater prospect variability analysis with spatio-temporal changes in Ranchi City, Jharkhand, India using geospatial technology. International Journal of Earth Sciences and Engineering 10, 616–625. Available from: https://doi.org/10.21276/ijee.2017.10.0320.

Ranjith, S., Shivapur, A.V., Shiva Keshava Kumar, P., Hiremath, C.G., Dhungana, S., 2019. Water quality evaluation in term of WQI river tungabhadra, Karnataka, India. International Journal of Innovative Technology and Exploring Engineering 89 (Special Issue 2), 247–253. Available from: https://doi.org/10.35940/ijitee.I1051.0789S219.

Rasool, A., Xiao, T., Farooqi, A., Shafeeque, M., Masood, S., Ali, S., et al., 2016. Arsenic and heavy metal contaminations in the tube well water of Punjab, Pakistan and risk assessment: a case study. Ecological Engineering 95, 90–100. Available from: https://doi.org/10.1016/j.ecoleng.2016.06.034.

Rastegari Mehr, M., Keshavarzi, B., Moore, F., Sharifi, R., Lahijanzadeh, A., Kermani, M., 2017. Distribution, source identification and health risk assessment of soil heavy metals in urban areas of Isfahan province, Iran. Journal of African Earth Sciences 132, 16–26. Available from: https://doi.org/10.1016/j.jafrearsci.2017.04.026.

Ravanipour, M., Hadi, M., Rastkari, N., Hemmati Borji, S., Nasseri, S., 2021. Presence of heavy metals in drinking water resources of Iran: a systematic review and *meta*-analysis. Environmental Science and Pollution Research 2821, 26223–26251. Available from: https://doi.org/10.1007/s11356-021-13293-y.

Raveesh, A., Mona, C., Jayveer, S., 2015. Waste management initiatives in India for human well being. European Scientific Journal 7881, 1857–7881. Available from: http://home.iitk.ac.in/~anubha/H16.pdf.

Raza, W., Saeed, S., Saulat, H., Gul, H., Sarfraz, M., Sonne, C., et al., 2021. A review on the deteriorating situation of smog and its preventive measures in Pakistan. Journal of Cleaner Production 279, 123676. Available from: https://doi.org/10.1016/j.jclepro.2020.123676.

Ren, Q., Wang, X., Li, W., Wei, Y., An, D., 2020. Research of dissolved oxygen prediction in recirculating aquaculture systems based on deep belief network. Aquacultural Engineering 90, 102085. Available from: https://doi.org/10.1016/j.aquaeng.2020.102085. September 2019.

Ren, B., Zhao, Y., Bai, H., Kang, S., Zhang, T., Song, S., 2021. Eco-friendly geopolymer prepared from solid wastes: a critical review. Chemosphere. Available from: https://doi.org/10.1016/j.chemosphere.2020.128900. Elsevier Ltd.

Reza, R., Singh, G., 2010. Heavy metal contamination and its indexing approach for river water. International Journal of Environmental Science and Technology 74, 785–792. Available from: https://doi.org/10.1007/BF03326187.

Rezaei, A., Hassani, H., Hassani, S., Jabbari, N., Fard Mousavi, S.B., Rezaei, S., 2019. Evaluation of groundwater quality and heavy metal pollution indices in Bazman basin, southeastern Iran. Groundwater for Sustainable Development 9, 100245. Available from: https://doi.org/10.1016/j.gsd.2019.100245.

Safiur Rahman, M., Shafiuddin Ahmed, A.S., Rahman, M.M., Omar Faruque Babu, S.M., Sultana, S., Sarker, S.I., et al., 2021. Temporal assessment of heavy metal concentration and surface water quality representing the public health evaluation from the Meghna River estuary, Bangladesh. Applied Water Science 117, 1–16. Available from: https://doi.org/10.1007/s13201-021-01455-9.

Sahoo, S., Dhar, A., Debsarkar, A., Kar, A., 2018. Impact of water demand on hydrological regime under climate and LULC change scenarios. Environmental Earth Sciences 779, 1–19. Available from: https://doi.org/10.1007/s12665-018-7531-2.

Saleh, H.N., Panahande, M., Yousefi, M., Asghari, F.B., Oliveri Conti, G., Talaee, E., et al., 2019. Carcinogenic and non-carcinogenic risk assessment of heavy metals in groundwater wells in Neyshabur Plain, Iran. Biological Trace Element Research 1901, 251–261. Available from: https://doi.org/10.1007/s12011-018-1516-6.

Salvo, A., Loredana, G., Torre, L., Mangano, V., Erminia, K., Bartolomeo, G., et al., 2018. Toxic inorganic pollutants in foods from agricultural producing areas of Southern Italy: level and risk assessment. Ecotoxicology and Environmental Safety 148, 114–124. Available from: https://doi.org/10.1016/j.ecoenv.2017.10.015. May 2017.

Sánchez-Navarro, A., Gil-Vázquez, J.M., Delgado-Iniesta, M.J., Marín-Sanleandro, P., Blanco-Bernardeau, A., Ortiz-Silla, R., 2015. Establishing an index and identification of limiting parameters for characterizing soil quality in Mediterranean ecosystems. Catena 131, 35–45. Available from: https://doi.org/10.1016/j.catena.2015.02.023.

Savassi, L.A., Paschoalini, A.L., Arantes, F.P., Rizzo, E., Bazzoli, N., 2020. Heavy metal contamination in a highly consumed Brazilian fish: immunohistochemical and histopathological assessments. Environmental Monitoring and Assessment 192 (8), 1–14.

Selvam, S., Jesuraja, K., Venkatramanan, S., Chung, S.Y., Roy, P.D., Muthukumar, P., et al., 2020. Imprints of pandemic lockdown on subsurface water quality in the coastal industrial city of Tuticorin, South India: a revival perspective. Science of the Total Environment 738, 139848. Available from: https://doi.org/10.1016/j.scitotenv.2020.139848.

Shahradnia, H., Chamani, A., Zamanpoore, M., Jalalizand, A., 2021. Heavy metal pollution in surface sediments of Ghareh-Aghaj River, one of the longest perennial rivers in Iran. Environmental Earth Sciences 803, 1–12. Available from: https://doi.org/10.1007/s12665-021-09384-1.

Sharma, D., Kansal, A., 2011. Water quality analysis of River Yamuna using water quality index in the national capital territory, India 2000–2009. Applied Water Science 13–4, 147–157. Available from: https://doi.org/10.1007/s13201-011-0011-4.

Sharma, B., Kumar, M., Denis, D.M., Singh, S.K., 2019. Appraisal of river water quality using open-access earth observation data set: a study of river Ganga at Allahabad (India). Sustainable Water Resources Management 5 (2), 755–765. Available from: https://doi.org/10.1007/s40899-018-0251-7.

Shekhar, S., Ghosh, M., Pandey, A.C., Tirkey, A.S., 2017. Impact of geology and geomorphology on fluoride contaminated groundwater in hard rock terrain of India using geoinformatics approach. Applied Water Science 76, 2943–2956. Available from: https://doi.org/10.1007/s13201-017-0593-6.

Sheykhi, V., Samani, N., 2020. Assessment of water quality compartments in Kor River, IRAN. Environmental Monitoring and Assessment 1928. Available from: https://doi.org/10.1007/s10661-020-08464-2.

Sigdel, B., 2017. Water quality measuring station: pH, turbidity and temperature measurement (pp. 1–31). PhD thesis.

Singh, U.K., 2016. Water quality assessment using physico-chemical parameters of kanke dam Ranchi, Jharkhand. International Journal of Civil Engineering and Technology 7 (4), 269–275.

Singh, L.N., Mandhyan, M., 2017. Seasonal variation in water quality of Kharkai and Subarnarekha river of Jamshedpur, Jharkhand. International Journal of Research in Engineering, Technology and Science vii, 1–6.

Singh, K.R., Dutta, R., Kalamdhad, A.S., Kumar, B., 2020. Review of existing heavy metal contamination indices and development of an entropy-based improved indexing approach. Environment, Development and Sustainability 228, 7847–7864. Available from: https://doi.org/10.1007/s10668-019-00549-4.

Singh, G., Kamal, R.K., 2017. Heavy metal contamination and its indexing approach for groundwater of Goa mining region, India. Applied Water Science 73, 1479–1485. Available from: https://doi.org/10.1007/s13201-016-0430-3.

Singh, A.K., Giri, S., 2018. Subarnarekha River: The Gold Streak of India, pp. 273–285. Available from: https://doi.org/10.1007/978-981-10-2984-4_22.

Singhal, A., Pandey, S., Kumari, N., Chauhan, D.K., Jha, P.K., 2021. Impact of Climate Change on Soil Microbes Involved in Biogeochemical Cycling, pp. 63–94. Available from: https://doi.org/10.1007/978-3-030-76863-8_5.

Sun, J., Pan, L., Tsang, D.C.W., Zhan, Y., Zhu, L., Li, X., 2018. Organic contamination and remediation in the agricultural soils of China: a critical review. Science of the Total Environment 615, 724–740. Available from: https://doi.org/10.1016/j.scitotenv.2017.09.271.

Tahmasebi, P., Taheri, M., Gharaie, M.H.M., 2020. Heavy metal pollution associated with mining activity in the Kouh-e Zar region, NE Iran. Bulletin of Engineering Geology and the Environment 792, 1113–1123. Available from: https://doi.org/10.1007/s10064-019-01574-3.

Tirgar, A., Aghalari, Z., Sillanpää, M., Dahms, H.-U., 2020. A glance at one decade of water pollution research in Iranian environmental health journals. International Journal of Food Contamination 71, 1–8. Available from: https://doi.org/10.1186/s40550-020-00080-9.

Tirkey, P., Bhattacharya, T., Chakraborty, S., Baraik, S., 2017. Assessment of groundwater quality and associated health risks: a case study of Ranchi city, Jharkhand, India. Groundwater for Sustainable Development 5, 85–100. Available from: https://doi.org/10.1016/j.gsd.2017.05.002.

Tiwari, A.K., Lavy, M., Amanzio, G., De Maio, M., Singh, P.K., Mahato, M.K., 2017. Identification of artificial groundwater recharging zone using a GIS-based fuzzy logic approach: a case study in a coal mine area of the Damodar Valley, India. Applied Water Science 78, 4513–4524. Available from: https://doi.org/10.1007/s13201-017-0603-8.

Uddin, M.J., Jeong, Y.K., 2021. Urban river pollution in Bangladesh during last 40 years: potential public health and ecological risk, present policy, and future prospects toward smart water management. Heliyon 72, e06107. Available from: https://doi.org/10.1016/j.heliyon.2021.e06107.

Ullah, A., Khan, D., Khan, I., Zheng, S., 2018. Does agricultural ecosystem cause environmental pollution in Pakistan? Promise and menace. Environmental Science and Pollution Research 2514, 13938–13955. Available from: https://doi.org/10.1007/s11356-018-1530-4.

US EPA, 2020. Best Practices for Solid Waste Management: Best Practices for Solid Waste Management: A Guide for Decision-Makers in Developing Countries, pp. 1–166.

Valenzuela-Diaz, M.J., Navarrete-Calvo, A., Caraballo, M.A., McPhee, J., Garcia, A., Correa-Burrows, J.P., et al., 2020. Hydrogeochemical and environmental water quality standards in the overlap between high mountainous natural protected areas and copper mining activities Mapocho river upper basin, Santiago, Chile. Journal of Hydrology 588, 125063. Available from: https://doi.org/10.1016/j.jhydrol.2020.125063.

Varol, M., 2020. Use of water quality index and multivariate statistical methods for the evaluation of water quality of a stream affected by multiple stressors: a case study. Environmental Pollution 266, 115417. Available from: https://doi.org/10.1016/j.envpol.2020.115417.

Verma, P., Singh, P.K., Sinha, R.R., Tiwari, A.K., 2020. Assessment of groundwater quality status by using water quality index WQI and geographic information system GIS approaches: a case study of the Bokaro district, India. Applied Water Science 101, 1–16. Available from: https://doi.org/10.1007/s13201-019-1088-4.

Wan, X., Yang, J., Song, W., 2018. Pollution status of agricultural land in China: impact of land use and geographical position. Soil and Water Research 134, 234–242. Available from: https://doi.org/10.17221/211/2017-SWR.

Wang, H., Liu, T., Tsang, D.C.W., Feng, S., 2017. Transformation of heavy metal fraction distribution in contaminated river sediment treated by chemical-enhanced washing. Journal of Soils and Sediments 174, 1208–1218. Available from: https://doi.org/10.1007/s11368-016-1631-y.

Wu, J., Li, P., Wang, D., Ren, X., Wei, M., 2020. Statistical and multivariate statistical techniques to trace the sources and affecting factors of groundwater pollution in a rapidly growing city on the Chinese Loess Plateau. Human and Ecological Risk Assessment 266, 1603–1621. Available from: https://doi.org/10.1080/10807039.2019.1594156.

Xia, Y., Zhang, M., Tsang, D.C.W., Geng, N., Lu, D., Zhu, L., 2020. Recent advances in control technologies for non-point source pollution with nitrogen and phosphorous from agricultural runoff: current practices and future prospects. Applied Biological Chemistry . Available from: https://doi.org/10.1186/s13765-020-0493-6.

Xu, X., Shrestha, S., Gilani, H., Gumma, M.K., Siddiqui, B.N., Jain, A.K., 2020. Dynamics and drivers of land use and land cover changes in Bangladesh. Regional Environmental Change 20 (2). Available from: https://doi.org/10.1007/s10113-020-01650-5.

Xueman, Y., Wenxi, L., Yongkai, A., Weihong, D., 2020. Assessment of parameter uncertainty for non-point source pollution mechanism modeling: a Bayesian-based approach. Environmental Pollution 263, 114570. Available from: https://doi.org/10.1016/j.envpol.2020.114570.

Yalçın, P., Kam, E., Yümün, Z.Ü., Kurt, D., 2018. An investigation of anthropogenic pollution in soil samples from residential areas in Erzincan city center and its vicinity by evaluating chemical factors. Arabian Journal of Geosciences 1122. Available from: https://doi.org/10.1007/s12517-018-4007-3.

Yang, Q., Li, Z., Lu, X., Duan, Q., Huang, L., Bi, J., 2018. A review of soil heavy metal pollution from industrial and agricultural regions in China: pollution and risk assessment. Science of the Total Environment 642, 690–700. Available from: https://doi.org/10.1016/j.scitotenv.2018.06.068.

Yu, R., He, L., Cai, R., Li, B., Li, Z., Yang, K., 2017. Heavy metal pollution and health risk in China. Global Health Journal 11, 47–55. Available from: https://doi.org/10.1016/s2414-64471930059-4.

Yue, H., Liu, Y., Li, Y., Lu, Y., 2019. Eco-environmental quality assessment in china's 35 major cities based on remote sensing ecological index. IEEE Access 7, 51295–51311. Available from: https://doi.org/10.1109/ACCESS.2019.2911627.

Yulistia, E., Fauziyah, S., Hermansyah, H., 2018. Assessment of ogan river water quality Kabupaten OKU SUMSEL by NSFWQI method. Indonesian Journal of Fundamental and Applied Chemistry 32, 54–58. Available from: https://doi.org/10.24845/ijfac.v3.i2.54.

Zahra, P., 2015. Ecological Crisis: Underground Water Resource Depletion in Ranchi, p. 13.

Zamora-Ledezma, C., Negrete-Bolagay, D., Figueroa, F., Zamora-Ledezma, E., Ni, M., Alexis, F., et al., 2021. Heavy metal water pollution: a fresh look about hazards, novel and conventional remediation methods. Environmental Technology and Innovation 22, 101504. Available from: https://doi.org/10.1016/j.eti.2021.101504.

Zerizghi, T., Yang, Y., Wang, W., Zhou, Y., Zhang, J., Yi, Y., 2020. Ecological risk assessment of heavy metal concentrations in sediment and fish of a shallow lake: a case study of Baiyangdian Lake, North China. Environmental monitoring and assessment 192 (2), 1–16.

Zhang, C., Shan, B., Zhao, Y., Song, Z., Tang, W., 2018. Spatial distribution, fractionation, toxicity and risk assessment of surface sediments from the Baiyangdian Lake in northern China. Ecological Indicators 90, 633–642. Available from: https://doi.org/10.1016/j.ecolind.2018.03.078. December 2017.

Zhao, Y., Duan, X., Han, J., Yang, K., Xue, Y., 2018. The main influencing factors of soil mechanical characteristics of the gravity erosion environment in the dry-hot valley of Jinsha river. Open Chemistry 161, 796–809. Available from: https://doi.org/10.1515/chem-2018-0086.

Zhitkovich, A., 2011. Chromium in drinking water: sources, metabolism, and cancer risks. Chemical Research in Toxicology 2410, 1617–1629. Available from: https://doi.org/10.1021/tx200251t.

Zhou, Q., Yang, N., Li, Y., Ren, B., Ding, X., Bian, H., et al., 2020. Total concentrations and sources of heavy metal pollution in global river and lake water bodies from 1972 to 2017. Global Ecology and Conservation 22, e00925. Available from: https://doi.org/10.1016/j.gecco.2020.e00925.

CHAPTER

9

Airborne heavy metals deposition and contamination to water resources

Harshbardhan Kumar[1,2], *Gurudatta Singh*[3], *Virendra Kumar Mishra*[3], *Ravindra Pratap Singh*[4] *and Pardeep Singh*[5]

[1]Academy of Scientific and Innovative Research (AcSIR), Ghaziabad, Uttar Pradesh, India [2]CSIR-National Institute of Oceanography, Dona Paula, Goa, India [3]Institute of Environment and sustainable Development, Banaras Hindu University, Varanasi, Uttar Pradesh, India [4]Central Public Works Department, New Delhi, India [5]Department of Environmental Science, PGDAV College University of Delhi, New Delhi, Delhi, India

9.1 Introduction

The Earth's system is mainly composed of four spheres: atmosphere, biosphere, hydrosphere, and lithosphere which continuously interact and exchange energy and matter across the spheres to produce conducive environment for the life to exist. In the same environment of this blue planet's humans, animals, plants and numerous other lives exist and flourish together. During the last 100 or 150 years, anthropogenic activities have dominated on the planet which has changed that intricate balance within the spheres in terms of both mass and energy. The major consequence of that misbalance is the problem of environmental pollution. Our very surrounding environment has been seriously polluted by several types of pollutants, which has a well-established adverse cause and effect relation with the living organisms. Among several other pollutants, here in this chapter we will be more concerned about the airborne heavy metal deposition and associated pollution (Manisalidis et al., 2020).

The term heavy metal is often used for a wide range of metallic elements which are toxic to the environment and human health. Generally, the elements having high atomic weight and density at least five times greater than that of water are considered under this category (Yedjou et al., 2012). However, exceptions exist wherein some lighter metals such as arsenic and aluminum are toxic to the environment, while some heavy metals such as

Metals in Water
DOI: https://doi.org/10.1016/B978-0-323-95919-3.00019-7

element gold are typically nontoxic in nature. The term heavy metal and its definition, is still a matter of ongoing discussion (Briffa et al., 2020). The toxicity and health effect of heavy metals such as mercury (Hg), lead (Pb), cadmium (Cd), chromium (Cr), arsenic (As), Zinc (Zn), copper (Cu), and nickel (Ni) are extensively studied (Ahamad et al., 2021). However, the potential toxic metals or metalloids name lists are not limited to only these metals (Ahamad et al., 2021; Madhav et al., 2020).

Metals or their speciated composition originate naturally from volcanism, bedrock weathering, soil erosion, sea-spray as well as anthropogenically from industrial activities, agricultural and medical associated drainage, metal mining, smelting, and refining (Sezgin et al., 2004; Ahmed and Ishiga, 2006; Banerjee, 2003; Faiz et al., 2009; Wei and Yang, 2010). The entrainment of these produced metals (metalloids) to the environmental system occurs via deposition of atmospheric particulates, discharge of metal enriched sewage, run off from acid mine drainage and leaching (Ahamad et al., 2021; Madhav et al., 2020). The latter all sources directly introduce the metal concentration into the receptor environment, while atmospheric transport and deposition is an indirect way. Since the Earth's history heavy metals are present in the natural surrounding at background concentration, having a lesser adverse effect to the life. A notable number of studies have concluded that potential toxic elements pollution not only degrades atmospheric quality, water bodies, soil, and food crops but also threaten the whole environment through their interaction with the food chain (Dong et al., 2011; Nabulo et al., 2010; Li et al., 2014).

In the modern era of rapid industrialization and urban expansion, it could not be possible to completely avoid any toxic metal exposure. Even though if individual did not occupationally exposed. Srivastava et al. (2018) highlighted the existence of heavy metals in ambient air and its adverse impact on human health due to both acute and chronic exposure. Several common cases of lead, cadmium, arsenic, and mercury exposure has been reported across the globe (Fernandez-Luqueno et al., 2013). Caussy et al. (2003), summarizes and consolidates the knowledge about the metal exposure, bioavailability, and associated risk factor. And their study also presents unified approach to investigate and control the outbreak due to metal poisoning.

This chapter will discuss more on the sources, emission, transportation, and subsequent deposition of heavy metals through the atmospheric pathway. The transportation and deposition of the metal-bound particles or their vapor phase, distant from their origin, depends on the circulation and strength of the prevailing air masses. Atmospheric particle may fall-out directly into water resources (such as river, lake, stream, community pond, or uncovered well), or indirectly through storm water run-off.

9.2 Pathways and fate of airborne heavy metals

The global race for economic development and associated anthropogenic activities has significantly changed the emission rate of metal bound particles or gases. The recent changed emission scenario has become a great cause of concern at all scales (Jaishankar et al., 2014). Despite the measures taken to reduce emissions of pollutants, their release into the atmosphere and deposition onto the terrestrial ecosystems remains significant (Al-Khashman et al., 2013; Bacardit and Camarero, 2009).

Surface based emissions of metal bound particulates or their vapor to the atmosphere is well known source of heavy metals into air. Metal bound fine particulate matter is ubiquitous in the atmosphere, while the gaseous or vapor phase emission of these metals is limited due to their low volatility. However, anthropogenic processes involving high temperature such as: fossil fuel combustion, smelting and refining, substantially increased the emission rate of heavy metals in gaseous phase too. This latter under the atmospheric dynamics and chemical kinetic processes stabilizes as crystalline particle through gas-to-particle conversion mechanism of nucleation. The physical and chemical property of both, directly emitted as well as nucleated particles, decides their atmospheric residence time and concentration. Since our atmospheric system is under regular turbulence and diffusive motion, particles cannot stay there forever. Metal-bound particulates return to the surface via a variety of dry and wet depositional processes (Tsuda et al., 2013; Herrmann et al., 2015). Settling process of particles under the influence of gravitation, impaction and diffusion is known as dry deposition, while wet deposition occurs through precipitation or rain wash. Wet deposition is an efficient mechanism for the removal of both vapor phase and particulate bound matter via rain out or wash out from the atmosphere. In rain-out process, particles are scavenged through the participation as cloud condensation nuclei for cloud droplet formation, whereas wash out is the collection and deposition by falling rain droplets (Wu et al., 2018).

The atmospheric deposition process is a function of physical characteristics of metals (or metalloids) such as size, vapor pressure, solubility, etc., and size of the metal-bound particles. All these parameter and atmospheric dynamics cumulatively favor the kind of deposition: that is, wet or dry. Finer fraction and gaseous particles are lofted high into the troposphere where they participate in the formation of raindrops, and are scrubbed out by precipitation. While the larger portion, as coarse soil materials or dust particles comparatively do not reach higher altitudes and thus undergo dry deposition.

Atmospheric heavy metals concentration entering aquatic system first trapped within surface microlayer (i.e., typical thickness of $1-1000\ \mu m$). As surface microlayer serves as source and sinks for materials through active interaction between aquatic surface and atmosphere (Chance et al., 2018). This subsurface microscopic layer has presence and abundancy of active molecules and microorganism that act as active trapper and mobilizer for the input heavy metals flux. Further, chemical interaction with several oxides, sulfides, carbonates, and ion exchange, sorption by minerals and scavenging by organic matter, make it available for biosorption (Chance et al., 2018). The degradation of organic matter and subsequent release of bound metals favors rapidly in aerobic condition whereas it is slow in anaerobic condition. Primarily the bryophytes, small nonvascular plant (such as mosses) efficiently uptake the released metals into their cell walls. Metal pollutants are get absorbed into the tissues of bivalve's mollusks (Gupta and Singh, 2011) and in gills and other organ of variety of fish species (Zeitoun and Mehana, 2014). Heavy metal pollutant also has profound impacts on the aquatic community structure in terms of drastic reduction in intolerable species and abruption of the tolerable one, mainly in the case of highly contaminated stream (Fatima et al., 2014). The induced changes in the aquatic species behavior, growth and diversity, adversely affects the food-web and ecosystem functioning.

Also, a significant fraction of heavy metal concentration deposited to the bottom, as primary sediment layer, where it is subject to diagenesis modification. The redox potential of sediment and pH of overlying water further determine the fixing and remobilization of metals from the sediment to above water (Fatima et al., 2014). Where the chemistry of water such as hydrogen ion activity, hardness and salinity govern the speciation of metals in aqueous solution. It has been recognized that the chemical speciation of heavy metals determines its bioavailability, environmental mobility and biogeochemical behavior (Fytianos, 2001). Potential risk of the living organism is strongly dependent on the available chemical species of the metals in the system. Studies (Rao et al., 1992; Connan et al., 2013; Ye et al., 2018) pertaining to chemical characterization of funnel-based collection of dry particulate matter and bulk rain-water chemistry confirms the presence of heavy metals (or metalloids), in both dry and wet atmospheric deposition.

9.3 Anthropogenic sources of heavy metals emission

According to the World Health Organization (WHO, 2011), guidelines on heavy metals compiles a list of 20 major metals that are of serious concern with respect to the health of all lives. In increasing order of their atomic number list includes: Beryllium (Be), Aluminum (Al), Chromium (Cr), Manganese (Mn), Iron (Fe), Cobalt (Co), Nickel (Ni), Copper (Cu), Zink (Zn), Arsenic (As), Selenium (Se), Molybdenum (Mo), Silver (Ag), Cadmium (Cd), Tin (Sn), Barium (Ba), Mercury (Hg), Thallium (Ti) and lead (Pb). Some of the listed metals are categorized as essential while other as nonessentials nutrients based on the need and nutritional value for living organism. However, the excessive levels or certain oxidation state of even essential nutrient become detrimental for the organism and ecosystem, thus listed as metal of concern. Nonnutritional metals such as Ar, Be, Pb and Hg are recognized toxic at all levels and form. The metals toxicity and health-based assessment must include both essential and nonessential nutrients metals. These metals (metalloids) are ubiquitous in every sphere of this planet and exposed to the living organism via air, water and food. Their sources and emission include both natural and anthropogenic processes that are explicit in varying spatio-temporal concentration. However, in most of the cases anthropogenic emission of metals exceeds natural background levels. Here some of the heavy metals' anthropogenic processes and subsequent emission sources has been listed below in Table 9.1. Among all anthropogenic sources, ore processing, smelting and refining operations are categorized as major emission sources that causes huge deposition of heavy metals such as Pd, Zn, Cu, As and vapor Hg. Local community scale open metallurgical activities are another potent source, that usually get unnoticed. Metals are emitted to the atmosphere either as vapor or particulate matter or in case both. Some volatile metals such as mercury and selenium are mostly emitted in vapor form that later condense onto available particles, while other such as cadmium and lead emitted in both form. The condensation of metal vapor mostly happens onto the process along generated ash particles. Some metals are rigid and nonvolatile such as chromium, that does not vaporize and thus emitted as particle along ash emission.

Along all the mentioned anthropogenic emissions in Table 9.1, natural sources (such as surface dust, volcano and forest fires) also contribute a significant fraction of heavy metals to the atmosphere. Apart from the direct emission of particulate as primary matter to the atmosphere, secondary particles also formed through physiochemical conversion mainly favored by the meteorological factors. However, it is still unknown and has to be

TABLE 9.1 Some typical anthropogenic sources of heavy metals emission to the atmosphere (heavy metals column represented as atomic number, name and symbol).

	Heavy metal	Sources of emission
Essential nutrient metals (Harmless at minute level)	26-Iron (Fe)	Ore-smelting and refining, metal casting, alloy processing, construction, urban transportation and manufacturing
	27-Cobalt (Co)	Alloy processing, open mining activities, coal combustion, pharma alloys manufacturing
	28-Nikel (Ni)	Metallurgical processes using nickel, electroplating with nickel, incineration, fuel combustion (coal or oil)
	29-Copper (Cu)	Iron and steel processing, metal polishing and plating, fly ash, fertilizers, mine discharge, disposal of municipal waste
	30-Zinc (Zn)	Oil refineries, galvanizing process, brass manufacturing unit, urban traffic, plumbing material manufacturing
Nonessential metals(Toxic at all levels)	24-Chromium (Cr)	Dyeing catalyst in textile industry, steel fabrication unit, electroplating, metallurgy, chemical industry, cement production unit
	33-Arsenic (As)	Coal-fired power plant, arsenic containing waste incineration, metal smelting, mining activities, agricultural pesticides and herbicides
	48-Cadmium (Cd)	Nonferrous metal production, municipal incinerators, welding, electroplating, metal refining units, fertilizer, nuclear plant
	80-Mercury (Hg)	Fuel combustion, paper and paint industry, metal refining, cement production, mercury based pesticides, mercury containing biomass burning
	82-Lead (Pb)	Automobile emission, lead smelters, high temperature fossil combustion, paint industry, mining activities

determined that whether the metal-bound secondary particle formation get affected or not, under the same meteorological influence. Because the secondary particle formation occurs from their atmospheric precursors (as in case of other gaseous molecules such as sulfate, etc.), favors readily scavenging of gaseous molecules from the atmosphere via participating in cloud droplet formation process. As hygroscopic air borne particles act as cloud condensational nuclei or as cloud seed nuclei for the vapor condensation and droplet growth. In the same process the dissolution and adsorption of the metals bound particles in cloud water droplet, make it efficiently rain out from the atmosphere.

9.4 Emission and atmospheric transportation of heavy metals

In fact, the anthropogenic emission of heavy metals, bound with other pollutants, started very early with the domestication of fire by human beings. As tracing the history of global metal pollution, Prof. J. O. Nriagu, in his popular article (Nriagu, 1996) has well regarded the release and deposition of metals concentration in the cave environment during burning of firewood. Later the industrial revolution powered by technological advancement and high temperature combustion process has exponentially increased the emission of metal concentration. The above mentioned literature well presented the historical change in the metal mining and emission of Cu, Pb and Zn level to the atmosphere since 1850–1990. The metal production and emission trend during 1850–1950 were almost consistent or stable except for Pb emission that dramatically increased after 1920. In the Ancient, less sophisticated metal mining and ore processing techniques constrain the production as well as consumption that actually limited the past emissions. After 1950, onset of the technology geared industrial revolution, brought about unprecedented production, consumption and consequent emission of the heavy metals to the atmosphere.

A schematic diagram of potential in-land sources of heavy metals, emission and their atmospheric long-range transportation and deposition pathways has been illustrated in Fig. 9.1. In this representation only the atmospheric pathways of deposition have been highlighted excluding storm water run-off or other surface discharge sources, though, they too contribute a significant fraction to water resources contamination. Storm water or localized sewage discharge sources are mainly categorized as point-sources, while open atmospheric deposition is considered as nonpoint sources. The metals input from the point sources longitudinally disperse across the water column and later settled down to bottom sediment layer via sorption with sinking material. In aquatic environment,

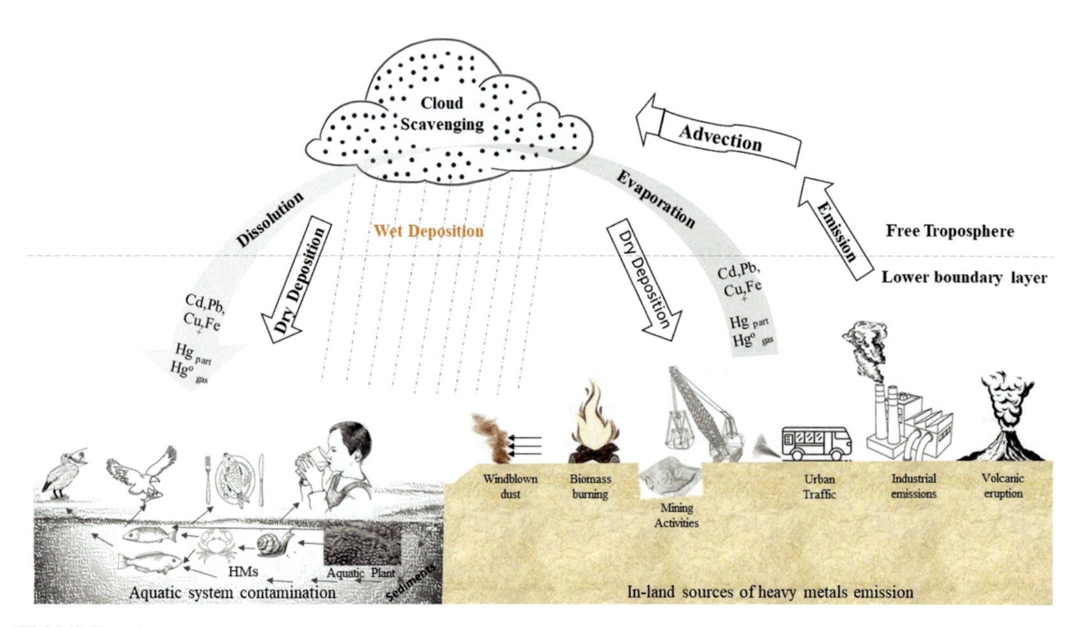

FIGURE 9.1 Schematic representation of heavy metals emission, transportation and deposition via atmospheric pathways to the aquatic system.

sediment materials act as scavenging agent for heavy metals. Thus, the metals concentration in sediment has been reported many times greater than in water column level. While the direct atmospheric deposition of the metals to the aquatic system and subsequent dispersion across column, their speciation and settling process has not been comprehensively understood so far.

The potential in-land sources mainly include ore extraction process, smelting and refining process that usually operated on high temperature in combustion chamber and releases substantial number of toxic metals such as: Co, Cu, Zn, Pb, As, Cd, Hg, etc., to the atmosphere. Apart from metal processing industry, agricultural pesticides, fertilizer, tanneries, batteries, electroplating, paint material, welding and pharma-lab discharge are recognized as other anthropogenic potent sources of heavy metal emission. The urban traffic emission, trapped within street canopies, is another major hotspot of the heavy metals such as: Cu, Ni and Zn from the tearing tire and brake wear, Pb from tailpipe exhaust and resuspension of metal bearing dust particles. It is important to note that whether its urban or remote location, the wind-blown metal-bound particle and their resuspension to the atmosphere, cannot be disregarded and they must be accounted in cumulative concentration assessment of metals. In both remote and urban locations, recent land-use and land-cover pattern has over exploited and exposed the soils surface. And the ongoing temperature rise trend has decreased the soil moisture content over increased evapotranspiration (Jeričević et al., 2012). As a consequence, higher natural emission and resuspension of crustal soil dust particles to the atmosphere is witnessed.

Regardless of the source of emission at the surface, once the metal-bound particles or in vapor-phase emitted to the atmospheric system, their fate comes under the purview of large-scale atmospheric processes and dynamics. The pollutants released close to the earth surface, do not travel far due to the surface topography, hindrance and confined air flow. While the stack/chimney emission, released the pollutant at a height, that directly come under the influence of atmospheric vertical mixing and consistent air circulation. Though the varying height of stack/chimney and prevailing weather condition, make a difference about how far the pollutants be advected. Meteorological conditions such as wind speed and direction, temperature, relative humidity, seasonal solar insolation and misty and clear days greatly influence the atmospheric accumulation, transportation and scavenging of the metal-bound particulates (Nimir, 2006).

During dry winter months lower temperature, moderate humidity and minimal solar insolation leads to high atmospheric stability that further supports to frequent formation of low-level temperature inversion. In inversion condition, lower tropospheric boundary layer mixing shifted low compared to summer, that significantly restrict the thorough mixing and dilution of the released pollutants. This condition is worst case scenario for the air-borne particulate accumulation and near surface deposition or direct exposure to the living beings. During summer in high temperature and humid condition, boundary layer deepens enough to favor thorough mixing and ambient ventilation of the pollutants. However, in summer prolonged dryness of upper soils and strong winds also increase the particulate bound metal emissions via erosion and resuspension of crustal elements such as: Al, Si, Ca, Ti, Mn, and Fe. From top soil surface emitted particles lofted to the upper tropospheric circulation system, through which long range transportation occur. Later progression of monsoon precipitation check out and control the atmospheric loading of pollutants via wet scavenging process. Thus, it can be concluded that during winter air borne particulate concentration will accumulate and cannot be advected far away from their origin (Morawska and Zhang, 2002). While in summer and monsoon both season witness strong wind flow favors long range transportation of the particles, however monsoonal rain efficiently washes out in due course of time. Another reason of particle-bound metals long range transportation is their atmospheric chemistry and residence time. Unlike organics, heavy metals are nondegradable and chemically inactive, even under the complex atmospheric chemistry process that involves the oxides, radicals and active ozone molecules. Neither these pollutants dissociate under photodissociation. This causes them to be persistent for long and thus they can be potentially long range transported (Morawska and Zhang, 2002). In this way air borne heavy metals affect even the most remote regions from the origin in a shorter span of time.

An assessment of transboundary pollution of heavy metals over the European region, presents clear evidence of long-range transportation of metal-bound particles (Travnikov et al., 2020). With an intention to protect the humans and their surrounding against harmful atmospheric pollutant, in the year 1983, a Convention on Long-range Transboundary Air Pollution signed and came into force. That directs guidelines pertaining to reduce and prevent the emission and their long-range transboundary transportation. The heavy metals addressed by the convention priority also include Pb, Cd and vapor Hg, that are characterized as longer residence time atmospheric pollutants. Atmospheric physiochemical processes and circulation dynamics play a critical role in dispersing and depositing these particles to the ground either through wet or dry mechanism. Eventually the atmospheric laden particle-bound metal (or metalloids) or condensed particle wash out in rain or get scavenged in hydrometeors. As wet deposition is more efficient mechanism of scavenging of atmospheric particles via precipitation, the receiving water resources both (above surface or underground) will be directly contaminated by the water-soluble elements. Whereas the dry deposition of particulates directly or indirectly through storm water runoff eventually input to nearby water bodies.

Past several studies has provided details insights into discharge of heavy metals via storm run-off and consequent water resource contamination (Gunawardena et al., 2013; Weerasundara et al., 2015; Sankhla et al., 2016; Liu et al., 2018; Rani et al., 2022). Experimental results have demonstrated that atmospheric deposition of metals persistently damage the aquatic ecosystem, whether it is at urban or rural locations (Rao et al., 1992; Mehra et al., 1998; Banerjee, 2003; Thakur et al., 2004; Karar et al., 2006; Sharma et al., 2008; Kulshrestha et al., 2009; Meena et al., 2014; Arshad et al., 2015). Analysis of water soluble or insoluble fraction of deposited atmospheric aerosols contains diverse range of metals concentration (Sah et al., 2019; Kumari et al., 2021), that even exceeds the permissible limit in cases of urban pollution. However, the collocated simultaneous measurements of atmospheric deposition of dry and wet fraction are still lacking at higher spatio-temporal Scale. The separate measurements and analysis of air borne metal's fraction enables to examine the metal's chemical fractionation, bioavailability, and associated health risk factor. Also, the deposition pattern of particles indirectly determines its physical characteristics such as size range, vapor pressure, and solubility. The separate depositional measurement will infer the comparative fraction (dry vs wet) of heavy metals concentrations.

9.5 Heavy metals deposition and contamination to water resources

Heavy metal contamination to the water resources is an ancient (due to natural release) problem that is now more severe (anthropogenic release) at global scale (Meybeck, 2013; Rani et al., 2022). Studies have been undertaken illustrating the metal's sources, dispersal and contamination to aquatic resources (Singh et al., 2020). Based on the sediment archives and depositional trajectories analysis at period of time such information regarding the origin and dispersal pathways of metals are inferred. Direct atmospheric deposition (wet and dry) or indirectly via surface storm water run-off of heavy metals produce changes in the physio-chemical conditions of the receiving water body (Hossain et al., 2021). The metals toxicity and lethal or sublethal deterioration impact to the organism, is mainly depends on the type of speciation and concentration. In distant locations far from any direct sources, the reported metal concentration in aquatic environment is low that causes sublethal or chronic exposure damages. While in or nearby the urban polluted environment, the toxic metal concentration is high enough to cause acute exposure and death of the several organisms.

The aquatic organisms respond to such alteration in terms of changes in species composition or decrease in their diversity or density. Based on the metals metabolization and speciation in receiving aquatic environment, initially up taken by the microinvertebrate phyto communities. The other higher microinvertebrate organism uptake the metal concentration either by direct food ingestion or indirectly through absorption/adsorption via tissues. Due to persistent nature of metals species, get enriched by organisms and converted to organic complexes which becomes even more toxic than earlier species. Gradually the metal concentration increases with accumulation and transferring processes through interlinked vertical (trophic levels) and horizontal (diet at a particular trophic level) position within food webs. At particular trophic levels, species based on their ability to cope with increased metals level and exposure pathways changes their abundances and diversity. And consequently, the aquatic food web structure gets changed by the replacement of tolerable abundant species. Although, some unicellular species has ability to adapt with higher metal levels through forming resistant spores. But due to poor knowledge about their specific life history and other intrinsic factors such as size, nutritional status, it becomes difficult to analyze the response of these species with respect to metal concentrations. Due to accumulation and magnification characteristics, diet-borne metals transferred along the food chain that eventually produce the greatest risk to upper trophic level organisms.

The cumulative assessment of heavy metals in aquatic system is typically done by examining the presence of heavy metals in water column, under-lying sediments layer and in biotic species (Gaikwad and Kamble, 2014; Pandiyan et al., 2021; Rubalingeswari et al., 2021). Dallinger et al. (1987), well reviewed the concept of bioavailability and uptake of heavy metals by the fish species. As fish are among the top consumer of aquatic trophic level, they are highly exposed and vulnerable species due to metal uptake via alimentary tract. Fishes are also major food items (along with mollusk, prawn, crab, frog, and others) for local or migratory species of birds and animals, that may get heavily affected with their dietary preferences (Gaikwad and Kamble, 2014; Pandiyan et al., 2021). Eventually humans at highest trophic level, having particular interest in fishes and other aqua-food products as their nutritional diet, get exposed to the heavy metals.

9.5.1 Contamination to surface water resources

Since the early civilization water body such as: lakes, ponds and river basins serve as main fresh water resources for the nearby residing community. They are highly vulnerable one because of the regular uptake and consumption of fresh water and food from such contaminated resources. The uptakes of metals contaminated water by the communities are associated with range of illness and disorders (Mohod and Dhote, 2013). In past it has been reported that long term exposure of metals levels via contaminated water resources to local communities have resulted in severe illness like Alzheimer's and Parkinson's disease, muscular dystrophy and multiple sclerosis. Acute exposure of heavy metals to humans have shown symptom of damaging their central nervous system, cardiovascular and gastro-intestinal system, kidneys, liver and bones (Mohod and Dhote, 2013). River basins are considered as reservoir or sink of atmospheric deposition or land run-off. Rivers have been regarded as cradle of human civilization since the immemorial (Rani et al., 2022). Millions depends on this resource for their daily water need and livelihoods, while countless others connect it to their physical and spiritual sustenance. But the recent trend of anthropogenic activities and consequent pollutants emission has led to deterioration in river water quality, especially in terms of heavy metals load. Indian main-land is blessed with many major and minor river basins. The Himalayan Rivers are perennial (fed by glacier melting), while the peninsular plateau rivers are rain fed basin. The basin catchment area receives a considerable number of pollutants via precipitation or through run-off. Heavy metal contamination in these river basins is increasingly becoming common in India. Paul (2017), presents a comprehensive review of the past experimental works that has been done by the several researcher to assess the metals level in one of the India's sacred river the Ganga. Tomar and Upadhyay (2018), attempt to review the assessments of heavy metals in river Yamuna.

All these attempts to assess the metals level across the Indian River is solely based on the past experimental work that is completely heterogeneous with time. The recent edition of 2019 on status of toxic metals in Indian river prepared by the Central Water Commission (Central Water Commission, 2019) herald to an alarming future pertaining to Indian river water quality. At the same its demands to ponder for remedial measures to check the pollution at the earliest. Based to the same report, out of the 424 chosen water quality sampling stations spread across the Indian rivers, 287 stations water quality were beyond the acceptable BIS standard mark, due to presence of toxic metals. At most of the sample location Iron concentration was reported beyond the acceptable limit. In the below table four common metals concentration analyzed and reported in eight major river basin of India has been presented, apart these four other metals levels were reported within the limit. However, more concerned Speciation of the Chromium (+3 & +6) and Arsenic (+3 & +5) need to be further assessed (Table 9.2).

So far it has been concluded that water quality of most of the Indian main-land river basin, particularly at some stretch, have been adversely affected by the anthropogenic activities. And all these has been understood based only on the water sample analysis, though it's not a complete picture. Comprehensive investigation must include analysis of samples from atmospheric deposition, metal fractionation from below sediment and in water column. Several past studies (Wong et al., 2003; Connan et al., 2013; Weerasundara et al., 2015; Liu et al., 2018; Gafur et al., 2018; Ye et al., 2018; Pandiyan et al., 2021) have reported the heavy

TABLE 9.2 Pronounced metals concentrations: Cd, Cr, Pb (in μg/L) and Fe (in mg/L) reported along the major river basin of India (Hussain and Rao, 2018).

River basin	Cd	Cr	Pb	Fe
Ganga	0.001–3.936	0.080–205.82	0.020–36.910	0.002–1.53
Yamuna	0.002–9.166	0.010–36.370	0.010–22.670	0.001–0.613
Brahmaputra	0.002–1.314	0.040–53.100	0.020–21.480	0.008–9.872
Narmada	0.002–1.201	0.080–26.660	0.080–21.930	0.002–1.312
Krishna	0.006–2.708	0.050–98.350	0.260–14.340	0.008–0.396
Cauvery	0.008–1.609	0.050–35.360	0.010–16.670	0.004–0.416
Godavari	0.009–1.489	0.010–21.510	0.020–22.870	0.008–0.670
Mahanadi	0.015–0.836	0.260–31.620	0.040–7.180	0.009–0.557

BIS (2012): Cd: 3 μg/L, Cr: 50 μg/L, Pb: 10 μg/L, Fe: 0.3 mg/L. http://cgwb.gov.in/Documents/WQ-standards.pdf.
Compiled from Hussain, J., Rao, N.P., 2018. Status of trace and toxic metals in Indian Rivers. Ministry of Water Resources, Government of India, River Data Compilation-2. Directorate Planning and Development Organization, New Delhi, 110066.

metals atmospheric load and their bulk deposition (both wet and dry) and consequent contamination to soil and water resources. Either via wet or dry deposition, it consists diverse range of heavy metal concentrations. It is mainly influenced by the local meteorology such as prevailing wind direction and strength of atmospheric wash out pattern (dry or wet) and distance from the emission sources (Meena et al., 2014). Although wet deposition has higher percentage of dissolved metal fraction than dry (Weerasundara et al., 2015), that can directly affect the water resources. Even the deposition of metals on soil or impermeable surfaces will be channelized through storm water to nearby receiving water body. Therefore, more in-depth studies, especially collocated simultaneous dry and wet deposition sampling and analysis of heavy metals are necessary for quantitative estimation of atmospheric deposition of heavy metals.

9.5.2 Contamination to groundwater resources

After open surface water resources (such as river, lake or community ponds) groundwater is second most vital source of fresh water supply, especially in arid and semiarid regions, where surface water sources are mostly depleted or dried-up (Singh et al., 2016). Recent ongoing trend of overexploitation and contamination of groundwater sources has posed another threat for safe water availability, even for drinking and household uses at large. Industrial release of nontreated effluents and subsequent seepage of these pollutants through sediment pores into deep aquifers significantly alter the physiochemical properties of water (Annapoorna and Janardhana, 2015; Ehya and Mosleh, 2018; Mahmoudi et al., 2017; Alexander et al., 2017). Although groundwater usually contains minor contamination of metals from the overlying parent rocks or minerals along the aquifers. But human mediated variation in surface flux to deep aquifer percolation has increased the concentration of heavy metals into the underlying water (Amalraj and Pius 2018; Madhav et al., 2021).

As discussed earlier the tendency of heavy metals to bioaccumulate in the food chain is the most dangerous characteristics of them, thus the knowledge of groundwater quality is vital to determine its fitness for use (Ahamad et al., 2018).

Through atmospheric pathways the deposition of air-borne heavy metals or metalloids via precipitation to groundwater usually occurs in less acidic condition (pH ~6.5 to 8.5) that become insoluble. Thus the increase in heavy metals concentration into deeper aquifer is not a serious threat because of its immobility (Tiller, 1989). While in shallow and acidic condition metals are more soluble and mobile thus poses greater risk of exposure. Current stage of metal bound particles or vapor emission and their long-range transportation has increased chances of deposition and subsequent percolation to the shallow groundwater resources. Although its takes longer to contaminate the aquifer compared to direct deposition into open surface water resources. Recent assessment of human health's cause and effect relationship with respect to groundwater quality and metal contamination has invited greater attention across the globe, as it continue exceed the permissible standard limits (Adamu et al., 2014; Jiménez-Ballesta et al., 2017). Past studies have concluded that high temperature combustion processes in thermal power plants considerably release a number of toxic metals either bound with fly-ash or in gaseous phase with sulfur and nitric oxides into the atmosphere (Usmani and Kumar, 2017; Sarode et al., 2010; Gao et al., 2013; Nanos et al., 2015). Fly-ash emitted from the thermal plants contains several metals like arsenic, mercury, cadmium, lead, selenium, chromium and nickel that vaporized during combustion from parent coal (Lemly, 2018). A fraction of emitted ash is captured and disposed-off through settling ponds with effluent outlet that directly cause pollution to the nearby soil or water bodies. While the uncaptured ash fraction and gaseous vapor travel long away from the source region and eventually get precipitated and infused to the soil and between rocks where groundwater table are located. Thermal power plants are characterized as elevated point source, that emit particles or gases at certain height through stack or chimney, that have greater possibility to long-range transportation. On the other hand, emission of heavy metals from urban traffic density have lesser chance to drift far away, due to near surface emission and obstruction by urban built-up. Vehicular emission and resuspension of metal bound solid particles mostly drifts along the urban street canopies that later channelize through storm water run-off and infuse to the underlying bottom sediment (Nawrot et al., 2020).

9.6 Caveats in monitoring and modeling of air-borne heavy metals

Heavy metals contamination owing to increasing anthropogenic sources is ancient and now is persistent global issue, having environmental and human health implications. A global assessment by Galloway et al. (1982), presents the detailed information about nineteen metals' atmospheric levels, their transportation, and modes of deposition. The spatio-temporal trend in atmospheric loading and subsequent deposition was assessed based on the available global study to infer whether these concentrations threaten human or organism health. It also highlights the cases of dry fallout of metals found to be significant relative to the wet scavenging. In the last couple of decades, the global anthropogenic activities and emission scenario has considerably changed. Particulate-bound or in vapor phase heavy metals emissions and their long transportation and deposition, distant from their source region, has become issue of paramount concern. Suvarapu and Baek (2017), presents an extensive global review regarding the determination of particulate-bound heavy metals levels in ambient air at different global region. That clearly pointed the need to determine the total as well as speciated concentration of air borne metals, as most. Because total concentration doesn't sufficient to establish cause and effect relation with respect to human health. For example, trivalent Cr is of less concern than carcinogenic hexavalent Cr. Most of the past studies has been done with the total bulk concentration assessment that cannot be sufficient to infer about their differentiation and consequent toxic impacts.

Apart instrumental measurement and characterization of particulate-bound metals or their speciated composition there is need of deterministic modeling approach for source apportionment, deposition rate prediction and health risk assessment. On the other hand, atmospheric borne particles undergo several changes during their course of transportation, such as dissolution in condensed water, evaporation under solar radiation, photolysis and coating with different elements or complexes. So all these transformational processes need to be better understood through physical and mathematical based processes modeling. Although, the physiochemical transformation of heavy metals bound to air-borne particles or in condensed vapor phase has not been studied so far. A critical review by Popoola et al. (2018), presents the caveats in the present modeling framework of particle-bound metal's or metalloid's source apportionment, transportation and deposition rate prediction. Several assumptions made during the development of these models, is constraining the models, to make better rate prediction and source apportionment. There is need to develop and present more nonlinear processes than the over-simplified one in the existing models.

So far numerous literatures are available on various methodologies for assessing atmospheric deposition of metals. This enabled to determine the increasing order of heavy metals concentration in the atmospheric deposition at large scale. However, there is need to do comparative analysis of estimated concentration under different methodologies, this will enhance the capability of individual methodology in terms of metals levels quantification. Also, there is need for intensive regional as well as global monitoring network to acquire more metals data at higher spatio-temporal resolution to better assess the global atmospheric burden of metals (Li et al., 2013). The sampling procedure and analytical methods must be standardized for both wet and dry deposition through-out the network. Further studies must include more metals species including those that takes different forms in the atmosphere.

9.7 Conclusion and future possibility

The problem of atmospheric heavy metal pollution and deposition is not going to disappear in overnight. Because all the potent sources cannot be removed at one go, considering all these aspects various treatment methods have been introduced to check and mitigate the emission. Now-a-days several remediation techniques have been in practice for removal of heavy metals from the contaminated systems (air, water and soil), based on physical, chemical, and biological uptake processes. So far, no single technique has been recommended for full proof check and with cent percent efficiency. Every single methodology has certain limitation, in terms of operating cost, treatment time consumption, set-up and logistical problem (Vareda et al., 2019). The production of secondary waste is another major problem with the most conventional treatment methods that demands additional cost for remediate that secondary waste. Most of the remedial measures includes in-situ immobilization of the metals (metalloids) species through phytoremediation or biological techniques (Dhaliwal et al., 2020).

The conventional method and modified adsorbent procedure for the removal of heavy metals from the aquatic system has been well reviewed in Vardhan et al. (2019). That briefly cover the pros and cons aspects of both conventional and modified adsorbent processes and highlights the need for further improvement on low-cost adsorbent remediation method. Recent application of aerogel as remediator is another promising way to deal with heavy metals pollution and contamination. Aerogels are specific class of three-dimensional porous material with high porosity, high surface area and low density. That can be used as an efficient adsorption media for removal for several pollutants including metals ion from the environment (Maleki, 2016). So far, different material derivatives aerogel has been synthesized and employed to treat pollutants from the system. Mostly used aerogels are derivative of cellulose, chitosan, graphene oxide and silica (Franco et al., 2021). Silica based aerogel have been used to remove heavy metals such as copper, lead, cadmium, chromium, and nickel from the water bodies (Vareda and Durães, 2019).

This technique can also be extended to employ at industry stack emission to capture the particulates and vapor bound heavy metals. The atmospheric loading and deposition of toxic metals can only be controlled by reducing its emission at the source region. For example, the stringent regulation pertaining to unleaded petrol and employment of catalytic converter in exhaust manifold of tailpipe has significantly reduce the lead emission. Similarly, to reduce the particulate matters emission from industrial stacks Cyclone precipitator, wet scrubber, electrostatic precipitator, fabric filtration and adsorption and absorption are some widely applied techniques. Despite the scope of severe air-borne heavy metals problem at global scale, little and heterogeneous experimental assessment has been done so far. In future it would be recommended:

1. Extensive monitoring network should be established to acquire more and homogenous data, including the analysis of metals speciated forms, to make cumulative toxicity assessment.
2. Simultaneous and collocated atmospheric dry versus wet deposition fractional should be determined.
3. The experimental methodologies must be standardized to achieve more accuracy in assessment. In case of differences in methodologies, comparative study should be recommended to enhance individual robustness.
4. More nonlinear mathematical modeling framework should be developed by reducing over simplified assumptions. So that the metals source apportionment and depositional rate prediction can be made precisely.

References

Adamu, C.I., Nganje, T., Edet, A., 2014. Hydrochemical assessment of pond and stream water near abandoned barite mine sites in parts of Oban massif and Mamfe Embayment, Southeastern Nigeria. Environmental Earth Sciences 71 (9), 3793–3811.

Ahamad, A., Madhav, S., Singh, P., Pandey, J., Khan, A.H., 2018. Assessment of groundwater quality with special emphasis on nitrate contamination in parts of Varanasi City, Uttar Pradesh, India. Applied Water Science 8 (4), 1–13.

Ahamad, A., Raju, N.J., Madhav, S., Gossel, W., Ram, P., Wycisk, P., 2021. Potentially toxic elements in soil and road dust around Sonbhadra industrial region, Uttar Pradesh, India: source apportionment and health risk assessment, Environmental Research, 202. p. 111685.

Ahmed, F., Ishiga, H., 2006. Trace metal concentrations in street dusts of Dhaka city, Bangladesh. Atmospheric Environment 40 (21), 3835–3844.

Alexander, A.C., Ndambuki, J., Salim, R., Manda, A., 2017. Assessment of spatial variation of groundwater quality in a mining basin. Sustainability 9 (5), 823.

Al-Khashman, O.A., Jaradat, A.Q., Salameh, E., 2013. Five-year monitoring study of chemical characteristics of wet atmospheric precipitation in the southern region of Jordan. Environmental Monitoring and Assessment 185 (7), 5715–5727.

Amalraj, A., Pius, A., 2018. Assessment of groundwater quality for drinking and agricultural purposes of a few selected areas in Tamil Nadu South India: a GIS-based study. Sustainable Water Resources Management 4 (1), 1–21.

Annapoorna, H., Janardhana, M.R., 2015. Assessment of groundwater quality for drinking purpose in rural areas surrounding a defunct copper mine. Aquatic Procedia 4, 685–692.

Arshad, N., Hamzah, Z., Wood, A.K., Saat, A., Alias, M., 2015. Determination of heavy metals concentrations in airborne particulates matter (APM) from Manjung district, Perak using energy dispersive X-ray fluorescence (EDXRF) spectrometer, AIP Conference Proceedings, 1659. AIP Publishing LLC, p. 050008, 1.

Bacardit, M., Camarero, L., 2009. Fluxes of Al, Fe, Ti, Mn, Pb, Cd, Zn, Ni, Cu, and As in monthly bulk deposition over the Pyrenees (SW Europe): the influence of meteorology on the atmospheric component of trace element cycles and its implications for high mountain lakes. Journal of Geophysical Research: Biogeosciences 114 (G2).

Banerjee, A.D., 2003. Heavy metal levels and solid phase speciation in street dusts of Delhi, India. Environmental Pollution 123 (1), 95–105.

Briffa, J., Sinagra, E., Blundell, R., 2020. Heavy metal pollution in the environment and their toxicological effects on humans. Heliyon 6 (9), e04691.

Caussy, D., Gochfeld, M., Gurzau, E., Neagu, C., Ruedel, H., 2003. Lessons from case studies of metals: investigating exposure, bioavailability, and risk. Ecotoxicology and Environmental Safety 56 (1), 45–51.

Central Water Commission, 2019. Status of trace & toxic metals in Indian rivers. Ministry of Jal Shakti, Department of Water Resources, River Development, and Ganga Rejuvenation: New Delhi, India.1-302. http://cgwb.gov.in/Documents/WQ-standards.pdf.

Chance, R.J., Hamilton, J.F., Carpenter, L.J., Hackenberg, S.C., Andrews, S.J., Wilson, T.W., 2018. Water-soluble organic composition of the Arctic sea surface microlayer and association with ice nucleation ability. Environmental Science & Technology 52 (4), 1817–1826.

Connan, O., Maro, D., Hébert, D., Roupsard, P., Goujon, R., Letellier, B., et al., 2013. Wet and dry deposition of particles associated metals (Cd, Pb, Zn, Ni, Hg) in a rural wetland site, Marais Vernier, France. Atmospheric Environment 67, 394–403.

Dallinger, R., Prosi, F., Segner, H., Back, H., 1987. Contaminated food and uptake of heavy metals by fish:a review and a proposal for further research. Oecologia 73 (1), 91–98.

Dhaliwal, S.S., Singh, J., Taneja, P.K., Mandal, A., 2020. Remediation techniques for removal of heavy metals from the soil contaminated through different sources: a review. Environmental Science and Pollution Research 27 (2), 1319–1333.

Dong, J., Yang, Q.W., Sun, L.N., Zeng, Q., Liu, S.J., Pan, J., et al., 2011. Assessing the concentration and potential dietary risk of heavy metals in vegetables at a Pb/Zn mine site, China. Environmental Earth Sciences 64 (5), 1317–1321.

Ehya, F., Mosleh, A., 2018. Hydrochemistry and quality assessment of groundwater in Basht plain, Kohgiluyeh-va-Boyer Ahmad Province, SW Iran. Environmental Earth Sciences 77 (5), 1–21.

Faiz, Y., Tufail, M., Javed, M.T., Chaudhry, M.M., 2009. Road dust pollution of Cd, Cu, Ni, Pb and Zn along islamabad expressway, Pakistan. Microchemical Journal 92 (2), 186–192.

Fatima, M., Usmani, N., Hossain, M.M., 2014. Heavy metal in aquatic ecosystem emphasizing its effect on tissue bioaccumulation and histopathology: a review. Journal of Environmental Science and Technology 7 (1), 1.

Fernandez-Luqueno, F., López-Valdez, F., Gamero-Melo, P., Luna-Suárez, S., Aguilera-González, E.N., Martínez, A.I., et al., 2013. Heavy metal pollution in drinking water-a global risk for human health: a review. African Journal of Environmental Science and Technology 7 (7), 567–584.

Franco, P., Cardea, S., Tabernero, A., De Marco, I., 2021. Porous aerogels and adsorption of pollutants from water and air: a review. Molecules 26 (15), 4440.

Fytianos, K., 2001. Speciation analysis of heavy metals in natural waters: a review. Journal of AOAC International 84 (6), 1763–1769.

Gafur, N.A., Sakakibara, M., Sano, S., Sera, K., 2018. A case study of heavy metal pollution in water of Bone River by Artisanal Small-Scale Gold Mine Activities in Eastern Part of Gorontalo, Indonesia. Water 10 (11), 1507.

Gaikwad, S.S., Kamble, N.A., 2014. Heavy metal pollution of Indian river and its biomagnifications in the molluscs. Octa Journal of Environmental Research 2 (1).

Galloway, J.N., Thornton, J.D., Norton, S.A., Volchok, H.L., McLean, R.A., 1982. Trace metals in atmospheric deposition: a review and assessment. Atmospheric Environment (1967) 16 (7), 1677–1700.

Gao, K.Q., Chen, J., Jia, J., 2013. Taxonomic diversity, stratigraphic range, and exceptional preservation of Juro-Cretaceous salamanders from northern China. Canadian Journal of Earth Sciences 50 (3), 255–267.

Gunawardena, J., Egodawatta, P., Ayoko, G.A., Goonetilleke, A., 2013. Atmospheric deposition as a source of heavy metals in urban stormwater. Atmospheric Environment 68, 235–242.

Gupta, S.K., Singh, J., 2011. Evaluation of mollusc as sensitive indicator of heavy metal pollution in aquatic system: a review. IIOAB Journal 2 (1), 49–57.

Herrmann, H., Schaefer, T., Tilgner, A., Styler, S.A., Weller, C., Teich, M., et al., 2015. Tropospheric aqueous-phase chemistry: kinetics, mechanisms, and its coupling to a changing gas phase. Chemical Reviews 115 (10), 4259–4334.

Hossain, M., Ahmed, M., Liyana, E., Jolly, Y.N., Kabir, M.J., Akter, S., et al., 2021. A case study on metal contamination in water and sediment near a coal thermal power plant on the eastern coast of Bangladesh. Environments 8 (10), 108.

Hussain, J., Rao, N.P., 2018. Status of trace and toxic metals in Indian Rivers. Ministry of Water Resources, Government of India, River Data Compilation-2, Directorate Planning and Development Organization, New Delhi, 110066.

Jaishankar, M., Tseten, T., Anbalagan, N., Mathew, B.B., Beeregowda, K.N., 2014. Toxicity, mechanism and health effects of some heavy metals. Interdisciplinary Toxicology 7 (2), 60–72.

Jeričević, A., Ilyin, I., Vidič, S., 2012. Modelling of heavy metals: study of impacts due to climate change. In: Fernando, H., Klaić, Z., McCulley, J. (Eds.), *National Security and Human Health Implications of Climate Change. NATO Science for Peace and Security Series C: Environmental Security*. Springer, Dordrecht, pp. 175–189.

Jiménez-Ballesta, R., García-Navarro, F.J., Bravo, S., Amorós, J.A., Pérez-de-Los-Reyes, C., Mejías, M., 2017. Environmental assessment of potential toxic trace element contents in the inundated floodplain area of Tablas de Daimiel wetland (Spain). Environmental Geochemistry and Health 39 (5), 1159–1177.

Karar, K., Gupta, A.K., Kumar, A., Biswas, A.K., 2006. Characterization and identification of the sources of chromium, zinc, lead, cadmium, nickel, manganese and iron in PM 10 particulates at the two sites of Kolkata, India. Environmental Monitoring and Assessment 120 (1), 347–360.

Kulshrestha, A., Satsangi, P.G., Masih, J., Taneja, A., 2009. Metal concentration of PM2. 5 and PM10 particles and seasonal variations in urban and rural environment of Agra, India. Science of the Total Environment 407 (24), 6196–6204.

Kumari, S., Jain, M.K., Elumalai, S.P., 2021. Assessment of pollution and health risks of heavy metals in particulate matter and road dust along the road network of Dhanbad, India. Journal of Health and Pollution 11 (29), 210305. Available from: https://doi.org/10.5696/2156-9614-11.29.210305.

Lemly, A.D., 2018. Environmental hazard assessment of coal ash disposal at the proposed Rampal power plant. Human and Ecological Risk Assessment: An International Journal 24 (3), 627–641.

Li, Z., Ma, Z., van der Kuijp, T.J., Yuan, Z., Huang, L., 2014. A review of soil heavy metal pollution from mines in China: pollution and health risk assessment. Science of the Total Environment 468, 843–853.

Li, H., Qian, X., Wang, Q.G., 2013. Heavy metals in atmospheric particulate matter: a comprehensive understanding is needed for monitoring and risk mitigation. Environmental Science & Technology 47 (23), 13210–13211.

Liu, A., Ma, Y., Gunawardena, J.M., Egodawatta, P., Ayoko, G.A., Goonetilleke, A., 2018. Heavy metals transport pathways: the importance of atmospheric pollution contributing to stormwater pollution. Ecotoxicology and Environmental Safety 164, 696–703.

Madhav, S., Ahamad, A., Singh, A.K., Kushawaha, J., Chauhan, J.S., Sharma, S., et al., 2020. Water pollutants: sources and impact on the environment and human health. Sensors in Water Pollutants Monitoring: Role of Material. Springer, Singapore, pp. 43–62.

Madhav, S., Raju, N.J., Ahamad, A., Singh, A.K., Ram, P., Gossel, W., 2021. Hydrogeochemical assessment of groundwater quality and associated potential human health risk in Bhadohi environs, India. Environmental Earth Sciences 80 (17), 1–14.

Mahmoudi, N., Nakhaei, M., Porhemmat, J., 2017. Assessment of hydrogeochemistry and contamination of Varamin deep aquifer, Tehran Province, Iran. Environmental Earth Sciences 76 (10), 1–14.

Maleki, H., 2016. Recent advances in aerogels for environmental remediation applications: a review. Chemical Engineering Journal 300, 98–118.

Manisalidis, I., Stavropoulou, E., Stavropoulos, A., Bezirtzoglou, E., 2020. Environmental and health impacts of air pollution: a review. Frontiers in Public Health 14.

Meena, M., Singh, M.B., Chandrawat, U., Rani, A., 2014. Seasonal variations and sources of heavy metals in free Fall dust in an industrial city of western India. Iranica Journal of Energy & Environment 5 (2), 160–166.

Mehra, A., Farago, M.E., Banerjee, D.K., 1998. Impact of fly ash from coal-fired power stations in Delhi, with particular reference to metal contamination. Environmental Monitoring and Assessment 50 (1), 15–35.

Meybeck, M.I.C.H.E.L., 2013. Heavy metal contamination in rivers across the globe: an indicator of complex interactions between societies and catchments. Proceedings f H04 Understanding Freshwater Quality Problems in a Changing World 361, 3–16.

Mohod, C.V., Dhote, J., 2013. Review of heavy metals in drinking water and their effect on human health. International Journal of Innovative Research in Science, Engineering and Technology 2 (7), 2992–2996.

Morawska, L., Zhang, J.J., 2002. Combustion sources of particles. 1. Health relevance and source signatures. Chemosphere 49 (9), 1045–1058.

Nabulo, G., Young, S.D., Black, C.R., 2010. Assessing risk to human health from tropical leafy vegetables grown on contaminated urban soils. Science of the Total Environment 408 (22), 5338–5351.

Nanos, N., Grigoratos, T., Rodríguez Martín, J.A., Samara, C., 2015. Scale-dependent correlations between soil heavy metals and As around four coal-fired power plants of northern Greece. Stochastic Environmental Research and Risk Assessment 29 (6), 1531–1543.

Nawrot, N., Wojciechowska, E., Rezania, S., Walkusz-Miotk, J., Pazdro, K., 2020. The effects of urban vehicle traffic on heavy metal contamination in road sweeping waste and bottom sediments of retention tanks. Science of the Total Environment 749, 141511.

Nimir, S.A., 2006. Environmental impact of some trace elements emission from the work environment of Atbara Cement Factory and Atbara Railway Foundry, 1–196.

Nriagu, J.O., 1996. History of global metal pollution. Science 272 (5259), 223–224.

Pandiyan, J., Mahboob, S., Govindarajan, M., Al-Ghanim, K.A., Ahmed, Z., Al-Mulhm, N., et al., 2021. An assessment of level of heavy metals pollution in the water, sediment and aquatic organisms: a perspective of tackling environmental threats for food security. Saudi Journal of Biological Sciences 28 (2), 1218–1225.

Paul, D., 2017. Research on heavy metal pollution of river Ganga: a review. Annals of Agrarian Science 15 (2), 278–286.

Popoola, L.T., Adebanjo, S.A., Adeoye, B.K., 2018. Assessment of atmospheric particulate matter and heavy metals: a critical review. International Journal of Environmental Science and Technology 15 (5), 935–948.

Rani, L., Srivastav, A.L., Kaushal, J., Grewal, A.S., Madhav, S., 2022. Heavy metal contamination in the river ecosystem. Ecological Significance of River Ecosystems. Elsevier, pp. 37–50.

Rao, P.P., Khemani, L.T., Momin, G.A., Safai, P.D., Pillai, A.G., 1992. Measurements of wet and dry deposition at an urban location in India. Atmospheric Environment. Part B. Urban Atmosphere 26 (1), 73–78.

Rubalingeswari, N., Thulasimala, D., Giridharan, L., Gopal, V., Magesh, N.S., Jayaprakash, M., 2021. Bioaccumulation of heavy metals in water, sediment, and tissues of major fisheries from Adyar estuary, southeast coast of India: an ecotoxicological impact of a metropolitan city. Marine Pollution Bulletin 163, 111964.

Sah, D., Verma, P.K., Kandikonda, M.K., Lakhani, A., 2019. Chemical fractionation, bioavailability, and health risks of heavy metals in fine particulate matter at a site in the Indo-Gangetic Plain, India. Environmental Science and Pollution Research 26 (19), 19749–19762.

Sankhla, M.S., Kumari, M., Nandan, M., Kumar, R., Agrawal, P., 2016. Heavy metals contamination in water and their hazardous effect on human health-a review. Int. J. Current Microbiology and Applied Sciences 5 (10), 759–766.

Sarode, D.B., Jadhav, R.N., Khatik, V.A., Ingle, S.T., Attarde, S.B., 2010. Extraction and leaching of heavy metals from thermal power plant fly ash and its admixtures. Polish Journal of Environmental Studies 19 (6), 1325–1330.

Sezgin, N., Ozcan, H.K., Demir, G., Nemlioglu, S., Bayat, C., 2004. Determination of heavy metal concentrations in street dusts in Istanbul E-5 highway. Environment International 29 (7), 979–985.

Sharma, R.K., Agrawal, M., Marshall, F.M., 2008. Atmospheric deposition of heavy metals (Cu, Zn, Cd and Pb) in Varanasi city, India. Environmental Monitoring and Assessment 142 (1), 269–278.

Singh, S., Raju, N.J., Gossel, W., Wycisk, P., 2016. Assessment of pollution potential of leachate from the municipal solid waste disposal site and its impact on groundwater quality, Varanasi environs, India. Arabian Journal of Geosciences 9 (2), 1–12.

Singh, J., Yadav, P., Pal, A.K., Mishra, V., 2020. Water pollutants: origin and status. Sensors in Water Pollutants Monitoring: Role of Material. Springer, Singapore, pp. 5–20.

Srivastava, A., Siddiqui, N.A., Koshe, R.K., Singh, V.K., 2018. Human health effects emanating from airborne heavy metals due to natural and anthropogenic activities: a review. Advances in Health and Environment Safety 279–296.

Suvarapu, L.N., Baek, S.O., 2017. Determination of heavy metals in the ambient atmosphere: a review. Toxicology and Industrial Health 33 (1), 79–96.

Thakur, M., Deb, M.K., Imai, S., Suzuki, Y., Ueki, K., Hasegawa, A., 2004. Load of heavy metals in the airborne dust particulates of an urban city of central India. Environmental Monitoring and Assessment 95 (1), 257–268.

Tiller, K.G., 1989. Heavy metals in soils and their environmental significance. Advances in soil Science. Springer, New York, NY, pp. 113–142.

Tomar, D., Upadhyay, R., 2018. Heavy metals in Yamuna River: a review. International Journal of Advanced Scientific Research and Management 11 (3), 339–348.

Travnikov, O., Batrakovam, N., Gusev, A., Ilyin, I., Kleimenov, M., Rozovskaya, O., et al., 2020. Assessment of transboundary pollution by toxic substances: heavy metals and POPs. EMEP Status Report 2, 2020.

Tsuda, A., Henry, F.S., Butler, J.P., 2013. Particle transport and deposition: basic physics of particle kinetics. Comprehensive Physiology 3 (4), 1437.

Usmani, Z., Kumar, V., 2017. Characterization, partitioning, and potential ecological risk quantification of trace elements in coal fly ash. Environmental Science and Pollution Research 24 (18), 15547–15566.

Vardhan, K.H., Kumar, P.S., Panda, R.C., 2019. A review on heavy metal pollution, toxicity and remedial measures: current trends and future perspectives. Journal of Molecular Liquids 290, 111197.

Vareda, J.P., Durães, L., 2019. Efficient adsorption of multiple heavy metals with tailored silica aerogel-like materials. Environmental Technology 40 (4), 529–541.

Vareda, J.P., Valente, A.J., Durães, L., 2019. Assessment of heavy metal pollution from anthropogenic activities and remediation strategies: a review. Journal of Environmental Management 246, 101–118.

Weerasundara, L., Vithanage, M., Ziyath, A.M., Goonetilleke, A., 2015. Heavy metals in atmospheric deposition in Kandy City; implications for urban water resources. 6th International Conference on Structural Engineering (SECM/15/094) .

Wei, B., Yang, L., 2010. A review of heavy metal contaminations in urban soils, urban road dusts and agricultural soils from China. Microchemical Journal 94 (2), 99–107.

WHO, 2011. Adverse Health Effect of Heavy Metals in Children, 2011. World Health Organization, Geneva. Available from: http://www.who.int/ceh/capacity/heavy_metals.pdf.

Wong, C.S.C., Li, X.D., Zhang, G., Qi, S.H., Peng, X.Z., 2003. Atmospheric deposition of heavy metals in the Pearl River Delta, China. Atmospheric Environment 37 (6), 767–776.

Wu, Y., Liu, J., Zhai, J., Cong, L., Wang, Y., Ma, W., et al., 2018. Comparison of dry and wet deposition of particulate matter in near-surface waters during summer. PLoS One 13 (6), 0199241.

Ye, L., Huang, M., Zhong, B., Wang, X., Tu, Q., Sun, H., et al., 2018. Wet and dry deposition fluxes of heavy metals in Pearl River Delta Region (China): characteristics, ecological risk assessment, and source apportionment. Journal of Environmental Sciences 70, 106–123.

Yedjou, C.G., Patlolla, A.K., Sutton, D.J., 2012. Heavy Metals Toxicity and the Environment Paul B Tchounwou. Published in final edited form as: EXS, 101, 133–164.

Zeitoun, M.M., Mehana, E.E., 2014. Impact of water pollution with heavy metals on fish health: overview and updates. Global Veterinaria 12 (2), 219–231.

CHAPTER

10

Metal pollution in marine environment: sources and impact assessment

Rahul Mishra[1], *Ekta Singh*[1], *Aman Kumar*[1], *Akshay Kumar Singh*[2], *Sughosh Madhav*[3], *Sushil Kumar Shukla*[4] *and Sunil Kumar*[1]

[1]Waste Reprocessing Division, CSIR-National Environmental Engineering Research Institute, Nagpur, Maharashtra, India [2]Department of Transport Science and Technology, Central University of Jharkhand, Ranchi, Jharkhand, India [3]Department of Civil Engineering, Jamia Millia Islamia, New Delhi, India [4]Department of Environmental Sciences, Central University of Jharkhand, Ranchi, Jharkhand, India

10.1 Introduction

Water is a valuable natural resource that is essential for human civilization and agricultural sustainability. About 70% of the earth's crust is covered with water while it has been over-exploited and highly contaminated day by day as a result of various anthropogenic activities. Water resources (both surface water and groundwater) are frequently handled and studied separately, despite the fact that they are interconnected and interdependent (Obinna and Ebere, 2019). Surface water percolates into the soil and forms groundwater contamination and vice versa. As a result, the same pollution sources may contaminate both groundwater and surface water sources. The ecosystem is damaged from these sources which contribute organic contaminants or heavy metals to marine waters. Heavy metal pollutants have been linked to carcinogenic, teratogenic and mutagenic changes and are persistent in the environment (Zhang et al., 2017; Jiang et al., 2018).

Heavy metal is elucidated as metalloids and metals with a specific gravity greater than 5 and atomic mass greater than 20 like zinc (Zn), nickel (Ni), chromium (Cr), lead (Pb), mercury (Hg), arsenic (As), copper (Cu) and cadmium (Cd). From a biological standpoint, heavy

Metals in Water
DOI: https://doi.org/10.1016/B978-0-323-95919-3.00006-9

metals refer to a group of metals, including metalloids, which can be hazardous to animals and plants even at low concentrations (Rascio and Navari-Izzo, 2011). Heavy metals produced by anthropogenic activities have surpassed natural sources' contribution over the previous two centuries (Lohani et al., 2008). These contaminants are dumped into the marine water, either indirectly or directly, particularly in underdeveloped nations. As per World Health Organization (WHO) and the US Environmental Protection Agency (US-EPA), the acceptable concentration bounds for all of these heavy metals in natural water vary between 0.01 and 0.05 mg/L (World Health Organization, 2018; EPA US, 2009). Furthermore, multiple researches have demonstrated the existence of heavy metal residues in ecosystems and marine waters (Mishra et al., 2022). Their consequences on animal and aquatic ecosystems health were also described by Fakhri et al. (2018). Water contamination as a result of these factors is a significant socioeconomic and environmental issue (World Health Organization, 2018). Heavy metal pollution, unlike organic pollutants, is irreversible, persistent and undetectable. It not only contaminates the food crops' quality, water bodies and atmosphere but also causes a serious hazard to living organisms' health and well-being via accumulation in the food chain (Jaishankar et al., 2014). Heavy metal contamination of marine water may cause hazards to the humans and ecosystem via the drinking contaminated groundwater, ingestion and food chain (McLaughlin et al., 2000; Ling et al., 2007). Consequently, even at very low concentrations, these contaminants are poisonous, nonbiodegradable, and typically resistant to traditional elimination procedures. As a result, they have the potential to significantly damage the drinking water's resource quality.

These heavy metals exposure can cause several human diseases like cancer, neurological disorders, kidney pathology, and respiratory problem. For instance, respiratory issues and skin lesions can be caused by Cr as it is carcinogenic (Mohammadi et al., 2020). Also, high blood pressure and kidney problems can be caused by Cd (Satarug et al., 2005), while neurological disorders and reduction in fertility can be caused by Pb (Landrigan et al., 2000).

The probable heavy metal pollutants' sources and the associated risk to the marine environment as well as human health have been insighted in the initial sections of the chapter whereas, in the last section, the assessment index has been given for water quality assessment. The detailed explanation of each topic related to heavy metals pollution of marine water in this chapter will help readers gain a background idea and knowledge in the field of marine pollution.

10.2 Source of heavy metals

Heavy metals naturally exist in the soil and eventually enter into marine water because of parent materials' pedogenetic weathering actions with trace quantities of 1000 mg/kg which is rarely harmful (Kabata-Pendias and Pendias, 2001). Most soil in the urban and rural areas may gather one or more than one type of heavy metal above permissible limits given by different agencies. This is plenty to cause harm to the environment as a result of man's disturbance of nature's steadily occurring geochemical cycle of metals (D'amore et al., 2005). Heavy metals primarily become contaminants in the environment due to (1) in comparison to natural process, the anthropogenic cycles have more generation rate of heavy metals (2) the metals become more bioavailable in the environment due to the chemical compounds in which metal is found (3) In comparison to those in the receiving

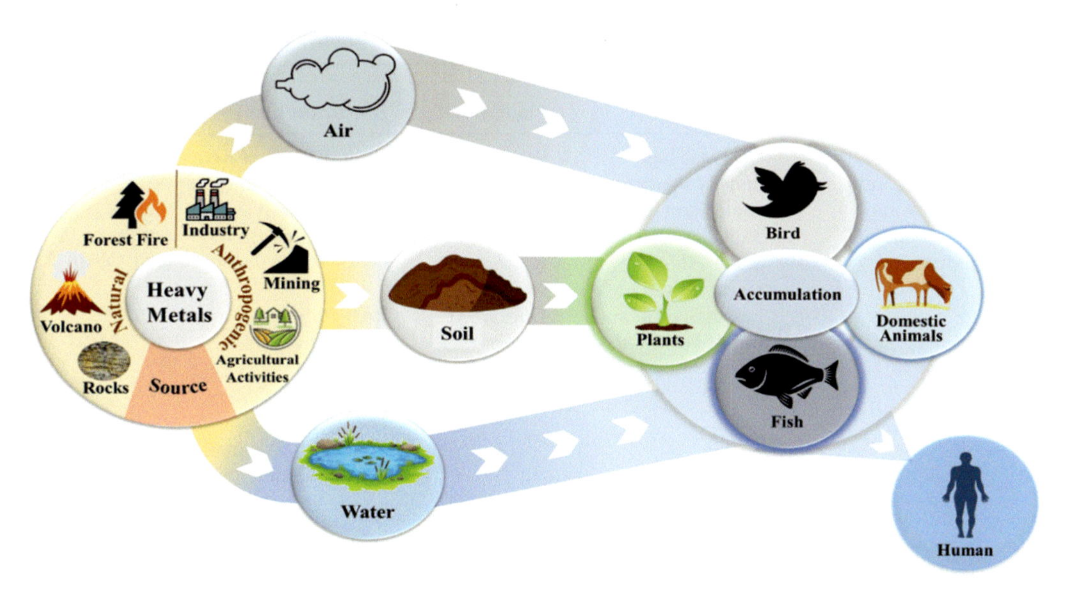

FIGURE 10.1 Anthropogenic and natural sources of heavy metals and its accumulation in food chain.

environment, metal concentrations in waste items are considerably large (4) transportation from mines to various sites in the environment where there is a greater risk of direct interaction (D'amore et al., 2005; Mishra et al., 2022). In the following section the natural activities and different anthropogenic activities have been discussed (Fig. 10.1) (Table 10.1).

10.2.1 Natural activities

Emissions of natural heavy metal emissions arise under multiple environmental conditions. Wind-borne soil particles, biogenic sources, rock weathering, forest fires, sea-salt sprays and volcanic eruptions are examples of such emissions. Heavy metals can be released from their endemic regions into various environmental areas as a result of natural weathering processes. Organic compounds, silicates, phosphates, sulfates, sulfides, hydroxides, and oxides are forms of heavy metals. Cu, Zn, Hg, As, Cd, Cr, Ni, Pb are the most common forms of heavy metals. Major health issues in humans and other mammals are posed by these above-mentioned heavy metals even after identifying their traces (Herawati et al., 2000; Mishra et al., 2022).

10.2.1.1 Forest fire

Many key studies on environmental emissions and their effects from wildfires were ignored. Weinhold (2011) recognized a variety of chemicals from wildfires but reported only four elements, that is, silicon (Si), sulfur (S), potassium (K), and chlorine (Cl). The hazardous metals Hg and Pb were notable emissions. Pb has been recognized as one of the most widespread and harmful elements to human beings (Patterson 1965). Many researchers have looked into how metals are remobilized after wildfires (Odigie and

TABLE 10.1 Heavy metals source and their impact on human health and plant.

Heavy metal	Sources	Impact on human health	Plant name	Impact on plant	References
Zn	Biosolids, mining, smelting and refining and electroplating industry	Decrease high-density lipoprotein cholesterol, damage pancreas, anemia, attacks the respiratory, hematological and digestive system.	Water lettuce (*Pistia stratiotes L.*) Water hyacinth (*Eichhornia crassipes*)	Chlorosis, decreased root length, plant height as well as stunted growth.	Lone et al. (2008); ATSDR (2005a,b); Sricoth et al. (2018); Radić et al. (2010)
			Duckweed (*Lemna minor L.*)	Reduction in chlorophyll, decreased photosynthetic energy conversion efficiency, enzymatic activity declined.	
Pb	Paints, industrial wastes, sewage, leaded gasoline burning, smelting and mining of ores.	Strokes, impact on joint pains and muscles, abortion, renal tubular dysfunction, colic, anorexia, ocular and musculoskeletal systems, renal gastrointestinal, impair neurodevelopment in children, impact on the central nervous system.	honey mesquite (*Prosopis sp.*) *Glomus deserticola*	Inhibits chlorophyll synthesis, plant growth, seed germination and phytochelatins.	Lone et al. (2008); ATSDR (2019); Pinho and Ladeiro (2012)
Ni	Weathering of soil and rock, bursting of the bubble, exchange of gas in the ocean, forest fire, landfill, volcanic eruptions.	Dermatitis, allergicreaction and chronic bronchitis	*Cajanus cajan*	leaf area diminished by 40%, irregular size of guard cells	Lone et al. (2008); ATSDR (2005a,b); Amari et al. (2017)
			Oryza sativa	Alternation of phospholipid composition of the plasma membrane	
Hg	Wood, peat and coal burning, emissions from caustic soda industries, forest fire, volcano eruptions.	Damage to the nervous system.	*Zea mays*	Gravimetric response of the seedling's inhibition, elongation of primary roots reduced	Lone et al. (2008); Azevedo and Rodriguez (2012)
			Vigna radiata	Declined respiration rates of seedling	

(*Continued*)

TABLE 10.1 (Continued)

Heavy metal	Sources	Impact on human health	Plant name	Impact on plant	References
Cu	Municipal sewage or biosolids, mining, smelting and refining, electroplating industry.	Immunotoxic, damage of kidney and liver as well as death.	Rhodes grass (*Chloris gayana*)	Root growth decreased	Lone et al. (2008); ATSDR (2004); Asati et al. (2016)
			black bindweed (*Polygonum convolvulus*)	Reduction in seed and biomass production, plant mortality	
			bean (*Phaseolus vulgaris*)	Reduction and malformation of the root	
Cr	Tanneries, solid waste, sludge and electroplating industry.	Affect the cardiovascular, respiratory, urinary and immune system, cause low birth weight and allergic dermatitis.	Water spinach (*Ipomonea aquatica*)	Reduced root length, increased root size	Lone et al. (2008); ATSDR (2012a,b); Chen et al. (2010)
Cd	Sewage sludge, phosphate fertilizers, fossil fuel burning, smelting and refining of metal, geogenic sources.	Affects, respiratory, reproductive, urinary, nervous, digestive, development and cardiovascular systems, lung damage, glomerular, renal tubular and osteoporosis.	Water lettuce (*Pistia stratiotes L.*) Water hyacinth (*Eichhornia crassipes*) Isoetaceae (*Isoetes taiwaneneses D.*) Iridaceae (*Gladiolous*) Duckweed (*Lemna minor L.*)	Chlorosis, reduce root length and plant height as well as stunted growth.	Lone et al. (2008); ATSDR (2012a,b); Razinger et al. (2008); Li et al. (2005); Sricoth et al. (2018)
As	Mining and smelting, volcanoes, herbicides, coal power plants, animal feed additives, wood preservatives, petroleum refining andsemiconductors	Dermatitis and bronchitis.	Brake fern (*Pteris vita*)	Decreased photosynthetic energy conversion efficiency, enzymatic activity declined.	Lone et al. (2008); Ma et al. (2001)

Flegal 2011; Nriagu 1989; Finley et al., 2009). During wildfires, considerable amounts of hazardous and harmless metals are released into the environment, according to these researches. Young and Jan (1977) discovered that smoke from a wildfire in California in 1975 released Zn, silver (Ag), Ni, Manganese (Mn), Pb, iron (Fe), Cu, Cr and Cd up to 100 km away from wildfire region. The wildfire's polluting effects on local marine waters with Mn, Fe and Pb surpassed the contaminating impacts of the local wastewater, which was the major metals' source.

10.2.1.2 Volcanic eruption

Active volcanoes are cracks in the Earth's crust, which are key suppliers of heavy metals and compounds from the deep layers of the crust, which expand in the nearby environment through gas, ash and lava emissions. Metals found in volcanic emissions enter into surface and groundwater supplies in the area. It can be transported across long distances due to local causes (rivers, winds), despite the fact that the nearby environment is always the most contaminated region, wherein increased metal concentrations can induce biological problems owing to their toxic effects. Heavy metal contamination from volcanic eruptions is far less well-known, despite the fact that more than 500 million humans reside in volcanic zones globally. This is owing to most volcanoes' irregular action, which makes it hard to describe volcanic-related contamination in the environment over a medium- or long-term timeframe. Furthermore, volcanoes differ with respect to chemical makeup, volcanic rock type and structure. Many heavy metals are observed in higher concentrations in volcanic soil (Favalli et al., 2004; Andronico et al., 2009). About 57-ton per year is the predicted global flux of mercury from volcanic eruption, whereas about 37.6-ton per year is the flux from degassing activities (Nriagu and Becker, 2003).

10.2.1.3 Rock weathering

Heavy metals can be found in varying quantities in different soils. Metals in natural soils are mostly derived from the lithosphere or the soil mineral component that produces minerals and rocks and make up the Earth's crust. Due to pedogenetic processes of parent materials weathering, naturally occurring elements enter the soil at quantities, which are considered trace and rarely harmful (Alloway, 2013; Yanagi, 2011). The 10 elements, that is, P, Ti, Mg, K, Na, Ca, Fe, Al, O, Siaccount for nearly 99% of the total crust composition, while the remaining elements are known as "trace elements," (concentrations <1000 mg/kg) (Hawkesworth and Kemp, 2006; Yanagi, 2011). Metals can be found in trace amounts in all soils. As a result, the metals' occurrence in the soil does not indicate pollution. The geology of the parent material in which the soil was created has a strong influence on the metal's concentration in unpolluted soil.

10.2.2 Anthropogenic processes

During uncontrolled smelting and mining of huge quantities of ores or metal in open fires, low concentrations of heavy metals are discharged. Metals are processed in industries after collecting from natural resources, where heavy metals were released into the atmosphere, with the industrial revolution. Likewise, heavy metal traces pollute the environment as a result of garbage disposal from both agricultural and household waste as well as motor exhaust. Several human activities which produce heavy metals in the environment are pesticides utilization that contains salts of heavy metals, auto exhausts release, domestic wastes discharge, agricultural waste discharge, fossil fuels such as kerosene oil, petrol, coal burning and mining (Armah et al., 2014). Some metals (Zn, Cu, Fe)'s low concentrations are important for organisms and these are called trace metals. In contrary, metals such as Cd, Pb, Hg are toxic to living beings above a certain limit. Mining activity is a major source of heavy metal contamination, although it is not the only one in the processes of industry. Pollution from Heavy metal can be produced by a variety of industrial operations in multiple ways.

10.2.2.1 *Mining activities*

Heavy metals can reach water resources via natural processes since they are found in the geological structures of the earth. Flowing water or heavy rains, for instance, can drain heavy metals from geological formations. When this geology is affected by mining, these activities are intensified. These operations expose the mined-out area to air and water, which can result in acid mine drainage and other problems. Heavy metals, especially radionuclides, are mobilized by the low pH conditions raised with acid mine drainage (Sankhla et al., 2016). Significant heavy metal pollution can also be led by mineral extraction via both direct extraction processes as well as leaching from tailings stockpiles and ore.

10.2.2.2 *Electronic waste*

Manufacturing companies of electronic goods must properly guarantee that their product's disposal options are mentioned in the user manual. Because E-waste is a recognized main source of heavy metals, toxic compounds, and carcinogens, disorders related to respiratory, skin, digestive, immunological, endocrine, and neurological systems, as well as malignancies, can all be eliminated by effective E-waste management and disposal. When ICT wastes are not managed scientifically, there is a worrying impact on human and the environment because of the exponential development in the usage of electrical and electronic equipment (EEE) to overcome the digital gap (Sankhla et al., 2016). It generates enormous amounts of effluents containing metals, making it a more possible polluter in comparison to the industry based on food processing. However, it is also a fact that it is the best economic interests in the electroplating industry to reduce discharges of metal, as they are inversely proportional to the efficiency of resources. Lower metal discharges can be achieved by decreasing losses by limiting plating baths. Another example of a sector that can produce metal-rich effluents and also airborne lead pollutants that might end up in surface water resources is the lead-acid battery manufacturing industry. Apparently, when heavy metals are used as important input materials in a manufacturing process, pollution hazards increase (Bagul et al., 2015).

10.2.2.3 *Power generation plants*

Among the other sources of heavy metal contamination, coal-fired power generation is a significant nonpoint source, and Hg discharge from boiler flues can affect water supplies. The sector also produces a lot of ash which includes a lot of heavy elements, like uranium. Metals released from coal-fired power stations are commonly recognized to have a substantial impact on the local environment's quality (Khillare et al., 2012). Heavy metals in coal can be found as gaseous and solid, eventually depositing as coal ash (Ćujić et al., 2017). Some ash is discharged into water and soil via dry or wet deposition, while some are deposited into the air via stacks.

10.2.2.4 *Agricultural activities*

The ever-increasing global population necessitates extensive land use for producing food that demands the utilization of soil conditioners, insecticides, and fertilizers on a regular basis. Fertilizers are put in the soil to supply more nutrients to crops. Pesticides are utilized for the protection of crops. Dredged sediments, animal manure and sewage sludge from harbors and rivers are all common sources of soil amendments. During dredging, heavy metals from these soils can be mobilized (Bradl, 2005) because the reducing environment shifts to an oxidizing one, allowing heavy metals to be remobilized. Metals put into sediments can impact surface water or aquifers via infiltration because the ground, surface water and soil are all interrelated systems (Mishra et al., 2022).

10.2.2.4.1 Fertilizers

Plants require not just macronutrients (Mg, Ca, S, K, P and N) but also vital micronutrients to grow. For healthy plant growth in some soils, some heavy metals are needed and these can be added to the soil (Wuana and Okieimen, 2011). Cereal crops are periodically supplemented with Cu as a soil amendment when cultivated on Cu deficient soils, while root and cereal crops can also be given Mn. In intensive farming methods, large amounts of fertilizer are applied on a regular basis to supply appropriate N, P, and K for the growth of crop. Heavy metals (Pb and Cd) are contaminants present in the fertilizer which is utilized in the agricultural field. If this fertilizer is used for a prolonged time, it may increase their level in the soil. Metals like Pb and Cd are recognized to have no physiological function. Some phosphatic fertilizers unintentionally contribute Cd and other possibly harmful substances to the soil such as Pb, Hg and F (Wuana and Okieimen, 2011).

10.2.2.4.2 Pesticides

Many conventional pesticides, which were historically widely utilized in horticulture and agriculture, had high levels of metal content. In high-production agriculture, pesticides are employed to manage insects and diseases and can be used as seed treatments by dusting, spraying or soil application. Though metal-based pesticides are no longer utilized, their previous utilization resulted in the larger heavy metal deposition in soils and groundwater, particularly Hg from methyl mercurials, Pb and As from lead arsenate. For several years, to prevent insects, lead arsenate was applied in orchards. In Australia and New Zealand, compounds of As were also commonly utilized to prevent pests in bananas. There are now many uninhabited sites where soil concentration levels of these components surpass concentrations of background (Wuana and Okieimen, 2011).

10.3 Adverse impact of heavy metals

Many prior studies have looked into the heavy metals' harmful impact on humans and the environment (Table 10.1) (Pinho and Ladeiro, 2012; Sricoth et al., 2018; Radić et al., 2010; Amari et al., 2017; Azevedo and Rodriguez, 2012). Elevated concentrations of heavy metal pollutants in water have a negative impact on water's ecological functions, such as primary nutrient generation and recycling. The health of wildlife and humans is also impacted by the bioaccumulation of metals in the food chain with the long-term effects of metal resistant evolution in some organisms. Additionally, metals' detrimental ecological effects may include info-disruption, which affects inter- and intraspecies interactions among freshwater fauna and microorganisms (Verla et al., 2019). Moreover, the implications of heavy metal contamination in water will be examined in terms of humans, aquatic fauna and plants. The toxicity of a specific heavy metal on human, aquatic flora and fauna is determined by the presence of other ions that hinder their bioavailability, pH as well as the metals' solubility. These ions may further affect the outcome of interaction with the component (de Souza and Silva, 2019).

10.3.1 Aquatic fauna

Fishes are a key indicator of heavy metal toxicity in the water habitats. Metals have a significant impact on them, despite their economic value. Fish species may be exposed to heavy metals indirectly through the food chain or through direct exposure. According to numerous researches, long-term exposure has been associated with the death of fish larvae and a reduction in the ability of adults to reproduce (Verla et al., 2016; Okoumassoun et al., 2002; Moon et al., 2010). At the organ level, chronic stress, DNA, and microscopic cellular, toxicology can produce structural changes in the organs, which can result in alterations in the function systems and eventually suppress the growth (Al-Kahtani, 2009). Heavy metal contamination of soil affects benthic organisms such as insects, crustaceans and worms. It is because of their nature to change the feeding habits of these organisms as well as lowering the availability of food for larger species like fishes (Kumar and Edward, 2009).

10.3.2 Aquatic flora

Aquatic flora requires some heavy metals for growth and maintenance. But, when the levels reach too high, the plant may be exposed to the heavy metals' toxicity. Heavy metals' high concentration in plants can disrupt metabolic functions, such as degeneration of major cell organelles, respiration, inhibition, photosynthesis and oxidative stress from the reactive oxygen species production, even leading to the plant's death (Fig. 10.2) (Sooksawat et al., 2013; Sneller et al., 2000). Other particular impacts are growth reduction, leaf necrosis and chlorosis accompanied by senescence and abscission traces, all of which alterations contribute to lower nutrient intake and impede the biomass obtained (Brunetto et al., 2017). Fig. 10.2 shows the direct and indirect impacts of heavy metals on the plant. Table 10.1 presents the impact of heavy metals on plants.

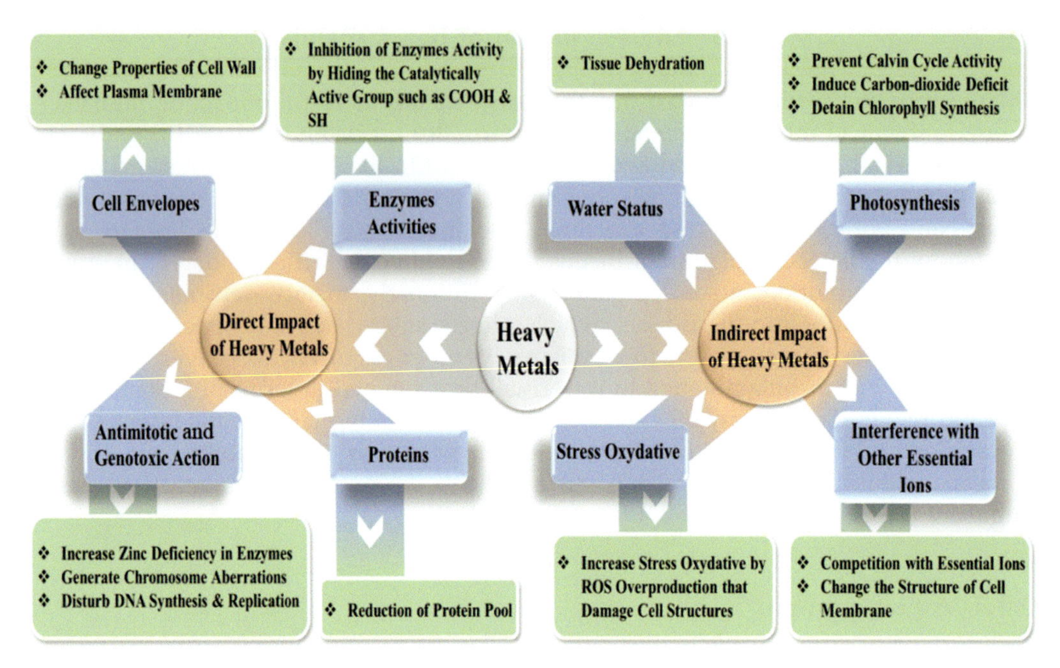

FIGURE 10.2 Direct and indirect impacts of heavy metal on plants.

The toxicity of heavy metal has multiple impacts on aquatic plants depending on the types of plants and interaction of water and heavy metals. Water hyacinth (*Eichhornia crassipes*) was exposed to an excess concentration of As (6 mg/L) for 3–8 days, which lead to plant death, whereas the plant appeared sick even after 3 days of exposure (Jasrotia et al., 2017). At the concentration of 2.5 and 6 mg/L, Eichhornia crassipes was capable to tolerate Cr (II) and Zn (II) sorption, respectively (Hasan et al., 2007). Moreover, without exhibiting any toxicity symptoms, the Brake fern (*Pteris vita*) absorbed As (about 7500 mg/kg) (Ma et al., 2001) whereas at such concentration, water hyacinth (*Eichhornia crassipes*) survives. In aquatic plants, unfavorable effects have been found for As, Hg, Cd and Pb at extremely low concentrations in the growth medium, according to the study. Furthermore, the presence or combination of various metals in the medium may amplify or lessen effects. When either two heavy metals were present, Wiafe et al. (2019) stated that the quantity of metals uptake (Pb, Cd, Hg and As) by *Typha capensis* was suppressed. Heavy metals cause some plants to endure or reverse their effects, whereas others gain in nutrition and size at certain concentrations. When *E. camaldulensis* was subjected to 45 mol/L Cd, a rise in root endoderm thickness, epidermis and carotenoids (associated to oxidative stress resistance) was detected (de Souza and Silva, 2019; Marques et al., 2011). Some phyto-compounds including antioxidant enzymes, thiols and anthocyanins may be responsible for the tolerance (Leão et al., 2014). Moreover, tomato plants with 50 mg/kg of Co had a higher nutritional content while the increase in antioxidant enzyme activities (catalase), biochemical content, nutrient content and plant growth in radish and mung bean (Jayakumar et al., 2013). After 14 days of exposure to elevated Cr^{3+} with the concentration of 10 mg/L in polluted water (in a hydroponic system), the *Ipomonea aquatica* (water spinach)'s root grew fatter instead of longer (Chen et al., 2010). After being exposed to high concentrations of heavy metals, certain aquatic flora has the potential to heal within days. Duckweed, for example, healed within days following significant exposure to Cd, Ni and Cu toxicity as per the study performed by Drost et al. (2007). It is fair to say that if a greater amount of exposure to stress or a toxicant is survived by a plant, it has a chance to recover completely (Ekperusi et al., 2019).

10.3.3 Human health

Metals can be found in water as a complex mixture of separate mineral stages. On the other hand, the effects on human health are dictated by metal bioavailability (as measured by metal speciation). Many researchers have looked into methods of water exposure, such as oral intake and dermal contact (Charity et al., 2018; Wirnkor et al., 2018; Diatta and Grzebisz, 2011). Individuals and concentrations (amount) ingested are the primary determinants of adverse health effects on humans (Diatta and Grzebisz, 2011). In general, assessing the harmful metals' health risk entails a quantitative valuation of the likelihood of negative consequences in a given set of circumstances (Charity et al., 2018).

The influence and toxicity of some significant heavy metals like Ni, Cd, Pb, Cr, Hg on the environment and human health have been given in the sections below and also presented in Table 10.1.

10.3.3.1 Impact and toxicity of heavy metals

Presently, metal trace elements (MTEs) comprises both toxic and less toxic (metalloids) heavy metals. The term MTEs is now commonly used instead of heavy metals. Most MTEs are hazardous once they reach a specific concentration. Zn, Ni, Hg, Pb, Cu, Cr and Cd at low quantities (less than 0.01 mg/L) are extremely dangerous to human beings. MTEs chronic exposure, most commonly by exposure to contaminated soils and industrial zones, polluted food and drinking water ingestion and air contaminants inhalation can have major impacts on animal and human health (Wuana and Okieimen, 2011). Metals accumulated in the surrounding water and soil can contaminate food supplies such as fish, fruits, grains and vegetables.

10.3.3.1.1 Impact and toxicity of nickel

Prolonged contact of Ni with the mucous membranes and skin can induce irritation and allergic reactions. The exposure of Ni causes dermatitis, which is the most dominant side effect. Furthermore, consumption of Ni salts can cause diarrhea, vomiting and nausea, while persistent inhalation of Ni in monoxide or metallic form can cause tubular dysfunction and asthma in some people (Coogan et al., 1989; Denkhaus and Salnikow 2002; Mishra et al., 2022). Furthermore, some organic Ni compounds, like tetra-carbonyl-nickel, are considered to be extremely poisonous. As a result, the International Agency for Research on Cancer (IARC) has classed Ni as a potentially carcinogenic chemical. While Lock and Janssen (2002) confirmed the Ni's chronic toxicity on three invertebrates, that is, *Folsomia candida, Enchytraeus albidus and Eisenia fetida*, it is not generally known because research on invertebrates has been studied infrequently. High dosages of Ni, as per to these authors, are hazardous to the reproductive of these invertebrates.

10.3.3.1.2 Impact and toxicity of cadmium

Cd is extremely hazardous in comparison to Hg and Pb. Cd was designated as a group 1 carcinogenic agent for humans by IARC in 1993 (Huff et al., 2007). Furthermore, several workplace investigations correlating to exposures by inhalation have demonstrated a considerable rise in mortality due to lung cancer when cadmium is present (Othumpangat et al., 2005). Cd can raise blood pressure and also cause kidney issues in people. Furthermore, inhaling it is hazardous. Chronic cadmium exposure produces a disease known as "itai-itai," that develops permanent nephropathy and may lead to failure of the kidney (Othumpangat et al., 2005; Waisberg et al., 2003). It is harmful to several fauna and flora species for both terrestrial and aquatic environment at low levels. It can, for example, distort the bones of minnows (small freshwater fish) when exposed to 7.5 g Cd/L, a dose 5200 times lower than the LC50 (96 h) (Bengtsson et al., 1975). Furthermore, cadmium can disrupt seed germination in some plants and cause oxidative stress in algae (Sethy and Ghosh, 2013; Pinto et al., 2003).

10.3.3.1.3 Impact and toxicity of lead

One of the most harmful heavy metals is Pb. Pb poisoning is the term for its long-term toxicity. It has the potential to harm the cardiovascular, hematopoietic and neurological, systems of humans. It can induce renal, hematological and neurological problems at large levels. It can also cause brain development problems in youngsters, as well as learning challenges and psychological issues. Pb poisoning causes spontaneous abortion, fetus death and reduced fertility as well as carcinogenic and mutagenic effects (Ahamed et al., 2008; Apostoli et al., 2000; Mishra et al., 2022). It is very poisonous for several invertebrates, particularly freshwater invertebrates. On the other hand, marine invertebrates withstand it better, and in comparison, to other metals, some invertebrates are more tolerant of Pb (Offem and Ayotunde, 2008). Due to Pb precipitation, Pb salts are difficult to dissolve in hard water and saltwater, and the existence of other salts lowers the Pb existence for organisms. The toxicity of Pb in aqueous media differs by fish species. Pb is the most susceptible in young fish and eggs. Pb poisoning in birds causes the blackening of the caudal area and spinal malformation (De Francisco et al., 2003). Birds can be poisoned by lead salts, which contain roughly 100 mg of Pb/kg of diet. Furthermore, exposing quail to diet containing 10 mg of Pb/kg can have a deleterious impact on the production of the egg (De Francisco et al., 2003).

10.3.3.1.4 Impact and toxicity of chromium

The Cr's toxicity differs greatly depending on the chemical form in which it is found, that is, valence 2–6, hydroxide, oxide, ion, nanoparticle and particle. For example, Cr^{6+} ions are 1000 times more poisonous than Cr^{3+} ions, making them exceedingly dangerous (Cohen et al., 1993). Through the ion transport system, Cr(VI) ions can pass via the cell and are becoming a growing health problem in the marine ecosystem. Cr(VI) ions have been studied extensively for their genotoxicity and cytotoxicity in medaka fish cells (Goodale et al., 2008). Glutathione can interact with Cr(VI) ions to create Cr(V) and Cr(IV) ions in organelle membranes and plasma as well as it can also reduce to Cr (III). DNA double-strand breaks can be mediated by reduced chromium species and can activate DNA damage, likely serving as a genotoxic agent at the end of the chain (Chiu et al., 2010).

The chromium consumption has been controlled by a system of plants. Though, changes in genetics of plants appear, when the chromium concentrations available in the soil rises to a high level. Plants chromium consumption can also be impacted by soil acidification (Hayat et al., 2012). Generally, accumulation of chromium has not been seen in the fish body, however high chromium concentration can harm the gills of fish near the discharge points. The formation of tumors, infertility, genetic defects, low potential to fight diseases and respiratory problems can be caused by the exposure of chromium in animals. Respiratory issues even bronchopulmonary cancers, skin lesions can be caused in humans (Ahmad et al., 2012).

10.3.3.1.5 Impact and toxicity of mercury

The humans' nervous system particularly in young children can be damaged by Hg poisoning (Ralston and Raymond, 2018; Counter and Buchanan, 2004). It is one of the most toxic metals as many diseases even death can be caused for a long exposure time at low doses. The toxicity of mercury under the vapor form is first entered through the respiratory tract and after that solubilized with hemoglobin, blood and plasma. It can attack the nervous system, the brain and the kidneys in these transportation circumstances. It can simply reach the fetus through the placenta in pregnant women. Human breast milk is also contaminated which creates risk after birth too. Hg is very neurotoxic under vapor form, particularly in children. On the central nervous system cells, it has a cytotoxic even at low concentrations. The unicellular organisms (fungi, bacteria and algae)'s growth and fishes (rainbow trout) can be inhibited as Hg causes larval mortality as well as an excess of embryo production (Tan et al., 2009; Xun et al., 2017). Furthermore, Hg has an endocrine effect in animals and humans as it interacts negatively with an essential hormone, that is, dopamine. The duration of exposure and the contamination level affect this interaction (Tan et al., 2009).

10.4 Water quality assessment

To identify the contaminants' effect on the environment and health of human beings, several organization and researchers focuses on water quality assessment. The development of several approaches is illustrated by the hydrological study to recognize the origin source and overall access the water quality. Heavy metal pollution index is an efficient and convenient approach for assessing water quality as per the literatures.

10.4.1 Heavy Metal Pollution Index

The total quality of water is characterized by Heavy Metal Pollution Index (HPI). The index is assessed by using following Eqs. (10.1) and (10.2).

$$Q_i = \sum_{i=1}^{n} \frac{[M_i(-)I_i]}{S_i - I_i} \times 100 \tag{10.1}$$

Ideal or highest desirable value for ith parameter is represented by I_i. M_i is represented as measured value for the ith parameter. Permissible or standard or permissible value for ith parameter is represented by S_i. The two values' numerical differences which ignore the algebraic sign is denoted by (-) sign (Giri and Singh, 2014; Edet and Offiong, 2002).

$$HPI = \frac{\sum_{i=1}^{n} W_i Q_i}{\sum_{i=1}^{n} W_i} \quad (10.2)$$

Subindex calculated and weight assigned for the ith parameter is represented by Q_i and W_i, respectively.

The parameters importance which varies from zero to one determines the samples' weight. Inversely proportional to the standard value can also be considered for each element (Giri and Singh, 2014; Edet and Offiong, 2002; Moghaddam et al., 2014). High heavy metal pollution ($HPI > 100$), threshold risk ($HPI = 100$) and low heavy metal pollution ($HPI < 100$)are categories of water quality related to heavy metal pollution index (Edet and Offiong, 2002).

10.4.2 Water Quality Index

Tyagi et al. (2013) described the WQI method to calculate the quality of water sources for consumption and irrigation. The following equations were applied to calculate the index values:

Firstly, the water quality parameters were selected for quality rating scale (Q_i). Eq. (10.3) was used for computing Qi of each selected parameter.

$$Q_i = \frac{(V_{\text{actual}} - V_{\text{ideal}}) \times 100}{(V_{\text{standard}} - V_{\text{ideal}})} \quad (10.3)$$

Where, the computed value of the ith parameter in the examined water sample is V_{actual}. In pure water, the ideal value of this parameter is V_{ideal}. The standard value of ith parameter is V_{standard}.

Eq. (10.4) was used for determining the unit weight (W_i) for each water quality parameter.

$$W_i = \frac{K}{V_{standard}} \quad (10.4)$$

Eq. (10.5) is used for calculating K (proportionality constant)

$$K = \frac{1}{\sum 1/V_{\text{standard}}} \quad (10.5)$$

Finally, Eq. (10.6) is used for calculating WQI.

$$WQI = \frac{\sum(Q_i \times W_i)}{\sum W_i} \quad (10.6)$$

The ratings of the index values were excellent; good; poor; very poor and unsuitable, that varies between 0–25, >25–50, >50–75, >75–100 and >100, respectively.

10.5 Conclusion

The knowledge of heavy metals' source and their impact on the marine environment is necessary for future researches regarding water pollution with respect to heavy metals. Therefore, in this chapter, generalities of environmental impact and toxicity, as well as source of heavy metals, have been briefly presented. Natural and anthropogenic actions are two sources of contamination in the marine environment but, anthropogenic sources contribute more than the former. Therefore, it is a critical concern for the general public, scientists, governmental and nongovernmental organizations to tackle. After going through the kinds of literature that quantify heavy metals in water and soil, it was observed that nearby areas where heavy metals are processed (smelting) or extracted (mine) are heavily polluted by these elements. Moreover, Heavy metal contamination is also caused by regular anthropogenic activity such as mixed industrial activities and road traffic. The toxicity and risk of heavy metals on the marine environment and eventually human health through the food chain need to be discussed with a focus on different heavy metals toxicity. The heavy metals' concentration in the marine environment is beyond the limitation prescribed by several organizations, which poses disease in marine as well as terrestrial organisms. These can be nonfatal like physical weakness to fatal like the disorder of brain nervous system. It is necessary to impose strict legislation on the maximum concentration of heavy metals in cultivated sediments, irrigation water, and drinking water to prevent heavy metal poising. For the safeguard of the environment, it is necessary to identify the source, impact of heavy metals as well as quality of water using efficient techniques.

References

Ahamed, M., Fareed, M., Kumar, A., Siddiqui, W.A., Siddiqui, M.K.J., 2008. Oxidative stress and neurological disorders in relation to blood lead levels in children. Redox Report 13 (3), 117–122.

Ahmad, A., Muneer, B., Shakoori, A.R., 2012. Effect of chromium, cadmium and arsenic on growth and morphology of HeLa cells. Journal of Basic & Applied Sciences 8, 53–58.

Al-Kahtani, M.A., 2009. Accumulation of heavy metals in tilapia fish (Oreochromis niloticus) from Al-Khadoud Spring, Al-Hassa, Saudi Arabia. American Journal of Applied Sciences 6 (12), 2024–2029.

Alloway, B.J., 2013. Sources of heavy metals and metalloids in soils. In Heavy Metals in Soils (pp. 11–50). Springer, Dordrecht.

Amari, T., Ghnaya, T., Abdelly, C., 2017. Nickel, cadmium and lead phytotoxicity and potential of halophytic plants in heavy metal extraction. South African Journal of Botany 111, 99–110.

Andronico, D., Spinetti, C., Cristaldi, A., Buongiorno, M.F., 2009. Observations of Mt. Etna volcanic ash plumes in 2006: an integrated approach from ground-based and polar satellite NOAA–AVHRR monitoring system. Journal of Volcanology and Geothermal Research 180 (2–4), 135–147.

Apostoli, P., Bellini, A., Porru, S., Bisanti, L., 2000. The effect of lead on male fertility: a time to pregnancy (TTP) study. American Journal of Industrial Medicine 38 (3), 310–315.

Armah, F.A., Quansah, R., Luginaah, I., 2014. A systematic review of heavy metals of anthropogenic origin in environmental media and biota in the context of gold mining in Ghana. International Scholarly Research Notices 2014.

Asati, A., Pichhode, M., Nikhil, K., 2016. Effect of heavy metals on plants: an overview. International Journal of Application or Innovation in Engineering & Management 5 (3), 56–66.

ATSDR, 2004. Toxicological Profile for Copper. https://www.atsdr.cdc.gov/toxprofiles/tp.asp?id = 206&tid = 37.

ATSDR, 2005a. Toxicological Profile for Nickel. https://www.atsdr.cdc.gov/toxprofiles/tp.asp?id = 245&tid = 44.

ATSDR, 2005b. Toxicological Profile for Zinc. https://www.atsdr.cdc.gov/toxprofiles/tp.asp?id = 302&tid = 54.
ATSDR, 2012a. Toxicological Profile for Cadmium. https://www.atsdr.cdc.gov/toxprofiles/tp.asp?id = 48&tid = 15.
ATSDR, 2012b. Toxicological Profile for Chromium. https://www.atsdr.cdc.gov/toxprofiles/tp.asp?id = 62&tid = 17.
ATSDR, 2019. Toxicological Profile for Lead. https://www.atsdr.cdc.gov/ToxProfiles/tp.asp?id = 96&tid = 22.
Azevedo, R., Rodriguez, E., 2012. Phytotoxicity of mercury in plants: a review. Journal of Botany 2012.
Bagul, V.R., Shinde, D.N., Chavan, R.P., Patil, C.L., Pawar, R.K., 2015. New perspective on heavy metal pollution of water. Journal of Chemical and Pharmaceutical Research 7 (12), 700–705.
Bengtsson, B.E., Carlin, C.H., Larsson, Å., Svanberg, O., 1975. Vertebral damage in minnows, Phoxinus phoxinus L., exposed to cadmium. Ambio 166–168.
Bradl, H.B., 2005. Sources and origins of heavy metals. In Interface Science and Technology (6, pp. 1–27). Elsevier.
Brunetto, G., Ferreira, P.A.A., Melo, G.W., Ceretta, C.A., Toselli, M., 2017. Heavy metals in vineyards and orchard soils. Revista Brasileira de Fruticultura 39.
Charity, L.K., Wirnkor, V.A., Emeka, A.C., Isioma, A.A., Ebere, C.E., Ngozi, V.E., 2018. Health risks of consuming untreated borehole water from Uzoubi Umuna Orlu, Imo State Nigeria. Journal of Environmental Analytical Chemistry 5 (250), 2380–2391.
Chen, J.C., Wang, K.S., Chen, H., Lu, C.Y., Huang, L.C., Li, H.C., et al., 2010. Phytoremediation of Cr (III) by Ipomonea aquatica (water spinach) from water in the presence of EDTA and chloride: effects of Cr speciation. Bioresource Technology 101 (9), 3033–3039.
Chiu, A., Shi, X.L., Lee, W.K.P., Hill, R., Wakeman, T.P., Katz, A., et al., 2010. Review of chromium (VI) apoptosis, cell-cycle-arrest, and carcinogenesis. Journal of Environmental Science and Health, Part C 28 (3), 188–230.
Cohen, M.D., Kargacin, B., Klein, C.B., Costa, M., 1993. Mechanisms of chromium carcinogenicity and toxicity. Critical Reviews in Toxicology 23 (3), 255–281.
Coogan, T.P., Latta, D.M., Snow, E.T., Costa, M., Lawrence, A., 1989. Toxicity and carcinogenicity of nickel compounds. CRC Critical Reviews in Toxicology 19 (4), 341–384.
Counter, S.A., Buchanan, L.H., 2004. Mercury exposure in children: a review. Toxicology and Applied Pharmacology 198 (2), 209–230.
Ćujić, M., Dragović, S., Đorđević, M., Dragović, R., Gajić, B., 2017. Reprint of Environmental assessment of heavy metals around the largest coal fired power plant in Serbia. Catena 148, 26–34.
De Francisco, N., Ruiz Troya, J.D., Agüera, E.I., 2003. Lead and lead toxicity in domestic and free living birds. Avian Pathology 32 (1), 3–13.
Denkhaus, E., Salnikow, K., 2002. Nickel essentiality, toxicity, and carcinogenicity. Critical Reviews in Oncology/Hematology 42 (1), 35–56.
de Souza, C.B., Silva, G.R., 2019. Phytoremediation of Effluents Contaminated with Heavy Metals by Floating Aquatic Macrophytes Species. IntechOpen.
Diatta, J.B., Grzebisz, W., 2011. Simulative evaluation of Pb, Cd, Cu, and Zn transfer to humans: the case of recreational parks in Poznan, Poland. Polish Journal of Environmental Studies 20 (6), 1433.
Drost, W., Matzke, M., Backhaus, T., 2007. Heavy metal toxicity to Lemna minor: studies on the time dependence of growth inhibition and the recovery after exposure. Chemosphere 67 (1), 36–43.
D'amore, J.J., Al-Abed, S.R., Scheckel, K.G., Ryan, J.A., 2005. Methods for speciation of metals in soils: a review. Journal of Environmental Quality 34 (5), 1707–1745.
Edet, A.E., Offiong, O.E., 2002. Evaluation of water quality pollution indices for heavy metal contamination monitoring. A study case from Akpabuyo-Odukpani area, Lower Cross River Basin (southeastern Nigeria). GeoJournal 57 (4), 295–304.
Ekperusi, A.O., Sikoki, F.D., Nwachukwu, E.O., 2019. Application of common duckweed (Lemna minor) in phytoremediation of chemicals in the environment: State and future perspective. Chemosphere 223, 285–309.
EPA US, 2009. National Primary Drinking Water Regulations. Washington, DC
Fakhri, Y., Saha, N., Miri, A., Baghaei, M., Roomiani, L., Ghaderpoori, M., et al., 2018. Metal concentrations in fillet and gill of parrotfish (Scarus ghobban) from the Persian Gulf and implications for human health. Food and Chemical Toxicology 118, 348–354.
Favalli, M., Mazzarini, F., Pareschi, M.T., Boschi, E., 2004. Role of local wind circulation in plume monitoring at Mt. Etna volcano (Sicily): insights from a mesoscale numerical model. Geophysical Research Letters 31, 9.

Finley, B.D., Swartzendruber, P.C., Jaffe, D.A., 2009. Particulate mercury emissions in regional wildfire plumes observed at the Mount Bachelor Observatory. Atmospheric Environment 43 (38), 6074–6083.

Giri, S., Singh, A.K., 2014. Assessment of surface water quality using heavy metal pollution index in Subarnarekha River, India. Water Quality, Exposure and Health 5 (4), 173–182.

Goodale, B.C., Walter, R., Pelsue, S.R., Thompson, W.D., Wise, S.S., Winn, R.N., et al., 2008. The cytotoxicity and genotoxicity of hexavalent chromium in medaka (Oryzias latipes) cells. Aquatic Toxicology 87 (1), 60–67.

Hasan, S.H., Talat, M., Rai, S., 2007. Sorption of cadmium and zinc from aqueous solutions by water hyacinth (Eichchornia crassipes). Bioresource Technology 98 (4), 918–928.

Hawkesworth, C.J., Kemp, A.I.S., 2006. Evolution of the continental crust. Nature 443 (7113), 811–817.

Hayat, S., Khalique, G., Irfan, M., Wani, A.S., Tripathi, B.N., Ahmad, A., 2012. Physiological changes induced by chromium stress in plants: an overview. Protoplasma 249 (3), 599–611.

Herawati, N., Suzuki, S., Hayashi, K., Rivai, I.F., Koyama, H., 2000. Cadmium, copper, and zinc levels in rice and soil of Japan, Indonesia, and China by soil type. Bulletin of Environmental Contamination and Toxicology 64 (1), 33–39.

Huff, J., Lunn, R.M., Waalkes, M.P., Tomatis, L., Infante, P.F., 2007. Cadmium-induced cancers in animals and in humans. International Journal of Occupational and Environmental Health 13 (2), 202–212.

Jaishankar, M., Tseten, T., Anbalagan, N., Mathew, B.B., Beeregowda, K.N., 2014. Toxicity, mechanism and health effects of some heavy metals. Interdisciplinary Toxicology 7 (2), 60.

Jasrotia, S., Kansal, A., Mehra, A., 2017. Performance of aquatic plant species for phytoremediation of arsenic-contaminated water. Applied Water Science 7 (2), 889–896.

Jayakumar, K., Rajesh, M., Baskaran, L., Vijayarengan, P., 2013. Changes in nutritional metabolism of tomato (Lycopersicon esculantum Mill.) plants exposed to increasing concentration of cobalt chloride. International Journal of Food Nutrition and Safety 4 (2), 62–69.

Jiang, C., Chen, H., Zhang, Y., Feng, H., Shehzad, M.A., Wang, Y., et al., 2018. Complexation Electrodialysis as a general method to simultaneously treat wastewaters with metal and organic matter. Chemical Engineering Journal 348, 952–959.

Kabata-Pendias, A., Pendias, H., 2001. Trace Elements in Soils and Plants. CRC Press Inc., *Boca Raton, FL, USA*.

Khillare, P.S., Jyethi, D.S., Sarkar, S., 2012. Health risk assessment of polycyclic aromatic hydrocarbons and heavy metals via dietary intake of vegetables grown in the vicinity of thermal power plants. Food and Chemical Toxicology 50 (5), 1642–1652.

Kumar, S.P., Edward, J.K., 2009. Assessment of metal concentration in the sediment cores of Manakudy estuary, south west coast of India. Indian Journal of Marine Sciences 38, 235–248.

Landrigan, P.J., Boffetta, P., Apostoli, P., 2000. The reproductive toxicity and carcinogenicity of lead: a critical review. American Journal of Industrial Medicine 38 (3), 231–243.

Leão, G.A., de Oliveira, J.A., Felipe, R.T.A., Farnese, F.S., Gusman, G.S., 2014. Anthocyanins, thiols, and antioxidant scavenging enzymes are involved in Lemna gibba tolerance to arsenic. Journal of Plant Interactions 9 (1), 143–151.

Ling, W., Shen, Q., Gao, Y., Gu, X., Yang, Z., 2007. Use of bentonite to control the release of copper from contaminated soils. Soil Research 45 (8), 618–623.

Li, H., Cheng, F., Wang, A., Wu, T., 2005, September. Cadmium removal from water by hydrophytes and its toxic effects. In Proceeding of the International Symposium of Phytoremediation and Ecosystem Health, Hangzhou, China.

Lock, K., Janssen, C.R., 2002. Ecotoxicity of nickel to Eisenia fetida, Enchytraeus albidus and Folsomia candida. Chemosphere 46 (2), 197–200.

Lohani, M.B., Singh, A., Rupainwar, D.C., Dhar, D.N., 2008. Seasonal variations of heavy metal contamination in river Gomti of Lucknow city region. Environmental Monitoring and Assessment 147 (1), 253–263.

Lone, M.I., He, Z.L., Stoffella, P.J., Yang, X.E., 2008. Phytoremediation of heavy metal polluted soils and water: progresses and perspectives. Journal of Zhejiang University. Science. B 9 (3), 210–220.

Marques, M., Aguiar, C.R.C., Silva, J.J.L.S.D., 2011. Technical challenges and social, economic and regulatory barriers to phytoremediation of contaminated soils. Revista Brasileira de Ciência do Solo 35 (1), 1–11.

Ma, L.Q., Komar, K.M., Tu, C., Zhang, W., Cai, Y., Kennelley, E.D., 2001. A fern that hyperaccumulates arsenic. Nature 409 (6820), 579–579.

McLaughlin, M.J., Hamon, R.E., McLaren, R.G., Speir, T.W., Rogers, S.L., 2000. A bioavailability-based rationale for controlling metal and metalloid contamination of agricultural land in Australia and New Zealand. Soil Research 38 (6), 1037–1086.

Mishra, R., Kumar, A., Singh, E., Kumar, S., Tripathi, V.K., Jha, S.K., et al., 2022. Current status of available techniques for removal of heavy metal contamination in the river ecosystem. Ecological Significance of River Ecosystems. Elsevier, pp. 217–234.

Moghaddam, M.H., Lashkaripour, G.R., Dehghan, P., 2014. Assessing the effect of heavy metal concentrations (fe, Pb, Zn, Ni, Cd, As, Cu, Cr) on the quality of adjacent groundwater resources of Khorasan Steel Complex. International Journal of Plant, Animal and Environmental Sciences 4 (2), 511–518.

Mohammadi, A.A., Zarei, A., Esmaeilzadeh, M., Taghavi, M., Yousefi, M., Yousefi, Z., et al., 2020. Assessment of heavy metal pollution and human health risks assessment in soils around an industrial zone in Neyshabur, Iran. Biological Trace Element Research 195 (1), 343–352.

Moon, H.B., Kim, H.S., Choi, M., Choi, H.G., 2010. Intake and potential health risk of polycyclic aromatic hydrocarbons associated with seafood consumption in Korea from 2005 to 2007. Archives of Environmental Contamination and Toxicology 58 (1), 214–221.

Nriagu, J., Becker, C., 2003. Volcanic emissions of mercury to the atmosphere: global and regional inventories. Science of the Total Environment 304 (1–3), 3–12.

Nriagu, J.O., 1989. A global assessment of natural sources of atmospheric trace metals. Nature 338 (6210), 47–49.

Obinna, I.B., Ebere, E.C., 2019. A review: Water pollution by heavy metal and organic pollutants: brief review of sources, effects and progress on remediation with aquatic plants. Analytical Methods in Environmental Chemistry Journal 2 (03), 5–38.

Odigie, K.O., Flegal, A.R., 2011. Pyrogenic remobilization of historic industrial lead depositions. Environmental Science & Technology 45 (15), 6290–6295.

Offem, B.O., Ayotunde, E.O., 2008. Toxicity of lead to freshwater invertebrates (Water fleas; Daphnia magna and Cyclop sp) in fish ponds in a tropical floodplain. Water, Air, and Soil Pollution 192 (1), 39–46.

Okoumassoun, L.E., Brochu, C., Deblois, C., Akponan, S., Marion, M., Averill-Bates, D., et al., 2002. Vitellogenin in tilapia male fishes exposed to organochlorine pesticides in Ouémé River in Republic of Benin. Science of the Total Environment 299 (1–3), 163–172.

Othumpangat, S., Kashon, M., Joseph, P., 2005. Eukaryotic translation initiation factor 4E is a cellular target for toxicity and death due to exposure to cadmium chloride. Journal of Biological Chemistry 280 (26), 25162–25169.

Patterson, C.C., 1965. Contaminated and natural lead environments of man. Archives of Environmental Health: An International Journal 11 (3), 344–360.

Pinho, S., Ladeiro, B., 2012. Phytotoxicity by Lead as Heavy Metal Focus on Oxidative Stress. Journal of Botany .

Pinto, E., Sigaud-kutner, T.C., Leitao, M.A., Okamoto, O.K., Morse, D., Colepicolo, P., 2003. Heavy metal–induced oxidative stress in algae 1. Journal of Phycology 39 (6), 1008–1018.

Radić, S., Babić, M., Škobić, D., Roje, V., Pevalek-Kozlina, B., 2010. Ecotoxicological effects of aluminum and zinc on growth and antioxidants in Lemna minor L. Ecotoxicology and Environmental Safety 73 (3), 336–342.

Ralston, N.V., Raymond, L.J., 2018. Mercury's neurotoxicity is characterized by its disruption of selenium biochemistry. Biochimica et Biophysica Acta (BBA)-General Subjects 1862 (11), 2405–2416.

Rascio, N., Navari-Izzo, F., 2011. Heavy metal hyperaccumulating plants: how and why do they do it? And what makes them so interesting? Plant Science 180 (2), 169–181.

Razinger, J., Dermastia, M., Koce, J.D., Zrimec, A., 2008. Oxidative stress in duckweed (Lemna minor L.) caused by short-term cadmium exposure. Environmental Pollution 153 (3), 687–694.

Sankhla, M.S., Kumari, M., Nandan, M., Kumar, R., Agrawal, P., 2016. Heavy metals contamination in water and their hazardous effect on human health-a review. International Journal of Current Microbiology and Applied Sciences 5 (10), 759–766.

Satarug, S., Nishijo, M., Ujjin, P., Vanavanitkun, Y., Moore, M.R., 2005. Cadmium-induced nephropathy in the development of high blood pressure. Toxicology Letters 157 (1), 57–68.

Sethy, S.K., Ghosh, S., 2013. Effect of heavy metals on germination of seeds. Journal of Natural Science, Biology, and Medicine 4 (2), 272.

Sneller, F.E.C., Van Heerwaarden, L.M., Schat, H., Verkleij, J.A., 2000. Toxicity, metal uptake, and accumulation of phytochelatins in Silene vulgaris exposed to mixtures of cadmium and arsenate. Environmental Toxicology and Chemistry: An International Journal 19 (12), 2982–2986.

Sooksawat, N., Meetam, M., Kruatrachue, M., Pokethitiyook, P., Nathalang, K., 2013. Phytoremediation potential of charophytes: bioaccumulation and toxicity studies of cadmium, lead and zinc. Journal of Environmental Sciences 25 (3), 596–604.

Sricoth, T., Meeinkuirt, W., Saengwilai, P., Pichtel, J., Taeprayoon, P., 2018. Aquatic plants for phytostabilization of cadmium and zinc in hydroponic experiments. Environmental Science and Pollution Research 25 (15), 14964– – 14976.

Tan, S.W., Meiller, J.C., Mahaffey, K.R., 2009. The endocrine effects of mercury in humans and wildlife. Critical Reviews in Toxicology 39 (3), 228–269.

Tyagi, S., Sharma, B., Singh, P., Dobhal, R., 2013. Water quality assessment in terms of water quality index. American Journal of Water Resources 1 (3), 34–38.

Verla, A.W., Verla, E.N., Enyoh, C.E., 2016. Petroleum hydrocarbons and heavy metals risk of consuming fish species from oguta lake, Imo State, Nigeria. In Proceedings of the Sixth International Science Congress, ISCA-ISC-8EVS-08-Poster, Pune, Maharastra, India.

Verla, A.W., Verla, E.N., Enyoh, C.E., Leizou, K., Peter, N.O., 2019. Using physicochemical properties in assessment of river water for consumption and irrigation in Nigeria. Eurasian Journal of Analytical Chemistry 5, 14–23.

Waisberg, M., Joseph, P., Hale, B., Beyersmann, D., 2003. Molecular and cellular mechanisms of cadmium carcinogenesis. Toxicology 192 (2–3), 95–117.

Weinhold, B., 2011. Fields and forests in flames: vegetation smoke and human health. Environmental Health Perspectives 119, A386–A393.

Wiafe, S., Buamah, R., Essandoh, H., Darkwah, L., 2019. Assessment of Typha capensis for the remediation of soil contaminated with As, Hg, Cd and Pb. Environmental Monitoring and Assessment 191 (6), 1–14.

Wirnkor, V., Ngozi, V., Emeka, A., 2018. Water Pollution scenario at river Uramurukwa flowing through owerri metropolis, IMO State, Nigeria. International Journal of Environment and Pollution Research 6, 38–49.

World Health Organization, 2018. WHO Water, Sanitation and Hygiene strategy 2018–2025 (No. WHO/CED/PHE/WSH/18.03). World Health Organization.

Wuana, R.A., Okieimen, F.E., 2011. Heavy metals in contaminated soils: a review of sources, chemistry, risks and best available strategies for remediation. International Scholarly Research Notices 2011.

Xun, Y., Feng, L., Li, Y., Dong, H., 2017. Mercury accumulation plant Cyrtomium macrophyllum and its potential for phytoremediation of mercury polluted sites. Chemosphere 189, 161– – 170.

Yanagi, T., 2011. Chemical composition of continental crust and the primitive mantle. In Arc Volcano of Japan (pp. 9 – 17). Springer, Tokyo.

Young, D.R., Jan, T.K., 1977. Fire fallout of metals off California. Marine Pollution Bulletin 8 (5), 109– – 112.

Zhang, T., Lu, Q., Su, C., Yang, Y., Hu, D., Xu, Q., 2017. Mercury induced oxidative stress, DNA damage, and activation of antioxidative system and Hsp70 induction in duckweed (Lemna minor). Ecotoxicology and Environmental Safety 143, 46–56.

CHAPTER

11

Marine environmental chemistry and ecotoxicology of heavy metals

Shriya Garg[1,2] *and Mangesh Gauns*[1,2]

[1]Academy of Scientific and Innovative Research (AcSIR), Gaziabad, India [2]CSIR- National Institute of Oceanography, Goa, India

11.1 Introduction

Heavy metals (HM) are defined as naturally occurring metallic elements with density (relative to water) greater than $5\,g/cm^3$ (Walker et al., 2005; Koller and Saleh, 2018; Ali and Khan, 2018). In the periodic table, elements with atomic number greater than 20 can be classified as HM which includes metal groups such as transition elements, metalloids, lanthanides, etc. Most of the HM are found at very low concentrations (in the range of parts per billion) in the natural environment and are also sometimes referred to as trace metals. Certain HM are essential for living organisms as they are required for carrying out metabolic activities inside the cells. Metals such as zinc (Zn), iron (Fe), copper (Cu) and molybdenum (Mo) form integral part of metal protein complexes in enzymes and aid in the enzymatic activity (Tchounwou et al., 2012). Iron and copper are constituents of oxygen transport proteins (hemoglobin and hemocyanin) inside the living organisms (Camaschella et al., 2020). Many others such as arsenic (As), mercury (Hg), lead (Pb), etc., have no reported role in biological functions. HM are mostly talked in context of environmental pollution caused due to human intervention and their toxic effect on living organisms.

HM enter the marine system through various natural and anthropogenic sources. Natural inputs are mainly from riverine discharges which carry with them weathered continental rock sediments containing HM. Volcanic eruptions and atmospheric deposition contribute minor fraction of HM in the oceans. With the advent of increasing developmental activities, HM input in the oceans has increased manifold. The coasts have become busier, with higher number of construction, mining, industrial and transport activities, not to forget the enormous amount of waste generated by these activities which is eventually discharged in the coastal seas. Nevertheless, our main focus in this chapter is to discuss the fate of the HM once they enter into the marine environment.

Metals in Water
DOI: https://doi.org/10.1016/B978-0-323-95919-3.00011-2

11.2 Heavy metals in the marine environment

Marine environment comprises of mainly three domains: seawater, marine sediments and marine living organisms. HM concentrations in seawater and sediments are discussed in the following sections. HM concentrations in living organisms is discussed with the ecotoxicological point of view later in this chapter.

11.2.1 Composition of heavy metals in seawater

HM concentration in seawater can vary both temporally and spatially. For instance, near the estuaries and coasts, there are greater inputs from anthropogenic activities and continental weathering, which leads to higher concentrations of HM as compared to the open ocean. Besides, processes such as primary production, upwelling, advection, etc., may also contribute to variability in HM concentration. These processes cycle HM within different spheres of the ocean (biosphere, hydrosphere and lithosphere). Residence time of the metal in the ocean is also an important factor regulating its concentration in seawater. Metals having higher residence time stays longer in the ocean resulting in their higher concentration in seawater. Mercury has an average concentration of 1.5 picomoles (pM) in the world ocean (Lamborg et al., 2002; Gworek et al., 2016). Northern Atlantic recorded higher mercury concentrations (2.0 pM) compared to Antarctic (0.8 pM) may be due to anthropogenic influence. A detailed review has been done on mercury in oceanic water by Gworek et al. (2016) in which they have discussed variability of mercury in different parts of the world ocean. For cadmium, average concentration in surface waters of the open-ocean ranges between 1 to about 100 ng/L (Neff, 2002). Lower concentrations of cadmium are recorded mainly in subtropical and central gyres while higher concentrations are recorded in areas of upwelling. In clean coastal waters, cadmium concentrations may come around 200 ng/L. In the continental shelf regions, cadmium is released from bottom sediments in the overlying waters, increasing cadmium concentrations of surface waters. Higher levels are also observed as a consequence of coastal pollution due to anthropogenic inputs. In the open ocean, cadmium concentration is very low in the surface waters (< 1 ng/L), however its concentration increases with depth below the euphotic zone. Maximum cadmium concentrations (~ 145 ng/L) is found at the depth of nutrient maximum (which varies from 900 to 1000 m) (Neff, 2002). Zinc concentrations ranges between 0.05 to 9 nmol/kg in the world ocean (Bruland and Lohan, 2003). Like cadmium, it also has nutrient like profile with lower surface concentrations and higher concentrations with increasing depth. Lead is found at trace levels in the ocean, its concentration is in the range of picomoles (10^{-12} moles) per kilogram (Cullen and McAlister, 2017). Maximum lead concentrations are found not at the surface but at the subsurface. This is due to sinking of surface water which was previously mixed with lead during periods of high anthropogenic lead emissions. Over time due to reduction in anthropogenic emissions of leaded gasoline along with seawater ventilation and mixing as well as scavenging losses of lead, average maximum concentration of lead has reduced from ~ 170 to ~ 35 pmol/kg and is found at deeper depths of around 1000–2000 m. Indian Ocean has higher concentrations of lead (40–80 pmol/kg) in surface waters due to emissions from the surrounding countries which had delayed phase out of leaded gasoline and are in rapidly developing stage. Southern Ocean, which has lesser anthropogenic influence, has low concentrations of lead (5–12 pmol/kg) (Cullen and McAlister, 2017). Arsenic concentration in the open-ocean ranges from 0.5 to 3 μg/L, with an average of around 1.7 μg/L (Neff, 2002). In the estuarine regions, concentration of arsenic increases downstream towards the mouth of the estuary as the arsenic poor freshwater mixes with arsenic rich seawater. Total nickel ranges from 0.2 to 2 mg/L in the open ocean (Glass and Dupont, 2017). In coastal areas, however, due to anthropogenic activities, Ni concentrations reach up to 250 mg/L. Vertical profile of nickel is similar to that of nutrient, with low surface concentrations (2–4 nM) that increase to around 6 nM in Pacific and 10 nM in Atlantic ocean at depths >1000 m. The nutrient-like profile indicates that nickel undergoes biological uptake in the surface ocean while at greater depths it is remineralized back into dissolved form. Elevated Ni concentrations is found below 500 m depth in Pacific Ocean due to its very old age water mass compared to other oceans (Glass and Dupont, 2017). For chromium, surface water concentration is around 2.8 nmol/kg. At deeper depths (800–2500 m) chromium concentration may become slightly higher around 3 nmol/kg. Below 2500 m, concentrations of chromium increase to about 3.5–4.4 nmol/kg in the Atlantic while around 4.5–6 nmol/kg in the Pacific Ocean (Geisler and Schmidt, 1992).

11.2.2 Heavy metals in marine sediments

Marine sediments are the largest sink of HM. Adsorbed on the suspended particulate matter, HM sink and get deposited in the sediments. Here, they may get remobilized into the overlying water column through biochemical processes or may form part of the deposited metals pool in the sediments. Cadmium concentrations in marine sediments ranges from 0.1 to 0.6 μg/g dry wt (Neff, 2002). For arsenic, concentrations in marine and estuarine sediments are in the range of 5–15 μg/g dry weight (Moore and Ramamoorthy, 2012; Neff, 2002). The average concentration of arsenic in deep sea sediments is around 40 μg/g. Sediments in the deep sea region of equatorial Pacific Ocean may contain higher concentrations (>400 μg/g) of total arsenic. Continental shelf sediments may also record higher concentrations of arsenic due to greater inputs from natural and anthropogenic sources. Chromium concentration in marine sediments is similar to that of crustal rocks and is in the range of 60–100 mg/kg (Geisler and Schmidt, 1992). Coastal and estuarine sediments influenced by industrial effluent discharges may record higher chromium levels, in the range of several hundred mg/kg.

11.2.3 Residence time of heavy metals in the ocean

The residence time of mercury in the ocean on an average is around 20–30 years (Gworek et al., 2016). The residence times of lead is very low in low productivity surface waters (around 2 years) but it is higher in the ocean interior (around $10–10^2$ years). Cadmium's residence time in the ocean is about 450 years. Nickel's residence time in the ocean is higher, around 10^4 years. Residence time of chromium in ocean water is very high and ranges from 25000 to 40000 years. Table 11.1 gives the concentration of HM in seawater and marine sediments with their residence time.

TABLE 11.1 Concentration and residence time of heavy metals in the ocean.

Heavy metal	Concentration (seawater)	Concentration (sediments) μg/g dry wt	Residence time (years)	Reference
Mercury	0.8–1.5 pmol/kg	–	20–30	Gworek et al. (2016)
Cadmium	1–145 ng/L	0.1 – 0.6	450	Neff (2002)
Zinc	0.05 – 9 nmol/kg	–	–	Bruland and Lohan (2003)
Lead	2–170 pmol/kg	–	2	Cullen and McAlister (2017)
Arsenic	0.5–3 μg/L	40	–	Neff (2002)
Nickel	0.2–250 mg/L	-	10^4	Glass and Dupont (2017)
Chromium	3–6 nmol/kg	60–100 mg/kg	25,000–40,000	Geisler and Schmidt (1992)

11.3 Biogeochemistry of heavy metals in the ocean

In the natural environment, metals do not exist in one form. They may exist in different oxidation states or bind with different anions or inorganic/organic ligands based on the environmental conditions present (Ansari et al., 2004; Valavanidis and Vlachogianni, 2010). Metal speciation refers to various physical and chemical forms in which metals may exist in the environment. It determines the mobility of the metal through various components of the marine environment vis a vis water, sediment, or living organisms. It also determines the nature of toxicity of the metal in marine biota. Based on the reactivity, residence time and biogeochemical interaction of the metals in the marine environment, Whitfield and Turner (1987) classified them as:

1. Conservative HM which have very low reactivity and very long residence time ($>10^5$ years)
2. Recycled metals which participate in biological processes and get internally recycled
3. Scavenged metals which have very high reactivity and short residence time ($<10^3$ years).

In seawater, mercury exists in inorganic as well as organic forms as Hg, Hg^{+2}, methyl mercury and dimethyl mercury (Morel et al., 1998). Hg^{+2} is unstable and may either undergo reduction to Hg^0 or adsorption on suspended particulate matter or methylation based on the surrounding conditions (Strode et al., 2007). Mercury in sediments usually undergoes methylation and forms methyl mercury. Cadmium in seawater is mostly in dissolved form, very less fraction of the total cadmium exists in association with the suspended particulate matter as particulate cadmium. The distribution of dissolved and particulate cadmium in the ocean usually follows that of phosphate. The ration of cadmium to phosphate is constant in almost all parts of the ocean but may vary with the age of water mass (Bruland et al., 1994). Similarity in cadmium and phosphate distribution points towards the control of biological activity (esp. that of phytoplankton) on cadmium speciation. Cadmium forms complexes with organic ligands, however stability of such complexes is usually low. In the open ocean, particulate matter content is very low, however in the coastal and estuarine regions, higher levels of suspended matter is present which can provide a medium for surface complexation of cadmium leading to a significant amount of scavenging of cadmium in coastal waters. The low stability of Cd-complexes with organic ligands (humic substances and amino acid) and low levels of dissolved organic matter (DOM)/particulate matter in the open ocean are the reasons why inorganic speciation of dissolved cadmium in the oceans holds more importance than its organic speciation. 97% of the inorganic cadmium in dissolved state usually form complexes with chloride ions in seawater, in which 10% exists as $CdCI^+$ and 87% as $CdCl_2$ (Mart and Nürnberg, 1986). Lead forms inorganic complexes with Cl^- (46%–75%), $CO_3{}^{2-}$ (11%–42%), and OH^- (8%–9%) in seawater (Cullen and McAlister, 2017). Higher proportion of lead is present in more labile form, that is, Pb^{+2} or as inorganic complexes towards the open seas compared to coastal ocean. Arsenic can occur in four oxidation states, +5, +3, 0, and −3 in the marine environment (Moore and Ramamoorthy, 2012; Landner, 1998). Elemental arsenic is rarely found. As^{-3} is mainly found in excessively reducing environments. Arsenate ($AsO^{-3}{}_4$) and arsenite ($AsO^{-3}{}_3$) are the dominant inorganic forms of arsenic in the ocean. Arsenate dominates in oxygenated seawater while arsenite in relatively reducing conditions. In open-ocean water, the ratio of arsenate to arsenite is approximately 327 (Neff, 2002). In coastal waters due to greater natural as well as human inputs from the continent and higher primary production resulting in reducing conditions, the ratio of As (V) to As (III) is much lower. Organic forms of arsenic also exists in seawater as methyl arsenic acid (MMA) and dimethyl arsenic acid (DMA), however they are in much lower proportion compared to inorganic forms. Nickel, around 50% exists in free inorganic form as Ni^{2+} in seawater and 10% as chloride complexes. Only 10%–30% of Ni is bound to the organic ligands (Glass and Dupont, 2017). Chromium in marine waters occurs in + 3 and + 6 oxidation states (Richard and Bourg, 1991). The most common species of chromium present are $Cr\ (OH)\ _2{}^{+\ 4}$ and $CrO_4{}^{2-}$. Nakayama et al. (1981) found that seawater contains around 45% − 60% organic chromium species of the total dissolved chromium present in seawater. Table 11.2 summarizes different species of HM found in the marine environment.

TABLE 11.2 Speciation of heavy metals in seawater and marine sediments.

Heavy metal	Metal species
Mercury	Hg, Hg^{+2}, CH_3Hg, $(CH_3)_2Hg$
Cadmium	Cd^{+2}, $CdCI^+$, $CdCl_2$
Lead	Pb^{+2}, $PbCl_2$, $Pb(OH)_2$, $PbCO_3$
Arsenic	As^{-3}, As^0, As^{+3}, As^{+5}, MMA, DMA
Nickel	Ni^{+2}
Chromium	Cr^{+3}, Cr^{+6}, $Cr\ (OH)\ _2{}^{+\ 4}$, $CrO_4{}^{2-}$

MMA: Monomethyl arsenic acid; DMA: dimethyl arsenic acid.

11.3.1 Factors controlling heavy metals speciation

In the marine environment, HM interact with seawater which possess unique chemical characteristics compared to freshwater. Also, presence of biological agents such as microbial communities and phytoplankton have profound influence on heavy metal speciation. We will discuss these chemical and biological factors in the following section, and how they affect the speciation of HM in the ocean. Fig. 11.1 illustrates the major factors controlling heavy metal speciation in marine environment.

11.3.1.1 pH

Metal precipitates (carbonates and hydroxides of copper, cadmium and zinc) have very low solubility in the seawater due to its pH range of 8.1–8.3. Mercuric ion hydrolyzes at pH 2 − 6, forming $Hg(OH)_2$ at pH 6 (Moore and Ramamoorthy, 2012). Cadmium is found in the divalent state (Cd^{+2}) up to pH 8. Cadmium does not precipitate if there is absence of anions such as phosphate and sulfide in the surrounding medium, in such case it is available for adsorption and complexation onto suspended organic matter. Cadmium starts hydrolyzing at pH 9 and forms $Cd(OH)^+$ (Moore and Ramamoorthy, 2012). In estuaries, as salinity and pH increase downstream toward the coastal region, more than 95% of lead is in particulate form. Lead precipitates (of phosphates and sulfides) hydrolyze at pH above 6. In the pH range of 8.1 to 8.2, $Pb(OH)^+$ predominates over the other lead chloride complexes in seawater. Insoluble $Pb(OH)_2$ is formed above pH 10 (Moore and Ramamoorthy, 2012). Anionic forms of arsenate (98% $HAsO_4^{-2}$, traces of $H_2AsO_4^{-1}$ and AsO_4^{-3}) and neutral forms of arsenite ($As[OH]_3^0$) are most abundant inorganic forms of arsenic at estuarine and marine water pH (Neff, 2002). Zinc starts to hydrolyze at pH 7 to 7.5. It forms stable complex $Zn(OH)_2$ at pH greater than 8 (Moore and Ramamoorthy, 2012).

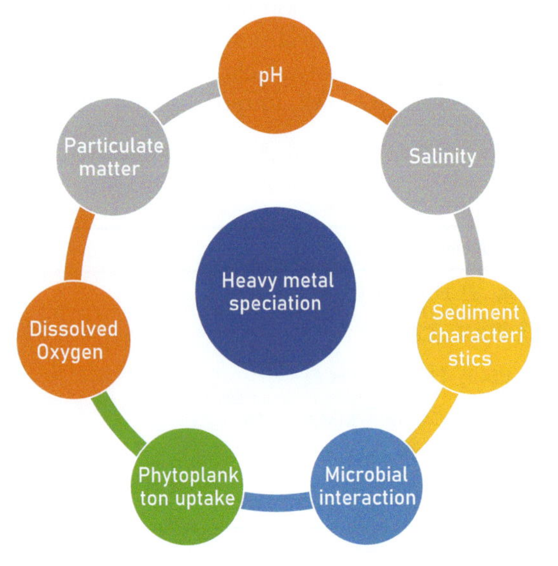

FIGURE 11.1 Flowchart showing factors affecting heavy metal speciation in marine environment.

11.3.1.2 Salinity

Anions such as sulfate, chloride, fluoride, bicarbonates/carbonates present in the seawater precipitate the dissolved metal cations. Chloride in seawater react with the HM to form metal chlorides. Metal chlorides brings the metal in dissolved form contrary to the conditions in freshwaters where usually metal oxides are formed which get precipitated. Chloride ions complex with mercury to form $[HgCl]^+$, $[HgCl_2]$, $[HgCl_3]^-$, $[HgCl]^{2-}$ or $HgBrCl^-$. Mercury forms soluble complexes of hydroxides and sulfides such as $[HgOH]^+$, and $[HgS_2]^{2-}$ in seawater (Gworek et al., 2016). 97% of dissolved cadmium exists as chloro-complexes. The dominant cadmium species in the dissolved state in ocean are $CdC1^+$, $CdC1_2$ and $CdC1_3^{-1}$ (Neff, 2002). Free ionic form of cadmium (Cd^{-2}), which is also the most toxic species of cadmium, is found in very less proportion (around 3 percent of the total dissolved cadmium) in the seawater. An inverse relation exists between seawater salinity and free ionic Cd^{+2}, with increase in salinity there is higher amount of chloride complexation leaving very less fraction of cadmium in free ionic state (Cd^{+2}) (Neff, 2002).

11.3.1.3 Dissolved oxygen

In the marine environment, many of the mercury forms depend on the redox conditions. $HgCl_4{}^{2-}$ and $HgOH^-$ are found mainly in oxygenated conditions, whereas sulfides (HgS, HgS_2H_2, HgS_2H^-, $HgS_2{}^2$, HgS^{2-}, and CH_3HgS^-) are formed in the reducing conditions. Mercuric sulfides are insoluble in water, they get associated with organic matter or iron oxides and get precipitated in the marine sediments (Gworek et al., 2016). Arsenate is the dominant form of arsenic in oxygenated marine waters, however significant amount of arsenite may also be found in oxygenated water as a result of arsenate reduction, atmospheric deposition, upwelling of anoxic waters, and waters entering from hypoxic river basins (Cutter, 1992; Landner, 1998; Neff, 2002). Similarly, arsenate present in anoxic bottom waters of estuaries and the deep ocean comes from sinking of organic matter containing arsenate adsorbed onto them, sinking of surface waters rich in dissolved oxygen into deeper anoxic zones and the slow rate of arsenate reduction (Cutter, 1992). Unlike other metals, dissolved Ni does not show much change to reducing conditions and increasing sulfide content in the bottom sea. Cr(VI) is the dominant form of chromium in oxygenated conditions while in reducing conditions and presence of H_2S, Cr(III) is the dominant species (Geisler and Schmidt, 1992).

11.3.1.4 Particulate matter

HM dissolved in river water converts to particulate form in the estuaries, since the colloidal particles in river water flocculates as they mix with the seawater. Particulate matter usually decreases as water moves closer to the estuarine mouth. Certain metals are readily adsorbed by the organic matter while others are not. In seawater, cadmium does not readily get adsorbed or form complexes with organic ligands such as humic or fulvic acids. Increase in salinity results in greater chloride complexation of cadmium, decreasing the affinity of cadmium for adsorption on organic matter. Cadmium is adsorbed on the particulate matter and is present in an exchangeable form. It readily gets desorbed back in the solution especially in lesser saline waters of the estuaries (Neff, 2002). Zinc forms strong stable complexes with humic and fulvic acids, which results in decrease in its free ionic form in seawater (Yang and Van den Berg, 2009; Sinoir et al., 2012). Similarly in places having higher concentrations of particulate matter like in estuarine and coastal waters, lead forms stable complexes with organic ligands, resulting in reduced concentration of freely available inorganic lead (Pb^{+2}). Chromium forms complexes with organic acids and this complexation is proportional to the amount of organic matter present in seawater. Cr (III) compounds have strong affinity to organic ligands and get readily adsorbed onto them (Nakayama et al., 1981).

11.3.1.5 Phytoplankton uptake

Cadmium concentrations get reduced in the surface waters of the ocean due to uptake by phytoplankton. Lane et al. (2005) reported cadmium in the enzyme carbonic anhydrase (CDCA1) (used by phytoplankton for acquiring inorganic carbon for photosynthesis) from the marine diatom *Thalassiosira weissflogii*. This provided explanation for the nutrient-like vertical profile of cadmium in the ocean. Lead's distribution and loss in the ocean is driven by processes involving scavenging of particle, in which Pb is adsorbed onto particle surfaces such as phytoplankton. Pb complexes which are lipophilic in nature can diffuse inside the cell membrane of the phytoplankton or else remain adsorbed onto their cell walls. The regions in the oceans having high productivity such as the coastal zones and areas of upwelling are the sites of Pb loss from dissolved phase in seawater through biological scavenging (Cullen and McAlister, 2017). Nickel also shows nutrient like vertical profile in the ocean with higher uptake in the surface waters compared to the greater depths. Phytoplankton growth is usually limited by the nitrogen (NO^{3-} and NH^{4+}) availability in the ocean. Ni is an important component of the enzyme urease which is involved in organic nitrogen assimilation (an alternative way of obtaining nitrogen in which urea is used as a substrate) in marine phytoplankton. Thus phytoplankton takes in large amount of nickel from the surface seawater to fulfill their demand for nitrogen in nitrogen limiting conditions (Glass and Dupont, 2017).

11.3.1.6 Microbial interaction

Microorganisms mediate transformation of HM into different species during uptake and detoxification process. They can both mobilize or immobilize the metals depending on the surrounding conditions. Mobilization of metals by microbes is attained through bioleaching or bioweathering of minerals, oxidation/reduction of metals and methylation. Immobilization of metals occurs due to biosorption of metals on microbial cell walls or extra cellular matrix, uptake of metals and their intracellular accumulation and oxidation or reduction of metals. For instance, microbes mediate four types of mercury transformation: (1) reduction of Hg (II) to Hg (0) (2) degradation of organic mercury compounds (3) methylation of Hg by sulfate reducing bacteria in anerobic conditions which leads to the formation of toxic methylmercury compounds in the marine sediments and (4) oxidation of Hg (0) to Hg (II) (Gworek et al., 2016; Batrakova et al., 2014).

11.3.1.7 Sediment characteristics

Physical and chemical characteristics of the sediments such as the size and texture, amount of organic matter present and the cation exchange capacity decides the rate of adsorption of HM on the sediments. Mercury can weakly bind through vander waal forces or strongly through covalent bonds with the sediments. It may also precipitate in the sediments along with ferromanganese oxides. Mercury adsorbed onto marine sediments is released in the overlying seawater. This process also depends on physical and chemical properties of the marine sediments such as pH, salinity (amount of chloride ions) and presence of complexing agents. Greater sulfate and methane content as well as persistence of anaerobic conditions in the sediments favors the formation of dimethyl mercury (Gworek et al., 2016). Sediment size greatly affects metals binding capacity. Finer the size of the sediments, greater the surface area for binding of metals. Thus in areas where finer sediments are found, there is higher probability that major proportion of metals will be present in bound state in the sediments. Highest cadmium concentrations are found in regions dominated by finer sediment fraction (Neff, 2002). In sediments where oxygenated conditions are present, cadmium forms complexes with carbonate and iron or manganese oxides which are water soluble in nature. With the increase in reducing conditions, manganese hydro-oxides get reduced and cadmium gets remobilized from the sediments into the overlying waters (Simpson, 1981). Cadmium in anoxic sediments form strong sulfide complexes. Arsenate is the dominant arsenic species in oxic sediments and associated pore water while arsenite dominates in the reduced sediments. Arsenite gets rapidly converted to arsenate, if present in oxidizing conditions. Arsenic gets adsorbed on manganese oxides in the oxygenated conditions or is present as $Fe_3\ (AsO_4)_2$ (Neff, 2002). Cr(VI) and Cr(III) ions get adsorbed onto minerals in the bottom sediments (Gorny et al., 2016). However, they also get desorbed quickly from the sediments. All the processes of metal adsorption and desorption on sediments are highly pH-dependent, thus heavy metal chemistry in marine sediments is characteristically different from that in freshwater sediments (Mart and Nürnberg, 1986).

11.4 Ecotoxicology of heavy metals in marine biota

HM form part of the marine ecosystem and are taken up through various routes by the organisms living in seawater and marine sediments. Inside the body of organisms, metals become part of their metabolic activities and go under various transformations. They may then either get accumulated inside the organism's body (process referred to as bioaccumulation) or may get excreted out. Metals may also be transferred from one trophic level to the next in the marine food chain and increase in concentration with each trophic level, phenomenon termed as biomagnification.

11.4.1 Bioavailability of heavy metals

Bioavailability of metal refers to the portion of total metal present in the environment, readily available for uptake into the biological system. To study the bioavailability of HM in the marine environment, the exchangeable fraction of metal in the seawater/sediments should be measured rather than its total concentration (Tomlinson et al., 1980). Bioavailability depends on the nature of metal species present in the environment (which further depends on environmental parameters such as pH, salinity and the type of organic and inorganic ligands present) as well as on the mode of uptake and physiology of the marine biota. For example, bioavailability of mercury in teleost fish, depends on the concentration of different species of mercury in the environment and also on how readily the fish can absorb these different mercury forms through gills, skin or gut. Cadmium has higher mobility and is more exchangeable and bioavailable in the marine environment because greater proportion of it is present in inorganic form, it binds weekly to the particulate matter, and also forms soluble sulfides under reducing conditions. Ionic cadmium (Cd^{+2}) is the most bioavailable species of cadmium, but it forms very less fraction of total cadmium in the seawater. Lead bioavailability and intracellular concentration in organisms depends on concentration of the free ion (Pb^{+2}) or its labile inorganic complexes in the seawater. However, lipophilic organic complexes of lead can passively diffuse through cell membrane into the organisms or may get adsorbed onto cell walls of marine phytoplankton and microbes. Arsenic undergoes chemical transformations in food chains of the marine ecosystems. These different chemical forms regulate the bioavailability of arsenic in marine organisms. Arsenate, the most dominant arsenic species in seawater gets converted to arsenite, mono- and dimethylarsinate or other organoarsenic compounds by the marine algae. Inorganic arsenic up taken from seawater is retained for longer time and more effectively in the organisms. Organoarsenic compounds are synthesized from inorganic arsenic inside the body of marine organisms or could be directly taken up through the food chain. Organoarsenic compounds are present at higher proportion (around 90%) compared to inorganic arsenic compounds at higher trophic levels. Bioavailability and toxicity of nickel changes as water chemistry changes from freshwater to seawater. Changes in water chemistry leads to changes in Ni transport pathways in the branchial and digestive epithelia of the marine organisms. For nickel too, the free ion (Ni^{+2}) is the most bioavailable as well as the most toxic chemical species as it easily enters through transporters on epithelial surfaces of the marine organisms (Wood, 2011). Complexation of nickel with inorganic/organic ligands and competition with other metal ions for uptake sites in organisms also impacts nickel's bioavailability and toxicity. Amount of dissolved organic carbon (DOC) (a complexing agent) in the seawater also alters bioavailability by reducing free Ni concentrations. In general, concentrations of DOC are lower in seawater thus it is of less importance in marine environment. Cr (VI) compounds mainly contribute to the toxic properties for chromium. These compounds are highly soluble in nature. They can pass through the cell membrane of the marine species as it is negatively charged. For mussels, most of the chromium is available in the dissolved state in the seawater. Chromium reaches to other marine invertebrates and fish through the food chain.

11.4.2 Toxicokinetics of heavy metals

Most of the metals (Cr, Cu, Hg, Ni and Zn) show higher toxicity in low salinity environment. This is due to the presence of greater free ions in freshwater as compared to saltwater where metals mostly bind with chlorides, carbonates and sulfates. Enrichment for some trace metals takes place when they are taken up from seawater by the phytoplankton, which are at the base of the marine food chain. For some metals (e.g., strontium), concentrations in the body of marine organisms are lower than that in seawater (Bernhard and Andreae, 1984).

Bioconcentration factor (BCF) is the ration of concentration of HM in the organism to that in its natural habitat, that is, water or sediments. It integrates the effects of a number of processes responsible for the concentration of an element in the organism relative to seawater/ marine sediments. It is calculated as (Ali et al., 2019):

$$BCF = C_{\text{organism tissue}} / C_{\text{abiotic medium}}$$

where $C_{\text{organism tissue}}$ is the metal concentration in the tissues of marine organisms and $C_{\text{abiotic medium}}$ is the metal concentration in the abiotic medium, that is, seawater/marine sediments.

Organism that are used to predict heavy metal pollution in the surrounding environment are called bioindicator organisms. We will now look at the fate of HM in the body of different groups of marine organisms.

11.4.2.1 *Marine plankton*

Effect of different HM on marine plankton has been seen in various studies. Amount of dissolved cadmium in the seawater and the ratio of cadmium to phosphate in the surrounding medium decides the bioconcentration of cadmium in the marine phytoplankton (Xu and Morel, 2013; Neff, 2002 and references therein). Radioisotope uptake experiments reported that the lead up taken by the phytoplankton mostly gets associated with structural components of the cells like the cell wall and cell membrane. Nickel was found to reduce chlorophyll-a concentration and rate of photosynthesis in the phytoplankton. Higher levels of nitrogen and phosphorus in the seawater results in greater bioaccumulation of Ni in the phytoplankton (Wang et al., 2007). In marine copepods (a group of marine zooplankton), nickel caused oxidative damage and modulated the gene expression. It also affected reproduction in copepods and decreased the viability of their eggs. Mechanisms of how exactly Ni is involved in the metabolic activities of the organisms is still not clear, however it is thought to affect the Ca^{+2} pathway inside the cells (Panneerselvam et al., 2018).

11.4.2.2 *Marine invertebrates*

Bioaccumulation of dissolved, ionic cadmium can take place by two means, either through passive or active transport through the gill epithelia of the marine invertebrates. In some crabs calcium pump is present in the posterior gills which aids in transport of cadmium through the gills. In the marine invertebrates, calcium and cadmium compete for the uptake site in the gills. The rate of uptake of cadmium in them is inversely proportional to the calcium concentration in the surrounding seawater. Bioaccumulation of cadmium and other HM leads to the production of metal-binding proteins, metallothioneins, in the tissues of marine organisms including the invertebrates. The metallothioneins bind with the cadmium inside the cell. This helps in maintaining low concentrations of cadmium (Cd^{+2}) in the cytoplasm. Glutathione also binds the metals inside the cells and serve similar function to metallothioneins. In bivalves and crabs, cadmium gets bound to plasma proteins in the hemolymph (Neff, 2002 and references therein). Binding of metals to organic compounds is a mechanism of detoxification in organisms to reduce the harmful effects of metals inside their bodies. The cadmium bound to organic chemicals such as metallothionein in the cytoplasm are bioavailable to the next trophic level in the marine food chain (Mart and Nürnberg, 1986; Neff, 2002). Lead is not taken up actively by marine organisms, but organisms which feed through filter feeding mechanism such as oysters take in dissolved or particulate lead directly from the water column or the sediments. Polychaete worms that live in marine sediments rich in metals, take in lead from the sediments and transfer it to the marine food web, however biomagnification of Pb is not reported (Cullen and McAlister, 2017). The integument (or the carapace) of the crustaceans strongly bind to nickel (Blewett and Leonard, 2017). Highest amount of Ni was reported in the carapace of green shore crabs exposed to water containing Ni (Blewett and Leonard, 2017 and references therein). In other marine invertebrates, Ni causes ion regulatory disruption. Ni was reported to cause disruption in Ca^{+2} influx in developing sea urchin embryos (Blewett and Leonard, 2017 and references therein). This further lead to skeletal malformations in embryos developing under Ni exposure.

11.4.2.3 Fish

Gills, digestive system and the skin are the major metal uptake sites in marine fish. Marine teleost fish drink seawater in order to replace water lost by them through osmosis. Thus gut plays a significant role in dissolved metals uptake and metal toxicity in marine fish. Mercury absorption and distribution in marine fish takes place as follows—mercury enters through branchial/digestive epithelia, it then dissolves in the blood and bind to its plasma proteins after which it is transported to various body tissues (Morcillo et al., 2017). In plasma, methyl mercury binds reversibly to cysteine (an amino acid) and consequently to sulfur-containing complexes, for example, glutathione. In the digestive tract, ingested methyl mercury is absorbed efficiently. Its distribution to the blood is complete within 30 hours of absorption from the gut. The blood may contain about 7% of the ingested dose of methyl mercury. The brain retains around 10% of the methyl mercury and is the most affected organ of the body. The remaining methyl mercury goes to the liver and kidney where it gets detoxified. After detoxification, it is excreted through bile and urine. Concentrations of methyl mercury get magnified within the food chain, reaching around 10,000- to 100,000-fold greater in fish than in the ambient water. Mercury primarily targets the central nervous system in the fish which results in numerous physiological dysfunctions such as brain lesions, cataracts, loss of appetite, impaired growth and development as well as reproductive anomalies (Morcillo et al., 2017). Nickel gets transported through cutaneous surface in some of the fish species such as Pacific hagfish (*Eptatretus stouti*) (Blewett and Leonard, 2017 and references therein). It gets accumulated in the intestine in killifish (Blewett and Leonard, 2017). Ni also gets accumulated in other body organs such as gills, intestine, stomach, liver and kidney in marine fish. Even at low concentrations, Ni can interact with Ca^{+2} and Mg^{+2} transport systems, bind to oxygen sensor proteins, bring changes in cellular homeostasis and produce oxidative stress in fish (Blewett and Leonard, 2017 and references therein). Toxicity of chromium for marine organisms is lower compared to other HM. This may be due to the fact that chromium ions have hard acid character. There has been no reported evidence for chromium bioaccumulation in any of the marine species. In marine species concentration of chromium in the muscle is around 0.5 mg/kg when collected near industrial areas (Moore and Ramamoorthy, 2012). Chromium is far less toxic to fish compared to marine invertebrates. Fig. 11.2 outlines the different toxicokinetic processes in marine organisms.

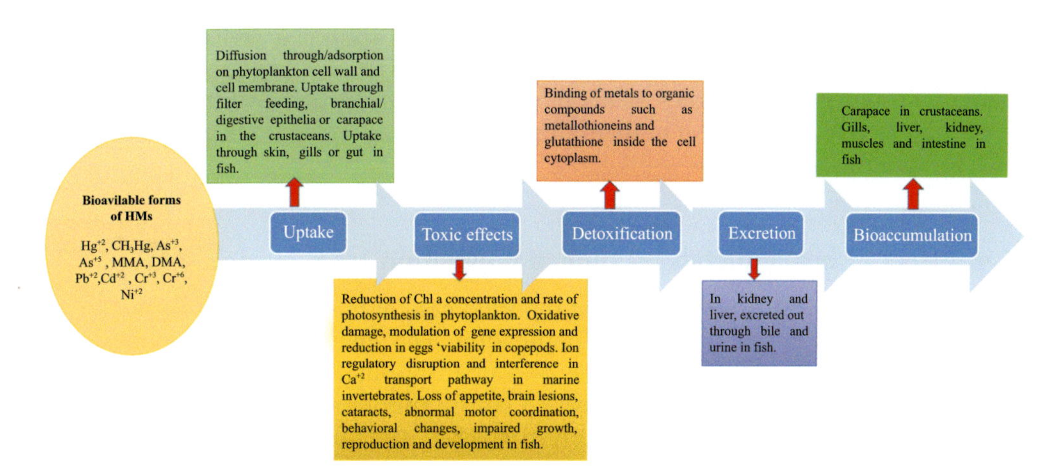

FIGURE 11.2 Overview of toxicokinetics of heavy metals (HMs) in marine biota. Bioavailable forms of HM are readily taken up by marine organisms through various entry points in their body. Inside the body, HM undergo various metabolic processes wherein they may get assimilated inside the body resulting in toxic side effects in the organisms. Organisms employ various detoxifying mechanisms to reduce the toxic effects of the metals. Part of HM is excreted outside the body with other waste products. Remainder gets accumulated inside the body tissues over a period of time. *DMA: Dimethyl arsenic acid; Chl: Chlorophyll; MMA: Monomethyl arsenic acid.*

11.4.3 Biomagnification of heavy metals along the marine food chain

Food chain in the marine environment starts with the seawater. Concentration of HM increases several times in phytoplankton as compared to seawater. The same trend is observed as one moves up the marine food chain from phytoplankton to copepods to plankton feeder fish such as anchovy, sardines, etc. (Amiard-Triquet et al., 1993; Wang, 2002). The increase in HM concentration is more at lower trophic levels as compared to higher trophic levels in the food chain as organisms at higher trophic levels are more efficient in regulating HM levels inside their bodies. However, the trend is not similar for all the HMs. Mercury increases along the food chain. Trend for As, Cd, Cu, Pb, and Zn is reverse and these metals decrease as they move higher up the food chain (Bernhard and Andreae, 1984). The organisms positioned at higher trophic levels have longer life span which also plays crucial role in concentration of HM. Phytoplankton divides within a few days, zooplankton (copepods) have a life cycle of 20–40 days, sardines live upto 5 years, majority of adult fish live from 1 to 3 years, tunas can reach upto 15 years (Bernhard and Andreae, 1984). Biomagnification of mercury in marine ecosystems is of greater importance compared to other heavy metals. Inorganic mercury has biological half-life of 60 days and gets completely eliminated in around 7 years. Efficient uptake and low elimination rate of methyl mercury compared to inorganic mercury, results in almost complete retention of the methyl mercury taken up by the organism. 10% of the inorganic mercury taken in is retained while 90% of methyl mercury is retained in the organism. Organisms at higher trophic levels feed on organisms at lower trophic level which have "preconcentrated" methyl mercury, with a higher relative methyl mercury content. Higher trophic level organisms retain methyl mercury for longer time in their bodies due to their longer life span (Bernhard and Andreae, 1984). HM biomagnification is a serious environmental concern as it is through this process, humans get exposed to greater concentrations of HM. Humans consume fish and other seafood which constitute higher trophic level in the marine food chain and thus have in them higher concentrations of HM accumulated inside their body tissues.

11.4.4 Effect of heavy metals on human population from marine sources

Minamata disease in 1950s was the first reported case of the effect of heavy metal pollution in the environment on the human population. The disease was result of poisoning caused by methyl mercury on large scale of human population in Japan. Humans were exposed to this toxic form of mercury through consumption of marine organisms that had accumulated methyl mercury in their body tissues. A similar case of mercury poisoning took place in Sweden (Zaib et al., 2015). Humans mainly get exposed to methyl mercury through seafood consumption (Bernhoft, 2012; Tamele and Loureiro, 2020). Mercury can cause brain damage especially in children who are in growing stage. Arsenic is suspected to cause carcinogenicity in humans (Gupta et al., 2017; Costa, 2019). Thus levels of arsenic in seawater from which seafood is harvested should be well monitored. Higher concentrations of Cd and Cu were found in the tissues of edible oyster along the west coast of India which pose a threat to human health (Shenai-Tirodkar et al., 2016). High concentrations of HM were found in the coastal marine sediments from the Red Sea which lead to severe environmental and human health concerns (Al-Mutairi and Yap, 2021). Five commercially important marine fish from Kutubdia Channel of the northern Bay of Bengal, Bangladesh were assessed for heavy metal concentration and its potential impact on human population. The study suggested that tropical marine fishes were free from heavy metal contamination and safe for human consumption (Rahman et al., 2019). At the cellular level, HM can affect cell growth and differentiation and processes such as apoptosis which is vital to prevent cancer. HM leads to generation of reactive oxidative species that causes DNA damage, gene toxicity and oxidative stress inside the cells (Kim et al., 2015; Balali-Mood et al., 2021). The extent of toxicity of HM in humans is mainly dependent on the amount of exposure or the dose intake. Higher the dose, more are the chances of developing severe health side effects.

11.5 Conclusion

HM naturally occur in the marine environment. HM concentrations may vary in space and time due to variations in physicochemical or biological processes occurring in the ocean. Human activities have changed the natural scenario and added an excess amount of HM in the marine system. Anthropogenic impact is more prominent in estuarine and coastal regions which usually results in greater concentrations of HM in these regions. Once HM enter into the ocean, unique seawater and sediment chemistry result in the formation of various metal species in the marine environment, a process known as metal speciation. Biological agents such as marine microbes and phytoplankton also play a major role in metal speciation. HM are taken up by different groups of marine organisms. In excess they show toxic effects on organisms and may get bioaccumulated in their different body tissues. We have looked over general toxicokinetics of HM on the marine biota. Some of the HM get biomagnified in the marine food chain and may even reach up to humans through marine fish or other seafood consumption wherein they become harmful to human health leading to DNA damage, gene toxicity, and oxidative stress inside the human cells as well as causing severe health effects such as brain damage and cancer.

References

Al-Mutairi, K.A., Yap, C.K., 2021. A review of heavy metals in coastal surface sediments from the Red Sea: health-ecological risk assessments. International Journal of Environmental Research and Public Health 18 (6), 2798.

Ali, H., Khan, E., 2018. What are heavy metals? Long-standing controversy over the scientific use of the term 'heavy metals'—proposal of a comprehensive definition. Toxicological & Environmental Chemistry 100 (1), 6–19.

Ali, H., Khan, E., Ilahi, I., 2019. Environmental chemistry and ecotoxicology of hazardous heavy metals: environmental persistence, toxicity, and bioaccumulation. Journal of Chemistry 2019.

Amiard-Triquet, C., Jeantet, A.Y., Berthet, B., 1993. Metal transfer in marine food chains: bioaccumulation and toxicity. Acta Biologica Hungarica 44 (4), 387–409.

Ansari, T.M., Marr, I.L., Tariq, N., 2004. Heavy metals in marine pollution perspective-a mini review. Journal of Applied Sciences 4 (1), 1–20.

Balali-Mood, M., Naseri, K., Tahergorabi, Z., Khazdair, M.R., Sadeghi, M., 2021. Toxic mechanisms of five heavy metals: mercury, lead, chromium, cadmium, and arsenic. Frontiers in Pharmacology 12.

Batrakova, N., Travnikov, O., Rozovskaya, O., 2014. Chemical and physical transformations of mercury in the ocean: a review. Ocean Science 10 (6), 1047–1063.

Bernhard, M. and Andreae, M.O., 1984. Transport of trace metals in marine food chains. In Changing Metal Cycles and Human Health (pp. 143-167). Springer, Berlin, Heidelberg.

Bernhoft, R.A., 2012. Mercury toxicity and treatment: a review of the literature. Journal of Environmental and Public Health 2012.

Blewett, T.A., Leonard, E.M., 2017. Mechanisms of nickel toxicity to fish and invertebrates in marine and estuarine waters. Environmental Pollution 223, 311–322.

Bruland, K.W., Lohan, M.C., 2003. Controls of trace metals in seawater. Treatise on Geochemistry 6, 625.

Bruland, K.W., Orians, K.J., Cowen, J.P., 1994. Reactive trace metals in the stratified central North Pacific. Geochimica et Cosmochimica Acta 58, 3171–3182.

Camaschella, C., Nai, A., Silvestri, L., 2020. Iron metabolism and iron disorders revisited in the hepcidin era. Haematologica 105 (2), 260.

Costa, M., 2019. Review of arsenic toxicity, speciation and polyadenylation of canonical histones. Toxicology and Applied Pharmacology 375, 1–4.

Cullen, J.T., McAlister, J., 2017. 2. Biogeochemistry of lead. Its release to the environment and chemical speciation. Lead: Its Effects on Environment and Health 21–48.

Cutter, G.A., 1992. Kinetic controls on metalloid speciation in seawater. Marine Chemistry 40 (1-2), 65–80.

Geisler, C.-D., Schmidt, D., 1992. An overview of chromium in the marine environment. Deutsche Hydrographische Zeitschrift 44, 185–196.

Glass, J.B., Dupont, C.L., 2017. Oceanic nickel biogeochemistry and the evolution of nickel use. The Biological Chemistry of Nickel 12–26.

Gorny, J., Billon, G., Noiriel, C., Dumoulin, D., Lesven, L., Madé, B., 2016. Chromium behavior in aquatic environments: a review. Environmental Reviews 24 (4), 503–516.

Gupta, D.K., Tiwari, S., Razafindrabe, B.H.N., Chatterjee, S., 2017. Arsenic contamination from historical aspects to the present. In Arsenic Contamination in the Environment (pp. 1-12). Springer, Cham.

Gworek, B., Bemowska-Kałabun, O., Kijeńska, M., Wrzosek-Jakubowska, J., 2016. Mercury in marine and oceanic waters—a review. Water, Air, & Soil Pollution 227 (10), 1–19.

Kim, H.S., Kim, Y.J., Seo, Y.R., 2015. An overview of carcinogenic heavy metal: molecular toxicity mechanism and prevention. Journal of Cancer Prevention 20 (4), 232.

Koller, M., Saleh, H.M., 2018. Introductory chapter: introducing heavy metals. Heavy Metals 1, 3–11.

Lamborg, C.H., Fitzgerald, W.F., Damman, A.W.H., Benoit, J.M., Balcom, P.H., Engstrom, D.R., 2002. Modern and historic atmospheric mercury fluxes in both hemispheres: global and regional mercury cycling implications. Global Biogeochemical Cycles 16 (4), 51-51.

Landner, L., 1998. Arsenic in the Aquatic Environment-Speciation and Biological Effects.

Lane, T.W., Saito, M.A., George, G.N., Pickering, I.J., Prince, R.C., Morel, F.M., 2005. A cadmium enzyme from a marine diatom. Nature 435 (7038), 42. -42.

Mart, L., Nürnberg, H.W., 1986. The distribution of cadmium in the sea. Cadmium in the Environment 28–40. Birkhäuser Basel.

Moore, J.W. and Ramamoorthy, S., 2012. Heavy Metals in Natural Waters: Applied Monitoring and Impact Assessment. Springer Science & Business Media.

Morcillo, P., Esteban, M.A., Cuesta, A., 2017. Mercury and its toxic effects on fish. AIMS Environmental Science 4 (3), 386–402.

Morel, F.M., Kraepiel, A.M., Amyot, M., 1998. The chemical cycle and bioaccumulation of mercury. Annual Review of Ecology and Systematics 29 (1), 543–566.

Nakayama, E., Tokoro, H., Kuwamoto, T., Fujinaga, T., 1981. Dissolved state of chromium in seawater. Nature 290 (5809), 768–770.

Neff, J.M., 2002. Bioaccumulation in Marine Organisms: Effect of Contaminants From Oil Well Produced Water. Elsevier.

Panneerselvam, K., Marigoudar, S.R., Dhandapani, M., 2018. Toxicity of nickel on the selected species of marine diatoms and copepods. Bulletin of Environmental Contamination and Toxicology 100 (3), 331–337.

Rahman, M.S., Hossain, M.S., Ahmed, M.K., Akther, S., Jolly, Y.N., Akhter, S., Kabir, M.J., Choudhury, T.R., 2019. Assessment of heavy metals contamination in selected tropical marine fish species in Bangladesh and their impact on human health. Environmental Nanotechnology, Monitoring & Management 11, 100210.

Richard, F.C., Bourg, A.C., 1991. Aqueous geochemistry of chromium: a review. Water Research 25 (7), 807–816.

Shenai-Tirodkar, P.S., Gauns, M.U., Ansari, Z.A., 2016. Concentrations of heavy metals in commercially important oysters from Goa, central-west coast of India. Bulletin of Environmental Contamination and Toxicology 97 (6), 813–819.

Simpson, W.R., 1981. A critical review of cadmium in the marine environment. Progress in Oceanography 10 (1), 1–70.

Sinoir, M., Butler, E.C., Bowie, A.R., Mongin, M., Nesterenko, P.N., Hassler, C.S., 2012. Zinc marine biogeochemistry in seawater: a review. Marine and Freshwater Research 63 (7), 644–657.

Strode, S.A., Jaeglé, L., Selin, N.E., Jacob, D.J., Park, R.J., Yantosca, R.M., et al., 2007. Air-sea exchange in the global mercury cycle. Global Biogeochemical Cycles 21, 1.

Tamele, I.J., Vázquez Loureiro, P., 2020. Lead, mercury and cadmium in fish and shellfish from the Indian Ocean and Red Sea (African Countries): public health challenges. Journal of Marine Science and Engineering 8 (5), 344.

Tchounwou, P.B., Yedjou, C.G., Patlolla, A.K., Sutton, D.J., 2012. Heavy metal toxicity and the environment. Molecular, Clinical and Environmental Toxicology 133–164.

Tomlinson, D.L., Wilson, J.G., Harris, C.R., Jeffrey, D.W., 1980. Problems in the assessment of heavy-metal levels in estuaries and the formation of a pollution index. Helgoländer Meeresuntersuchungen 33 (1-4), 566–575.

Valavanidis, A. and Vlachogianni, T., 2010. Metal pollution in ecosystems. Ecotoxicology Studies and Risk Assessment in the Marine Environment. Dept. of Chemistry, University of Athens University Campus Zografou, 15784.

Walker, C.H., Sibly, R.M., Peakall, D.B., 2005. Principles of Ecotoxicology. CRC press.

Wang, M.H., Wang, D.Z., Wang, G.Z., Huang, X.G., Hong, H.S., 2007. Influence of N, P additions on the transfer of nickel from phytoplankton to copepods. Environmental Pollution 148 (2), 679–687.

Wang, W.X., 2002. Interactions of trace metals and different marine food chains. Marine Ecology Progress Series 243, 295–309.

Whitfield, M., Turner, D.R., 1987. Aquatic Surface Chemistry; Chemical Processes at the Particle-Water Interface: The Role of Particles in Regulating the Composition of Seawater. *Wiley & Sons, New York.*

Wood, C.M., 2011. An introduction to metals in fish physiology and toxicology: basic principles. Fish Physiology 31, 1–51. Academic Press.

Xu, Y., Morel, F.M., 2013. Cadmium in marine phytoplankton. Cadmium: From Toxicity to Essentiality 509–528.

Yang, R., Van den Berg, C.M., 2009. Metal complexation by humic substances in seawater. Environmental Science & Technology 43 (19), 7192–7197.

Zaib, M., Athar, M.M., Saeed, A., Farooq, U., 2015. Electrochemical determination of inorganic mercury and arsenic—a review. Biosensors and Bioelectronics 74, 895–908.

CHAPTER

12

Various methods for the recovery of metals from the wastewater

Priya Mukherjee, Uttkarshni Sharma*, Ankita Rani*, Priyanka Mishra* and Pichiah Saravanan*

Environmental Nanotechnology Laboratory, Department of Environmental Science and Engineering, Indian Institute of Technology (ISM), Dhanbad, Jharkhand, India

12.1 Introduction

The rapid industrialization and technological advancement had increased the occurrence of metals in the aquatic environment. The waste generated from industrial operations like chemical manufacturing, electroplating, mining, and metallurgy are some of the anthropogenic causes that contaminate the waters with heavy metals. These metals are also a common occurrence in the water bodies due to natural sources. The runoff impacts the metallic composition in surface water bodies like ponds, rivers and lakes. The metals existing on the surface of the soil are crafted from their channel, which ends up within the wild and inside the lakes (Salem et al., 2000). Rainwater gets polluted as it scrubs the atmospheric gaseous and particulate pollutants. Where else the groundwater is polluted naturally due to the occurrence in the geochemistry and anthropogenically by the leachates originated from the industrial landfill, municipal landfill and so on (Oyeku and Eludoyin, 2010). Physical and chemical factors such as absorption, cation exchange, temperature, pH, dwelling organisms, evaporation, etc., also affect the leaching ability of the metals present in the waste (Selvi et al., 2019). The remediation of such metals from the aquatic environment is a vast research concern in today's world due to their poor degradability. Furthermore, these metals possess toxicity concerns to the living organisms leading to an insecure ecosystem. Metals like Pb(II), is known to cause carcinogenic, reproductive, genotoxic and neurological diseases. While high Cd intake can lead to lung cancer, prostatic proliferative lesions, renal disturbances and bone fractures. Cu(II) toxicity even in

* Authors have contributed equally.

DOI: https://doi.org/10.1016/B978-0-323-95919-3.00007-0

micrograms can damage the central nervous system and cause capillary, hepatic and renal problems (Dahaghin et al., 2017; Uddin, 2017). Hence various strategies to eliminate the metals from water bodies have been applied, which include precipitation, phytoremediation, algal decomposition, coagulation, filtration, etc. (Selvi et al., 2019). However, these technologies require large areas, extra operating costs and huge sludge generation, resulting in secondary pollution.

The evolution of the circular economy created a great demand for such metals. These metals with uneven distribution and low sustainability possess higher economic value (Moss et al., 2011). The critical metals are hence classified under different classes like rare earth elements, platinum group elements, and other metals, like cobalt (Co), gallium (Ga), indium (In), molybdenum (Mo), nickel (Ni), tellurium (Te), and tungsten (W), etc. (Hennebel et al., 2015; Won et al., 2014). Of them, the platinum group elements are precious with the least quantity. The growing demand for catalysts and rechargeable batteries, permanent magnets, computer hard drives, etc., enhanced the utilization of these platinum and rare metals, resulting in the generation of huge demand and the high price of these metals (Yu et al., 2020).

The solution to these problems is implementing the 3R principle, that is, reduce, reuse, and recycle the used metals. The recovery of metals is a classical practice that uses techniques like hydrometallurgy and pyrometallurgy for the extraction of metals from metal oxides and sulfides. However, these classical methods suffer limitations like high operating costs, low recovery yields and additional environmental pollution (Pathak et al., 2009). Moreover, these technologies are unrealistic for the extraction of critical metals when present in lower concentrations (Lo et al., 2014). Hence the operations like adsorption, bioleaching, biological organisms, precipitation, biosorption, electrochemical treatment, filtration were practiced (Fu and Wang, 2011; Carolin et al., 2017; Wang et al., 2021a). Fig. 12.1 illustrates the methods available for metal recovery from the aquatic stream. The earliest and vastly practiced method for metal recovery is precipitation, but it has a very low recovery percentage and selectivity. Hence adsorption method came into the picture but the drawback is its slow kinetics and metal concentrated sludge generation. Alternatively, the sorption was practiced on the biosorbent, but the process requires prolonged operation time and is mostly used for the removal of metals rather than recovery. The desorption of metals after adsorption is not much deliberated and there is a vacuum on the large-scale studies. In recent years, electrochemical and filtration technology are focused on metal recovery. Although electrochemical methods are good in metal recovery, they are not cost-effective and often require an external power source. The filtration technology has been used for a long time but it was mostly used for the removal process. On the other hand, recovery of metals via membrane filtration shows high recovery efficiency although the costs for membrane are high. Nowadays, various integrated systems are also applied for high recovery of the metals from the water bodies with low sludge generation.

The chapter focuses on the recent well-researched technologies and methods applied for the recovery of metals and heavy metals from polluted water. The basic principle and viability of the methods are focused along with integrated systems for metal recovery with high concentration. The chapter will also discuss the current challenges faced in the recovery, along with the type of technology best suited for specified metal and the commercially practiced methods.

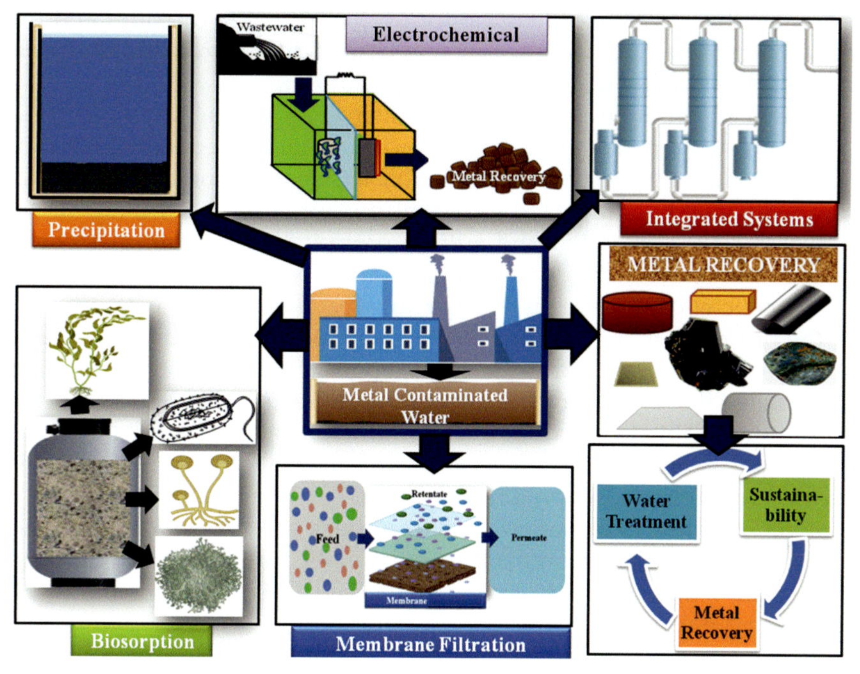

FIGURE 12.1 Methods available for metal recovery from the aquatic stream.

12.2 Chemical precipitation

The chemical precipitation method involves the precipitation of dissolved metal ions, led by the addition of precipitants in the form of chemicals to the aqueous medium. The precipitate thus formed is separated by sedimentation, floatation, or filtration processes (Dahman et al., 2017). The process being a chemical reaction between the dissolved metal ion and the precipitating agents it is controlled by the type and concentration of meta ion (pollutant) and precipitant, organic contaminants, pH and temperature. The reaction is pH sensitive and most often highly alkaline condition favors the process and aids in reducing the solubility of the metal. Each metal precipitates at a specific pH beyond which it solubilizes again. Therefore, a pH specified for a specific metal may not be the same for the other. This means that the process needs to be performed in multiple stages for the precipitation of every metal present in the waste stream. In order to reach a particular pH value, the precipitant dose added needs to be customized. A higher dose of precipitant may result in the formation of ultrafine precipitates with enhanced pH. These precipitates are usually available with the size <100 nm and take a longer time for sedimentation. Therefore, the size of precipitates formed should be large enough for rapid settling and easy separation (Chen et al., 2018). The presence of oil, grease and other organic contaminants in the medium interferes and slows down the process. Hydroxides as calcium and sodium while sulfides of sodium, ferrous and bisulfides of sodium are the major precipitants used in the wastewater treatment process (Yadav et al., 2020). By adding these metals are precipitated in the respective chemical forms, that is, the hydroxides and the sulfides. Advantages of the process include low-cost and ease of operation while excess sludge generation becomes the disadvantage (Adhikari et al., 2019).

Coprecipitation of multiple metals due to their precipitating characteristics within a common pH range often makes their separation problematic. Sequential addition of more than one precipitant was reported as a solution for treating complex metal wastewater. Wang and Chen applied a multistage treatment process where they integrated hydroxide precipitation with H_2O_2 oxidation and sulfide precipitation for the selective removal of Cu, Fe and Zn from an industrial wastewater sample (Wang and Chen, 2019). The release of H_2S gas in acidic circumstances is a major drawback associated with sulfide precipitation. Thus, the process is mostly carried out in a neutral environment. However, precipitation at a higher pH range thickens the sludge and eases its separation. Sulfide precipitation is also through biological agents (Xu and Chen, 2020). The sulfate-reducing bacteria (SRB) can precipitate metals in form of sulfides. The SRB is anaerobic heterotrophs and utilizes sulfate as the terminal electron acceptor. In the presence of sufficient organic constituents or hydrogen as electron donors, the SRBs oxidize these while at the same time reducing sulfate from its $+6$-oxidation state to sulfides of -2 oxidation state. Most of the time under a circumneutral environment it results in the formation of bisulfides which are highly reactive and reduce the metals into metal sulfides (Muyzer and Stams, 2008; Gadd, 2009; Ojuederie and Babalola, 2017).

12.3 Sorption by biological and chemical agents

Adsorption is one of the major separation techniques used for the recovery of metals from contaminated water sources. The term adsorption refers to the process where the pollutant either organic or inorganic (termed as adsorbate) are transferred from the bulk phase to the surface of the substrate added (termed as adsorbent). Regeneration of the adsorbent takes place by the addition of suitable regenerating agents added once the process reaches equilibrium. Widely used low-cost substrates applied for metal recovery from the contaminated source include the unicellular microbes, multicellular plants, natural surfactants, zeolites, natural clay and agricultural biowastes (Barakat, 2011).

In the biosorption process, biological living matter and dried biomass are applied as sorbents. Microbes including bacteria, algae and fungi and multicellular plants are also applied for the purpose of biosorption. The cell walls of bacteria and fungi are rich in negatively charged surface ligands which attract the positively charged metal ions through electrostatic force of attraction. The carboxyl, amino, phosphate and hydroxyl functional groups present on the surface are responsible for the electrostatic attraction (Kikuchi and Tanaka, 2012). It has been observed that the use of mixed consortium and immobilized forms of sorb are more efficient than the single strain and nonimmobilized one. The enhanced stability, rigidity and optimum porosity offered by the consortia and immobilized biomass make significant changes (Kader et al., 2007). *Bacillus, Pseudomonas, Micrococcus* and *Acientobacter* are examples of the bacterial genus; *Aspergillus, Saccharomyces* and *Sphaerotilus* are examples of the certain fungal genus; and *Spirogyra, Spirulina, Nostoc, Chlorella* are certain algae genus actively participating in metal biosorption (Igiri et al., 2018).

The adsorption by these is totally metabolism independent (Kikuchi and Tanaka, 2012). The process is influenced by several factors including temperature, pH, metal ion concentration, the bioavailability of metals and the presence of organic constituents. The presence of higher temperature makes the microorganism competent thus enhances its enzymatic activity and metabolism rate. At the same time, the availability of metals around the external boundary layer of microorganisms is also increased. In general, the possibility of metal sorption is always low at acidic conditions and this is because of the H^+ ion interaction with the metal sorption sites on the cell surface. The presence of low molecular weight organic acids and humic acids alter the speciation of metals and interfere with the transportation and bioavailability of metals. A high concentration of metals makes the process of biosorption ineffective in the presence of microorganisms alone due to the severe toxicity leading to poor competitiveness of microbes.

Application of surfactants also aids in the sorption of metals either directly by adsorbing the dissolved metal ions or indirectly by facilitating the growth of microorganisms on unfavorable fluids interface. The surfactants produced by microorganisms and plants are cheaper, possess diverse structures, biodegradable and nontoxic. These are naturally derived chemical moieties that are amphiphilic, that is, consist of a hydrophilic and a hydrophobic part. When these molecules are added to the water bodies at a concentration below critical micelle concentration (CMC), they stay as monomers while above it forms micelles thus reducing the interfacial surface tension. In both the monomeric or micelle form, they adsorb the metal ions which result in the formation of an inorganic-organic complex. Separation of the complex from water bodies and further separation of metals from the complexed surfactants is achieved by sequential operation of solvent extraction, centrifugation, evaporation, membrane filtration and precipitation processes. The most commonly applied biosurfactants are rhamnolipids produced by *Pseudomonas aeruginosa*, surfactin produced by *Bacillus* strain, and the sorpholipids produced by yeast *Torulopsis bombicola* (Akbari et al., 2018).

Plants play a major role in metals extraction from contaminated water through the process called phytofiltration. This process can take place in three ways: rhizofiltration, caulofiltration, and blastofiltration (Dixit et al., 2015; Rahman et al., 2016; Ojuederie and Babalola, 2017). In former roots are the major organs for the removal of metals whereas in the median excised shoots are the major organs and in later seedlings are the major organs for metal extraction from contaminated water. The phytofiltration process needs to be explored more to establish new organs effective for the filtration process. In the rhizofiltration process the plants are hydroponically grown first on clean water for achieving high root growth and biomass production, after this the clean water is substituted with contaminated water in order to acclimatize the plants with various levels of metal contamination. After their incubation on the acclimatized water, the plants are transferred to the actual metal-contaminated water or wetland. These plants can either adsorb the metals across their root surface or may absorb and translocate them to various organs. After the roots are totally saturated these are removed from the medium and harvested. Both wetland and terrestrial plants can be chosen for the purpose, however they should possess tolerance to metal toxicity, should be fast-growing in order to produce more biomass in less time, bear extensive root system and should be resistant to the plant-feeding insects in the metal-contaminated environment. Several plants like Indian mustard, water hyacinth, poplar and azola are reported to have good rhizofiltration capacities (Kikuchi and Tanaka, 2012; Yan et al., 2020).

Apart from the biological sorbents other commonly used inexpensive adsorbents include agricultural wastes, natural zeolites, clay minerals, etc. Zeolites are porous hydrated aluminosilicates with bound alkali and alkali earth cations. These are known for their reversible cation exchange ability. Thus, are used for metal ions removal from water and soil. The addition of zeolite increases the pH of the medium which improves the exchange of metals on the zeolite surface. Zeolites are both naturally existing and synthesized. Clinoptilolite and mordenite are some of the widely used zeolites for metal removal (Yuna, 2016). Natural clay is another that falls under low-cost adsorbent category consisting of aluminosilicates similar to that of zeolites. However, they possess a porous layered shrinkable crystalline structure, unlike zeolite which possesses a rigid structure. Their particle size lies below 2 μm, hence possessing a large specific surface area and net little negative charge which facilitates interaction and exchange of dissolved metallic species. Kaolinites, halloysite and montmorillonites are among the widely used ones (Gu et al., 2019). This is followed by agricultural waste that includes the application of pretreated fruit shells (coconut shell, hazelnut shell, pecan shell), peels (orange peel, potato peel), husks (rice, maize) for the removal of metals from contaminated resources. Pretreatment of these biowastes, or formation of activated carbon by heating at higher temperature improves the metal binding capacity (Farhadi et al., 2021). Among the various methods applied for the removal of metals from contaminated water, adsorption is one that was extensively studied. Biosorption is now prime focus because of the low cost, less sludge generation and excess availability. Table 12.1 compiles the adsorption characteristics of various classifications of adsorbent for the aquatic removal of metal ions.

TABLE 12.1 Enlists the studies on the various classification of adsorbents on metal removal.

Sl no.	Adsorbent (dose) (g/L)	Initial metal conc. (mg/L)	% removed	References
1.	Tangerine cortex (2)	10	Cr (88%), Cd (67%), Zn (43%)	Al-Qahtani (2016)
2.	Kiwi cortex (2)	10	Cr (91%), Cd (70%), Zn (40%)	Al-Qahtani (2016)
3.	Banana (10)	50	Zn^{2+} (92.57%)	Castro et al. (2021)
4.	Orange (10)	50	Zn^{2+} (97.13%)	Castro et al. (2021)
5.	Granadilla (10)	50	Zn^{2+} (88.21%)	Castro et al. (2021)
6.	Tannin immobilized coconut husk cellulose fiber (4)	Pb^{2+} (100) Cd^{2+} (100) Cu^{2+} (100)	Pb^{2+} (52%) Cd^{2+} (90%) Cu^{2+} (90%)	Taksitta et al. (2020)
7.	Palm Kernel shell (15)	25	Cr^{6+} (98.92%), Pb^{2+} (99.01%) Cd^{2+} (84.23%) Zn^{2+} (83.45%)	Baby et al. (2019)

(Continued)

TABLE 12.1 (Continued)

Sl no.	Adsorbent (dose) (g/L)	Initial metal conc. (mg/L)	% removed	References
8.	NaCl-HCl treated Natural Zeolite (Clinoptilolite) (12)	40	Co^{2+} (87%)	Rodríguez et al. (2019)
9.	ZIF 8 (0.1 for Pb^{2+}), (0.2 for Cu^{2+})	200	Pb^{2+} (99.4%) Cu^{2+} (97.81%)	Huang et al. (2018)
10.	ZIF 67 (0.1 for Pb^{2+}), (0.2 for Cu^{2+})	200	Pb^{2+} (100%), Cu^{2+} (99.43%)	Huang et al. (2018)
11.	Zeolite A and X prepared from Kaolin (0.4)	20	Pb^{2+} (>99%)	Jamil et al. (2010)
12.	Fe-Mn Clay (1.5)	20	As^{5+} (>98.82%)	Foroutan et al. (2019)
13.	Triazole modified clay (0.5)	20	Co^{2+} (91%), Pb^{2+} (89.5%), Zn^{2+} (90.25%)	Kakaei et al. (2020)
14.	*Geobacillus thermodenitrificans* dried biomass (2)	175	Fe^{3+} (91.31%), Cr^{3+} (80.8%), Co^{2+} (79.71%), Cu^{2+} (57.14%), Zn^{2+} (55.14%), Cd^{2+} (49.02%), Ag^{+} (43.25%), Pb^{2+} (36.86%)	Chatterjee et al. (2010)
15.	*Klebsiella* sp. (1)	Cd^{2+} (417) Mn^{2+} (1258.44)	Cd^{2+} (81.92%) Mn^{2+} (90.55%)	Hou et al. (2018)
16.	*Penicillium* sp. MRF1 (7.5)	639	Ni^{2+} (74.6%)	Sundararaju et al. (2020)
17.	*Penicillium fellutinum* dead biomass/ bentonite (50)	200	Ni^{2+} (80.5%) Zn^{2+} (39.25%)	Rashid et al. (2016)
18.	Acid treated *Spirulina platensis*	75	Al (95%), Ni (87%), Cu (63%)	Almomani and Bohsale (2021)
19.	Acid treated *Chlorella vulgaris*	75	Al (87%), Ni (79%), Cu (80%)	Almomani and Bohsale (2021)
20.	*Plectranthus amboinicius*	50	Pb^{2+}(75%)	Ignatius et al. (2014)
21.	Water hyacinth	B (14.4) Cu (20) Zn (8.13) Fe (1.51) Cr (19.25)	B (75%) Cu (88%) Zn (97%) Fe (99%) Cr (70%)	Elias et al. (2014)
23.	Rhamnolipid activated coal waste (250 surfactant)	100	Cd^{2+}(> 99%)	Shami et al. (2019)
24.	Surfactin (cassava wastewater) (2.5)	25	Cu^{2+} (40%)	Kummer et al. (2016)

Though the removal of metals from contaminated water is a primary goal, the studies on desorption of the spent adsorbent for the recovery of adsorbed metals and the reusability of the adsorbent are limited. Metal-loaded sorbents are the carriers for secondary pollution, harmful in the locations where dumped and may leach to the groundwater, slowly contaminate surface water and pose a health threat. Thus, there is a growing demand for regenerating the spent adsorbent and recovering the loaded metals for their further reuse and recycle (Lata et al., 2015). An adsorbent with good desorption properties reduces the overall cost of the adsorption process because of the regeneration and reuse. Hence an ideal adsorbent must have the ability to desorb the adsorbed element completely with negligible damage to the adsorbing property of the adsorbent. However, this depends on the eluent type chosen, eluent concentration, pH, contact time, adsorbed metal species, and adsorbent type and properties (Das, 2010). Chemical regeneration methods employing chemical agents including acids, bases, alcohols, chelating agents and metal salts were used. The eluents which are extensively practiced include: HCl, HNO_3, H_2O_2, NaOH, Na_2CO_3, EDTA (ethylene diamine tetraacetate), NTA (nitrilotriacetic acid), $NaNO_3$ and $CaCl_2$ (Chatterjee and Abraham, 2019). Acidic eluents are most widely applied for the recovery of metal of cationic nature where else the anionic metals are handled by alkaline eluents. Both the H^+ ions and OH^- ions of the acidic and basic eluents compete for the cationic and anionic species binding sites, thus resulting in the desorption of metal ions and adsorption of H^+ and OH^- ions on the sorbent surface. It was widely reported that the recovery efficiency by the alkali eluents was observed to be good for the regeneration of chemically synthesized adsorbents (metal oxide modified sand, nanoparticle suspension, nanozerovalent suspension, etc.). The presence of a more positive charge on sorbents surface attracts negatively charged hydroxyl ions. Regeneration capacities for acidic eluents were observed to be good for biowaste-derived adsorbents (coconut and rice husk, charcoal, activated carbon, cellulose-based wastes, etc.). The diverse functional groups present on the adsorbent surface with negative charge play a key role in such regeneration ability. On the other hand, EDTA has proved to be a potential eluent for the regeneration of biomass-based adsorbents and was due to its least toxicity effect causing less damage as compared to acidic and basic agents (Lata et al., 2015). Sonication of solution mixture of eluents and metal loaded sorbent can result in quick desorption of ions. The ultrasonic waves with optimum frequency result in the generation of cavitation bubbles that collapse with each other. The energy released while collapsing releases an enormous amount of pressure and temperature which is sufficient for breaking the bond between metal ions and sorbent surface. This causes metal desorption from the sorbent surface (Hamdaoui et al., 2005; Hokkanen et al., 2013).

The main challenge of adsorption process is the regeneration and the recovery of the adsorbate material. Sometimes maximizing the number of cycles of the applied adsorbent can also enhance the cost-benefit ratio of the adsorption process (Dotto and McKay, 2020). In biosorption the challenges being faced are uncontrolled bacterial growth leading to process instability while poor biological growth limits the biosorption. Extraction of metals from the phyto-remediated plants is also a major challenge faced today. Further finding a inexpensive and robust biosorbent is a severe drought.

12.4 Electrochemical methods for metal recovery

Electrochemical treatment of heavy metals is one of the most advanced controllable and eco-friendly methods for heavy metal remediation. It is an electric current-driven, nonspontaneous selective redox reaction process. The process employs the transfer of charge carriers at interfaces through metal-rich wastewater being used as electrolytes, under the applied potential difference between conductive electrodes (Tran et al., 2017). This system also offers many benefits over others comprising continuous running, ambient environmental operation and capability to sustain under corrosive composition of electrolyte (Ya et al., 2018). The efficiency of metal removal completely depends upon the selection of electrode materials and the liquid electric medium for charge kinetics. Some other factors like potential differences at the interface, applied current, current density, specific electrode area, current efficiency (CE) of the solution and charge transfer resistance can also impact the system performance. The significance of using electrochemical treatment is that it does not generate any harmful by-products and additional chemical pollutants. Electrochemical treatment follows several processes for recovery of the metals like electro deposition, precipitation, coagulation, flotation, ion-exchange, and electroplating. Owing to the applied potential difference between electrodes, oxidation of water molecules or electrolyte takes place at anode subsequently hydroxyl radicals are generated under strong oxidative conditions. Thus, generated radicals lead to the oxidation of metal complexes. Consequently, oxidized metal ions produced at the anode further move towards the cathode to be reduced as solid metals by getting the electrons from the outer circuit. These metals can be oxidized and precipitate in the form of oxide, sulfide and as pure solid metals (Kakakhel et al., 2007; Knápek et al., 2010). The chemical composition of final products or yielded bulk metal depends upon various factors like electrolyte (oxide or sulfide), the primitive composition of metal pollutants and selection of electrode materials. The principal steps of the mechanism are (1) oxidation of metal complexes at anode (2) reduction of metal ions at the cathode.

Electrochemical recovery reactions could be performed in the electrochemical reactors. The performance of the reactor is calculated by the space–time yield (YST) of the reactor, that is, the amount of product (solid metal) produced by the reactor volume in unit time:

$$\mathrm{YST} = (iaM/1000zF) \times \mathrm{CE} \tag{12.1}$$

Where M represented total amount of product or total molecular mass, "i" stand for current density, "a" is specific electrode area, "z" is charge number, "F" Faraday constant and "CE" is current efficiency (Beauchesne et al., 2014). Basically, space–time yield is the influence of specific electrode area "a" on the overall index of reactor performance.

The CE of the system depends upon the ratio of deposited metal (W_1) at cathode and theoretical weight of deposited metal (W_2) (Beauchesne et al., 2014).

$$\mathrm{CE}(\%) = \left(\frac{W_1}{W_2}\right) \times 100 \tag{12.2}$$

Where theoretical weight can be calculated from the Eq. (12.3) given above.

$$W_2 = MQ/ZF \tag{12.3}$$

Where "M" is the molecular weight of the metal, "Z" is the valence state of metal "Q" is applied current intensity and "F" is Faraday's constant (Beauchesne et al., 2014).

Metal extraction can also be calculated as Leaching Yield (LY) and the equation is as follows:

$$LY(\%) = (1 - n_f/n_i) \times 100 \tag{12.4}$$

Where "n_f" is the mole number of heavy metals present in treated water while "n_i" is the mole number of heavy metals in polluted water (Beauchesne et al., 2014).

These heavy metals such as Cr^{3+}, Cr^{6+}, Cu^{2+}, Cd^{2+}, Zn^{2+}, As^{3+}, As^{5+} and Ni^{+} could be recovered by altering the pH of the wastewater. Efficient electrorecovery of gold and cyanide from gold mine wastewater was achieved through electrogenerative methods. The soluble ions of gold were oxidized as solid gold onto the cathode of the electrochemical cell (Yap and Mohamed, 2007; Turygin et al., 2008; Acheampong et al., 2010). Recovery of chromium metal could be obtained by electroreduction of chromium (+6 and +3) by using scrap iron as reducing agents (Lakshmipathiraj et al., 2008; Basha et al., 2008; Nepel et al., 2020).

Another method for metal recovery under the flag ship of the electrochemical process is Electrodeposition. It is a reversible chemical reaction consisting of electrochemical oxidation and reduction phenomenon by which heavy metal ions get reduced at the cathode. However, anodic deposition of metals is also reported but they could be in the form of oxides. Therefore, the cathodic route was found to be more proficient for the metal's recovery.

The general equation of cathodic electrodeposition is written as follows.

$$M^{n+} + n^{e-} \leftrightarrow M(\text{Reduction}) \tag{12.5}$$

From the above equation, the recovery phenomenon is illustrating the positively charged metal ions getting reduced on the cathode and depositing over as solid metal.

However, this is a reversible reaction thus it follows the Nernst equation where equilibrium potential (E^{eq}) is directly proportional to normal electrode potential (E^{0}) of the single electrode which could be dependent on the ratio of metallic ions and atoms. The Nernst equation is given below.

$$E^{eq} = E^{0} + \frac{RT}{nF} \times \ln(a_{Mn+}/a_{M}) \tag{12.6}$$

Where, "R", "T" and "F" represent the ideal gas constant, temperature (K) and Faraday constant respectively. However "α_{Mn+}" is the metal ion is the (Product after deposition) and "α_{M}" is the metal atoms (reactants) (Zhu et al., 2019).

Similarly, another method under electrochemical processes is Electrocoagulation. It was first developed and patented by A. E. Dietrich in 1906 for metal pollutants elimination from wastewater. Later J. T. Harries in 1909 proposed wastewater treatment by electrocoagulation method employing aluminum and iron anodes (Emamjomeh and Sivakumar, 2009). Electrocoagulation cell contains two electrodes one is anode also called a sacrificial electrode (owing to the release of metal cations to perform the oxidation of contaminants) and the other is a cathode submerged into wastewater that acts as charge transporting solution (Akbal and Camcı, 2010). This process was carried out with anode-cathode cells where charged metal ions present in the colloid suspension of wastewater are destabilized by the agglomeration of counter ions. Hereby, neutralization of heavy metal ions takes upon by release of metal cations anode under applied direct current. Moreover, metal cations released from the anode can form metal hydroxide precipitates which also act as a coagulating agent. The further destabilized pollutants by coagulant eventually will be collected over the surface of wastewater by flocculation. And simultaneously at cathode water splitting can be occurred where oxygen, hydrogen gases and hydroxide ions will be formed (Yadav et al., 2020). In electrocoagulation of heavy metals including aluminum (Al), chromium (Cr), copper (Cu), Iron (Fe), and zinc (Zn) were used as anode. The electrocoagulation process comprises of various mechanisms such as adsorption, coprecipitation, oxidation, and reduction. General equations are given below (Un and Ocal, 2015).

At Anode

$$M_{(s)} \rightarrow M^{n+}_{(aq)} + ne^- \tag{12.7}$$

(If Iron anode is used)

$$Fe_{(s)} \rightarrow Fe^{2+}_{(aq)} + 2e^- \tag{12.8}$$

$$2H_2O \rightarrow O_{2(g)} + 4H^+ + 4e^- \tag{12.9}$$

Coagulant formation

$$Fe^{2+} + O_2 + 2H_2O \rightarrow Fe^{3+} + 4OH^- \tag{12.10}$$

$$Fe^{2+} + 2OH^- \rightarrow Fe(OH)_2 \tag{12.11}$$

At cathode

$$M^{n+}_{(aq)} + ne^- \rightarrow M_s \tag{12.12}$$

$$2H_2O + 2e^- \rightarrow H_{2(g)} + 2OH^- \tag{12.13}$$

The next method under the electrochemical process is electrodialysis. This technique employs the selective separation of soluble metal ions via ion-exchange membranes from wastewater streams under applied potential differences. Electrodialysis system is configured as a two-compartment system separated by a selectively permeable membrane that allows only ions with the appropriate charge to pass through with anode to perform oxidation and cathode for reduction reaction under external electrical driving force (Gering and Scamehorn, 1988). The selectivity of the membrane is dependent on the ions required to be recovered either from cations and anion exchange membrane (Smara et al., 2007; Babilas and Dydo, 2018). However, ion exchange is the heart of this system therefore new generation ion-exchange membranes have been introduced to accelerate the system performance. Thus, to develop an ion exchange membrane with high selectivity, stability, durability, and low resistivity towards specific ions numerous researches are being done. However, this deficiency can be overcome by new generation nanomaterial through membrane development.

12.5 Membrane-based metal recovery

Pressure-driven membrane processes are a well-known strategy for metal recovery. These systems can meet the environmental criteria and provide high efficiency, ease in operation and a low ecological footprint (Kehrein et al., 2020). The large-scale deposition of metal nanoparticles on the membrane surface as well signifies their removal and recovery potential from wastewater. The first process among membrane filtration is ultrafiltration (UF). This process assists in the removal of dissolved as well as colloidal particles under low transmembrane pressure. Their usage is often limited by their large pore size which ranges from 2 to 50 nm (Singh and Hankins, 2016). Micelles/ surfactants and polymers are usually added to UF to enhance metal recovery which is referred to as Micellar enhanced UF (MEUF) or polymer enhanced UF (PEUF) respectively. MEUF technique has high removal efficiency for wastewater with a low level of heavy metal concentration. They are characterized as high flux, selectivity, and lower energy consumption process (Rahmati et al., 2017). The surfactant can be ordered into four categories: nonionic, anionic, cationic, and amphoteric. These must be added above the CMC to form larger amphiphilic transparent micelles having a hydrodynamic diameter larger than the pore diameter of the UF membrane. (Zeng et al., 2008; Chung et al., 2009; Zaghbani et al., 2009). The CMC, type of solute, the charge of surfactant and the polymeric material of the membranes are important parameters for determining the performance of these membranes. Surfactants having an electric charge opposite to that of contaminants have shown high efficiency (Sum et al., 2021). PEUF methods involve the integration of a water-soluble polymer as a complexing agent. The common functional groups of these polymers are carboxylate, sulfonate, phosphonic, amine (Biersack et al., 2009; Rivas et al., 2011). They form macromolecules with the metal ions via ionic or chelating bonds allowing easy retention through ultrafiltration membranes (Sánchez et al., 2018).

The next in the membrane category is the nanofiltration (NF) membranes. It is composed of asymmetric polymers having a pore size of less than 1–2 nm (Carvalho et al., 2011). The operating range of NF is between UF and reverse osmosis (RO) and can concentrate solutes with molecular weights ranging from 200 to 1000 Da. They are preferred due to the higher flux rate and low operating pressure. The pH-dependent dissociation of functional groups at the topmost selective layer gives charge to the membranes (Tajuddin et al., 2019; Ye et al., 2019a). Overall, their separation procedure works as a summation of convection and diffusion transport mechanisms, that is, sieving effects, together with electromigration as a result of membrane charge (Carvalho et al., 2011). The sieving is the major factor determining separation for uncharged molecules, which is mainly dependant on the molecular size of the solute. However, both sieving and electromigration are the major factors for separation in the case of ionic species (Childress and Elimelech, 2000; Qin et al., 2004). Metal ion recovery is also significantly impacted by membrane charge which is a factor of feed pH (Mullett et al., 2014).

The next classification is the RO membranes. These are nonporous asymmetric or thin-film composites with a pore size of 0.5–1.5 nm and molecular weight cut-off of approximately 100 Da. They act as an effective barrier (99%) to all dissolved inorganic salt as well as charged organic molecules compared to all other processes (Algieri et al., 2021). The separation mechanism is governed by the application of hydrostatic pressure against the osmotic pressure through the semipermeable membrane. Though the process is compact and efficient, it involves high energy consumption, wastage of processed water and large capital investment. Besides the dense nature of the RO membranes also leads to quick fouling (Wenten, 2016). A way to reduce it is to pretreat the feed using UF or NF processes. Future research focusing on developing membranes with targeted extraction of some metals from the concentrate can have multiple environmental and economic benefits.

The forward osmosis (FO) is another membrane filtration technique and a potential green technology that can effectively retain and concentrate valuable substances (Awad et al., 2019). It is a naturally driven osmotic process, across a semipermeable membrane. Very simply it is opposite to the RO process where the draw solution (DS) is more concentrated than the feed solution (FS) which forms the concentration gradient. The osmotic pressure thus developed drives the water molecules from the FS to DS side (Suwaileh et al., 2018). FO does not require hydraulic pressure, hence low fouling, more energy-saving and environmentally friendly. However, FO is often limited by concentration polarization, regeneration of DS and reverse salt flux (RSF) (Cui et al., 2014; He et al., 2020). Studies have now been focused on researching suitable DS to minimize RSF and reduce energy consumption. Table 12.2 summarizes the performance of various pressure-driven membranes- for metal recovery.

Electrospun Nanofibre Membranes are also used for metal recovery. The adsorption capacity of such nanofibers depends upon the nature of the metal, solution pH, particle size, adsorbate dose, agitation, initial metal concentration and temperature. Besides, the kinetics of ion sorption/desorption is an important parameter in designing optimum operational conditions (Krishnani et al., 2008). Further, the recovery of metals from water can be challenging when the concentration level drops from ppm to ppb where it requires complex and multiple filtration units. Nanofiber membrane developed by electrospinning polyacrylonitrile (PAN) polymer along with amorphous WO_3 nanoparticle proved to be very effective in terms of metal recovery and membrane reusability. This specific modification reduces the membrane under solar light irradiation upon immersion in wastewater, it successfully lowered the amount of metal ions by two to four orders of magnitude in ppb level within an hour. The negatively charged membrane surface promoted strong electrostatic attraction. Overall the metal reduction and deposition process prevented desorption of metal ions in wastewater and allowed 99% effective removal/ recovery. Nobel metals such as Ag, Au, Pt and Pd and toxic metals

TABLE 12.2 Performance of various pressure-driven membranes for metal recovery.

Process	Metal	Membrane	Surfactant/ polymer	pH	Efficiency (%)	References
MEUF	Cd^{2+}	Polyethersulfone (HFUF20)	Rhamnolipid—RHL	7.8	98%	Verma and Sarkar (2020)
MEUF	Hg^{2+}	Polyacrylo nitrile	SDS	7.0	95.75% using MEUF; 96.83% using MEUF-ACF	Yaqub and Lee (2020)
PEUF	Cr^{3+}, Cr^{6+}	Regenerated cellulose	Sodium alginate (SA)	2–5	100%	Butter et al. (2021)
PEUF	Cd^{2+}	Flat sheet and hollow fiber UF membrane	Chitosan, polyvinyl alcohol, and polyacrylate sodium	46–10	100%	Meng et al. (2021)

(Continued)

TABLE 12.2 (Continued)

Process	Metal	Membrane	Surfactant/ polymer	pH	Efficiency (%)	References
PEUF	Zn^{2+} and Pb^{2+}	Polyethersulfone (PES)	Carboxymethyl chitosan	4–10	80% and 90% respectively	Edward et al. (2021)
NF	Cu^{2+} and Ni^{2+}	Polyimide/polyethersulphone dual layer hollow fibers	–	9 and 7.1	99.1% and 91.2%, respectively	Wang et al. (2021b)
NF	Cu^{2+}, Zn^{2+} and Ni^{2+}	Sulfonated polysulfone (SPSf) on polyethersulfone (PES)	–	2	> 90.0% for single cation and >95% for the mixture	Gao et al. (2020)
NF	Mg^{2+}, Ni^{2+}, Pb^{2+}, Zn^{2+} and Ca^{2+}	Carboxylated cardo poly (arylene ether ketone)s (PAEKCOOH)	–	7	>92.0%, >92.0%, >92.0%, >92.0%, >92.0%, respectively	Liu et al. (2017)
NF/ RO	Cr^{6+}	NF-HL-2514 Spiral wound RO-SG-2514		8	85%–99.9%	Mnif et al. (2017)
RO	Cu^{2+}, Ni^{2+}	Aromatic polyamide-urea advanced composite membrane	EDTA	3.3–5.5	99%	Rodrigues Pires da Silva et al. (2016)
RO	As (V)	Polyamide (Thin Film Composite)	–		99%	Abejón et al. (2015)
RO	Ni^{2+} and Sb^{3+}	Polyamide	–	7–9	95% and 85% respectively	Samaei, et al. (2020)
FO	Cd^{2+}, Pb^{2+}, and Cu^{2+}	Thin film composite	–	7	> 99.4%	Zhao and Liu (2018)
FO	Zn^{2+}, Cu^{2+}, Co^{2+}, Fe^{3+}, Mg^{2+}, Ca^{2+}, Si, Al^{3+}, and Mn^{2+}	Thin-Film Composite	–	–	99.5%, 98.9%, 99.4%, 100.0%, 99.6%, 99.3%, 97.1%, 99.8%, and 98.9%, respectively	Vital et al. (2018)
FO	Cd^{2+}, Pb^{2+}, Cu^{2+} and Zn^{2+}	Nanoporous thin-film inorganic membrane	–	–	94%	You et al. (2017)
VRO-LPM	Mn, As, Cd and Pb Fe, Cu, Zn	Thin Film Composite	–	–	80%–100%	Choi et al. (2019)

such as Hg, Cu and Pb were recovered at ppb level. The membranes were regenerated and were stable for up to five cycles without a significant drop in performance (Wei et al., 2018). The regeneration of the modified adsorbent and the stability of any moiety attached to the surface of the nanofiber are very important for their reusability. Grafting high-density functional groups with hyperbranched structures increase the affinity and selectivity of nanofibers for metal ions recovery (Yu et al., 2019; Zeng et al., 2019). The progress in the modification approach of nanofibers is currently an emerging field and requires further research and innovations for developing more efficient and selective adsorbents.

The recently focused filtration method for metal recovery is Nanohybrid membranes. Nanohybrid membranes are subdivided into two types based upon the property of nanoparticles incorporated, that is, nanocomposite membranes (NCM) and mixed matrix membranes (MMM) (Pendergast and Hoek, 2011). NCM generally refers to a mixed matrix containing nanoparticles that are not permeable or have no sieving effect. The presence of nanoparticles like TiO_2 and Ag improve hydrophilicity, antibacterial surface characteristics of the membranes and make the diffusion path longer. They also provide better affinity/selectivity for specific metals (Yin and Deng, 2015; Kedchaikulrat et al., 2020). On the other hand, the hybrid membranes composed of porous materials with molecular sieving effects (such as MOFs and Zeolites) are called MMM. In addition to providing desired surface properties, these form selective and additional pathways for water molecules, thereby enhancing water permeability and ion rejection (Liu et al., 2013; Yurekli, 2016). For instance, when the inside surface of polyvinylchloride hollow fiber UF membrane was modified with carboxylated multiwalled CNTs (f-MWCNTs) this prevented leaching of nanoparticles and displayed 70% removal of Zn^{2+} from wastewater at pH 7. The adsorption/desorption study further revealed that these membranes can efficiently regenerate even after multiple working cycles and the Zn^{2+} can be easily recovered (Ali et al., 2019).

Metal-organic frameworks (MOF) are also studied for metal recovery. Zirconia-based MOF was found to be ideal for noble metal recovery due to their high surface area, chemical stability, and strong interaction between the organic ligand of MOF and polymer. However, adsorbent dispersibility in polymers and metal selectivity is a major issue often leading to additional costs due to secondary separation procedures (Lin et al., 2017). To overcome these challenges electrospinning method was adopted to disperse UiO-66-NH_2 in polyurethane (PU). The results showed uniform distribution without any leak of MOF. The fabricated membranes showed selective removal of precious metals (palladium (Pd) and platinum (Pt)) from a mixed metal solution containing Ni, Cu, Pd and Pt ions. The Density Functional Theory calculations further revealed that Pt/Pd anions had very high binding energy with MOFs. In contrast, Cu/Ni cations had negligible binding energy. The MOF-based fibrous membranes were regenerated by an acidified thiourea solution and were stable up to five cycles of adsorption-desorption. The super acid resistance of PU/UiO-66-NH_2 fibrous membranes was attributed to the strong $\pi-\pi$ interactions between PU and MOF nanoparticles (Liu et al., 2019).

12.6 Integrated methods of metal recovery

The above-discussed physical, chemical, and biological methods for metal recovery serve the demand but are still limited by the aforementioned (Selvi et al., 2019). So, the researchers have upgraded the situation by process integration thus exhibiting higher efficiency of metal recovery with commercialization. The integration involves two or more different methods for metal recovery combined for the successful and vast recovery of metals. They have pros like large-scale treatment options, versatility, effectiveness, short duration, economic feasibility, and eco-friendliness (Selvi et al., 2019). However such integrated systems require in-depth knowledge of the processes and their scalability for feasible operation under large-scale application. Various such integrated processes have been successfully tested and operated which include the precipitation-oxidation, filtration-reduction, electrolysis-adsorption processes, etc.

12.6.1 Precipitation and oxidation method

Precipitation-oxidation based metal recovery procedure is applied for the recovery of Cu, Fe, and Zn from wastewater (Wang and Chen, 2019). Altering the process pH to 9 and followed by the addition of 0.1% (v/v) H_2O_2 an effective metal recovery. The integrated process however achieved more than 90% recovery in all three metals with Fe (99.8%), Cu (94%), and Zn (96%). Thus, proved to be an eco-friendly and economical solution for long turn operation. Though it is seen that hydroxide precipitation is preferred in industries for metal treatment, but it generates surplus metal hydroxide sludge. This sludge thus requires treatment before disposal.

12.6.2 Adsorption and crystallization method

Hoang and coworkers experimented with the ability of the integrated adsorption crystallization method for recovery of lead (Pb) from the aquatic environment (Hoang et al., 2019). The adsorption was carried out using citric acid-modified sugarcane bagasse followed by desorption. Sodium chloride was employed as the regenerating reagent. The desorption liquid was then subjected to crystallization in a pellet reactor. The Pb undergoes both precipitation and crystallization in the pellet reactor, where lead carbonate is formed from the reaction of supersaturated carbonate ions with Pb. This lead carbonate was then subjected to nucleation in granular seed followed by water reaction leading to the formation of cerussite ($PbCO_3$) which is recovered from the system on garnet sand. The method showed 99% lead recovery at alkaline conditions.

12.6.3 Electrodialysis, electrolysis and adsorption method

This specific process involves the integration of electrodialysis, electrolysis and adsorption together for recovery. This method was successfully applied for the Pb^{2+} removal from aqueous solutions in the laboratory-scale study (Abou-Shady et al., 2012). A feedwater with 600 mg/L of Pb^{2+} concentration was selected for recovery testing and was constantly supplied to the electrodialysis process at 7 L/h. The electrodialysis process separated the feedwater into two categories one as a concentrated solution with (Pb^{2+}) 2600 mg/L and a second diluted solution with (Pb^{2+}) 15–17 mg/L. The concentrated lead solution was then used as FS for the electrolysis process while the dilute lead solution was directed to adsorption by a cation exchange membrane. The electrolysis process reduced the Pb^{2+} concentration to 223–242 mg/L with a recovery of 89.4%–91.6%. The discharge of the electrolysis solution was again used as an influent for electrodialysis for separating into concentrated and dilute solution. The dilute solution was subjected to CEM, for further reducing the lead concentration to 1.0–1.3 mg/L. This formed a continuous integrated system achieving a 90%–91% recovery of Pb^{2+}.

12.6.4 Integrated two-stage electrode ionization method

An integrated two-stage electrode-ionization (EDI) method was also applied for metal recovery. In a typical study, nickel was recovered from electroplating rinse discharge (Lu et al., 2015). The authors used a voltage range of 15 V for the first stage and achieved a Ni^{2+} removal of 94%. The second stage EDI was operated at 25 V achieving deep purification and high-quality water generation by 96.7% Ni^{2+} removal. The first stage of the EDI treatment was based on an enhanced transfer regime while the second stage was operated in the electro regeneration regime with an overall treatment capacity of 1.0 m^3/h. The integrated system was expected to reduce the annual wastewater discharge by 7200 m^3 and sludge disposal by 12.8 tons. Thus, can enhance the overall profit of the electroplating industry by 208,400 Yuan RMB annually as compared to one-step chemical precipitation processes.

12.6.5 Ion exchange and reduction-precipitation

An integrated Ion-exchange and reduction-precipitation method were studied for chromium (Cr) recovery from wastewater generated from the electroplating industry (Ye et al., 2019b). The ion exchange process was operated with silica-based pyridine resin (SiPyR-N4) and showed 99.3% removal efficiency. The adsorption equilibrium for SiPyR-N4 was achieved in 5 min with an excellent adsorption capacity under acidic conditions. This was followed by the reduction-precipitation process where 96.67% of the chromium was removed by reductive desorption, and 98.6% purity of chromium was recovered by precipitation. It was obvious from the study that the integrated process obtained high-quality pure chromium compared to conventional reduction precipitation, along with low reagent consumption (Ye et al., 2019b).

12.7 Recent developments in technology for metal recovery process

In recent years some new technologies have been implemented for the effective recovery of the metals present in water bodies. One such is the use of gamma radiation to detect and recover noble metals from wastewater. The gamma radiation associates radiolytic reduction to recover the Pd particles from wastewater (Kang et al., 2018). In this method, firstly basic orange 2 (BO) was utilized as agent to detect and to stabilize Pd^{2+} nanoparticles. The Pd^{2+} forms a coordination complex with BO via the nitrogen functional group in BO, producing a change in color from yellow to red. Thus, confirming the presence of the

Pd^{2+} in the water. Furthermore, the Pd^{2+} forms a square-planar structured complex with the BO which helped in the formation of Pd as nanoparticles during radiolytic reduction. The gamma rays exposure produces H, H^+, OH, H_2, H_2O_2, e^-_{aq}, etc., species in the aqueous solution. Off them, e^-_{aq} and H are highly reactive which converts the higher valence metal ions in to zero-valent. Thus, converting the Pb^{2+} to Pb^0 for recovery. The size of the nanoparticle was controlled by altering the gamma-ray intensity from 10 to 50 kGy. The Pb particles thus recovered varied in size ranging from 5 to 400 nm.

On the other the photo-electrochemical osmotic system was also used for the recovery of metals from metal-containing wastewater along with electricity production as a new generation technology (Wang et al., 2020). In this method, Ni functionalised Titanium nanowire was used as photoanode and Carbon fiber as the cathode. A FO membrane was used as the separator between the electrodes. A recovery rate of 51 $g/h/m^2$ of copper was reported. The system proved to be an effective metal recovering system along with electrical energy and freshwater generation with the help of sunlight.

Recently sulfate reducing bacteria have also gained wide popularity for recovering heavy metals from wastewater (Kumar et al., 2021). SRB are anaerobic prokaryotes found in various habitats like marshes, lakes, paddy fields, industrial wastewater, petroleum deposits, etc. (Xu and Chen, 2020). These bacteria can tolerate various pH ranges and extracts the metals in aqueous media following precipitation technique. The bioprocess is preferred now-a-days owing to its sustainable nature and low cost.

12.8 Commercial systems

Some commercial recovery systems were readily available, one such is THIOPAQ technology, developed and marketed by Paques Bio Systems B.V., Balk, the Netherlands, for metal sulfide precipitation (Van Lier et al., 1999). The technology was first carried out in 1992 for the treatment of contaminated groundwater at the Budelco zinc refinery in the Netherlands. The technique consisted of two-stage biological treatment where the first one converts sulfate to sulfide (also known as sulfate reduction or anaerobic stage) while the second stage converts sulfide to elemental sulfur (also designated as sulfide oxidation or aerobic stage). The recovered sulfide and sulfur are biogenic in nature and can be used as precipitation agents. The THIOPAQ techniques treat water containing metal-sulfur species including sulfate, sulfite, thiosulfate, and (poly) sulfide. The sulfide precipitation process also separates metals like As from Cu, Cu from Zn, Fe from Ni, etc., in various stages of the reaction process and at distinct pH. This allows selective recovery of metals from the reactor. The THIOPAQ system at Budelco zinc refinery, processed $\sim$300 m^3/h of polluted groundwater since 1992 and was increased to 400 m^3/h in 1998. The THIOPAQ system is now operational in industries like metallurgy, chemical and electronics.

The system developed by Paques has grown vastly and now a new unit dedicated to metal recovery known as THIOTEQ Metal technology was evolved. The system now operates on two-stage, the first being chemical and the second being biological. The metal recovery is mainly derived from the formation of high-purity metal sulfides. The system can work under liquids with a dissolved metal concentration between 50 and 5000 mg/L. The metal sulfides formed are separated by gravity settling and dewatered for metal recovery.

The THIOTEQ system was operated by The Pueblo Viejo Gold Mine for copper recovery. The Pueblo Viejo Mine operates on 24,000 tons/day of polymetal ores containing copper, silver, gold, and zinc. Since its commission, it has recovered 12,000 tons of copper annually (Paques, 2021).

12.9 Conclusion

The main focus of the circular economy and the ongoing industrial revolution is to enhance the value of the wastes generated. Hence the recovery and reuse of the metals from any waste serve sustainability and green economy purpose. This chapter clearly presented the treatment of metals in contaminated water as well as recovery of those metals for reuse. The various technologies applied for the recovery of metals from contaminated water bodies were revealed. The integrated systems and ongoing successful commercial-scale systems for metal recovery were deliberated.

Acknowledgment

The authors are grateful for the financial support under IMPRINT with grant code IMP/2019/00028 from Science and Engineering Research Board (DST-SERB).

References

Abejón, A., Garea, A., Irabien, A., 2015. Arsenic removal from drinking water by reverse osmosis: minimization of costs and energy consumption. Separation and Purification Technology 144, 46–53.

Abou-Shady, A., Peng, C., Bi, J., Xu, H., 2012. Recovery of Pb (II) and removal of NO_3^- from aqueous solutions using integrated electrodialysis, electrolysis, and adsorption process. Desalination 286, 304–315.

Acheampong, M.A., Meulepas, R.J., Lens, P.N., 2010. Removal of heavy metals and cyanide from gold mine wastewater. Journal of Chemical Technology & Biotechnology 85 (5), 590–613.

Adhikari, S., Mandal, S., Kim, D.H., Mishra, A.K., 2019. An overview of treatment technologies for the removal of emerging and nanomaterials contaminants from municipal and industrial wastewater. Emerging and Nanomaterial Contaminants in Wastewater 3–40.

Akbal, F., Camcı, S., 2010. Comparison of electrocoagulation and chemical coagulation for heavy metal removal. Chemical Engineering & Technology 33 (10), 1655–1664.

Akbari, S., Abdurahman, N.H., Yunus, R.M., Fayaz, F., Alara, O.R., 2018. Biosurfactants—a new frontier for social and environmental safety: a mini review. Biotechnology Research and Innovation 2 (1), 81–90.

Algieri, C., Chakraborty, S., Candamano, S., 2021. A way to membrane-based environmental remediation for heavy metal removal. Environments 8 (6), 52.

Ali, S., Rehman, S.A.U., Shah, I.A., Farid, M.U., An, A.K., Huang, H., 2019. Efficient removal of zinc from water and wastewater effluents by hydroxylated and carboxylated carbon nanotube membranes: behaviors and mechanisms of dynamic filtration. Journal of Hazardous Materials 365, 64–73.

Almomani, F., Bohsale, R.R., 2021. Bio-sorption of toxic metals from industrial wastewater by algae strains *Spirulina platensis* and *Chlorella vulgaris*: application of isotherm, kinetic models and process optimization. Science of the Total Environment 755, 142654.

Al-Qahtani, K.M., 2016. Water purification using different waste fruit cortexes for the removal of heavy metals. Journal of Taibah University for Science 10 (5), 700–708.

Awad, A.M., Jalab, R., Minier-Matar, J., Adham, S., Nasser, M.S., Judd, S.J., 2019. The status of forward osmosis technology implementation. Desalination 461, 10–21.

Babilas, D., Dydo, P., 2018. Selective zinc recovery from electroplating wastewaters by electrodialysis enhanced with complex formation. Separation and Purification Technology 192, 419–428.

Baby, R., Saifullah, B., Hussein, M.Z., 2019. Palm Kernel Shell as an effective adsorbent for the treatment of heavy metal contaminated water. Scientific Reports 9 (1), 1–11.

Barakat, M.A., 2011. New trends in removing heavy metals from industrial wastewater. Arabian Journal of Chemistry 4 (4), 361–377.

Basha, C.A., Ramanathan, K., Rajkumar, R., Mahalakshmi, M., Kumar, P.S., 2008. Management of chromium plating rinsewater using electrochemical ion exchange. Industrial & Engineering Chemistry Research 47 (7), 2279–2286.

Beauchesne, I., Drogui, P., Seyhi, B., Mercier, G., Blais, J.F., 2014. Simultaneous electrochemical leaching and electrodeposition of heavy metals in a single-cell process for wastewater sludge treatment. Journal of Environmental Engineering 140 (8), 04014030.

Biersack, B., Diestel, R., Jagusch, C., Sasse, F., Schobert, R., 2009. Metal complexes of natural melophlins and their cytotoxic and antibiotic activities. Journal of Inorganic Biochemistry 103 (1), 72–76.

Butter, B., Santander, P., Pizarro, G.D.C., Oyarzún, D.P., Tasca, F., Sánchez, J., 2021. Electrochemical reduction of Cr (VI) in the presence of sodium alginate and its application in water purification. Journal of Environmental Sciences 101, 304–312.

Carolin, C.F., Kumar, P.S., Saravanan, A., Joshiba, G.J., Naushad, M., 2017. Efficient techniques for the removal of toxic heavy metals from aquatic environment: a review. Journal of Environmental Chemical Engineering 5 (3), 2782–2799.

Carvalho, A.L., Maugeri, F., Pradanos, P., Silva, V., Hernández, A., 2011. Separation of potassium clavulanate and potassium chloride by nanofiltration: transport and evaluation of membranes. Separation and Purification Technology 83, 23–30.

Castro, D., Rosas-Laverde, N.M., Aldás, M.B., Almeida-Naranjo, C.E., Guerrero, V.H., Pruna, A.I., 2021. Chemical modification of agro-industrial waste-based bioadsorbents for enhanced removal of Zn (II) ions from aqueous solutions. Materials 14 (9), 2134.

Chatterjee, A., Abraham, J., 2019. Desorption of heavy metals from metal loaded sorbents and e-wastes: a review. Biotechnology Letters 41 (3), 319–333.

Chatterjee, S.K., Bhattacharjee, I., Chandra, G., 2010. Biosorption of heavy metals from industrial waste water by *Geobacillus thermodenitrificans*. Journal of Hazardous Materials 175 (1–3), 117–125.

Chen, Q., Yao, Y., Li, X., Lu, J., Zhou, J., Huang, Z., 2018. Comparison of heavy metal removals from aqueous solutions by chemical precipitation and characteristics of precipitates. Journal of Water Process Engineering 26, 289–300.

Childress, A.E., Elimelech, M., 2000. Relating nanofiltration membrane performance to membrane charge (electrokinetic) characteristics. Environmental Science & Technology 34 (17), 3710–3716.

Choi, J., Im, S.J., Jang, A., 2019. Application of a volume retarded osmosis–low pressure membrane hybrid process for treatment of acid whey. Chemosphere 219, 261–267.

Chung, Y.S., Yoo, S.H., Kim, C.K., 2009. Effects of membrane hydrophilicity on the removal of a trihalomethane via micellar-enhanced ultrafiltration process. Journal of Membrane Science 326 (2), 714–720.

Cui, Y., Ge, Q., Liu, X.Y., Chung, T.S., 2014. Novel forward osmosis process to effectively remove heavy metal ions. Journal of Membrane Science 467, 188–194.

Dahaghin, Z., Mousavi, H.Z., Sajjadi, S.M., 2017. Trace amounts of Cd (II), Cu (II) and Pb (II) ions monitoring using Fe_3O_4@ graphene oxide nanocomposite modified via 2-mercaptobenzothiazole as a novel and efficient nanosorbent. Journal of Molecular Liquids 231, 386–395.

Dahman, Y., Deonanan, K., Dontsos, T., Lammatteo, A., 2017. Nanopolymers. In: Dahman, Y. (Ed.), Nanotechnology and Functional Materials for Engineers. Elsevier, pp. 121–144.

Das, N., 2010. Recovery of precious metals through biosorption—a review. Hydrometallurgy 103 (1–4), 180–189.

Dixit, R., Malaviya, D., Pandiyan, K., Singh, U.B., Sahu, A., Shukla, R., et al., 2015. Bioremediation of heavy metals from soil and aquatic environment: an overview of principles and criteria of fundamental processes. Sustainability 7, 2189–2212.

Dotto, G.L., McKay, G., 2020. Current scenario and challenges in adsorption for water treatment. Journal of Environmental Chemical Engineering 8 (4), 103988.

Edward, K., Dalmia, M., Samuel, A.M., Sivaraman, P., Rajesh, M.P., 2021. Optimization of process conditions for the removal of zinc and lead by ultrafiltration using biopolymer. Chemical Papers 1–15.

Elias, S.H., Mohamed, M.A.K.E.T.A.B., Ankur, A.N., Muda, K., Hassan, M.A.H.M., Othman, M.N., et al., 2014. Water hyacinth bioremediation for ceramic industry wastewater treatment-application of rhizofiltration system. Sains Malaysiana 43 (9), 1397–1403.

Emamjomeh, M.M., Sivakumar, M., 2009. Review of pollutants removed by electrocoagulation and electrocoagulation/flotation processes. Journal of Environmental Management 90 (5), 1663–1679.

Farhadi, A., Ameri, A., Tamjidi, S., 2021. Application of agricultural wastes as a low-cost adsorbent for removal of heavy metals and dyes from wastewater: a review study. Physical Chemistry Research 9 (2), 211–226.

Foroutan, R., Mohammadi, R., Adeleye, A.S., Farjadfard, S., Esvandi, Z., Arfaeinia, H., et al., 2019. Efficient arsenic (V) removal from contaminated water using natural clay and clay composite adsorbents. Environmental Science and Pollution Research 26 (29), 29748–29762.

Fu, F., Wang, Q., 2011. Removal of heavy metal ions from wastewaters: a review. Journal of Environmental Management 92 (3), 407–418.

Gadd, G., 2009. Heavy metal pollutants: environmental and biotechnological aspects. Encyclopedia of Microbiology. Elsevier, pp. 321–334.

Gao, J., Wang, K.Y., Chung, T.S., 2020. Design of nanofiltration (NF) hollow fiber membranes made from functionalized bore fluids containing polyethyleneimine (PEI) for heavy metal removal. Journal of Membrane Science 603, 118022.

Gering, K.L., Scamehorn, J.F., 1988. Use of electrodialysis to remove heavy metals from water. Separation Science and Technology 23 (14–15), 2231–2267.

Gu, S., Kang, X., Wang, L., Lichtfouse, E., Wang, C., 2019. Clay mineral adsorbents for heavy metal removal from wastewater: a review. Environmental Chemistry Letters 17 (2), 629–654.

Hamdaoui, O., Djeribi, R., Naffrechoux, E., 2005. Desorption of metal ions from activated carbon in the presence of ultrasound. Industrial & Engineering Chemistry Research 44 (13), 4737–4744.

He, M., Wang, L., Lv, Y., Wang, X., Zhu, J., Zhang, Y., et al., 2020. Novel polydopamine/metal organic framework thin film nanocomposite forward osmosis membrane for salt rejection and heavy metal removal. Chemical Engineering Journal 389, 124452.

Hennebel, T., Boon, N., Maes, S., Lenz, M., 2015. Biotechnologies for critical raw material recovery from primary and secondary sources: R&D priorities and future perspectives. New Biotechnology 32 (1), 121–127.

Hoang, M.T., Pham, T.D., Nguyen, V.T., Nguyen, M.K., Pham, T.T., Van der Bruggen, B., 2019. Removal and recovery of lead from wastewater using an integrated system of adsorption and crystallization. Journal of Cleaner Production 213, 1204–1216.

Hokkanen, S., Repo, E., Sillanpää, M., 2013. Removal of heavy metals from aqueous solutions by succinic anhydride modified mercerized nanocellulose. Chemical Engineering Journal 223, 40–47.

Hou, Y., Cheng, K., Li, Z., Ma, X., Wei, Y., Zhang, L., et al., 2018. Correction: biosorption of cadmium and manganese using free cells of Klebsiella sp. isolated from waste water. PLoS One 13 (5), e0198309.

Huang, Y., Zeng, X., Guo, L., Lan, J., Zhang, L., Cao, D., 2018. Heavy metal ion removal of wastewater by zeolite-imidazolate frameworks. Separation and Purification Technology 194, 462–469.

Igiri, B.E., Okoduwa, S.I., Idoko, G.O., Akabuogu, E.P., Adeyi, A.O., Ejiogu, I.K., 2018. Toxicity and bioremediation of heavy metals contaminated ecosystem from tannery wastewater: a review. Journal of Toxicology 2018.

Ignatius, A., Arunbabu, V., Neethu, J., Ramasamy, E.V., 2014. Rhizofiltration of lead using an aromatic medicinal plant *Plectranthus amboinicus* cultured in a hydroponic nutrient film technique (NFT) system. Environmental Science and Pollution Research 21 (22), 13007–13016.

Jamil, T.S., Ibrahim, H.S., Abd El-Maksoud, I.H., El-Wakeel, S.T., 2010. Application of zeolite prepared from Egyptian kaolin for removal of heavy metals: I. Optimum conditions. Desalination 258 (1–3), 34–40.

Kader, J., Sannasi, P., Othman, O., Ismail, B.S., Salmijah, S., 2007. Removal of Cr (VI) from aqueous solutions by growing and non-growing populations of environmental bacterial consortia. Global Journal of Environmental Research 1 (1), 12–17.

Kakaei, S., Khameneh, E.S., Hosseini, M.H., Moharreri, M.M., 2020. A modified ionic liquid clay to remove heavy metals from water: investigating its catalytic activity. International Journal of Environmental Science and Technology 17 (4), 2043–2058.

Kakakhel, L., Lutfullah, G., Bhanger, M.I., Shah, A., Niaz, A., 2007. Electrolytic recovery of chromium salts from tannery wastewater. Journal of Hazardous Materials 148 (3), 560–565.

Kang, S.M., Kwak, C.H., Rethinasabapathy, M., Jang, S.C., Choe, S.R., Roh, C., et al., 2018. Gamma radiation mediated green technology for Pd nanoparticles recovery from wastewater. Separation and Purification Technology 197, 220–227.

Kedchaikulrat, P., Vankelecom, I.F., Faungnawakij, K., Klaysom, C., 2020. Effects of colloidal TiO_2 and additives on the interfacial polymerization of thin film nanocomposite membranes. Colloids and Surfaces A: Physicochemical and Engineering Aspects 601, 125046.

Kehrein, P., Van Loosdrecht, M., Osseweijer, P., Garfí, M., Dewulf, J., Posada, J., 2020. A critical review of resource recovery from municipal wastewater treatment plants—market supply potentials, technologies and bottlenecks. Environmental Science: Water Research & Technology 6 (4), 877–910.

Kikuchi, T., Tanaka, S., 2012. Biological removal and recovery of toxic heavy metals in water environment. Critical Reviews in Environmental Science and Technology 42 (10), 1007–1057.

Knápek, J., Komárek, J., Novotný, K., 2010. Determination of cadmium, chromium and copper in high salt samples by LA-ICP-OES after electrodeposition—preliminary study. Microchimica Acta 171 (1), 145–150.

Krishnani, K.K., Meng, X., Christodoulatos, C., Boddu, V.M., 2008. Biosorption mechanism of nine diferent heavy metals onto biomatrix from rice husk. Journal of Hazardous Materials 153, 1222–1234.

Kumar, M., Nandi, M., Pakshirajan, K., 2021. Recent advances in heavy metal recovery from wastewater by biogenic sulfide precipitation. Journal of Environmental Management 278, 111555.

Kummer, L., Cosmann, N.A.J., Pastore, G.M., Simiqueli, A.P.R., Gomes, S.D., 2016. Adsorption of copper, zinc and lead on biosurfactant produced from cassava wastewater. African Journal of Biotechnology 15 (5), 110–117.

Lakshmipathiraj, P., Raju, G.B., Basariya, M.R., Parvathy, S., Prabhakar, S., 2008. Removal of Cr (VI) by electrochemical reduction. Separation and Purification Technology 60 (1), 96–102.

Lata, S., Singh, P.K., Samadder, S.R., 2015. Regeneration of adsorbents and recovery of heavy metals: a review. International Journal of Environmental Science and Technology 12 (4), 1461–1478.

Lin, S., Reddy, D.H.K., Bediako, J.K., Song, M.H., Wei, W., Kim, J.A., et al., 2017. Effective adsorption of Pd (ii), Pt (iv) and Au (iii) by Zr (iv)-based metal–organic frameworks from strongly acidic solutions. Journal of Materials Chemistry A 5 (26), 13557–13564.

Liu, C., Bi, W., Chen, D., Zhang, S., Mao, H., 2017. Positively charged nanofiltration membrane fabricated by poly (acid–base) complexing effect induced phase inversion method for heavy metal removal. Chinese Journal of Chemical Engineering 25 (11), 1685–1694.

Liu, X., Jin, H., Li, Y., Bux, H., Hu, Z., Ban, Y., et al., 2013. Metal–organic framework ZIF-8 nanocomposite membrane for efficient recovery of furfural via pervaporation and vapor permeation. Journal of Membrane Science 428, 498–506.

Liu, Y., Lin, S., Liu, Y., Sarkar, A.K., Bediako, J.K., Kim, H.Y., et al., 2019. Super-stable, highly efficient, and recyclable fibrous metal–organic framework membranes for precious metal recovery from strong acidic solutions. Small (Weinheim an der Bergstrasse, Germany) 15 (10), 1805242.

Lo, Y.C., Cheng, C.L., Han, Y.L., Chen, B.Y., Chang, J.S., 2014. Recovery of high-value metals from geothermal sites by biosorption and bioaccumulation. Bioresource Technology 160, 182–190.

Lu, H., Wang, Y., Wang, J., 2015. Recovery of Ni^{2+} and pure water from electroplating rinse wastewater by an integrated two-stage electrodeionization process. Journal of Cleaner Production 92, 257–266.

Meng, Q., Nan, J., Mu, Y., Zu, X., Guo, M., 2021. Study on the treatment of sudden cadmium pollution in surface water by a polymer enhanced ultrafiltration process. RSC Advances 11 (13), 7405–7415.

Mnif, A., Bejaoui, I., Mouelhi, M., Hamrouni, B., 2017. Hexavalent chromium removal from model water and car shock absorber factory effluent by nanofiltration and reverse osmosis membrane. International Journal of Analytical Chemistry 2017.

Moss, R.L., Tzimas, E., Kara, H., Willis, P., Kooroshy, J., 2011. Critical metals in strategic energy technologies. JRC-scientific and strategic reports, European Commission Joint Research Centre Institute for Energy and Transport. Publications Office of the European Union, Luxembourg.

Mullett, M., Fornarelli, R., Ralph, D., 2014. Nanofiltration of mine water: impact of feed pH and membrane charge on resource recovery and water discharge. Membranes 4 (2), 163–180.

Muyzer, G., Stams, A.J., 2008. The ecology and biotechnology of sulphate-reducing bacteria. Nature Reviews. Microbiology 6 (6), 441–454.

Nepel, T.C.D.M., Costa, J.M., Vieira, M.G.A., Almeida Neto, A.F.D., 2020. Copper removal kinetic from electroplating industry wastewater using pulsed electrodeposition technique. Environmental Technology 1–9.

Ojuederie, O.B., Babalola, O.O., 2017. Microbial and plant-assisted bioremediation of heavy metal polluted environments: a review. International Journal of Environmental Research and Public Health 14 (12), 1504.

Oyeku, O.T., Eludoyin, A.O., 2010. Heavy metal contamination of groundwater resources in a Nigerian urban settlement. African Journal of Environmental Science and Technology 4 (4).

Paques, B.V., 2021. Paques, leading in biological wastewater and gas treatment [Online]. Available: http://www.paques.nl (accessed 10.09.21).

Pathak, A., Dastidar, M.G., Sreekrishnan, T.R., 2009. Bioleaching of heavy metals from sewage sludge: a review. Journal of Environmental Management 90 (8), 2343–2353.

Pendergast, M.M., Hoek, E.M., 2011. A review of water treatment membrane nanotechnologies. Energy & Environmental Science 4 (6), 1946–1971.

Qin, J.J., Oo, M.H., Lee, H., Coniglio, B., 2004. Effect of feed pH on permeate pH and ion rejection under acidic conditions in NF process. Journal of Membrane Science 232 (1–2), 153–159.

Rahman, M.A., Reichman, S.M., De Filippis, L., Sany, S.B.T., Hasegawa, H., 2016. Phytoremediation of toxic metals in soils and wetlands: concepts and applications. In: Hasegawa, H., Rahman, M.M., Rahman, I. (Eds.), Environmental Remediation Technologies for Metal-Contaminated Soils. Springer, Tokyo, Japan, pp. 161–195.

Rahmati, N.O., Chenar, M.P., Namaghi, H.A., 2017. Removal of free active chlorine from synthetic wastewater by MEUF process using polyethersulfone/titania nanocomposite membrane. Separation and Purification Technology 181, 213–222.

Rashid, A., Bhatti, H.N., Iqbal, M., Noreen, S., 2016. Fungal biomass composite with bentonite efficiency for nickel and zinc adsorption: a mechanistic study. Ecological Engineering 91, 459–471.

Rivas, B.L., Maureira, A., Guzmán, C., Contreras, D., Kaim, W., Geckeler, K.E., 2011. Poly (L-lysine) as a polychelatogen to remove toxic metals using ultrafiltration and bactericide properties of poly (L-lysine)–Cu^{2+} complexes. Polymer bulletin 67 (5), 763–774.

Rodrigues Pires da Silva, J., Merçon, F., Guimarães Costa, C.M., Radoman Benjo, D., 2016. Application of reverse osmosis process associated with EDTA complexation for nickel and copper removal from wastewater. Desalination and Water Treatment 57 (41), 19466–19474.

Rodríguez, A., Sáez, P., Díez, E., Gómez, J.M., García, J., Bernabé, I., 2019. Highly efficient low-cost zeolite for cobalt removal from aqueous solutions: characterization and performance. Environmental Progress & Sustainable Energy 38 (s1), S352–S365.

Salem, H.M., Eweida, E.A., Farag, A., 2000. September. Heavy metals in drinking water and their environmental impact on human health. In *International Conference on the Environment Hazards Mitigation, Cairo University Egypt* (pp. 542–556).

Samaei, S.M., Gato-Trinidad, S., Altaee, A., 2020. Performance evaluation of reverse osmosis process in the post-treatment of mining wastewaters: case study of Costerfield mining operations, Victoria, Australia. Journal of Water Process Engineering 34, 101116.

Sánchez, J., Espinosa, C., Pooch, F., Tenhu, H., Pizarro, G.D.C., Oyarzún, D.P., 2018. Poly (*N*, *N*-dimethylaminoethyl methacrylate) for removing chromium (VI) through polymer-enhanced ultrafiltration technique. Reactive and Functional Polymers 127, 67–73.

Selvi, A., Rajasekar, A., Theerthagiri, J., Ananthaselvam, A., Sathishkumar, K., Madhavan, J., et al., 2019. Integrated remediation processes toward heavy metal removal/recovery from various environments-a review. Frontiers in Environmental Science 7, 66.

Shami, R.B., Shojaei, V., Khoshdast, H., 2019. Efficient cadmium removal from aqueous solutions using a sample coal waste activated by rhamnolipid biosurfactant. Journal of Environmental Management 231, 1182–1192.

Singh, R., Hankins, N.P., 2016. Introduction to membrane processes for water treatment. Emerging Membrane Technology for Sustainable Water Treatment 15–52.

Smara, A., Delimi, R., Chainet, E., Sandeaux, J., 2007. Removal of heavy metals from diluted mixtures by a hybrid ion-exchange/electrodialysis process. Separation and Purification Technology 57 (1), 103–110.

Sum, J.Y., Kok, W.X., Shalini, T.S., 2021. The removal selectivity of heavy metal cations in micellar-enhanced ultrafiltration: a study based on critical micelle concentration. Materials Today: Proceedings.

Sundararaju, S., Manjula, A., Kumaravel, V., Muneeswaran, T., Vennila, T., 2020. Biosorption of nickel ions using fungal biomass Penicillium sp. MRF1 for the treatment of nickel electroplating industrial effluent. Biomass Conversion and Biorefinery 1–10.

Suwaileh, W.A., Johnson, D.J., Sarp, S., Hilal, N., 2018. Advances in forward osmosis membranes: altering the sub-layer structure via recent fabrication and chemical modification approaches. Desalination 436, 176–201.

Tajuddin, M.H., Yusof, N., Wan Azelee, I., Wan Salleh, W.N., Ismail, A.F., Jaafar, J., et al., 2019. Development of copper-aluminum layered double hydroxide in thin film nanocomposite nanofiltration membrane for water purification process. Frontiers in Chemistry 7, 3.

Taksitta, K., Sujarit, P., Ratanawimarnwong, N., Donpudsa, S., Songsrirote, K., 2020. Development of tannin-immobilized cellulose fiber extracted from coconut husk and the application as a biosorbent to remove heavy metal ions. Environmental Nanotechnology, Monitoring & Management 14, 100389.

Tran, T.K., Chiu, K.F., Lin, C.Y., Leu, H.J., 2017. Electrochemical treatment of wastewater: selectivity of the heavy metals removal process. International Journal of Hydrogen Energy 42 (45), 27741–27748.

Turygin, V.V., Smirnov, M.K., Smetanin, A.V., Zhukov, E.G., Fedorov, V.A., Tomilov, A.P., 2008. Electrochemical arsenic extraction from nonferrous metals industry waste. Inorganic Materials 44 (9), 946–953.

Uddin, M.K., 2017. A review on the adsorption of heavy metals by clay minerals, with special focus on the past decade. Chemical Engineering Journal 308, 438–462.

Un, U.T., Ocal, S.E., 2015. Removal of heavy metals (Cd, Cu, Ni) by electrocoagulation. International Journal of Environmental Science and Development 6 (6), 425.

Van Lier, R.J.M., Buisman, C.J.N., Piret, N.L., 1999. THIOPAQ® technology: versatile high-rate biotechnology for the mining and metallurgical industries. In Proceedings of the TMS Fall Extraction and Processing Conference (Vol. 3, pp. 2319–2328).

Verma, S.P., Sarkar, B., 2020. Analysis of flux decline during rhamnolipid based micellar-enhanced ultrafiltration for simultaneous removal of Cd^{2+} and crystal violet from aqueous solution. Journal of Water Process Engineering 33, 101048.

Vital, B., Bartacek, J., Ortega-Bravo, J.C., Jeison, D., 2018. Treatment of acid mine drainage by forward osmosis: heavy metal rejection and reverse flux of draw solution constituents. Chemical Engineering Journal 332, 85–91.

Wang, L.P., Chen, Y.J., 2019. Sequential precipitation of iron, copper, and zinc from wastewater for metal recovery. Journal of Environmental Engineering 145 (1), 04018130.

Wang, C., Sun, M., Zhao, Y., Huo, M., Wang, X., Elimelech, M., 2020. Photo-electrochemical osmotic system enables simultaneous metal recovery and electricity generation from wastewater. Environmental Science & Technology 55 (1), 604–613.

Wang, N., Qiu, Y., Hu, K., Huang, C., Xiang, J., Li, H., et al., 2021a. One-step synthesis of cake-like biosorbents from plant biomass for the effective removal and recovery heavy metals: effect of plant species and roles of xanthation. Chemosphere 266, 129129.

Wang, Z.Y., Wang, Y.C., Wang, W.J., Tao, S.N., Chen, Y.F., Tang, M., et al., 2021b. Designing scalable dual-layer composite hollow fiber nanofiltration membranes with fully cross-linked ultrathin functional layer. Journal of Membrane Science 628, 119243.

Wei, J., Jiao, X., Wang, T., Chen, D., 2018. Fast, simultaneous metal reduction/deposition on electrospun a-WO_3/PAN nanofiber membranes and their potential applications for water purification and noble metal recovery. Journal of Materials Chemistry A 6 (30), 14577–14586.

Wenten, I.G., 2016. Reverse osmosis applications: prospect and challenges. Desalination 391, 112–125.

Won, S.W., Kotte, P., Wei, W., Lim, A., Yun, Y.S., 2014. Biosorbents for recovery of precious metals. Bioresource Technology 160, 203–212.

Xu, Y.N., Chen, Y., 2020. Advances in heavy metal removal by sulfate-reducing bacteria. Water Science and Technology 81 (9), 1797–1827.

Ya, V., Martin, N., Chou, Y.H., Chen, Y.M., Choo, K.H., Chen, S.S., et al., 2018. Electrochemical treatment for simultaneous removal of heavy metals and organics from surface finishing wastewater using sacrificial iron anode. Journal of the Taiwan Institute of Chemical Engineers 83, 107–114.

Yadav, M., Gupta, R., Arora, G., Yadav, P., Srivastava, A., Sharma, R.K., 2020. Current status of heavy metal contaminants and their removal/recovery techniques. Contaminants in Our Water: Identification and Remediation Methods. American Chemical Society, pp. 41–64.

Yan, A., Wang, Y., Tan, S.N., Mohd Yusof, M.L., Ghosh, S., Chen, Z., 2020. Phytoremediation: a promising approach for revegetation of heavy metal-polluted land. Frontiers in Plant Science 11, 359.

Yap, C.Y., Mohamed, N., 2007. An electrogenerative process for the recovery of gold from cyanide solutions. Chemosphere 67 (8), 1502–1510.

Yaqub, M., Lee, S.H., 2020. Micellar enhanced ultrafiltration (MEUF) of mercury-contaminated wastewater: experimental and artificial neural network modeling. Journal of Water Process Engineering 33, 101046.

Ye, C.C., An, Q.F., Wu, J.K., Zhao, F.Y., Zheng, P.Y., Wang, N.X., 2019a. Nanofiltration membranes consisting of quaternized polyelectrolyte complex nanoparticles for heavy metal removal. Chemical Engineering Journal 359, 994–1005.

Ye, Z., Yin, X., Chen, L., He, X., Lin, Z., Liu, C., et al., 2019b. An integrated process for removal and recovery of Cr (VI) from electroplating wastewater by ion exchange and reduction–precipitation based on a silica-supported pyridine resin. Journal of Cleaner Production 236, 117631.

Yin, J., Deng, B., 2015. Polymer-matrix nanocomposite membranes for water treatment. Journal of Membrane Science 479, 256–275.

You, S., Lu, J., Tang, C.Y., Wang, X., 2017. Rejection of heavy metals in acidic wastewater by a novel thin-film inorganic forward osmosis membrane. Chemical Engineering Journal 320, 532–538.

Yu, D., Wang, Y., Wu, M., Zhang, L., Wang, L., Ni, H., 2019. Surface functionalization of cellulose with hyperbranched polyamide for efficient adsorption of organic dyes and heavy metals. Journal of Cleaner Production 232, 774–783.

Yu, Z., Han, H., Feng, P., Zhao, S., Zhou, T., Kakade, A., et al., 2020. Recent advances in the recovery of metals from waste through biological processes. Bioresource Technology 297, 122416.

Yuna, Z., 2016. Review of the natural, modified, and synthetic zeolites for heavy metals removal from wastewater. Environmental Engineering Science 33 (7), 443–454.

Yurekli, Y., 2016. Removal of heavy metals in wastewater by using zeolite nano-particles impregnated polysulfone membranes. Journal of Hazardous Materials 309, 53–64.

Zaghbani, N., Hafiane, A., Dhahbi, M., 2009. Removal of Direct Blue 71 from wastewater using micellar enhanced ultrafiltration. Desalination and Water Treatment 6 (1–3), 204–210.

Zeng, H., Wang, L., Zhang, D., Yan, P., Nie, J., Sharma, V.K., et al., 2019. Highly efficient and selective removal of mercury ions using hyperbranched polyethylenimine functionalized carboxymethyl chitosan composite adsorbent. Chemical Engineering Journal 358, 253–263.

Zeng, G.M., Xu, K., Huang, J.H., Li, X., Fang, Y.Y., Qu, Y.H., 2008. Micellar enhanced ultrafiltration of phenol in synthetic wastewater using polysulfone spiral membrane. Journal of Membrane Science 310 (1–2), 149–160.

Zhao, X., Liu, C., 2018. Efficient removal of heavy metal ions based on the optimized dissolution-diffusion-flow forward osmosis process. Chemical Engineering Journal 334, 1128–1134.

Zhu, Y., Fan, W., Zhou, T., Li, X., 2019. Removal of chelated heavy metals from aqueous solution: a review of current methods and mechanisms. Science of the Total Environment 678, 253–266.

CHAPTER

13

Processes of decontamination and elimination of toxic metals from water and wastewaters

Sylvester Chibueze Izah[1], *Clement Takon Ngun*[2], *Paschal Okiroro Iniaghe*[3], *Ayobami Omozemoje Aigberua*[4] *and Tamaraukepreye Catherine Odubo*[1]

[1]Department of Microbiology, Faculty of Science, Bayelsa Medical University, Yenagoa, Bayelsa State, Nigeria [2]Department of Biochemistry and Biophysics, Saratov State University, Saratov, Saratov Region, Russian Federation [3]Department of Chemistry, Faculty of Science, Federal University Otuoke, Bayelsa State, Nigeria [4]Department of Environment, Research and Development, Anal Concept Limited, Elelenwo, Rivers State, Nigeria

13.1 Introduction

Human operations in various industries such as petroleum, agriculture, nuclear power, research, medical and health care, usually generate wastes that pollute the environment hence, increasingly becoming one of today's significant environmental issues. Some of these wastes are infectious, carcinogenic, and neurotoxic to living species, particularly humans. Reactive wastes from nuclear activity, for example, are volatile and can react with water and air to produce hazardous vapors or even explosions. Ignitable wastes from construction and petrochemical industries become ignitable at low temperatures, posing a fire threat. Corrosive wastes from pesticides, herbicides, paints, fertilizers and various industrial solvents can harm materials or tissue when they make contact. If infectious waste from biological research institutions, laboratories, and hospitals are not appropriately managed, it can threaten human health and the environment (Eskander and Saleh, 2017).

Before the mid-1950s, organic pollution of water bodies was not a severe issue. However, in recent times, water pollution remains an issue of global environmental worry.

Metals in Water
DOI: https://doi.org/10.1016/B978-0-323-95919-3.00003-3

This is largely a spill-over effect of growing industrialization, population expansion, and massive urbanization (Dhall et al., 2012). With the increased volume of various hazardous wastes generated each year, the negative consequences of poor waste management have been on the rise, and the health effects have become far-reaching. Landfilling is still one of the most frequently adopted waste management procedures. This particularly applies to municipal solid wastes. Other typical strategies for solid waste reduction have also been examined, but in some instances, storage has been compromised, resulting in contamination of groundwater resources. This is frequent in areas where the water table is high.

Waters, which frequently contain toxic materials, are harmful to the ecosystem and biota that live in them. When toxicant concentrations surpass the permitted limit for discharge into the receiving environment, both acute and chronic impacts may occur. Although, this is largely dependent on exposure duration and characteristics of the toxicant in question. The effects of various toxicants on living creatures vary. For instance, hydrocarbons from crude oil, polychlorinated biphenyls (PCBs), dye, pesticides, fertilizers, and trace metals have different environmental effects.

Trace metals are metalloids with a density approximately five times that of water (Izah et al., 2016, 2017c; Izah and Aigberua, 2020). Trace metals are frequently classed according to their level of toxicity. They are also divided into essential (chromium, iron, zinc, copper, manganese, cobalt, molybdenum, selenium, nickel, amongst others) and non-vital (mercury, cadmium, lead, arsenic, etc.). In humans and other living beings, vital metals perform biochemical and physiological roles (Tchounwou et al., 2012; Izah et al., 2016, 2017c). When the level of critical elements exceeds the appropriate amount for a particular organism, it becomes poisonous. Suggested levels for the various metals are different. For example, compared to the acceptable limit for chromium, the concentration of zinc and copper allowed is much higher (Izah et al., 2016, 2017c; Ahamad et al., 2020; Madhav et al., 2020).

Trace metals are recalcitrant pollutants found in a variety of environmental medium including: soil, ground/surface waters, sediment, atmospheric particulates and food crops (Izah et al., 2016, 2017b,c, 2021a,b,c; Izah and Aigberua, 2020; Aigberua et al., 2021; Ogamba et al., 2017a,b, 2021; Uzoekwe et al., 2021; Izah and Aigberua, 2017). Several wastewaters have been shown to contain trace metals which is challenging to remediate. Techno-economic, environmental, and societal factors contribute to the successful extermination of potentially harmful metal ions from wastewaters (Diep et al., 2018). As a result, there are numerous processes that can be utilized for the efficient removal of trace metal-laden wastewater. Trace metals have been successfully removed from wastewater using a variety of chemical-based approaches. These methods include chemical precipitation, coagulation or flocculation, membrane filtering, ion-exchange, photocatalysis, and adsorption of inorganic matter (Diep et al., 2018). Some of the benefits of utilizing chemical methods in the extrication of trace metals contained in wastewaters include quick processing time, resistance to a high concentration of toxic metals, operational ease and well-established molecular model (Fu and Wang, 2011; Gunatilake, 2015; Le and Nunes, 2016; Carolin et al., 2017). However, there are several other ways to remove trace metals from wastewater.

Some of the most frequently employed strategies for eliminating toxicants from contaminated environments are bioaccumulation, biosorption, immobilization, biotransformation, bioprecipitation, biocrystallization, and metal bioleaching. Biodegradation breaks down organic molecules into minor compounds by microorganisms, regardless of the mechanism used. Some microorganisms can utilize chemical substances as an energy source and/or growth stimulus.

Several microorganisms, including bacteria, algae, and fungi (molds and yeasts), have been efficient for the extraction and recovery of trace metals in general. The efficacy, relatively low cost, and vast availability of these trace metals recovery techniques have gotten them more attention (Wang and Chen, 2009).

Trace metals, if significantly present in the environment pose reasonable threat to biodiversity, while long-term exposure to these elements can result in liver and kidney toxicity, intestinal damage and even cancer. Therefore, this review aims to determine the various chemical and microbial-based trace metal removal techniques in wastewater. In addition, the research will focus on the factors that influence trace metal removal from wastewater.

13.2 Environmental pollutants and methods of degradation

13.2.1 Hydrocarbons from petroleum

Crude oil spills cause severe environmental problems worldwide, particularly in Nigeria's oil-producing Niger Delta states (Izah et al., 2017a; Aigberua et al., 2016a,b, 2017a; Singh et al., 2017). Physical methods such as scooping, trenching, booming, and application of dispersant and surfactant, can be used to remove crude oil from the environment. In these contaminated locations, hydrocarbons from crude oil serve as organic carbon sources for indigenous microorganisms (Marinescu et al., 2009). Several bacteria can degrade hydrocarbons in aerobic and anaerobic environments (Yakimov et al., 2007). Biotechnology and molecular techniques like genetic sequencing are useful for screening and identifying hydrocarbon degraders from hydrocarbon-contaminated areas. Over 70 genera which possess the tendency for decomposing petroleum hydrocarbons have been discovered during recent investigations (Tremblay et al., 2017; Kafilzadeh et al., 2011). *Saccharomyces, Saccharomycopis, Trichoderma, Neurospora, Mortierella, Cladosporium, Absidia, Candida, Pichia*, and *Rhodotorulla* are some yeast species that can degrade hydrocarbons. *Sporobolomyces, Penicillium, Mucor, Aspergillus, Fusarium, Trichosporon* (Balba et al., 1998; Milic et al., 2009), as well as bacteria like *Nocardia, Pseudomonas, Acinetobacter, Flavobacterium, Micrococcus, Arthrobacter, Rhodococcus, Corynebacterium, Achromobacter, Alcaligenes, Bacillus, Staphylococcus, Brevibacterium* (Balba et al., 1998; Milic et al., 2009), algae such as *Amphora, Porphyrindium, Petalonia, Nostoc, Oscillatoria, Chlamydomonas, Ulva* (Cerniglia et al., 1980), *Chlorella, Scenedesmus, Selenanatrum* (Lei et al., 2007), among others can also degrade hydrocarbons.

13.2.2 Polychlorinated biphenyls

PCBs are commercially used as a synthetic aromatic chemical that does not occur naturally in the environment. PCBs are used in various industrial processes, including plastics, paints, dyes, paper, electrical, hydraulic equipment (Seeger et al., 2010), lubricants, dielectric fluids, and plasticizers (Seeger et al., 2010; Aken and Bhalla, 2011). As a recognized carcinogen, PCBs are grouped as persistent organic contaminants with significant toxicity for both the environment and humans (Lallas, 2001). These compounds could bioconcentrate along the food chain, particularly after being discharged into the environment by burning PCB-containing wastes, landfill leaks, or poorly discarded industrial wastes. Its presence has been linked to chronic immunological damage, bronchitis, and impaired pulmonary functioning (Van Gerven et al., 2004; Schecter et al., 2006).

With reference to Jing et al. (2018), PCB contamination can be remedied by utilizing phytoremediation, microbial degradation, chemical reagent dehalogenation, and the use of activated carbon. Some of the physical and chemical methods of removing PCB in the environment include supercritical oxidation of water, accelerated radiation, bimetal systems, iron-centered diminutive dehalogenation on a nanoscale, etc. Bacteria has been documented to enhance the breakdown of PCBs (Anyasi and Atagana, 2011; Bedard, 2008). Through oxidation reactions and chlorine elimination, aerobic and anaerobic bacteria participates in the biotransformation of PCBs (dehalogenation). *Pseudomonas, Burkholderia, Sphingomonas, Rhodococcus, Bacillus, Micrococcus, Dehalococcoides, Ochrobactrum*, and *Phenylobacterium* species are examples of bacteria engaged in oxidation reactions and chlorine elimination (Li et al., 2016; Krumins et al., 2009; Petric et al., 2007; Jing et al., 2018).

13.2.3 Pesticides

Pesticides are chemical formulations utilized for the prevention, control, or elimination of pests. Some pesticides can be retained in the environment, thereby exacerbating significant risk to humans and the environment (Hassaan and El Nemr, 2020). Pesticides have been observed to be degraded by bacteria utilizing it as a primary source of energy. The use of the *Agrobacterium radiobacter* strain for the degradation of atrazine was reported by Struthers et al. (1998). Rani et al. (2008) identified a chlorpyrifos-degrading *Providencia stuartii* strain. Kanade et al. (2012) have documented the utilization of *Azospirillum lipoferum* to degrade malathion, one of the world's most widely used organophosphorus insecticides. Huang et al. (2018) found that bacteria (*Pseudomonas, Bacillus, Alcaligenes, Flavobacterium*), *Actinomycetes* (*Micromonospora, Actinomycetes* like *Nocardia, Streptomyces*), and certain species of fungi that belong to the genera *Rhizopus, Cladosporium, Aspergillus, Penicillium, Aspergillus, Fusarium*, among others are often utilized for hydrocarbon degradation.

13.2.4 Dyes

Dyes are widely used in various products such as pharmaceuticals, textiles, rubber, printing, and cosmetics. Textile dyes, such as azo dyes, that are released into nearby water bodies, cause significant pollution effects that is toxic to aquatic life and harmful to humans through bioaccumulation. Microbial, physical and chemical approaches can be used to breakdown dyes. Physicochemical procedures like counting filtration, congelation, and chemical gelation can be used to remediate azo dyes (Ajaz et al., 2020). Dyes can also be removed using physical and chemical methods such as Fenton's reagent, ion interchange, irradiation, electrokinetic congelation, ozone treatment, photocatalysis, sodium hypochlorite, methylene-linked glycoluril, electrocatalytic scarring, activated coke, marsh, wood fragments, vitreous silica and partition infiltration (Robinson et al., 2001). Several studies have demonstrated that fungi (mold and yeasts) and bacteria can be used for removing colors from wastewater (Robinson et al., 2001; Ajaz et al., 2020).

13.2.5 Trace metal

Trace metals are a prevalent contaminant affecting the environment and human food supplies (Izah et al., 2016, 2017c, 2021a,b; Aigberua et al., 2021). Ions in the form of OH^-, O^{2-}, S^{2-}, PO_4^{3-}, SiO_2, and organic complexes are all examples of trace metals (Masindi and Muedi, 2018). Trace metals have been found in soil (Izah et al., 2017a), water (Izah et al., 2016; Ogamba et al., 2017a, 2021), air (Uzoekwe et al., 2021), foods (Izah et al., 2017c), vegetables (Izah and Aigberua, 2017), and fish tissues (Aghoghovwia et al., 2016; Aigberua et al., 2021; Aigberua and Izah, 2018; Izah et al., 2016; Ogamba et al., 2015, 2016a,b, 2017b), meat (Seiyaboh et al., 2018), edible oil (Aigberua et al., 2017b), etc. The toxicity and bioavailability of trace metals are affected by thermal condition, phase coalition, surface assimilation and physical condition, species identification at thermodynamic balance, complexation mobility, fats and oil solvability, synthetic environment, species features, interactions across the energy pyramid, and biological status (Tchounwou et al., 2012).

For trace metals removal in environmental matrices, microorganisms (bacteria, fungi, and algae) have shown promising results. The procedures used to decontaminate trace metals from environmental matrices is dependent on the bioreactor type and the status of its operating system, metal toxicity and bioavailability, and the physical conditions and qualities of metallic ions.

13.2.6 Wastewater

Many industrial wastewaters, whether emanating from manufacturing, agricultural, or food processing industries, are contaminated with potentially toxic ionic species such as toxic metals and microorganisms. As a result, they pose serious threat to human health and the environment. Several chemical procedures, such as precipitation produce sludge effluent, activated coke and ion interchange adhesives which can decontaminate trace metals in wastewater. Additionally, the application of microbial organisms or biomass for decontamination of micropollutant metal ions contained in wastewater is cost-effective, ecologically friendly and simple to maneuver. Therefore, it remains a long-term solution (Diep et al., 2018).

13.3 Sources of trace metals in water

There is a growing worry about potentially toxic metals in the environment and human dietary sources. Trace metals are found in various industrial, agricultural, and domestic effluents and technological applications of trace metal-containing materials. Potentially toxic metal pollution is prevalent due to increased geogenic and atmospheric deposition. This section highlights the divergent origin of trace metals contamination in the ecosystem.

13.3.1 Anthropogenic sources

Increased human activities or pressure is one of the leading causes of potentially toxic metal pollution incidents in the environment. Industrial release from tanneries, smelting, fertilizer, metal coating of surfaces, and paint industries, aerial emissions from leaded fuel and stack emissions, surface metal coating and polishing, fabric manufacturing, batteries cache, lead amalgam, mineral excavation, polishing, pottery, glass pest killers, fertilizer applications, ore mining, electronic waste, wood preservatives, municipal waste sludge and wastewater, personal care products and cosmetics, pharmaceutical and food processing effluents have caused the exacerbation of trace metallic ions loading on the habitat (Foroughi et al., 2011; Izah et al., 2016). Automobile wastes are another source of release of potentially toxic metal releases into the habitat (Foroughi et al., 2011; Izah et al., 2016; David et al., 2013).

Individual trace metal sources in the environment have been identified through research. For example, improper discharge of cans of arsenic-containing compounds such as insecticides, wood preservers, dyes, and drugs (arsenic trioxide) used for the treatment of acute promyelocytic leukemia (Tchounwou et al., 2012) can lead to environmental pollution. The inappropriate disposal of empty cans of old pesticides, particularly herbicides, is relatively common in Nigeria's Niger Delta region. This action poses a reasonable danger to the abundant aquatic biodiversity that inhabits the region's vast surface water resources. Because of its raised water table, most wastes from human activities are emptied into recipient water bodies, either straightforward or accidentally, via effluent channels and rainwater runoffs.

Cadmium (Cd) is a very poisonous metal that can pose environmental and human health risks to exposed persons. It is used in the manufacturing of alloys, pigments, batteries, mining and smelting processes (Tchounwou et al., 2012). Incinerator emissions and fuel combustion are examples of anthropogenic activities that lead to cadmium pollution (Medfu Tarekegn et al., 2020). Meanwhile, improperly disposed battery waste, e-waste, paint sludge materials could also result in cadmium pollution in the environment.

Chromium (Cr) is a micropollutant metal with variable valences spanning from Cr^{3+} to Cr^{6+}. The toxicity and stability of the various oxidative states vary depending on the prevailing environmental conditions. Metal purification, tannery mills, chromate generation, cast iron welding, ferrochromium and chromium plating, color manufacturing dyestuff and tints, curing of hide and skin (leather), and wood varnish production (Tchounwou et al., 2012; Madhav et al., 2018), mining, industrial coolants, chromium salts production, and leather tanning are all examples of anthropogenic activities that release chromium into the environment (Medfu Tarekegn et al., 2020). Chromium is also discharged within environmental matrices via wastewater from metallurgical, refractory, and chemical industries (Tchounwou et al., 2012).

Another naturally occurring element is lead. Lead (Pb) pollution can occur from a variety of sources, including the use of lead-acidified batteries, bullets, soldered metal pipes, radiogram protecting devices, paints and ceramic products, caulking, pipe solder (Tchounwou et al., 2012; Madhav et al., 2021), electronic wastes, smelting processes, coke-powered thermal engines and potteries (Medfu Tarekegn et al., 2020).

Mercury (Hg) is a peculiar trace element found in its elemental form (as liquid with high vapor pressure). Also, it exists both as inorganic and organic species. Mercury's valence state is either +1 (mercurous) or +2 (mercuric) (Tchounwou et al., 2012; Guzzi and La Porta, 2008). Their presence and oxidative states determine the toxicity of mercury compounds. According to Dopp et al. (2004) and Tchounwou et al. (2012), the most prevalent organic semblance of mercury in the ecosystem is methyl-mercury, generated by bacteria methylating mercuric forms. When materials containing the compounds are not adequately maintained, mercury contamination might result. Electrical parts production, dental synthesis, and a variety of industrial processes (caustic soda production, nuclear equipment, antifungal additives for wood varnishing, solvent for reactive and precious metals, pharmaceutic drug preservatives) are just a few examples of the application of mercury (Tchounwou et al., 2003, 2012). Food contamination, preventive medical practices, industrial and agricultural operations (Tchounwou et al., 2012), Chlor-alkali and heat-powered equipment, fluorescent bulbs, medical garbage also contribute to the release of mercury into the environment (Medfu Tarekegn et al., 2020).

Zinc (Zn) has been found in various effluents, including cassava mill effluents (Izah et al., 2017a) and oil palm processing wastewater (Ohimain et al., 2012). Zinc is released as a waste product through manufacturing activities like metal surface coating, amalgamation, battery production, galvanization, as well as metallurgical industries (Plum et al., 2010; Medfu Tarekegn et al., 2020).

Another primary metal that has been found in various effluents is copper. Copper (Cu) is a significant trace metal that pollutes the environment. The principal human activities that culminate in the release of copper to the surroundings, according to Medfu Tarekegn et al. (2020), is waste dumpsite leachates, mining processes, electroplating, smelting operations, fossil fuel combustion, and domestic or municipal wastewaters (Madhav et al., 2020).

Nickel (Ni) is another significant metal that pollutes the environment and can build up in the food chain. It can be discharged into the environment from human activities involving nickel-containing substances. According to Medfu Tarekegn et al. (2020), nickel carbonyl $Ni(CO)_4$, an intermediate product of nickel refining activities, can damage the atmospheric environment. Furthermore, smelting activities, thermal power plants, fuel-burning (Medfu Tarekegn et al., 2020), and mining processes can pollute the environment with toxic metals like arsenic. Also, molybdenum can be released into the environment when used catalysts are discarded (Medfu Tarekegn et al., 2020).

13.3.2 Natural sources

Micropollutant metals are found naturally in crustal formations of the earth (Tchounwou et al., 2012). Metal corrosion, air deposition, leaching of potentially toxic metals in surfaces under certain conditions, suspension of underlying sediment and vaporization of metallic ions from surface water supplies to soil, as well as groundwater are among native sources of potentially toxic metal contamination (Nriagu, 1989; Tchounwou et al., 2012). Conversely, weathering and volcanic eruptions are two of the most important natural origination of trace metal loading (Nriagu, 1989; Tchounwou et al., 2012; He et al., 2005; Shallari et al., 1998; Madhav et al., 2021).

13.4 Methods for decontaminating toxic metals from water and wastewaters

Many trace metals are regarded as critical macro- and microelements, particularly at levels with no adverse consequences. However, when their concentrations are relatively high, as is the case for polluted environments, they become harmful to human health. Other metals, including lead, cadmium, nickel, and mercury, depict inherent toxicity at minute concentration.

Trace metals are becoming more common in aquatic ecosystems as a result of rising industrial activities and urban sprawling. The presence of metallic ions in the environment is a significant source of concern due to its toxicity, bioaccumulation potential in aquatic creatures, and biomagnification tendency across the food chain. As a result, decontaminating trace metal ions from water is critical for environmental sustainability and human health. Trace metals can be removed from water using a variety of methods. However, environmental and social factors, cost analysis, and economic issues must be considered when selecting an appropriate technique. Trace metals can be decontaminated from water using various physical and chemical techniques (Fig. 13.1). This section highlights some of the standard physicochemical methods utilized for decontamination of residual metal ions in water and wastewater.

13.4.1 Electrocoagulation

The on-site production of coagulant in water-based medium through the passing of electric charge through one or more submerged, sacrificial electrodes is known as electrocoagulation. It has the advantages of being safe and straightforward to use, producing less sludge, and being automated (Mouedhen et al., 2008).

13.4.2 Precipitation of chemicals

Chemical precipitation involves using hydroxide and sulfide to generate an insoluble metal ion residue that can easily be removed using a screen filter or sedimentation. Due to its reduced cost and ease of operation, this technique is widely utilized by most companies in conventional water treatment plants. Its usefulness is enhanced when trace metal

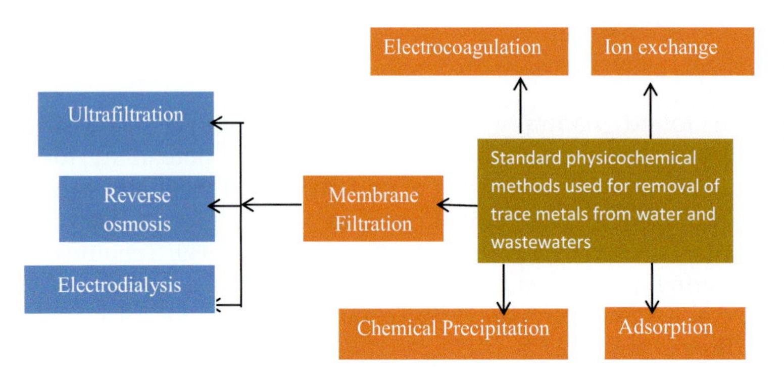

FIGURE 13.1 Some standard physicochemical methods utilized for the extrication of metallic ions from water and wastewaters.

concentration is high but limited when the metal ion concentration is at low levels (Renu and Singh, 2017). Another disadvantage of this method is sludge production.

13.4.3 Ion-exchange

Ion interchangers are natural or chemical adhesives possessing active binding points on their exterior for sulfonic ($-SO_3H$) and carboxylic acid groups ($-COOH$). Synthetic resins are more effective and commonly used at removing trace metals from wastewater than natural resins (Ince and Ince, 2019). The hydrogen ions of the resin's ion-exchanging functional group-swap electrons with the metal ion in the water.

This approach frequently removes metals from wastewater because of various advantages, including high metals removal efficiency, low energy requirements, high treatment capability, and rapid kinetics.

13.4.4 Membrane filtration

Using this procedure, water is passed through a screen with mesh size spanning between 0.1 and 0.5 mm (Ince and Ince, 2019). The method is simple to use, requires minimum workspace and entails several steps, including:

1. Ultrafiltration: This process employs low transmembrane pressures for removing metal ions and other suspended solids.
2. Reverse osmosis: This technique is also called hyperfiltration, as it utilizes a semipermeable membrane, which allows metal ions to be ejected as liquids flow through. It is regarded as the most economical approach because of the "ultrapure" state of treated water produced.
3. Electrodialysis: This method involves the separation of contaminants by using a charged membrane. The membrane is charged, with an electric current serving as the catalyst. Ion interchange membranes are popularly utilized across process applications.

13.4.5 Adsorption

Adsorbents are utilized in the adsorption technique. Metal ions adhere to adsorbents because they have porous high surface areas. Due to its economic and technical feasibility, low cost, nontoxic tendency, and local availability of adsorbents, this technique is commonly utilized in wastewater remediation studies (Renu and Singh, 2017). Additionally, the method is useful for decreasing biological and chemical sludge formation, which allows for the redevelopment of adsorbents and improved metal recovery (Renu and Singh, 2017). The physicochemical attributes of adsorbent, metals in solution, and operating conditions (pH value, temperature, quantity of adsorbent used, preliminary metal amount in solution, adsorption interval, particle diameter and quality of adsorbent) tend to limit the efficiency of its use (Qasem et al., 2021) (Fig. 13.2). Adsorption research has led to a broad spectrum of alternative adsorbents.

Carbon-based surface-active agents, including activated coke, graphene, and nanotubes from charcoal (Karnib et al., 2014), have been reported to be utilized. Electrochemical

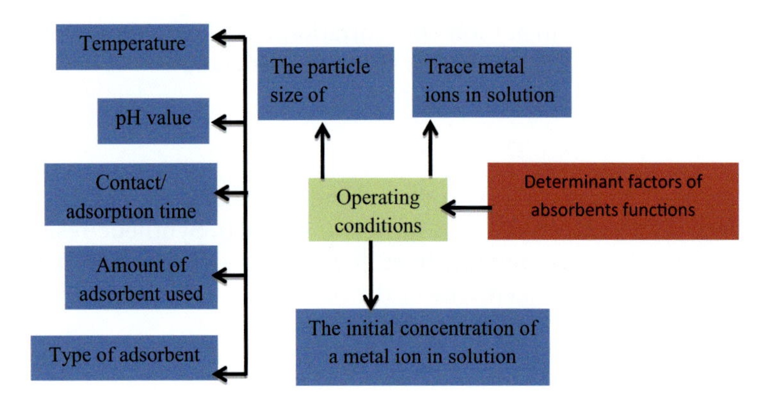

FIGURE 13.2 Some factors that determine the efficiency of absorbents.

treatment, agricultural materials, and mineral adsorbents have also been employed across different treatment systems.

Materials with reasonable coke content can be utilized for the production of activated charcoal. They feature vast surface area and well-built pores (Rao et al., 2006). However, recently, efforts are being focused towards obtaining activated carbon from agricultural waste, as coal-based activated carbon is prohibitively expensive for trace metal removal, limiting its use. Renu and Singh (2017) described graphene as a nanomaterial with good stability, a large specific surface area and a two-dimensional structure that has been vastly utilized in exterminating chromium ions in effluent waters in the form of graphene oxide (Yang et al., 2014).

Carbon nanotubes have become increasingly popular because of their stability, good adsorbent properties, and large surface areas (Renu and Singh, 2017). There have been reports of chemical modification to increase their adsorption capabilities (Gupta et al., 2011). Watermelon peel has been reported to form magnetic carbon nanocomposites, more of a biosorbent, with outstanding adsorption capabilities.

The biosorption process removes trace metals from agricultural commodities by utilizing solid agricultural/biological materials (called biosorbents/bio-adsorbents) (Mahmoud et al., 2010). The existence of a variety of atomic moieties or functional groupings on their surfaces (thiol, carboxyl, amino, phosphate, and hydroxyl) speeds up the biosorption process by creating complexes with metals (Qasem et al., 2021). For example, cellulose is a significant component of agricultural products and it is made up of a linear chain of multiple D-glucose units containing various functional groups. Water hyacinth, rice chaff, surfactant-modified effluent, sugarcane residue, wheat bran, coconut hard-shell, orange skin, wood dust, and eggshell are standard agricultural products utilized as biosorbents (Renu and Singh, 2017). The main benefits of biosorption over other previously listed options include: high efficiency, cost-effectiveness, regeneration of biosorbent and high metal recoverability, minimized sludge formation, absence of additional nutrient requirement, etc. (Das et al., 2008; Ahalya et al., 2003). With reference to Ahluwalia and Goyal (2005) and Das et al. (2008), the disadvantages of this technique include timely saturation (whereby metal-interactive sites are full, and metal desorption is required) and limited possibility for biological improvement (through genetic engineering of unmetabolized cells, and inability to biologically alter metal valency state).

Mineral adsorbents such as activated zeolites, quartz, and modified kaolin are typically utilized. Activated zeolites are made of hydrated aluminosilicate minerals. Hence, they possess high surface area, good exchange properties, and are hydrophilic. In addition, metal salts (Al_2SO_4 and $FeCl_3$) can concentrate metal ions in wastewater since aluminum ions are present in dissolved form. However, the main disadvantage of using mineral adsorbents is their decreasing efficiency after a few cycles, and modifications may not be environmentally friendly or cost-effective. Because of its significant exchange capacity and selectivity for cations, as well as its surface electronegativity, clay has been widely used, either in its natural state or after modification (Zhang et al., 2021).

Many of the techniques listed above have been reported with limitations in terms of in-situ treatment and failure in large-scale operations (Selvi et al., 2015). As a result, integrated processes that combine two different methods to achieve greater synergy and effectiveness in trace metals removal are becoming increasingly popular (Chen et al., 2013). Howbeit, integrating two processes necessitates a thorough understanding of the process goals. That is, ensuring the manner of combination is practically feasible, economically viable, and more efficient in large-scale applications than individual processes (Selvi et al., 2019). The different processes utilized for the decontamination of trace metals from water are as summarized by Selvi et al. (2019) and includes chemical and biological attenuation, electrokinetic-microbiological, electrokinetic coupled with phytoremediation, enhanced phyto-remedial and electrokinetic-geosynthetic approaches.

13.5 Elimination of trace metals in water and wastewater using microorganisms

Removing trace metals from water by physical and/or chemical techniques requires huge capital as these methods are primarily expensive and produce sizeable chemical waste volumes. However, the deployment of biosensor bacteria in the elimination of metals in polluted environments represents a more ecologically friendly and economical method. These bacteria have a variety of ways for sequestering metals, thereby allowing for greater metal bioabsorption.

Parameterization is a spontaneous process wherein microbiological processes degrade toxic compounds into less or nontoxic forms, eliminating contaminants from the environment (United States Environmental Protection Agency US EPA, 2010). Microorganisms can exploit these pollutants as energy sources through their metabolic activities throughout the biological process.

Many microorganisms are being identified as efficient, cheap options for removing metals from water and soil (Ahirwar et al., 2016). Recently, techniques for bioremediation have advanced, with the definitive aim of restoring the contaminated habitat in an ecologically sustainable way and at a minimal cost. Indigenous microorganisms present in contaminated sites or water bodies are crucial in tackling major biodegradation and bioremediation difficulties (Azubuike et al., 2016). This is because microorganisms have a means of accumulating potentially toxic metals in environmental matrices. Furthermore, microorganisms are known to effectively restore trace metal-polluted water bodies in one of the following ways (Fig. 13.3).

Microorganisms can naturally repel chemicals up to a specific concentration, after which the chemicals become harmful, putting the microbe in danger. According to Diep

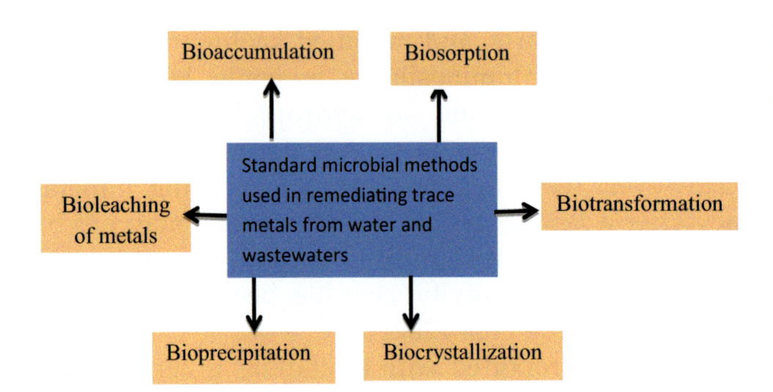

FIGURE 13.3 Common techniques involving microbial elimination of trace metals in water and wastewaters.

et al. (2018), bioaccumulation of trace metals is a natural process wherein microorganisms utilize proteins and peptide ligands in the assimilation and segregation of elements in cellular activities (balancing charge sites on biological molecules, enzyme catalysis, signaling, etc.). The authors further reported the metals-bioaccumulative tendency of microbial biomass as a performance measure often described as "micromol of X" or "milligram of X per gram" dry weight, where X is the element. The sensitivity of microorganisms to chemical species varies with respect to participating chemical species and microorganisms (Mishra and Malik, 2013). For instance, *Escherichia coli* can bioaccumulate trace metals like cobalt and nickel (Raghu et al., 2008; Krishnaswamy and Wilson, 2000; Diep et al., 2018; Deng and Wilson, 2001).

13.5.1 Biosorption

Microorganisms (dead or alive) use their biomass to absorb and eliminate trace metals and other pollutants from water bodies or polluted environments (Kisielowska et al., 2010). Hence, biosorption can take place with microorganisms through surface adsorption which is the collection of metals on cell surfaces and binding them with extra-cellular polymers. The outer cell shield of bacteria confers them with sorption qualities, and metals are connected by the active groups in the cell's surface layers. According to Diep et al. (2018), biosorption is the adsorption of particles onto a biological-microbial medium with the aid of many linking mechanisms, including physical (electrostatic forces), chemical (ion/proton displacement), and chelation or complexation. Some of the shared binding sites that biosorbents contain include: amino, amide, sulphydryl, phenol, imidazole, sulfate, carboxyl, phosphate, carbonyl, thioether and hydroxyl moieties (Wang and Chen, 2009). This reaction frequently involves the transfer of ions between the positively charged ions and active compounds with negative cell structure potential, as well as microorganisms classified under divergent systematic units characterized by a variety of chemical groupings with mobile metallic ion associations connecting the outer structures.

For most microorganisms, biosorption of trace metals occurs at neutral pH. Studies have demonstrated that the extra-cellular exterior of microorganisms possesses negatively charged atomic particles that become binding sites for positively charged micropollutant elements at pH of 7 (Michalak et al., 2013; Fomina and Gadd, 2014). The cell structure and surface components found in many fungi and algae cellular wall linings, such as cellulose, are critical in metal binding (Wang and Chen, 2009). Several microorganisms depict adsorption tendencies with trace metals (Volesky, 2007; Srivastava et al., 2015; Vijayaraghavan and Yun, 2008; Wang and Chen, 2009; Ayangbenro and Babalola, 2017; Ilyas et al., 2017).

Frequently, dissimilarities in pH and ionic strength of wastewater (effluents) affect microorganisms' biosorption of trace metals. Furthermore, biosorbents made of dead microbial biomass tend to degrade, resulting in fouling and loss of molecular bonding points (Gadd, 2009; Fomina and Gadd, 2014). Biosorbents are less expensive and more effective at removing trace elements (Wang and Chen, 2009). In a review study, Kanamarlapudi et al. (2018) reported *Pichia guilliermondii, Aspergillus niger, Aspergillus flavus, Bacillus circulans, Bacillus megaterium, Saccharomyces cerevisiae, Drepanomonas revolute, Uronema nigricans, Euplotes* species as some microbial biosorbents that can be utilized for the bioaccumulation of trace metals including copper, lead and chromium.

Abioye et al. (2018) researched on the bio-adsorption of trace elements in tannery effluent using different bacterial species, including *Bacillus megaterium, B. subtilis, Penicillium* species and *A. niger*. Pb reduction was found to be greatest in *B. megaterium* and *B. subtilis; A. niger* and *Penicillium* species were shown to reduce the level of Cr, while *B. subtilis* and *B. megaterium* greatly reduced Cd levels after 20 days. Kim et al. (2015) used a zeolite-immobile *Desulfovibrio desulfuricans* batch system to remove Cr^{6+}, Ni and Cu, with extrication capabilities of 99.8%, 90.1% and 98.2%, concurrently.

13.5.2 Biotransformation

Microorganisms convert trace metals through a sequence of reactions such as oxidation, reduction, methylation, and demethylation (Medfu Tarekegn et al., 2020), and the microorganism's enzymatic systems participates in these reactions. Gram-positive bacteria isolated from tannery sewers has been shown to cause Cr^{6+} (a highly toxic metal) to be reduced to Cr^{3+} (a less toxic ionic form) (Kisielowska et al., 2010).

13.5.3 Biocrystallization and precipitation

Precipitation of trace metals may occur due to microorganisms' activities, whereby metals can be sparsely transformed into toxicity-diminishing forms. The processes of precipitation and biocrystallization play a key part in nutrient cycling and processing of microfossils, deposition of Fe and Mn, and fossilization of Ag and Mn. Discharge of trace metal ions on the external or inner cellular walls could have a detrimental effect on the direct mobility of enzymatic agents and the impact of secretion of secondary metabolites (Sklodowska, 2000).

13.5.4 Metals bioleaching

Bioleaching is a simple and effective technique of recovering precious metal ions from low grade minerals and its concentrated forms. The method could also be used to recover metals and detoxify wastewaters that have been contaminated with trace metals (Bosecker, 1997). It mainly entails the solubilization and extraction of metals by microorganisms. This method relies on the conversion of metallic mixtures found in the habitat from their insoluble forms (most commonly sulfides) into soluble species (Kisielowska et al., 2010). *Thiobacillus* is a Gram-negative, nonspore producing rod that grows under aerobic circumstances whilst being the most active bacteria in bioleaching (Bosecker, 1997). The bioleaching technique is performed by a succession in the growth of acidophobic and acidophilic *Thiobacilli*, with *Thiobacillus thioparus* being the essential less-acidophilic *thiobacilli*. In contrast, *Thiobacillus ferooxidans* and *Thiobacillus thiooxidans* are the most common *thiobacilli* species taking part in the acidophilic grouping of *Thiobacilli* (Blais et al., 1993).

When metal ions remain in solution, bacteria leaching occurs under acidic conditions (Bosecker, 1997); this makes acidophilic species very important in this technique. Kamizela and Worwag (2020) determined the bioleaching efficacy, pH and redox potential on sludge using *Acidithiobacillus thiooxidan* isolated strain. The authors reported high effectiveness of the process for most of the tested metals, with the maximum efficiency being recorded for Mn and Pb (99.5% removal), while the least was reflected for Cu (80.6%).

The ability of microorganisms like fungi to bioleach and mobilize metals from wastewater is primarily linked to two processes: the production of numerous organic acids (citric, gluconic, and oxalic acids) in the living environment and the secretion of cosmetic compounds. For such fungi, *Aspergillus* sp., *Penicillium* sp., *Rhizopus* sp., *Mucor* sp., *Alternaria* sp., and *Cladosporium* sp. may be included in metal percolation due to their biological and chemical potential, and corresponding strong opposition to unfavorable pH and temperature conditions. This option can be utilized mainly when it is not challenging to employ chemical or bacterial leaching protocols.

Environmental factors including temperature, pH, etc., could alter the conversion, mobility, valence state and absorptivity of trace metals in microorganisms inhabiting water bodies (Rasmussen et al., 2000).

13.6 Factors that determine toxic metal removal in water and wastewater

Despite the various efficiencies recorded by current trace metal removal techniques from waste and wastewater, such as physical approach (adsorption and biosorption), chemical approach (synthetic and natural), and phytoremediation methods. Certain factors exist which impede the effectiveness of trace metal removal from water and wastewater by microorganisms and these will be discussed.

High efficiency and low cost of microbial remediation of water and wastewater polluted with trace metals have led to a growing interest in applying microorganisms in treating large amounts of industrial wastewater and effluents (Pleshakova et al., 2021; Ngun et al., 2018). The activated sludge technique is often preferred for studies on trace metal removal in water and effluent water by microorganisms due to its increased efficiency and extensive variability of removal. Research studies conducted so far have identified varying levels of concerns and factors which serve as bottlenecks to the overall success of microbial treatment of water and effluent water (Kushner, 1971; Sterritt and Lester, 1981, 1982; Tynecka et al., 1975; Bitton and Freihofer, 1977; Brown and Lester, 1982). Some of the identified factors include influences of solutes and ions on microorganisms (Kushner, 1971), the influence of extra-cellular polysaccharides and polymers on cadmium and copper toxicity in pure bacterial cultures in activated sludges (Bitton and Freihofer, 1977; Brown and Lester, 1982), and also the effect of mean cell residence time (Brown and Lester, 1982). Other factors are the existence of plasmid dependent permeability barrier which blocks the absorption of certain trace metals (Tynecka et al., 1975), the impact of sludge maturation, which affects elemental speciation and controls the behavior of metals in activated sludge process (Sterritt and Lester, 1981; Sterritt and Lester, 1982), the dominance of process parameters (Rossin et al., 1982) and trace metal sensitivity (Pickett and Dean, 1979).

Removal or uptake of trace metals in water and wastewater by microorganisms occurs either as bioaccumulation or biosorption (Pleshakova et al., 2021). These mechanisms result from the interaction between microbial cell surfaces and metals in the aqueous solution phase (Ngun et al., 2018). Understanding these phases would provide a clear insight into the factors affecting trace metal uptake and removal in activated sludges (Nelson et al., 1981). Studied experimental factors which influence the efficiency of trace metal removal by microorganisms include; initial metal ion concentration, the type of metal, contact period, particle diameter, biosorbent dosage and bacterial genus play an important part in bio-adsorption because different genera have inconsistent cellular content. Ecological factors such as; pH, thermal condition, and the domination of low molecular weight humic and organic acids can affect the bioavailability of these trace metals and further impede their transportation and transformation (Sharma et al., 2016; Medfu Tarekegn et al., 2020).

13.6.1 Preliminary metal accumulation

Trace metals removal via microorganisms occurs in two stages: adsorption of the metals onto cellular surfaces and active transport of elements in the cellular interior. These processes utilize a living biosorbent whose initial capacity is usually optimal with increased metal ion concentrate resulting from functional binding sites and a subsequent reduction in efficiency when the binding sites are fully saturated. Microorganisms that do not have inhibitory process for accumulating large quantities of metals and can cope with high loads of metal ions perform better (Kanamarlapudi et al., 2018).

13.6.2 Contact time

Although the initial rate of microbial trace metal removal in water and wastewater is usually fast, with about 90% of the metal ions attaching themselves to all the available free active sites, the metal ions, type of biosorbent and the metal combinations are crucial in determining the time needed to achieve maximum biosorption. With the increase in time comes an elevation in the percentage concentration arising from trace metals that are bound to most active sites, leading to a reduction in the rate of metals removal from solution. Microbial species have different biosorption and bioaccumulation rates. Joo et al. (2010) studied the biosorption of Zn^{2+} and reported an equilibrium contact and biosorption time of 30 min for *Pseudomonas aeruginosa* and *Bacillus cereus* while Liu et al. (2006)

reported that *A. niger*, mostly depicted the absorption of dissolved cadmium and zinc ions within 6 h with no further biosorption after 24 h.

13.6.3 Biosorbent dose

For the microbial extrication of trace elements from water and liquid effluent setups such as activated sludge system, microorganisms serve as living biosorbent, therefore, providing binding sites for trace metal ions. Bacterial and fungal species possess excellent resistant mechanisms, which make them very effective biosorbents. Capsules on bacterial surfaces composed of charged polysaccharides play a key role in detoxifying metal ions from industrial effluents and wastewater. Glycoproteins and polysaccharides in fungal cell walls and features such as chitin, lipids, polyphosphates and the presence of metal-binding groups such as hydroxyls, carboxyls, phosphates and amines make fungal species good absorbents for trace metal ions in wastewater. Different bacterial and fungal species have been successfully utilized as bio-adsorbents for the extermination of trace elements, some of which include: *Enterococcus faecium* for Cu^{2+} ions from aqueous solutions with maximum bio-adsorptive efficiency of 106.4 mg/g of dry weight biomass (Yilmaz et al., 2010), *B. subtilis* for Cu^{2+}, Zn^{2+} (Sabae et al., 2006), *B. megaterium* for Fe^{2+} (Ngun et al., 2018), *S. cerevisiae* for Zinc (Gaensly et al., 2014; Ponce de León et al., 2002; Shet et al., 2011) and *Rhizopus cohnii* for cadmium (Luo and Xiao, 2010). Increasing the bio-adsorbent dose. That is, the microbial load in a treatment vessel, at an established metal dose. For example, an activated sludge provides more binding sites for bio-adsorption as a result of increased surface area for elements (Kanamarlapudi et al., 2018).

13.6.4 Age of biomass

According to studies depending on the type of microrganisms used, the age of biomass determines the biosorption efficiency of microbial cell walls. Although not well understood, it is believed that older microbial cultures could increase or decrease trace metal removal processes (Chojnacka, 2009).

13.6.5 Existence of supplementary ions in liquid mixtures

The presence of supplemental charged atoms in wastewater leads to a contest for active binding sites and often hinders the binding of trace metals. This affects the biosorption dynamics and inhibits the required trace metal (Chojnacka, 2009).

13.6.6 pH

pH plays an important part in determining the efficiency of bio-adsorption and bioaccumulation processes. It influences activity of important atomic moieties in bio-adsorbents, as well as affecting the chemical interactions of elements, thereby determining how metals are speciated. During high acidity, trace metals in solution form ionic species, promoting rapid saturation of trace metal bonding points resulting from increased availability of positively charged atoms. Low pH value of reaction medium can result in the competitiveness for bonding points by available metallic ions and positively charged particles. Active sites and functional groups become readily available at high pH, particularly resulting from reduced density of H^+ ions and consequently increasing attraction of positively charged trace metal ions. Metal precipitation can also occur at high pH due to the formation of hydroxides, thereby obstructing the trace metal removal process. According to recent studies, a rise in pH from 1 to 4.5 improved the bio-adsorption of chromium by *A. niger* and *S. cerevisiae* and cadmium by *R. cohnii* (Luo and Xiao, 2010; Mungasavalli et al., 2007; Hlihor et al., 2013; Kanamarlapudi et al., 2018).

13.6.7 Temperature

The thermal status of a reacting vessel largely controls the thermodynamics and kinetics of microbial trace metal removal processes. Temperature changes can positively or negatively influence the trace metal removal process at various reaction stages. According to studies, increased temperature results in greater solvability of trace elements, bioavailability and accelerates the biosorption process but can result in structural damage of the biosorbent (Medfu Tarekegn et al., 2020; Kanamarlapudi et al., 2018). For efficient trace metal removal, an optimum temperature for the microbial absorbent has to be selected to facilitate maximum binding of trace metal ions. In a work carried out by Goyal et al. (2003), *Streptococcus equisimilis* showed reasonable rise in Cr (VI) absorption rates due to an increased incubation temperature, rising from 25°C to 40°C (Goyal et al., 2003). Hlihor et al. (2013) also reported an 86% absorption of cadmium ions by *S. cerevisiae* at 40°C (Hlihor et al., 2013).

13.6.8 Agitation speed

Agitation of the bioreactor plays an important part in determining the efficiency of biosorption and bioaccumulation processes. The effect of agitation speed can be advantageous or lead to a disruption of the trace metal removal process (Kanamarlapudi et al., 2018). Enhanced metal ion sorption leading to increased bio-adsorption efficiency of the bio-adsorbent is amongst the positive effects of high agitation speed. Moderate agitation speed allows for improved homogeneity of the suspended solution, thereby resulting in increased biosorption rate (Kanamarlapudi et al., 2018). Negative effects of increased agitation rate include; loss of homogeneity in the suspension due to vortex phenomenon and a reduction in the interval of interaction between the bio-adsorbate and the bio-adsorbent, which consequently impact the biosorption process. The optimum speed for maximum biosorption can vary in microbial species and applying the right speed enhances the efficiency of the microbial trace metal removal process. Liu et al. (2006) reported a rise in biosorption efficiency from 32.4% to 65% in his study on biosorption of zinc and cadmium by *A. niger* (Liu et al., 2006; Kanamarlapudi et al., 2018).

13.6.9 Metal affinity to bio-adsorbents

Metal affinity can be enhanced through chemical and physical treatment of the bio-adsorbent. This influences the exterior charge, whilst improving the permeability of biomass, making it more accessible to trace metal binding groups and consequently improving metal uptake (Shamim, 2018).

There are other factors capable of affecting trace metal removal by microorganisms. Some of them are—surface area to volume ratio, which is a crucial phenomenon in biofilms, and the energy-consuming process of elemental adsorption. Also, the application of wild-type or genetically engineered microorganisms can drastically impact the trace metal removal process. Although wild-type microorganisms are adapted to the chemical composition of wastewater, thereby guaranteeing better survival than genetically engineered microorganisms, their efficiency is limited. According to studies, genetically engineered microorganisms that have been extensively studied are made of import storage systems composed of engineered and modified proteins with optimized trace metal specificity and selectivity (Diep et al., 2018).

13.7 Conclusion

In our world today, the human population growth has triggered the need for rapid industrialization, exploitation, and exploration of terrestrial and aquatic ecosystems. Most of the anthropogenic factors highlighted in the earlier parts of this book section, including pesticides, nuclear research, oil exploration and others, often lead to the release of various concentrations of trace metals into aquatic ecosystems in the form of wastewater, thereby contaminating scarce freshwater sources. The need for the application of efficient treatment methods aimed at complete removal of these trace metals either as elements or compounds cannot be overemphasized, considering the recalcitrant nature of trace metals and the health challenges they pose to humans. A variety of standard physical and chemical techniques currently used for efficient trace metal removal include electrocoagulation, chemical deposition, surface binding, membrane filtration, ion interchange, reverse diffusion, ultra-filtration, and electrodialysis. However, as a result of low operating costs and increased efficacy of biotechnological techniques, modern wastewater treatment methods currently prefer techniques such as bioaccumulation, biosorption, bioleaching of metals, biotransformation, precipitation and biocrystallization. In other cases, the application of both physical, chemical and biotechnological techniques are considered to offer the best results. Despite the efficiency of these techniques, a variety of extrinsic and intrinsic factors should be considered, including: temperature, hydrogen ion potential, interval of contact, type of adsorbent, preliminary accumulation of metallic ions, presence of other ions, age of biomass, biosorbent dose, agitation speed and metal affinity to bio-adsorbents. These factors largely determine the overall success of the trace elements extermination process.

References

Abioye, O.P., Oyewole, O.A., Oyeleke, S.B., Adeyemi, M.O., Orukotan, A.A., 2018. Biosorption of lead, chromium and cadmium in tannery effluent using indigenous microorganisms. Brazil Journal of Biological Sciences 5 (9), 25–32.

Aghoghovwia, O.A., Ohimain, E.I., Izah, S.C., 2016. Bioaccumulation of heavy metals in different tissues of some commercially important fish species from Warri River, Niger Delta, Nigeria. Biotechnological Research 2 (1), 25–32.

Ahalya, N., Ramachandra, T.V., Kanamadi, R.D., 2003. Biosorption of heavy metals. Research Journal of Chemistry and Environment 7 (4), 71–79.

Ahamad, A., Madhav, S., Singh, A.K., Kumar, A., Singh, P., 2020. Types of water pollutants: conventional and emerging. Sensors in Water Pollutants Monitoring: Role of Material. Springer, Singapore, pp. 21–41.
Ahirwar, N.K., Gupta, G., Singh, R., Singh, V., 2016. Isolation, identification and characterization of heavy metal resistant bacteria from industrial affected soil in central India. Indian Journal of Pure & Applied Biosciences 4 (6), 88–93.
Ahluwalia, S.S., Goyal, D., 2005. Removal of heavy metals by waste tea leaves from aqueous solution. Engineering in Life Sciences 5 (2), 158–162.
Aigberua, A.O., Ekubo, A.T., Inengite, A.K., Izah, S.C., 2016a. Evaluation of total hydrocarbon content and polycyclic aromatic hydrocarbon in an oil spill contaminated soil in Rumuolukwu community in Niger Delta. Journal of Environmental Treatment Techniques 4 (4), 130–142.
Aigberua, A.O., Ekubo, A.T., Inengite, A.K., Izah, S.C., 2016b. Seasonal variation of nutrient composition in an oil spill contaminated soil: a case of Rumuolukwu, Eneka, Port Harcourt, Nigeria. Biotechnological Research 2 (4), 179–186.
Aigberua, A.O., Ekubo, A.T., Inengite, A.K., Izah, S.C., 2017a. Assessment of some selected heavy metals and their pollution indices in an oil spill contaminated soil in the Niger Delta: a case of Rumuolukwu community. Biotechnological Research 3 (1), 11–19.
Aigberua, A.O., Ovuru, K.F., Izah, S.C., 2017b. Evaluation of selected heavy metals in palm oil sold in some markets in Yenagoa Metropolis, Bayelsa State, Nigeria. EC Nutrition 11, 244–252.
Aigberua, A.O., Izah, S.C., 2018. Evaluation of heavy metals in tissue of *Tympanotonus fuscatus* sold in some markets in Port Harcourt metropolis, Nigeria. MOJ Toxicology 4 (5), 334–338.
Aigberua, A.O., Izah, S.C., Richard, G., 2021. Hazard analysis of trace metals in muscle of *Sarotherodon melanotheron* and *Chrysichthys nigrodigitatus* from Okulu River, Rivers State, Nigeria. Journal of Environmental Health and Sustainable Development 6 (3), 1340–1356.
Ajaz, M., Shakeel, S., Rehman, A., 2020. Microbial use for azo dye degradation—a strategy for dye bioremediation. International Microbiology 23 (2), 149–159.
Aken, B.V., Bhalla, R., 2011. Microbial degradation of polychlorinated biphenyls. Reference module in earth systems and environmental sciences. Comprehensive Biotechnology 6, 151–166. second ed.
Anyasi, R.O., Atagana, H.L., 2011. Biological remediation of polychlorinated biphenyls (PCB) in the environment by microorganisms and plants. African Journal of Biotechnology 10 (82), 18916–18938.
Ayangbenro, A.S., Babalola, O.O., 2017. A new strategy for heavy metal polluted environments: a review of microbial biosorbents. International Journal of Environmental Research and Public Health 14 (1), 94. Available from: https://doi.org/10.3390/ijerph14010094.
Azubuike, C.C., Chikere, C.B., Okpokwasili, G.C., 2016. Bioremediation techniques–classification based on site of application: principles, advantages, limitations and prospects. World Journal of Microbiology and Biotechnology 32 (11), 1–18.
Balba, M.T., Al-Awadhi, N., Al-Daher, R., 1998. Bioremediation of oil-contaminated soil: microbiological methods for feasibility assessment and field evaluation. Journal of Microbiological Methods 32 (2), 155–164.
Bedard, D.L., 2008. A case study for microbial biodegradation: anaerobic bacterial reductive dechlorination of polychlorinated biphenyls—from sediment to defined medium. Annual Review of Microbiology 62, 253–270.
Bitton, G., Freihofer, V., 1977. Influence of extracellular polysaccharides on the toxicity of copper and cadmium toward *Klebsiella aerogenes*. Microbial Ecology 4, 119–125.
Blais, J.F., Tyagi, R.D., Auclair, J.C., 1993. Bioleaching of metals from sewage sludge: microorganisms and growth kinetics. Water Research 27 (1), 101–110.
Bosecker, K., 1997. Bioleaching: metal solubilization by microorganisms. FEMS Microbiology Reviews 20 (3–4), 591–604.
Brown, M.J., Lester, J.N., 1982. Role of bacterial extracellular polymers in metal uptake in pure bacterial culture and activated sludge—I. Effects of metal concentration. Water Research 16 (11), 1539–1548.
Carolin, C.F., Kumar, P.S., Saravanan, A., Joshiba, G.J., Naushad, M., 2017. Efficient techniques for the removal of toxic heavy metals from aquatic environment: a review. Journal of Environmental Chemical Engineering 5 (3), 2782–2799.
Cerniglia, C.E., Gibson, D.T., Van Baalen, C., 1980. Oxidation of naphthalene by cyanobacteria and microalgae. Microbiology (Reading, England) 116 (2), 495–500.
Chen, S., Chao, L., Sun, L.N., Sun, T.H., 2013. Plant-microorganism combined remediation for sediments contaminated with heavy metals, Advanced Materials Research, 610. Trans Tech Publications Ltd, pp. 1223–1228.

Chojnacka, K., 2009. Biosorption and Bioaccumulation in Practice. Nova Science Publishers, New York.

Das, N., Vimala, R., Karthika, P., 2008. Biosorption of heavy metals—an overview. Indian Journal of Biotechnology 7, 159–169.

David, T.W., Awoh, D.K., Essa, G.A., 2013. Investigation of heavy metals in drinking water (sachet and bottled) in ago-Iwoye and environs, Ijebu North lga, Ogun state, Nigeria. Scholarly Journals of Biotechnology 2 (1), 1–6.

Deng, X., Wilson, D., 2001. Bioaccumulation of mercury from wastewater by genetically engineered *Escherichia coli*. Applied Microbiology and Biotechnology 56 (1), 276–279.

Dhall, P., Kumar, R., Kumar, A., 2012. Biodegradation of sewage wastewater using autochthonous bacteria. The Scientific World Journal 2012. Available from: https://doi.org/10.1100/2012/861903.

Diep, P., Mahadevan, R., Yakunin, A.F., 2018. Heavy metal removal by bioaccumulation using genetically engineered microorganisms. Frontiers in Bioengineering and Biotechnology 6, 157. Available from: https://doi.org/10.3389/fbioe.2018.00157.

Dopp, E., Hartmann, L.M., Florea, A.M., Rettenmeier, A.W., Hirner, A.V., 2004. Environmental distribution, analysis, and toxicity of organometal (loid) compounds. Critical Reviews in Toxicology 34 (3), 301–333.

Eskander, S., Saleh, H.E.D., 2017. Biodegradation: process mechanism. Environmental Science and Engineering 8 (8), 1–31.

Fomina, M., Gadd, G.M., 2014. Biosorption: current perspectives on concept, definition and application. Bioresource Technology 160, 3–14.

Foroughi, M., Najafi, P., Toghiani, S., 2011. Trace elements removal from waster water by *Ceratophyllum demersum*. Journal of Applied Sciences and Environmental Management 15 (1), 197–201.

Fu, F., Wang, Q., 2011. Removal of heavy metal ions from wastewaters: a review. Journal of Environmental Management 92 (3), 407–418.

Gadd, G.M., 2009. Biosorption: critical review of scientific rationale, environmental importance and significance for pollution treatment. Journal of Chemical Technology & Biotechnology: International Research in Process, Environmental & Clean Technology 84 (1), 13–28.

Gaensly, F., Picheth, G., Brand, D., Bonfim, T., 2014. The uptake of different iron salts by the yeast *Saccharomyces cerevisiae*. Brazilian Journal of Microbiology 45 (2), 491–494.

Goyal, N., Jain, S.C., Banerjee, U.C., 2003. Comparative studies on the microbial adsorption of heavy metals. Advances in Environmental Research 7 (2), 311–319.

Gunatilake, S.K., 2015. Methods of removing heavy metals from industrial wastewater. Methods (San Diego, Calif.) 1 (1), 14.

Gupta, V.K., Agarwal, S., Saleh, T.A., 2011. Chromium removal by combining the magnetic properties of iron oxide with adsorption properties of carbon nanotubes. Water Research 45 (6), 2207–2212.

Guzzi, G., La Porta, C.A., 2008. Molecular mechanisms triggered by mercury. Toxicology 244 (1), 1–12.

Hassaan, M.A., El Nemr, A., 2020. Pesticides pollution: classifications, human health impact, extraction and treatment techniques. The Egyptian Journal of Aquatic Research 46 (3), 207–220.

He, Z.L., Yang, X.E., Stoffella, P.J., 2005. Trace elements in agroecosystems and impacts on the environment. Journal of Trace Elements in Medicine and Biology 19 (2–3), 125–140.

Hlihor, R.M., Diaconu, M., Fertu, D., Chelaru, C., Sandu, I., Tavares, T., et al., 2013. Bioremediation of Cr (VI) polluted wastewaters by sorption on heat inactivated *Saccharomyces cerevisiae* biomass. International Journal of Environmental Research 7, 581–594.

Huang, Y., Xiao, L., Li, F., Xiao, M., Lin, D., Long, X., et al., 2018. Microbial degradation of pesticide residues and an emphasis on the degradation of cypermethrin and 3-phenoxy benzoic acid: a review. Molecules (Basel, Switzerland) 23 (9), 2313. Available from: https://doi.org/10.3390/molecules23092313.

Ilyas, S., Kim, M.S., Lee, J.C., Jabeen, A., Bhatti, H.N., 2017. Bio-reclamation of strategic and energy critical metals from secondary resources. Metals 7 (6), 207.

Ince, M., Ince, O.K., 2019. Heavy metal removal techniques using response surface methodology: water/wastewater treatment. In: Biochemical Toxicology – Heavy metals and Nanomaterials. Ince, M., Ince, O.K., and Ondrasek, G., pp. 1–19. Available from: https://doi.org/10.5772/intechopen.88915.

Izah, S.C., Aigberua, A.O., 2017. Comparative assessment of selected heavy metals in some common edible vegetables sold in Yenagoa metropolis, Nigeria. Journal of Biotechnology Research 3 (8), 66–71.

Izah, S.C., Aigberua, A.O., 2020. Microbial and heavy metal hazard analysis of edible tomatoes (*Lycopersicon esculentum*) in Port Harcourt, Nigeria. Toxicology and Environmental Health Sciences 12 (4), 371–380.

Izah, S.C., Aigberua, A.O., Richard, G., 2021a. Concentration, source, and health risk of trace metals in some liquid herbal medicine sold in Nigeria. Biological Trace Element Research 1–14.

Izah, S.C., Richard, G., Aigberua, A.O., Ekakitie, O., 2021b. Variations in reference values utilized for the evaluation of complex pollution indices of potentially toxic elements: a critical review. Environmental Challenges 5, 100322.

Izah, S.C., Uzoekwe, S.A., Aigberua, A.O., 2021c. Source, geochemical spreading and risks of trace metals in particulate matter 2.5 within a gas flaring area in Bayelsa State, Nigeria. Advances in Environmental Technology 7 (2), 101–118.

Izah, S.C., Angaye, T.C., Aigberua, A.O., Nduka, J.O., 2017a. Uncontrolled bush burning in the Niger Delta region of Nigeria: potential causes and impacts on biodiversity. International Journal of Molecular Ecology and Conservation 7 (1), 1–15.

Izah, S.C., Bassey, S.E., Ohimain, E.I., 2017b. Assessment of heavy metal in cassava mill effluent contaminated soil in a rural community in the Niger Delta region of Nigeria. EC Pharmacology and Toxicology 4 (5), 186–201.

Izah, S.C., Inyang, I.R., Angaye, T.C., Okowa, I.P., 2017c. A review of heavy metal concentration and potential health implications of beverages consumed in Nigeria. Toxics 5 (1), 1–15.

Izah, S.C., Chakrabarty, N., Srivastav, A.L., 2016. A review on heavy metal concentration in potable water sources in Nigeria: human health effects and mitigating measures. Exposure and Health 8, 285–304.

Jing, R., Fusi, S., Kjellerup, B.V., 2018. Remediation of polychlorinated biphenyls (PCBs) in contaminated soils and sediment: state of knowledge and perspectives. Frontiers in Environmental Science 6, 79. Available from: https://doi.org/10.3389/fenvs.2018.00079.

Joo, J.H., Hassan, S.H., Oh, S.E., 2010. Comparative study of biosorption of Zn^{2+} by *Pseudomonas aeruginosa* and *Bacillus cereus*. International Biodeterioration & Biodegradation 64 (8), 734–741.

Kafilzadeh, F., Sahragard, P., Jamali, H., Tahery, Y., 2011. Isolation and identification of hydrocarbons degrading bacteria in soil around Shiraz Refinery. African Journal of Microbiology Research 5 (19), 3084–3089.

Kamizela, T., Worwag, M., 2020. Processing of water treatment sludge by bioleaching. Energies 13 (24), 6539.

Kanade, S.N., Ade, A.B., Khilare, V.C., 2012. Malathion degradation by *Azospirillum lipoferum* Beijerinck. Science Research Reporter 2 (1), 94–103.

Kanamarlapudi, S.L.R.K., Chintalpudi, V.K., Muddada, S., 2018. Application of biosorption for removal of heavy metals from wastewater. Biosorption 18 (69), 70–116.

Karnib, M., Kabbani, A., Holail, H., Olama, Z., 2014. Heavy metals removal using activated carbon, silica and silica activated carbon composite. Energy Procedia 50, 113–120.

Kim, I.H., Choi, J.H., Joo, J.O., Kim, Y.K., Choi, J.W., Oh, B.K., 2015. Development of a microbe-zeolite carrier for the effective elimination of heavy metals from seawater. Journal of Microbiology and Biotechnology 25 (9), 1542–1546.

Kisielowska, E., Hołda, A., Niedoba, T., 2010. Removal of heavy metals from coal medium with application of biotechnological methods. Górnictwo i Geoinżynieria 34, 93–104.

Krishnaswamy, R., Wilson, D.B., 2000. Construction and characterization of an *Escherichia coli* strain genetically engineered for Ni (II) bioaccumulation. Applied and Environmental Microbiology 66 (12), 5383–5386.

Krumins, V., Park, J.W., Son, E.K., Rodenburg, L.A., Kerkhof, L.J., Häggblom, M.M., et al., 2009. PCB dechlorination enhancement in Anacostia River sediment microcosms. Water Research 43 (18), 4549–4558.

Kushner, D.J., 1971. Influence of solutes and ions on micro-organisms. In: Hugo, W.B. (Ed.), Inhibition and Destruction of the Microbial Cell. Academic Press, London, p. 259.

Lallas, P.L., 2001. The Stockholm Convention on persistent organic pollutants. American Journal of International Law 95 (3), 692–708.

Lei, A.P., Hu, Z.L., Wong, Y.S., Tam, N.F.Y., 2007. Removal of fluoranthene and pyrene by different microalgal species. Bioresource Technology 98 (2), 273–280.

Le, N.L., Nunes, S.P., 2016. Materials and membrane technologies for water and energy sustainability. Sustainable Materials and Technologies 7, 1–28.

Liu, Y.G., Ting, F.A.N., Zeng, G.M., Xin, L.I., Qing, T.O.N.G., Fei, Y.E., et al., 2006. Removal of cadmium and zinc ions from aqueous solution by living *Aspergillus niger*. Transactions of Nonferrous Metals Society of China 16 (3), 681–686.

Li, G., Xiong, J., Wong, P.K., An, T., 2016. Enhancing tetrabromobisphenol A biodegradation in river sediment microcosms and understanding the corresponding microbial community. Environmental Pollution 208, 796–802.

Luo, J.M., Xiao, X.I.A.O., 2010. Biosorption of cadmium (II) from aqueous solutions by industrial fungus *Rhizopus cohnii*. Transactions of Nonferrous Metals Society of China 20 (6), 1104–1111.

Madhav, S., Ahamad, A., Singh, A.K., Kushawaha, J., Chauhan, J.S., Sharma, S., et al., 2020. Water pollutants: sources and impact on the environment and human health. Sensors in Water Pollutants Monitoring: Role of Material. Springer, Singapore, pp. 43–62.

Madhav, S., Ahamad, A., Singh, P., Mishra, P.K., 2018. A review of textile industry: wet processing, environmental impacts, and effluent treatment methods. Environmental Quality Management 27 (3), 31–41.

Madhav, S., Raju, N.J., Ahamad, A., Singh, A.K., Ram, P., Gossel, W., 2021. Hydrogeochemical assessment of groundwater quality and associated potential human health risk in Bhadohi environs, India. Environmental Earth Sciences 80 (17), 1–14.

Mahmoud, M.E., Osman, M.M., Hafez, O.F., Elmelegy, E., 2010. Removal and preconcentration of lead (II), copper (II), chromium (III) and iron (III) from wastewaters by surface developed alumina adsorbents with immobilized 1-nitroso-2-naphthol. Journal of Hazardous Materials 173 (1–3), 349–357.

Marinescu, M., Dumitru, M., Lăcătuşu, A.R., 2009. Biodegradation of petroleum hydrocarbons in an artificial polluted soil. Research Journal of Agricultural Science 41 (2), 157–162.

Masindi, V., Muedi, K.L., 2018. Environmental contamination by heavy metals. In: Saleh, H.E.M., Aglan, R.F. (Eds.), Heavy Metals. IntechOpen, pp. 115–133. Available from: http://doi.org/10.5772/intechopen.76082.

Medfu Tarekegn, M., Zewdu Salilih, F., Ishetu, A.I., 2020. Microbes used as a tool for bioremediation of heavy metal from the environment. Cogent Food & Agriculture 6 (1), 1783174.

Michalak, I., Chojnacka, K., Witek-Krowiak, A., 2013. State of the art for the biosorption process—a review. Applied Biochemistry and Biotechnology 170 (6), 1389–1416.

Milic, J.S., Beškoski, V., Ilić, M.V., Ali, S.A., Gojgić-Cvijović, G.D., Vrvić, M.M., 2009. Bioremediation of soil heavily contaminated with crude oil and its products: composition of the microbial consortium. Journal of the Serbian Chemical Society 74 (4), 455–460.

Mishra, A., Malik, A., 2013. Recent advances in microbial metal bioaccumulation. Critical Reviews in Environmental Science and Technology 43 (11), 1162–1222.

Mouedhen, G., Feki, M., Wery, M.D.P., Ayedi, H.F., 2008. Behavior of aluminum electrodes in electrocoagulation process. Journal of Hazardous Materials 150 (1), 124–135.

Mungasavalli, D.P., Viraraghavan, T., Jin, Y.C., 2007. Biosorption of chromium from aqueous solutions by pretreated *Aspergillus niger*: batch and column studies. Colloids and Surfaces A: Physicochemical and Engineering Aspects 301 (1–3), 214–223.

Nelson, P., Chung, A., Hudson, M., 1981. Factors affecting the fate of heavy metals in the activated sludge process. Journal (Water Pollution Control Federation) 53 (8), 1323–1333.

Ngun, C.T., Ragusina, D.A., Pleshakova, E.V., Reshetnikov, M.V.A., 2018. Study of iron-oxidizing microorganisms for possible use in biotechnology of water purification. Izvestiya of Saratov University. Series: Chemistry. Biology. Ecology 18 (2), 204–210. Available from: https://doi.org/10.18500/1816-9775-2018-18-2-204-210.

Nriagu, J.O., 1989. A global assessment of natural sources of atmospheric trace metals. Nature 338 (6210), 47–49.

Ogamba, E.N., Charles, E.E., Izah, S.C., 2021. Distributions, pollution evaluation and health risk of selected heavy metal in surface water of Taylor creek, Bayelsa State, Nigeria. Toxicology and Environmental Health Sciences 13 (2), 109–121.

Ogamba, E.N., Ebere, N., Izah, S.C., 2017a. Heavy metal concentration in water, sediment and tissues of *Eichhornia crassipes* from Kolo Creek, Niger Delta. Greener Journal of Environmental Management and Public Safety 6 (1), 001–005.

Ogamba, E.N., Ebere, N., Izah, S.C., 2017b. Levels of lead and cadmium in the bone and muscle tissues of *Oreochromis niloticus* and *Clarias camerunensis*. EC Nutrition 7 (3), 117–123.

Ogamba, E.N., Izah, S.C., Isimayemiema, F., 2016a. Bioaccumulation of heavy metals in the gill and liver of a common Niger Delta wetland fish, Clarias garepinus. British Journal of Applied Research 1 (1), 17–20.

Ogamba, E.N., Izah, S.C., Ofoni-Ofoni, A.S., 2016b. Bioaccumulation of chromium, lead and cadmium in the bones and tissues of *Oreochromis niloticus* and *Clarias camerunensis* from Ikoli creek, Niger Delta, Nigeria. Advanced Science Journal of Zoology 1 (1), 13–16.

Ogamba, E.N., Izah, S.C., Numofegha, K., 2015. Effects of dimethyl 2, 2-dichlorovinyl phosphate on the sodium, potassium and calcium content in the kidney and liver of *Clarias gariepinus*. Research Journal of Pharmacology and Toxicology 1 (1), 27–30.

Ohimain, E.I., Seiyaboh, E.I., Izah, S.C., Oghenegueke, V., Perewarebo, T., 2012. Some selected physico-chemical and heavy metal properties of palm oil mill effluents. Greener Journal of Physical Sciences 2 (4), 131–137.

Petric, I., Hrsak, D., Fingler, S., Voncina, E., Cetkovic, H., Kolar, A.B., et al., 2007. Enrichment and characterization of PCB-degrading bacteria as potential seed cultures for bioremediation of contaminated soil. Food Technology and Biotechnology 45 (1), 11–20.

Pickett, A.W., Dean, A.C., 1979. Cadmium and zinc sensitivity and tolerance in *Bacillus subtilis* subsp. niger and in a *Pseudomonas* sp. Microbios 24 (95), 51–64.

Pleshakova, E., Ngun, C., Reshetnikov, M., Larionov, M.V., 2021. Evaluation of the ecological potential of microorganisms for purifying water with high iron content. Water 13, 901. Available from: https://doi.org/10.3390/w13070901.

Plum, L.M., Rink, L., Haase, H., 2010. The essential toxin: impact of zinc on human health. International Journal of Environmental Research and Public Health 7 (4), 1342–1365.

Ponce de León, C.A., Bayón, M.M., Paquin, C., Caruso, J.A., 2002. Selenium incorporation into *Saccharomyces cerevisiae* cells: a study of different incorporation methods. Journal of Applied Microbiology 92 (4), 602–610.

Qasem, N.A., Mohammed, R.H., Lawal, D.U., 2021. Removal of heavy metal ions from wastewater: a comprehensive and critical review. Npj Clean Water 4 (1), 1–15.

Raghu, G., Balaji, V., Venkateswaran, G., Rodrigue, A., Maruthi Mohan, P., 2008. Bioremediation of trace cobalt from simulated spent decontamination solutions of nuclear power reactors using *E. coli* expressing NiCoT genes. Applied Microbiology and Biotechnology 81 (3), 571–578.

Rani, M.S., Devi, K.V.L.P.S., Madhuri, R.J., Aruna, S., Jyothi, K., Narasimha, G., et al., 2008. Isolation and characterization of a chlorpyrifos-degrading bacterium from agricultural soil and its growth response. African Journal of Microbiology Research 2 (2), 26–31.

Rao, M.M., Ramesh, A., Rao, G.P.C., Seshaiah, K., 2006. Removal of copper and cadmium from the aqueous solutions by activated carbon derived from Ceiba pentandra hulls. Journal of Hazardous Materials 129 (1–3), 123–129.

Rasmussen, L.D., Sørensen, S.J., Turner, R.R., Barkay, T., 2000. Application of a mer-lux biosensor for estimating bioavailable mercury in soil. Soil Biology and Biochemistry 32 (5), 639–646.

Renu, M.A., Singh, K., 2017. Methodologies for removal of heavy metal ions from wastewater: an overview. Interdisciplinary Environmental Review 18 (2), 124–142.

Robinson, T., McMullan, G., Marchant, R., Nigam, P., 2001. Remediation of dyes in textile effluent: a critical review on current treatment technologies with a proposed alternative. Bioresource Technology 77 (3), 247–255.

Rossin, A.C., Sterritt, R.M., Lester, J.N., 1982. The influence of process parameters on the removal of heavy metals in activated sludge. Water, Air, and Soil Pollution 17 (2), 185–198.

Sabae, S., Hazaa, M., Hallim, S., Awny, N., Daboor, S., 2006. Bioremediation of Zn, Cu and Fe using *Bacillus subtilis* d215 and Pseudomonas putida biovar ad 225. Bioscience Research 3, 189–204.

Schecter, A., Birnbaum, L., Ryan, J.J., Constable, J.D., 2006. Dioxins: an overview. Environmental Research 101 (3), 419–428.

Seeger, M., Hernández, M., Méndez, V., Ponce, B., Córdova, M., González, M., 2010. Bacterial degradation and bioremediation of chlorinated herbicides and biphenyls. Journal of Soil Science and Plant Nutrition 10 (3), 320–332.

Seiyaboh, E.I., Kigigha, L.T., Aruwayor, S.W., Izah, S.C., 2018. Level of selected heavy metals in liver and muscles of cow meat sold in Yenagoa Metropolis, Bayelsa State, Nigeria. International Journal of Public Health and Safety 3 (2).

Selvi, A., Das, D., Das, N., 2015. Potentiality of yeast *Candida* sp. SMN04 for degradation of cefdinir, a cephalosporin antibiotic: kinetics, enzyme analysis and biodegradation pathway. Environmental Technology 36 (24), 3112–3124.

Selvi, A., Rajasekar, A., Theerthagiri, J., Ananthaselvam, A., Sathishkumar, K., Madhavan, J., et al., 2019. Integrated remediation processes toward heavy metal removal/recovery from various environments-a review. Frontiers in Environmental Science 7, 66. Available from: https://doi.org/10.3389/fenvs.2019.00066.

Shallari, S., Schwartz, C., Hasko, A., Morel, J.L., 1998. Heavy metals in soils and plants of serpentine and industrial sites of Albania. Science of the Total Environment 209 (2–3), 133–142.

Shamim, S., 2018. Biosorption of heavy metals. In: Biosorption. Derco, J. and Vrana B (Eds.). pp. 21–49. Available from: https://doi.org/10.5772/intechopen.72099.

Sharma, S., Rana, S., Thakkar, A., Baldi, A., Murthy, R.S.R., Sharma, R.K., 2016. Physical, chemical and phytoremediation technique for removal of heavy metals. Journal of Heavy Metal Toxicity and Diseases 1 (2), 1–15.

Shet, A.R., Patil, L.R., Hombalimath, V.S., Yaraguppi, D.A., Udapudi, B.B., 2011. Enrichment of *Saccharomyces cerevisiae* with zinc and their impact on cell growth. Journal of Biotechnology, Bioinformatics and Bioengineering 1 (4), 523–527.

Singh, P., Jain, R., Srivastava, N., Borthakur, A., Pal, D.B., Singh, R., et al., 2017. Current and emerging trends in bioremediation of petrochemical waste: a review. Critical Reviews in Environmental Science and Technology 47 (3), 155–201.

Sklodowska, A., 2000. Biologiczne metody lugowania metali ciezkich-biohydrometalurgia. Postępy mikrobiologii 39 (1), 73–89.

Srivastava, S., Agrawal, S.B., Mondal, M.K., 2015. A review on progress of heavy metal removal using adsorbents of microbial and plant origin. Environmental Science and Pollution Research 22 (20), 15386–15415.

Sterritt, R.M., Lester, J.N., 1981. The influence of sludge age on heavy metal removal in the activated sludge process. Water Research 15 (1), 59–65.

Sterritt, R.M., Lester, J.N., 1982. Speciation of copper and manganese in effluents from the activated sludge process. Environmental Pollution Series A, Ecological and Biological 27 (1), 37–44.

Struthers, J.K., Jayachandran, K., Moorman, T., 1998. Biodegradation of atrazine by *Agrobacterium radiobacter* J14a and use of this strain in bioremediation of contaminated soil. Applied and Environmental Microbiology 64 (9), 3368–3375.

Tchounwou, P.B., Ayensu, W.K., Ninashvili, N., Sutton, D., 2003. Environmental exposure to mercury and its toxicopathologic implications for public health. Environmental Toxicology: An International Journal 18 (3), 149–175.

Tchounwou, P.B., Yedjou, C.G., Patlolla, A.K., Sutton, D.J., 2012. Heavy metal toxicity and the environment. Molecular, Clinical and Environmental Toxicology 101, 133–164.

Tremblay, J., Yergeau, E., Fortin, N., Cobanli, S., Elias, M., King, T.L., et al., 2017. Chemical dispersants enhance the activity of oil-and gas condensate-degrading marine bacteria. The ISME Journal 11 (12), 2793–2808.

Tynecka, Z., Zajac, J., Goś, Z., 1975. Plasmid dependent impermeability barrier to cadmium ions in *Staphylococcus aureus*. Acta Microbiologica Polonica. Series A: Microbiologia Generalis 7 (1), 11–20.

United States Environmental Protection Agency (US EPA), 2010. Assessment and remediation of contaminated sediments (ARCS) program, final summary report (EPA-905-S-94-001). EPA Chicago.

Uzoekwe, S.A., Izah, S.C., Aigberua, A.O., 2021. Environmental and human health risk of heavy metals in atmospheric particulate matter (PM10) around gas flaring vicinity in Bayelsa State, Nigeria. Toxicology and Environmental Health Sciences 13 (4), 323–335.

Van Gerven, T., Geysen, D., Vandecasteele, C., 2004. Estimation of the contribution of a municipal waste incinerator to the overall emission and human intake of PCBs in Wilrijk, Flanders. Chemosphere 54 (9), 1303–1308.

Vijayaraghavan, K., Yun, Y.S., 2008. Bacterial biosorbents and biosorption. Biotechnology Advances 26 (3), 266–291.

Volesky, B., 2007. Biosorption and me. Water Research 41 (18), 4017–4029.

Wang, J., Chen, C., 2009. Biosorbents for heavy metals removal and their future. Biotechnology Advances 27 (2), 195–226.

Yakimov, M.M., Timmis, K.N., Golyshin, P.N., 2007. Obligate oil-degrading marine bacteria. Current Opinion in Biotechnology 18 (3), 257–266.

Yang, S., Li, L., Pei, Z., Li, C., Lv, J., Xie, J., et al., 2014. Adsorption kinetics, isotherms and thermodynamics of Cr (III) on graphene oxide. Colloids and Surfaces A: Physicochemical and Engineering Aspects 457, 100–106.

Yilmaz, M., Tay, T., Kivanc, M., Turk, H., 2010. Removal of corper (II) Ions from aqueous solution by a lactic acid bacterium. Brazilian Journal of Chemical Engineering 27, 309–314.

Zhang, T., Wang, W., Zhao, Y., Bai, H., Wen, T., Kang, S., et al., 2021. Removal of heavy metals and dyes by clay-based adsorbents: from natural clays to 1D and 2D nano-composites. Chemical Engineering Journal 420, 127574.

CHAPTER

14

Heavy metal water pollution: an overview about remediation, removal and recovery of metals from contaminated water

Shobha Singh[1], *Sanjeet Kumar Paswan*[1], *Pawan Kumar*[2], *Ram Kishore Singh*[1] *and Lawrence Kumar*[1]

[1]Department of Nanoscience and Technology, Central University of Jharkhand, Ranchi, Jharkhand, India [2]Department of Physics, Mahatma Gandhi Central University, Motihari, Bihar, India

14.1 Introduction

The metals, which are required in a very minute amount and are considered toxic, are termed as heavy metals. Researchers have widely investigated and studied these metals due to their dangerous and harmful influence on health and the environment. Due to their ability to accumulate and toxic nature these are addressed to be a vital source of environmental contamination. Heavy metals have critically polluted the environment and its components. This has severely damaged its abilities to promote life and provide its intrinsic values. These are naturally available compounds and because of their anthropogenic origin they are commonly found in various environmental domain. This results in deterioration of the environment competence to support life and health of human, animals and plants becomes threatened. This takes place because of bioaccumulation of these heavy metals in the food chain which is a direct consequence of nondegradable state of the heavy metals. The surroundings within which the human life exists is referred as the environment. It basically comprises of the water, land, microorganisms, animal and plant life and the atmosphere of the earth. It is also represented by different spheres and that impacts its actions and intrinsic values. Biosphere is the most vital arena of the environment as it caters the

Metals in Water
DOI: https://doi.org/10.1016/B978-0-323-95919-3.00018-5

living organisms. This sphere provides the field of interaction among the living organism with their nonliving counter parts. But at the same time, environmental contamination and pollution is a major concern. However, pollution is different from contamination, but it is assumed that the contaminants are pollutants which helps and promotes harmful impact on the environment. Contamination in a simple word can be referred as rise in concentrations of compounds in the environment beyond the required level for the organisms naturally (Wong, 2012).

Heavy metals are found in varying concentrations in each environmental compartment as they are naturally occurring elements. They are observed in different chemical compounds as well as in elemental form. Heavy metals are generally characterized as elements having high atomic numbers, large atomic weights and higher densities almost five times larger than that of water. Heavy metals can be defined in different ways on the basis of aim and objective. Now a days the term "Metal trace elements (MTEs)" which is widely used instead of heavy metals. The continuous and long exposure to MTEs may have extreme consequences on plant, animal and human health usually by intake of contaminated air, polluted food and drinking water and close proximity through contaminated soil and industrial areas (Shen et al., 2019; Mohammadi et al., 2020). Some of them are quite volatile and they get easily attached to the fine particles and eventually migrate to large distances on large scales. There is no denying fact that the heavy metals in trace amounts holds prime importance. But their bio-toxic effects in human biochemistry is of grave concern. The rapid urbanization, industrialization and globalization have resulted in an enhancement of concentrations of heavy metals in different environmental compartments relative to their natural background levels. There are several factors playing in the easy mobility of these heavy metals with convenience. These are available in traces in the crust of the earth and are used in our daily lives. They conduct a magnificent role in the performance of the biological systems of the humans. Some of the heavy metals like copper and iron are regarded as quintessential to life. The most common heavy metals which can pollute the environment are arsenic, cadmium, nickel, lead, chromium, copper and mercury. Among all of these heavy metals the one which captures the most attention are mercury, cadmium and lead owing to their ability to migrate large distances in the atmosphere. Some of these heavy metals are quite toxic and harmful even in minute concentrations. It is quite interesting to note that, owing to their toxic nature the heavy metals have garnered a generous amount of attention from the researchers across the globe.

Since last decade, the industrialization and globalization have drastically affected the pure environment and have severely impaired it to promote life. The three super power countries namely, The United States, Germany and Russia are known to consume about 75% of the world's most extensively used metals with only 8% of the total population of the globe. The continuous utilization of heavy metals is becoming a critical crisis among the developing countries as a large part of the population still resides in the rural areas. As there is an enhancement in the geologic and anthropogenic activities the problems related to heavy metals are also emerging. They increased to an extent which can cause harmful effects on the environment (Chibuike and Obiora, 2014). It is very evident that excess of any metal in the body can be a potential origin of disease. The essential metals are quintessential to the body and both either deficiency or excess can lead to harmful influence on the human body. Deficiencies are generally led by malabsorption and by a diarrheal condition. On the other hand, the surplus of metal and its relative toxic impacts depends on certain parameters like intake of the metal, tissue distribution, entry rate of the metal, concentration achieved and the excretion rate of the metal. Metals possessing carcinogenic and toxic features can react with the nuclear proteins and DNA, which led to oxidative degradation of biological macromolecules.

Heavy metals are regarded as components of an ill-ordered and defined subgroup of elements that shows metallic features. Transition metals, lanthanides, some metalloids and actinides are included in the list. Heavy metals are the most commonly found transition metals like copper, zinc and lead (Nies, 1999). Heavy metals can be either metals or the metalloids (elements that possess features of both metals as well as nonmetals). The elements which receive the utmost amount of attention in this regard and are persistent in the environment includes: arsenic, cobalt, cadmium, manganese, mercury, nickel, lead, tin, chromium, copper and thallium. These cannot be degraded or destroyed and possess densities usually greater than $5\ g/cm^3$ (Garbarino et al., 1995). Ninty metals are present naturally, out of which fifty three are heavy metals (Weast, 1984). Several scientists have given different definitions of heavy metals with numerous insights on its properties and level of toxicity. The natural occurrence of heavy metals is due to the activity of thermal springs, volcanoes, erosion, infiltration, etc. There are measures to examine the level of contamination that incorporates contamination indices for example, contamination factor (Cf), pollution load index (PLI), Enrichment factor (EF) and geo-accumulation index (I-geo). Zinc, copper, boron, iron, nickel, molybdenum are some of the heavy metals that are vital for the progressive growth of the plants whereas, their concentration above permissible level exert hazardous effect on the marine organisms and plants. Other heavy metals like, lead, mercury, cadmium, and arsenic are not required for the growth and proper development of flora and fauna. Heavy metal can be categorized into three groups:

1. Toxic metals: Hg, Cr, Pb, Cu, Ni, Cd, As, Co, Sn, etc.
2. Precious metals: Au, Pt, Ru, Ag, Pd, etc.
3. Radioactive: Th, U, Ra, Am

Discharging of heavy metals in the form of polluted industrial water, sewage waste and fertilizers is responsible for soil pollution and soil degradation (Edelstein and Ben-Hur, 2018; Gupta et al., 2010; Liu et al., 2018). The food chain is considered as the prime routes for expose of human beings to soil pollution (Sang et al., 2018; Sharma et al., 2018). Along with promoting soil pollution it also severely impacts on the generation and quality of food and its well-being (Muchuweti et al., 2006). The most exceptionally imperative hot spot for the existence of life is water. But still a large part of the total population of the globe suffers from the paucity of clean drinking water. As there is rapid increment in the worldwide population the demand of water for the anthropogenic activities has increased manifolds (Pendergast and Hoek, 2011). As assessed and reported by the World Water Council, the population of around 3.9 billion people will suffer from acute scarcity of water by 2030 (Xu et al., 2018). It is very evident that the present and future urge of water supply is going to increase tremendously due to the spontaneous increase in the urban population and the industrial requirements. Thus, it has become very crucial to look for alternate arrangements of clean and pure drinking water. The untreated and poisonous effluents in the form of heavy metals are regularly discharged from the industries have been the major contributor in the increase of water contamination. It is astonishing to admit and assess that water pollution contribute almost 70%–80% of all the ongoing issues in the developing countries. These toxic heavy metals have also damaged the entire marine ecosystem by completely contaminating the water bodies, as the aquatic organisms are suffering by consuming them (Padilla-Ortega et al., 2013; Bhatnagar and Sillanpää, 2010; Sadeek et al., 2015; Zazycki et al., 2017; Kobielska et al., 2018; Li et al., 2018; Sahmoune, 2018). Numerous physical, chemical and biological processes come into picture when the heavy metals arrive in the aquatic ecosystem (Guo et al., 2018).

Water source with minute concentration of heavy metal having hazardous and toxic impact on human health and alter the other ecospheres. It is extremely paramount to consider that the metal toxicity level relies on specific parameters like the bio-organisms which are in close contact to it, its nature, its biological effect and duration of metal exposure with bio-organisms. The all organisms depict inter relationship through food chains and food webs (Lee et al., 2002). Owing to the harmful and severe effect of heavy metals on the ecosphere and human health, it continues to be one of the most explored and studied areas of research and investigation (Fairbrother et al., 2007). Because of the availability, toxicity and tenacity of the heavy metals they are of specific attention for ecosystems. It has been extensively studied that metal specification decides the ecotoxicological implications of the heavy metals. The partitioning of metal largely influences the metal bioavailability. In the aquatic ecosystem, plethora of parameters for example, pH, temperature, oxidation reduction potential, total concentration of metal and mineralogy, water transport and removal by rainfall performs vital and complicated role (Aung et al., 2008). These parameters are interrelated and changes continuously. Hence, the characterization of species of metal in the dynamic systems like rivers, streams or wetlands becomes ponderous and tiresome. For the proper and accurate understanding of the sources and origin of heavy metal contamination and its influence on environment is a challenge that is being addressed since decades. To stand out in the process of heavy metal remediation the research on heavy metal is significant as the elimination efficacy for various heavy metals differs with numerous elimination approaches (Fig. 14.1).

14.2 Heavy metals as environmental pollutants

The environmental contamination because of the extensive use of the heavy metals have come out as a global issue which needs utmost attention in the present day as it impacts the public health. Heavy metals among a large pool of environmental pollutants defines a vital role as its concentration in water bodies, soil, air is rapidly and continuously increasing with each passing day because of the anthropogenic activities. The parent rocks and metallic minerals dominates the natural source on the other hand the major anthropogenic sources involve agricultural activities and metallurgical activities. In general, heavy metal is considered as a metal which is harmful irrespective of their density and atomic mass (Singh, 2007). Even the low concentration of metal ions having lethal and toxic impact on human health (Kumar and Gayathri, 2009; Shyam et al., 2013; Zhou et al., 2015; Peng et al., 2017; Kyzas et al., 2018; Sherlala et al., 2018; Bibaj et al., 2019; Hemavathy

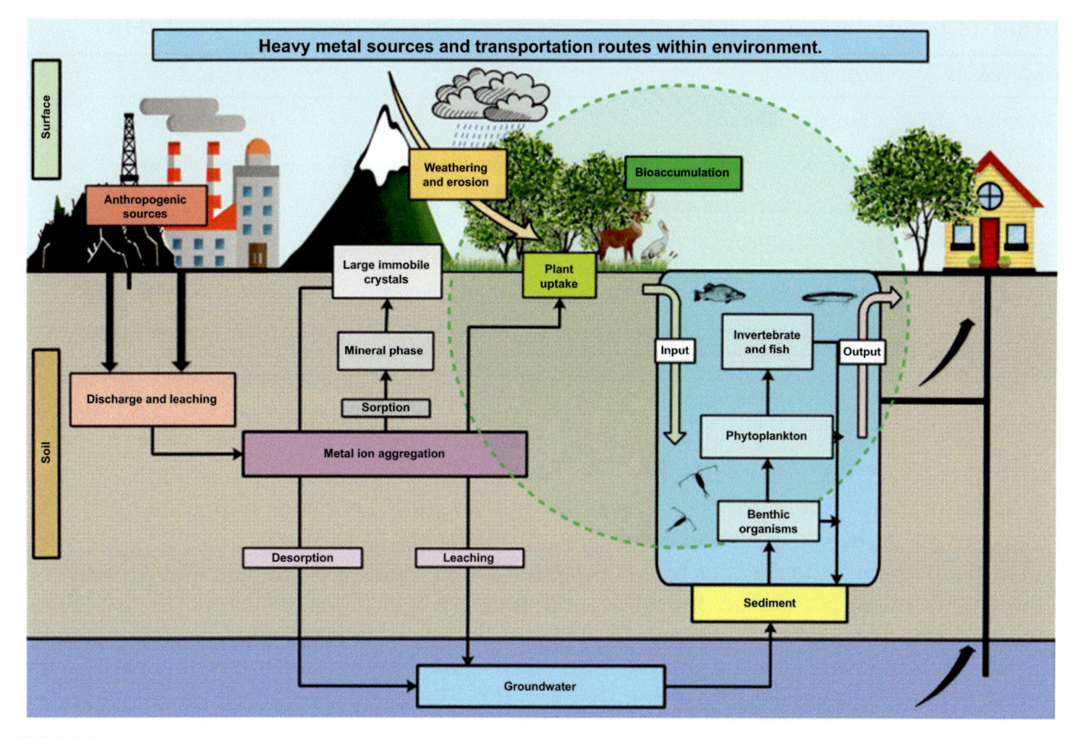

FIGURE 14.1 A flow-chart describing about redistribution of soil, fresh and groundwater system based heavy metals of anthropogenic origin. Source: *Reprinted with the permission Kobielska, P.A., Howarth, A.J., Farha, O.K., Nayak, S., 2018. Metal–organic frameworks for heavy metal removal from water. Coordination Chemistry Reviews, 358, 92–107 (Elsevier).*

et al., 2019; Lei et al., 2019; Li et al., 2019; Suganya, 2019; Saravanan et al., 2019). The two prime reasons for the increase in water contamination are urbanization and industrialization. Through industries discharged water the metals are transferring from the towns, cities and villages. They further assemble in the sediments of the water bodies. They might be quite toxic if consumed by the humans or the marine animals, they are even harmful for the entire aquatic ecosystem. It is very natural that even if single organism is being influenced then it will eventually influence the entire food chain. In view of the fact that in the food chain the humans are located at the pinnacle it is very obvious that human beings will be the most influenced one. The prime sources of discharge of heavy metals in large quantities into the water bodies through different industries are: chemical fertilizers and pesticides, leather training, ink and pigment manufacturing, nuclear power plants, pharmaceutical industries, metallurgical and mining, petroleum refineries, coal burning in power plants, microelectronics, gas and crystal industries (Bradl, 2005). Heavy metals are also available in the atmosphere in the forms of aerosols, particulates and gases. In atmosphere cadmium (Cd), zinc (Zn), copper (Cu) and lead (Pb) are abundantly present as metals. The heavy metals come in contact with the environment starts from the beginning of the production chain, when the ores are mined in addition to that, it continues during

TABLE 14.1 Major man-made sources of toxic heavy metals in the environment (Patra et al., 2020).

Toxic metals	Sources	References
Chromium	Chromite mine soil	Patra et al. (2019)
Nickel	Batteries of automobile and discharges from the industries	Tariq et al. (2006)
Copper	Chemical fertilizers and insecticides	Khan et al. (2007)
Cadmium	Dyes, electroplating	Salem et al. (2000)
Mercury	Surgical instruments, medical residue, burning of coal	Rodrigues et al. (2012)
Lead	Petrol burning and insecticides	Wuana and Okieimen (2011)

From Patra, D.K., Pradhan, C. and Patra, H.K., 2020. Toxic metal decontamination by phytoremediation approach: concept, challenges, opportunities and future perspectives. Environmental Technology & Innovation, 18, *100672.*

the utilization of products containing them and finally at the end of the production chain. The deterioration of heavy metals is just beyond imagination hence they accumulate in a huge amount in the ecosystem (Sarwar et al., 2010) (Table 14.1).

The pollution of natural water bodies is related to the rapid development and expansion of cities, towns, urban population, industries and intensive agriculture which makes use of various chemicals. This leads to exit and discharge of the toxic effluents which includes nitrates, phosphates, fertilizers into the natural ecosystems and hydrosystems. As reported by the United States Environment Protection Agency (US-EPA) and World Health Organization (WHO), the permissible concentration limits of all the heavy metals have been wide-ranging from 0.01 to 0.05 mg/L (EPA, US, 2009; WHO, 2019). The heavy metals possess majorly negative influence on the plants and animals too. They have exhibited and caused disturbance in the germination of seed, metabolism of plants, its growth and resistance mechanisms (Cheng, 2003; Aydinalp and Marinova, 2009). Due to the occurrence of the phenomena known as bioconcentration the heavy metals concentrations are relatively higher in the plants relative to the environment. The toxicity level of heavy metals in plants depends upon various factors like type of heavy metal, its concentration, and the affected species of plants. It has also been observed that this inhibits the chlorophyll concentration and hence reduction of photosynthesis. The cellular division of plants is also disrupted by the availability of heavy metals. As a result of bioaccumulation and biomagnification process the heavy metals are raising the level of risk at different trophic levels. To control the mobility of heavy metals speciation is considered to be a paramount factor. However, to safeguard and conserve the environmental standard of the sediment, soil, groundwater and surface water, thus it is very critical to move ahead with exact and precise assessment of the movement and toxicity of heavy metals. The sharp raise in the commercialization of the heavy metals in the modern industries have caused an environmental liability (Table 14.2).

14.2.1 Aluminum

Aluminum (Al) originally found in the bauxite ore having density of 2.7 g/cm^3. It is a lightweight and strong element. Aluminum is the part of several food sources and food

TABLE 14.2 Adverse influence of toxic heavy metals on human beings.

Heavy metals	Effect on humans	References
Nickel	Kidney, respiratory diseases, asthma, cancers, congenital malformations, pulmonary fibrosis	Das et al. (2019), Coogan et al. (2008), Denkhaus and Salnikow (2002)
Copper	Wilson disease, insomnia, convulsions and cramps	Gunatilake (2015)
Cadmium	Kidney dysfunction, bone weakening and brittleness, spinal and leg pain, renal diseases	Waisberg et al. (2003), Satoh et al. (2003)
Lead	Nervous system disordering, fertility related problems, liver and kidney disorders, dizziness, irritability, muscle weakening	Apostoli et al. (2000), Ahamed et al. (2008)
Arsenic	DNA breakdown followed by death, vascular diseases	Gunatilake (2015)
Zinc	Short term memory loss, depression, renal failure, lethargy	Gunatilake (2015)
Mercury	Chest pain, mental illness, neurological disruptions (memory and sensory functions)	Counter and Buchanan (2004), Ralston and Raymond (2018)
Chromium	Skin irritability, lung cancer, mental illness, anemia	Chiu et al. (2010)

products like, cheese, soft drinks sponge cakes, tea and coffee and several food preservatives and additives. A permicible limit of ≤ 0.2 mg/L for purified drinking water quality guidelines has been set by World Health Organization. They led to the damaging of central nervous system of the human beings. It also leads to dementia, dysfunction of liver and kidney, colitis, pulmonary fibrosis and lung damage. The raise in aluminum concentration particularly in adult cause a neurological diseases like Alzheimer's disease. The vital organs that are the major focus of the poisoning of aluminum are the central nervous system, lungs, and bone. It also causes two types of osteomalacia.

14.2.2 Copper

It is among the heavy metals that is utilized on regular basis for a number of applications. During its mining it enters the environment. Commercially it is widely used in the electronics industries such as wires, semiconductors, electronic chips, cell-phones, pulp and paper industries, insecticides and in metals processing industries. Cu is necessary micronutrient required by a human body in the formation of hemoglobin in red platelets. It also assists in the working of several enzymes. Although it is required for the basic healthy development but if consumed in high doses it can be very unsafe. For the amphibian life, Cu^{2+} is quite hazardous. Queasiness, respiratory challenges, gastric problems, gastrointestinal bleeding and many more. The side-effects related to intake of higher concentration of copper is visible in the ages between 6 and 40 years and if proper treatment is not done it may lead to liver failure. The lack of copper in the human body also led to numerous ailments. The allowed level of copper in drinking water is 0.05 mg/L according to the Bureau of Indian Standards (BIS).

14.2.3 Cadmium

The density of Cd is 8.69 g/cm^3. The existence of Cd on the earth's crust Cd is regarded as majorly toxic heavy metals for both plants and animals. It enters into the food chain through the agricultural soils. It diffuses into the soil mainly through the anthropogenic activities. The compounds of Cd are very much soluble as compared to other heavy metals, so it is swiftly taken up by the plants and assembles in various edible parts of the plants. Commercially, it is widely used in pesticides, corrosion-resistant plating, Nickel–Cadmium batteries, television screens, phosphate fertilizers and many more. It is also found in food sources like mushroom, shrimps, mussels, shellfish, etc. Kidneys, bones and lungs are quite prone to the harmful effects of Cd. Several studies have confirmed about the cause of apoptosis in organs due to Cd. It has harmful effect on humans such as Itai-Itai disease, renal dysfunction, alternation in metabolism of calcium, psychological and gastrointestinal disorders, impairment of DNA and many more. Cadmium is known as an intense poison which posses a nearly 20 years of a natural half-life. Cd present in the body is known to affect different enzymes. Cd discharges out of the body through urine but the rate of discharge is very low which results in high retention of Cd in the body. Tobacco smoking and food intake are the most common pathways through which Cd makes its entry inside the human body. If it is being accumulated chronically in the body then it may severely damage the functioning of the kidney (Wuana and Okieimen, 2011). Cd is believed to have significant capacity of disruption of endocrine system in the human body and increase the blood pressure. Moreover, oxidative stress has been observed in algae due to presence of Cd (Pinto et al., 2003) and in many plants it also alters the seed germination. Cd has been listed as "Carcinogenic agent for humans" by The International Agency for Research in Cancer (IARC) (Huff et al., 2007).

14.2.4 Nickel

Nickel (Ni) in its metallic form is considered as among the lightest heavy metals. It is popularly used in the modern trending technologies. Commercially, they are being widely utilized in the electroplating, Ni–Cd batteries and electronic equipment. As Ni is ubiquitous element hence its exposure is very natural but it is not among the essential elements required by a human body. The most common effect that occurs when Ni is exposed to skin is dermatitis. In addition that, intake of nickel salts may result in vomiting, nausea and diarrhea. On the other hand, when nickel is chronically exposed then it can cause asthma and tubular dysfunction. According to IARC, it has been listed as potentially carcinogenic substance as it is highly toxic and carcinogenic.

14.2.5 Lead

It is regarded as one of the most dangerous heavy metal. When it is exposed for a longer period of time then it can lead to lead poisoning. It is very harmful for the water invertebrates. In the marine ecosystem eggs and quite young fishes are at higher threat. It causes deformity in the spine of the bird and blackening of the caudal region among the birds. They can also affect their egg production. There are several adverse effects of lead poisoning in the humans such as decreased fertility, spontaneous abortion, neurological and cardiovascular disorders. At higher doses, it can lead to neurological and renal disruptions. It can affect the brain development and learning difficulties among the children.

14.2.6 Mercury

Mercury (Hg) is a heavy metal having density 13.6 g/cm^3 that is lethal to entire known living organisms. Mercury having adverse effect on the growth and development of unicellular life (algae, bacteria and fungi) and of marine animals like fishes. In case of humans and animals it causes an endocrine influence as it reacts negatively with the necessary hormone called as dopamine. The reaction precisely depends upon the contamination level and time of exposure. It has adverse effects on the pregnant women as it easily reaches out to the fetus by crossing the placenta. In infants the risk still persists as the human breast milk may also be contaminated. It has severe effects on the children as it is very neurotoxic and has a tendency to bioaccumulate. They cause cytotoxic effects on the nervous system even at minute concentration. It is one of those metals which can lead to death if exposed for a longer period of time even at small concentrations. In the vapor form it can majorly affect the respiratory system of the humans. Mercury may also lead to phenomena known as autoimmunity phenomena, in which the human cells are attacked by their own immunity systems.

14.2.7 Chromium

The toxicity of this heavy metal depends strongly on its chemical form (valence 3–6). As Cr (IV) can make their way through the ion transport system, this tends to increase the concern for the aquatic environment. As they are known for damaging the DNA hence, they are also genotoxic agents. If the level of chromium increases in the soil then genetic changes starts to appear. Soil acidification also drives the intake of chromium in the plants. If high concentration of chromium is present in the water bodies then it damages the fish gills near the discharge points. Cr cause respiratory problems, lowering of immunity, genetic defects and formation of tumor cells in the animals. Humans are also adversely affected by high levels and long time of exposure to chromium.

14.2.8 Zinc

Zinc (Zn) is not commonly found in nature although the extraction process from its ore is very simple. Since, 2000 years Zn has been known as a metal. ZnS is the principal ore and is available across the globe. It is used commercially owing to its chemical and metallurgical properties. Zinc is widely used in the process of galvanization of iron and steel items. It is also used in cosmetics, medicinal items, photocopy paper, paints and agricultural products. Usually, for human body the zinc act as essential micronutrients with toxicity. Disability of growth can be caused due to an extreme consumption of zinc. It is mainly discharged into the environment through the industrial venture like mining, coal burning and steel production and manufacturing. Diarrhea, nausea, abdominal cramps, vomiting, blood urine, stomach ache, etc., can be caused due to zinc poisoning. In accordance to BIS the permitted range of lead in drinking water is 5 mg/L.

14.2.9 Arsenic

Arsenic is among the most toxic and carcinogenic metals that exist in the form of either sulfides or oxides. In all its form of oxidation states it persists to be carcinogenic and at level of high exposure may even result in death. Intake of arsenic through drinking water exhibits high risk of lung, kidney and bladder cancer. Arsenic can also be cause of skin cancer including hyperkeratosis and pigmentation. It adversely attacks the sulphydryl group of cells, which lead to breakdown of cell enzymes and cell respiration. Autoimmune disorder can also take place due to the toxicity of arsenic. It causes nerve inflammation and weakness in muscles (Kalita and Baruah, 2020).

14.2.10 Nuclear waste

As the nuclear energy sector is swiftly developing to meet the emerging demand of clean energy a huge quantity of nuclear waste is also generated. Usually, the radionuclides are insoluble in water and geochemically inert. But they get oxidized and turn into pollutants when exposed to natural environment. However, these oxidized radionuclides are largely soluble in water and adversely affects the marine organisms. Uranium (U), Plutonium (Pt), Technetium (Tc), Neptunium (Np) exhibit high toxicity towards human health and marine ecosystem in their high oxidation state. The radionuclides contaminate the ground and surface water used for drinking and agricultural activities. They cause hazardous effects to humans such as neural disorders, infertility issues, birth defects, breast and lung cancer, stomach related problems. Uranium damages the functioning of kidney among the humans. Strontium causes leukemia and weakens the immune system of humans.

14.3 Application of nanotechnology for metals remediation from contaminated water

With the advent of rapid development and advancement of nanoscience and technology in the 21st century, plethora of possibilities has been explored to fabricate and develop better, cost effective, time saving and most significantly environment friendly remediation processes. Nanotechnology is a blooming science that has exhibited tremendous potential in humanizing various dimensions of life ranging from medicine to industrial materials. Among a large pool of applications, one such application is the remediation of contaminated groundwater. The contamination of groundwater and surface water is emerging as a serious threat for not only developing nations like India but also for the developed nations. Since the last decade the over-exploitation and contamination of groundwater has increased several folds. The quality of soil and groundwater is swiftly deteriorating. The current breakthrough developments in the nanotechnology have ventured new dimensions for water remediation. Nanomaterials can be utilized for refinement of water through pathways like adsorption of heavy metals and pollutants, deactivation of pathogens and finally the catalytic deterioration of toxic metals into not so harmful compounds (Adeleye et al., 2016). Despite of the fact that the nanostructured metal oxides possess certain chemical and biological side-effects, they have been so far considered as the best savior for the environmental conservation. The most popularly utilized nanosized metal oxides for water purification are TiO_2, ZnO, CuO, SnO_2, Iron oxide nanoparticles and other nanoparticles. Recently the elimination of hazardous heavy metal ions from contaminated water by using metal oxides have been explored widely. Numerous adsorbents like zeolites, nanocomposites, metal oxides are frequently exploited for the eradication of poisonous metal ions from polluted water (Fig. 14.2).

The utilization of TiO_2 nanoparticles for water purification have garnered a huge amount of attention and importance because of its physiochemical properties, catalytic efficiency, cost worthy, high photosensitivity and of course its nontoxic nature. Photocatalysis has proved to be a ground breaking technological advancement in the domain of water remediation and beneficial in the elimination of heavy metal ions. They have been studied and investigated fir the eradication of natural organic matters (Gora et al., 2018), pesticides (Yola et al., 2014) and dyes (Ali et al., 2017) from contaminated water. ZnO is known for its antibacterial properties and is cost-effective too. The photocatalytic properties of ZnO are very helpful in purification of water as it removes the dyes and bacterial decomposition. High photocatalytic properties, large oxidation capability and better free-excitation binding energy have made ZnO nanowires a potential candidate for purification of water and remediation of toxic waste. CuO nanoparticles behaves as a capable adsorbent regarding the elimination arsenic in samples of groundwater leaving no negative chemical impact on the sample. CuO_2 nanoparticle introduces better separation of charge carrier rendering the photocatalyst more potent than pure TiO_2. It has been reported that CuO has been used for catalytic degradation of organic dyes. SnO_2 nanoparticles are widely investigated for water purification particularly for degradation of

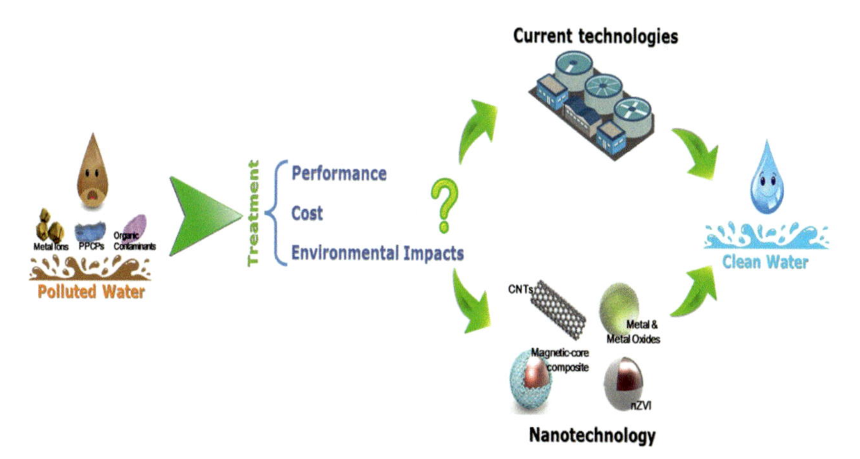

FIGURE 14.2 Schematic diagram, of use of nanotechnology for the remediation of wastewater. Source: *Reprinted with the permission Stefaniuk, M., Oleszczuk, P., Ok, Y.S., 2016. Review on nano zero valent iron (nZVI): from synthesis to environmental applications. Chemical Engineering Journal, 287, 618–632 (Elsevier).*

dye. Owing to low-cost synthesis of iron oxide nanoparticles they are utilized as an adsorbent to eliminate the harmful metal ions as they are environment friendly too. WO_3 possess a very good sensing properties so it can perform as a sensor for the identification and detection of toxic organic compounds. NiO nanosheets have exhibited porous structures which makes it a potent adsorbent for the removal of red dyes from the polluted water. The radioactive materials available in the nuclear waste are very toxic to all kinds of life on earth which also includes the human beings. The radionuclides have great mobility in the aqueous solution and it migrate to the soil. Thereafter the plants intake them and they make their path into the food chain. They also have very long half-lives which increases the threat of long-term exposure. Carbon nanotubes (CNTs), magnetic nanomaterials, silver nanomaterials, graphene-based nanomaterials are extensively utilized regarding the removal of nuclear waste. Large surface area and high reactivity are the two properties that pave the path for the utilization of nanomaterials for the remediation of nuclear waste.

Zero valent iron nanomaterials (nZVI) an environmentally friendly substance has appeared to be an appropriate alternative pathway for water remediation (Stefaniuk et al., 2016). It has exhibited numerous unique and efficient properties such as extremely less toxicity, cost-effective, large surface are-to-volume ratio which further results in good adsorption capacity and increased reactive sites (Fu et al., 2014; Ponder et al., 2000). As the nano iron possess magnetic properties so on the application of magnetic field they eases out their separation from soil and water (Boparai et al., 2011). While coming in close proximity with metal ions like Zn (II) and Cd (II) NZVI acts as an adsorbent. It also behaves as a suitable reductant while reacting with the oxidized metal ions like Cr (IV) (Zou et al., 2016) (Fig. 14.3).

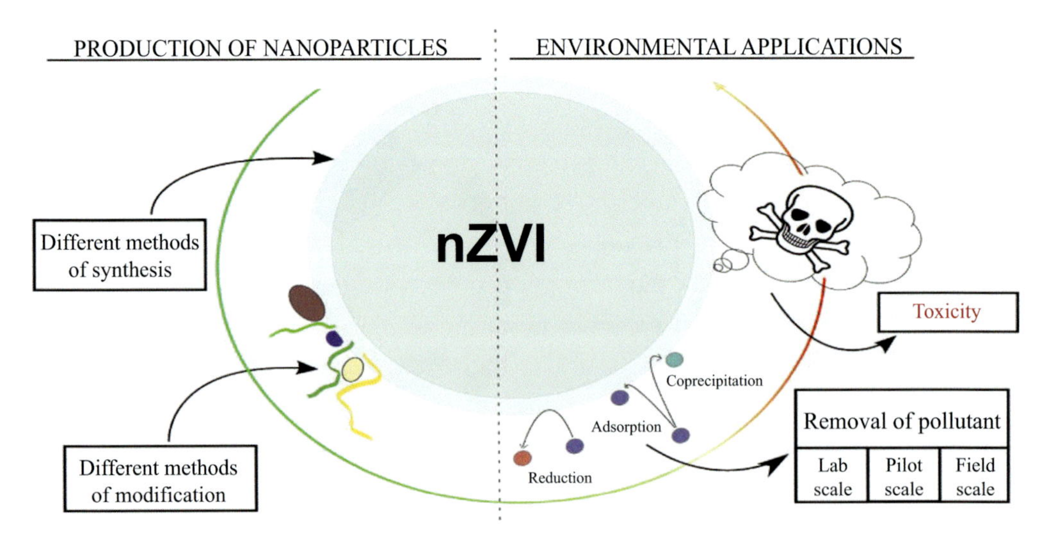

FIGURE 14.3 Schematic representation of working mechanism of nZVI. Source: *Reprinted with the permission Stefaniuk, M., Oleszczuk, P. and Ok, Y.S., 2016. Review on nano zero valent iron (nZVI): from synthesis to environmental applications. Chemical Engineering Journal, 287, 618–632.*

14.3.1 Nanoadsorbents

Nanomaterials which are chemically active and possess good adsorption capacity are refer to as nanoadsorbents. They are categorized on the basis of process of adsorption. Numerous adsorbents include the carbon-based nanoparticles (graphene and CNTs), metal and metal-oxide nanoparticles and magnetic nanoparticles. In addition, nanoclays, aerogels and polymer-based nanomaterials are also regarded as nanoadsorbents. Owing to their peculiar structural, mechanical and chemical properties carbon-based nanomaterials like CNTs (single-walled and multiwalled), graphene and fullerenes are widely utilized for water remediation. Graphene a two-dimensional which consists sp_2 hybridized carbon atoms ordered in a hexagonal lattice has exhibited tremendous potential in wastewater remediation. Fullerenes are another class of carbon-based nanomaterial which has sp_2 hybridized carbon atoms and has spherical closed-cage like structure is also a potential candidate regarding the wastewater remediation. The properties which make fullerenes appropriate candidate for the elimination of various species from the aqueous solution are large surface area-to-volume ratio, great electron affinity and hydrophobic surface. Because of their outstanding properties Fe_3O_4 nanoparticles have excellent adsorption capacity. A huge number of carboxyl groups are available on the surface of Fe_3O_4 nanoparticles that results in the attraction of surface negative charges through the positively charged heavy metal ions. Therefore, the predominant factor for the adsorption mechanism is the electrostatic attraction.

14.3.2 Nanocatalysts

Photocatalysts (Quintana et al., 2017), electrocatalysts (Sun et al., 2014), and Fenton-based catalysts (Tušar et al., 2012) are some of the nanocatalysts used for treatment of wastewater. When the catalytic activities are performed in the presence of light by the nanomaterials then it refereed as photocatalysis. The decline of organic pollutants like detergents, dyes, pesticides, etc., are facilitated by the metallic nanoparticles and light energy interaction. SnO_2 nanosheets behave as a photocatalyst for the deterioration of methylene blue and Congo red. Zirconium oxide nanoparticles has been explored as an appropriate photocatalyst for the deterioration of Rhodamine B dye. The phenomenon of the photocatalytic degradation in the catalyst is created upon the photoexcitation. As per the light is exposed on the catalyst it produces exited electrons and holes in the conduction band. The water molecules (H_2O) trap the generated holes (h^+) and led to the formation of hydroxyl radicals (^-OH) in the aqueous solution. One more scientific and modern pathway for wastewater treatment is microbial fuel cell established on electrocatalysis. Oxidation of organic contaminants through Fenton's reaction is another method of treatment of contaminated water. Although it performs efficiently only in acidic mediums.

14.3.3 Nanomembranes

One of the most promising and powerful methodologies implemented for the wastewater treatment by utilizing nanomaterials is the membrane filtration technique (Ho et al., 2012). Better catalytic activity, fouling resistance and high permeability are some of the novel functionalities that have been contributed to the treatment of wastewater through application of nanotechnology. As compared to the other traditional techniques membrane treatment methodology are highly economical, simple to handle and quite efficient. Particle separation and chemical decomposition of organic contaminants are prime roles performed by the nanomaterials in nanomembranes. Nanomembranes incorporates one-dimensional materials including the organic and inorganic ones like nanofibers, nanoribbons and nanotubes. Nowadays nanocomposites are also used for wastewater treatment where the reactivity of nanoparticles are utilized without release of free nanoparticles to the environment. Nanocomposites beads, membranes and 3D-nanocomposites are the three broad classification of nanocomposites.

14.4 Methods of recovery and elimination of metals from contaminated water

The entrance in the human food chain of the toxic heavy metals is through the bioaccumulation mechanism and the toxicity increases over the raise in metal ions concentration and time of exposure. The hazardous heavy metals are easily capable of penetrating into the marine ecosystem because of the industrial and agricultural waste. Several remediation techniques likes, electrochemical methods, ion exchange, chemical precipitation, flocculation and coagulation, absorption, membrane filtration and bioremediation have used to eliminate harmful and toxic heavy metals from contaminated water (exhibited in Fig. 14.4).

14.4.1 Chemical precipitation

Owing to its basic and affordable nature chemical precipitation is among the most extensively used water treatment methods. In the beginning the pH of the wastewater is adjusted thereafter the addition of precipitating agent takes place. Then to form insoluble it interacts with the heavy metal ions in the polluted water. Sedimentation or filtration technique is used for the separation of formed precipitates. Sulfide and hydroxide precipitation methods are the conventional chemical precipitation techniques. Hydroxide precipitation is of low cost, simple to operate and easy control of pH (8.0–11.0). Heavy metals contaminations like chromium and copper have been eliminated from polluted water by using hydroxide precipitating agents like $Ca(OH)_2$ and NaOH. Some of the hydroxide precipitates are amphoteric while the sulfide precipitates are not. By tuning the factors like, pH, initial concentration, temperature and charge of the ions, etc., removal percentage of metal ions in the solution can be enhanced. The heavy metals are generally in acidic condition so the sulfide precipitation takes place in neutral or basic environment because the sulfide precipitants can form harmful hydrogen sulfide vapors. A huge quantity of chemical is required to deduce the metals to permissible level to be released in the environment by the chemical precipitation method. There are also several advantages attached to this particular technique like formation of huge amount of sludge, slower precipitation of metal, less settling and the long-span negative impact (Yang et al., 2001).

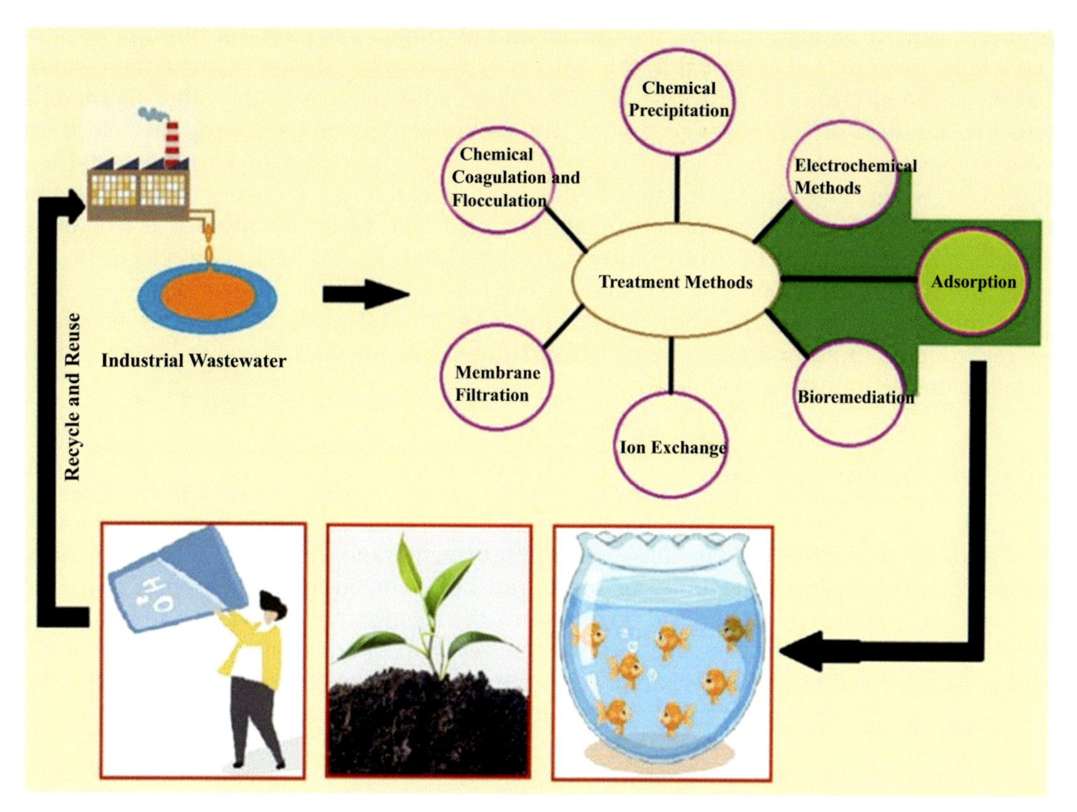

FIGURE 14.4 A schematic representation of treatment methods for the removal of heavy metals from the contaminated water. Source: *Reprinted with the permission Vardhan, K.H., Kumar, P.S., Panda, R.C., 2019. A review on heavy metal pollution, toxicity and remedial measures: current trends and future perspectives. Journal of Molecular Liquids, 290, 111197.*

14.4.2 Electrochemical methods

Regarding heavy metals removal particularly from the industrial contaminated water the electrochemical methods have been immensely suitable. This technique uses the electricity, as it incorporates the retrieval of heavy metals in the elemental metallic nature by utilizing cathode plate and an insoluble anode in an electrochemical cell. The generation of electricity is from the movement of one electron from one element to other. This method is not feasible economically as it need huge capital investment. In this research field many scientists and researchers are highly interested and motivated to work and explore because of the strict guidelines regarding the discharge of wastewater into the surroundings. Among the numerous electrochemical methods, electrocoagulation, electroflotation and electrodeposition techniques are the most widely used.

The process of electrocoagulation comprises electrodes that acts as an anode and a cathode where the oxidation and reduction reactions occur respectively (Widhiastuti et al., 2018). It is easy to handle and cost-effective process for industrial cable operations. Heavy metal dyes, nitrates, sulfates and pharmaceuticals waste have been removed through this technique where, anode produces metal ions and hydrogen gas pass out through cathode. There is no need to add chemical in the process of electrocoagulation because the essential reagent is the electron. This process is very eco-friendly relative to the other methods as very low amount of sludge is generated. By the utilization of an appropriate electrode, electrodeposition technique is extensively utilized in the recovery of harmful metal ions from wastewater. The major attraction of this method is that the recovery of metal is done in its pristine from and the cost of operation is also low. Electroflotation is a techniques used to remove toxic metal from industrial wastewater. In this technique electrolysis of water produce small bubbles of oxygen and hydrogen gases which cause floating of toxic heavy metal and help in separation (Casqueira et al., 2006). The electrochemical techniques are regarded as rapid and well-controlled treatment strategies for the elimination of toxic metals from the industrial polluted water.

14.4.3 Membrane filtration

Membrane filtration technique having exceptional property of heavy metal removal from industrial wastewater. It consists complex membrane structures with nanometer range. The membrane permeability of water and heavy metal ions elimination properties relies on the physical and chemical attributes of the membrane. The prime advantages of using this method are lesser space requirement, easy to operate and better removal efficiency. Reverse osmosis, nanofiltration, ultrafiltration and electrodialysis are some of the various types of membranes that have been tested and implemented successfully.

14.4.4 Ion-exchange

Owing to the exceptional rate of elimination, better treatment capacity and swift kinetics ion exchange mechanism has been popularly implemented for wastewater treatment of (Kang et al., 2004). In ion-exchange resin there is exchange of ions (cations or anions) present in solution and present in insoluble substances takes place. In the ion-exchange column the heavy metal wastewater enters from one end and is migrated through the surface bed which eventually filters out the heavy metals from the wastewater. The ion-exchange resin can either be natural one or synthetic. The synthetic ones are preferred as they are more appropriate and effective in the removal of heavy metals. This technique is quite expensive as huge amount of resin is required during the entire process and the chemical used produces secondary pollutants which are not desirable.

14.4.5 Chemical coagulation and flocculation

For removal of toxic heavy metal ions from wastewater chemical coagulation or flocculation technique has also used widely. This technique mainly destabilizes the colloidal particles and eventually causes sedimentation. Usually, the process of coagulation is succeeded by flocculation of the not so stable particles into huge floccules to enhance the size of the particle (Teh et al., 2016). Lastly, the huge floccules are precipitated in the sedimentation tank. Alum, ferrous sulfate, etc., are the coagulants used in this method. For coagulation process suspended solids and hydrophobic colloids are the major objects. Flocculation is a method of bond formation between the flocs and knot the particles by using polymer to form a large agglomerates. Poly-aluminum chloride, polyferric sulfate and polyacrylamide are the various types of flocculants utilized in this method. This method is relatively very expensive and it is not environment friendly as large amount of chemicals are involved in the entire process.

14.4.6 Bioremediation

Bioremediation process refers to the utilization of biological systems, plants and animals that incorporates the microorganisms to eradicate the harmful contaminants from the aquatic ecosystem (Head, 1998). For hazardous heavy metals removal microbial system has been widely used. Wetland ecosystems have proved to be potential candidates for the removal of heavy metals. The entire operation is very easy to handle and is of low cost. Algae which are microscopic organisms and are available in every kind of ecosystem are suitable for use in wastewater remediation. Different functional groups present on algal cell wall act as major binding sites for the elimination of heavy metals from the polluted water. Bacteria and fungi have also been observed as an appropriate candidate for the heavy metal ions removal from the wastewater.

14.4.7 Adsorption

The adsorption technique is utmost suitable, economical and precise treatment technique for the elimination of toxic heavy metals from the wastewater (Da'na, 2017). It is solid-liquid mass migration mechanism, where the heavy metal which is known as adsorbate is transferred from the contaminated water to the solid surface called the adsorbent and then bonding takes place on the surface due to either physical or chemical adsorption. The physical adsorption takes place because of the Van der Waals force of attraction whereas the chemical adsorption takes place because of presence of strong covalent bond between the adsorbate and the adsorbent. The entire mechanism is reversible. Owing to its greater surface area and better electron affinity activated carbon is considerably utilized as an adsorbent. There are numerous advantages of using this particular technique such as there is no sludge production, no extra need of nutrients, quite easy to operate and have quite high rate of efficiency.

14.5 Future perspectives and challenges

Water contamination due to harmful heavy metals is of prime environmental stress and anxiety and significant threat to human existence on Earth. The foremost concern is that the assembling of heavy metals in the soil sediments, road dust and water bodies should be regarded as chemical time bombs anticipating to explode. With the advancement of human development, it is very ironic that the use of heavy metals has also increased tremendously, which has triggered the existing problem. Although the complete elimination of the use of heavy metals is not possible at all but the judicious use and proper remediation methodologies can help in curbing the problem. The entire scientific coterie is suggesting a laboratory and field based eco-friendly solution for the eradication of toxic heavy metals from the polluted soil and water. There are several disadvantages and challenges attached to each existing technique for the wastewater treatment. Owing to their exceptional morphological, structural and magnetic properties, nanotechnology has stirred a technological revolution in different areas of research. Since long time it has been observed that development of technology has been intricately concerned with the design and configuration of a specific material with basic functions and enhanced performance. It is very necessary to highlight the significance of understanding and to control the properties of nanomaterials used in the water remediation process. Toward the sustainable move, nanotechnology can surely act as an intelligent weapon. From the perspective of science and technology, the upcoming future seems to be very fascinating and exciting. The advantages of nanotechnology will enable us to achieve things that we never contemplated before. There are five principal requirements that should be followed by an ideal method of wastewater treatment high adsorption rate, high specificity, time saving, cost-effective and simple operation. Across the globe one in every nine people face the scarcity of safe water, that amount to almost 844 million people. In every 2 min a child dies of waterborne disease. Clearly, these challenges need to be addressed.

14.6 Conclusion

There is need of an hour when it comes to manage and control the heavy metal contamination in the environment. Urbanization and globalization are persistent driving global phenomena that demands precise and appropriate knowledge of the interplay among the anthropogenic activities, related outflow of contaminants and the urban surroundings. In this chapter we have attempted to explain the various heavy metal contaminants and their toxic and hazardous impact on all forms existing on the earth. The different sources be it either the natural or man-made both are discussed briefly. The relevance of nanotechnology in the domain of wastewater treatment is very convincing and promising. Nonetheless, this is beyond doubt that the nanomaterials will enable the purification of wastewater like never before. Since the first research in the 1980s, nanomaterials have traveled a huge distance. In the present-day nanotechnology is poised to solve some of the world's most critical health issues while on the other hand creating entire novel industries. Nothing makes more sense than the application of such an exceptional technology to curb the huge global challenge of safe water supplies. There is no denying the fact that the quality of groundwater is deteriorating with each passing day and there is an indispensable need for novel and advanced methodology for the purification of contaminated water. The scientists across the globe are actively investigating and studying on the different advance and environmental-friendly techniques to eradicate the prevailing problem of heavy metal pollution. To conclude, developments and advancements in nanotechnology in the upcoming time will surely be a prominent headway in the domain of wastewater remediation to prevent diseases and epidemics caused by consumption of toxic heavy metals and offer a healthy and sound life to each and every living organism on the earth.

References

Adeleye, A.S., Conway, J.R., Garner, K., Huang, Y., Su, Y., Keller, A.A., 2016. Engineered nanomaterials for water treatment and remediation: costs, benefits, and applicability. Chemical Engineering Journal 286, 640–662.

Ahamed, M., Fareed, M., Kumar, A., Siddiqui, W.A., Siddiqui, M.K.J., 2008. Oxidative stress and neurological disorders in relation to blood lead levels in children. Redox Report 13 (3), 117–122.

Ali, M.B., Barras, A., Addad, A., Sieber, B., Elhouichet, H., Férid, M., et al., 2017. $Co_2\ SnO_4$ nanoparticles as a high performance catalyst for oxidative degradation of rhodamine B dye and pentachlorophenol by activation of peroxymonosulfate. Physical Chemistry Chemical Physics 19 (9), 6569–6578.

Apostoli, P., Bellini, A., Porru, S., Bisanti, L., 2000. The effect of lead on male fertility: a time to pregnancy (TTP) study. American Journal of Industrial Medicine 38 (3), 310–315.

Aung, N.N., Nakajima, F., Furumai, H., 2008. Trace metal speciation during dry and wet weather flows in the Tama River, Japan, by using diffusive gradients in thin films (DGT). Journal of Environmental Monitoring 10 (2), 219–230.

Aydinalp, C., Marinova, S., 2009. The effects of heavy metals on seed germination and plant growth on alfalfa plant (*Medicago sativa*). Bulgarian Journal of Agricultural Science 15 (4), 347–350.

Bhatnagar, A., Sillanpää, M., 2010. Utilization of agro-industrial and municipal waste materials as potential adsorbents for water treatment—a review. Chemical Engineering Journal 157 (2–3), 277–296.

Bibaj, E., Lysigaki, K., Nolan, J.W., Seyedsalehi, M., Deliyanni, E.A., Mitropoulos, A.C., et al., 2019. Activated carbons from banana peels for the removal of nickel ions. International Journal of Environmental Science and Technology 16 (2), 667–680.

Boparai, H.K., Joseph, M., O'Carroll, D.M., 2011. Kinetics and thermodynamics of cadmium ion removal by adsorption onto nano zerovalent iron particles. Journal of Hazardous Materials 186 (1), 458–465.

Bradl, H. (Ed.), 2005. Heavy Metals in the Environment: Origin, Interaction and Remediation. Elsevier.

Casqueira, R.G., Torem, M.L., Kohler, H.M., 2006. The removal of zinc from liquid streams by electroflotation. Minerals Engineering 19 (13), 1388–1392.

Cheng, S., 2003. Effects of heavy metals on plants and resistance mechanisms. Environmental Science and Pollution Research 10 (4), 256–264.

Chibuike, G.U., Obiora, S.C., 2014. Heavy metal polluted soils: effect on plants and bioremediation methods. Applied and Environmental Soil Science 2014.

Chiu, A., Shi, X.L., Lee, W.K.P., Hill, R., Wakeman, T.P., Katz, A., et al., 2010. Review of chromium (VI) apoptosis, cell-cycle-arrest, and carcinogenesis. Journal of Environmental Science and Health, Part C 28 (3), 188–230.

Coogan, T.P., Latta, D.M., Snow, E.T., Costa, M., Lawrence, A., 2008. Toxicity and carcinogenicity of nickel compounds. CRC Critical Reviews in Toxicology 19 (4), 341–384.

Counter, S.A., Buchanan, L.H., 2004. Mercury exposure in children: a review. Toxicology and Applied Pharmacology 198, 209–230. *Find this article online.*

Da'na, E., 2017. Adsorption of heavy metals on functionalized-mesoporous silica: a review. Microporous and Mesoporous Materials 247, 145–157.

Das, K.K., Reddy, R.C., Bagoji, I.B., Das, S., Bagali, S., Mullur, L., et al., 2019. Primary concept of nickel toxicity—an overview. Journal of Basic and Clinical Physiology and Pharmacology 30 (2), 141–152.

Denkhaus, E., Salnikow, K., 2002. Nickel essentiality, toxicity, and carcinogenicity. Critical Reviews in Oncology/Hematology 42 (1), 35–56.

Edelstein, M., Ben-Hur, M., 2018. Heavy metals and metalloids: sources, risks and strategies to reduce their accumulation in horticultural crops. Scientia Horticulturae 234, 431–444.

EPA, US, 2009. National primary drinking water regulations. Washington, DC.

Fairbrother, A., Wenstel, R., Sappington, K., Wood, W., 2007. Framework for metals risk assessment. Ecotoxicology and Environmental Safety 68 (2), 145–227.

Fu, F., Dionysiou, D.D., Liu, H., 2014. The use of zero-valent iron for groundwater remediation and wastewater treatment: a review. Journal of Hazardous Materials 267, 194–205.

Garbarino, J.R., Hayes, H., Roth, D., 1995. Contaminants in the Mississippi River US Geological survey circular, 1133. Virginia, USA.

Gora, S., Liang, R., Zhou, Y.N., Andrews, S., 2018. Settleable engineered titanium dioxide nanomaterials for the removal of natural organic matter from drinking water. Chemical Engineering Journal 334, 638–649.

Gunatilake, S.K., 2015. Methods of removing heavy metals from industrial wastewater. Methods 1 (1), 14.

Guo, Q., Li, N., Bing, Y., Chen, S., Zhang, Z., Chang, S., et al., 2018. Denitrifier communities impacted by heavy metal contamination in freshwater sediment. Environmental Pollution 242, 426–432.

Gupta, N., Khan, D.K., Santra, S.C., 2010. Determination of public health hazard potential of wastewater reuse in crop production. World Review of Science, Technology and Sustainable Development 7 (4), 328–340.

Head, I.M., 1998. Bioremediation: towards a credible technology. Microbiology 144 (3), 599–608.

Hemavathy, R.R.V., Kumar, P.S., Suganya, S., Swetha, V., Varjani, S.J., 2019. Modelling on the removal of toxic metal ions from aquatic system by different surface modified *Cassia fistula* seeds. Bioresource Technology 281, 1–9.

Ho, H.L., Chan, W.K., Blondy, A., Yeung, K.L., Schrotter, J.C., 2012. Experiment and modeling of advanced ozone membrane reactor for treatment of organic endocrine disrupting pollutants in water. Catalysis Today 193 (1), 120–127.

Huff, J., Lunn, R.M., Waalkes, M.P., Tomatis, L., Infante, P.F., 2007. Cadmium-induced cancers in animals and in humans. International Journal of Occupational and Environmental Health 13 (2), 202–212.

Kalita, E., Baruah, J., 2020. Environmental remediation. Colloidal Metal Oxide Nanoparticles. Elsevier, pp. 525–576.

Kang, S.Y., Lee, J.U., Moon, S.H., Kim, K.W., 2004. Competitive adsorption characteristics of Co^{2+}, Ni^{2+}, and Cr^{3+} by IRN-77 cation exchange resin in synthesized wastewater. Chemosphere 56 (2), 141–147.

Khan, M.A., Ahmad, I., Rahman, I.U., 2007. Effect of environmental pollution on heavy metals content of Withania somnifera. Journal of the Chinese Chemical Society 54 (2), 339–343.

Kobielska, P.A., Howarth, A.J., Farha, O.K., Nayak, S., 2018. Metal–organic frameworks for heavy metal removal from water. Coordination Chemistry Reviews 358, 92–107.

Kumar, P.S., Gayathri, R., 2009. Adsorption of Pb^{2+} ions from aqueous solutions onto bael tree leaf powder: isotherms, kinetics and thermodynamics study. Journal of Engineering Science and Technology 4 (4), 381–399.

Kyzas, G.Z., Deliyanni, E.A., Mitropoulos, A.C., Matis, K.A., 2018. Hydrothermally produced activated carbons from zero-cost green sources for cobalt ions removal. Desalination and Water Treatment 123, 288–299.

Lee, G., Bigham, J.M., Faure, G., 2002. Removal of trace metals by coprecipitation with Fe, Al and Mn from natural waters contaminated with acid mine drainage in the Ducktown Mining District, Tennessee. Applied Geochemistry 17 (5), 569–581.

Lei, S., Shi, Y., Qiu, Y., Che, L., Xue, C., 2019. Performance and mechanisms of emerging animal-derived biochars for immobilization of heavy metals. Science of the Total Environment 646, 1281–1289.

Li, J., Chen, J., Chen, S., 2018. Supercritical water treatment of heavy metal and arsenic metalloid-bioaccumulating-biomass. Ecotoxicology and Environmental Safety 157, 102–110.

Li, Y., Bai, P., Yan, Y., Yan, W., Shi, W., Xu, R., 2019. Removal of Zn^{2+}, Pb^{2+}, Cd^{2+}, and Cu^{2+} from aqueous solution by synthetic clinoptilolite. Microporous and Mesoporous Materials 273, 203–211.

Liu, L., Li, W., Song, W., Guo, M., 2018. Remediation techniques for heavy metal-contaminated soils: principles and applicability. Science of the Total Environment 633, 206–219.

Mohammadi, A.A., Zarei, A., Esmaeilzadeh, M., Taghavi, M., Yousefi, M., Yousefi, Z., et al., 2020. Assessment of heavy metal pollution and human health risks assessment in soils around an industrial zone in Neyshabur, Iran. Biological Trace Element Research 195 (1), 343–352.

Muchuweti, M., Birkett, J.W., Chinyanga, E., Zvauya, R., Scrimshaw, M.D., Lester, J.N., 2006. Heavy metal content of vegetables irrigated with mixtures of wastewater and sewage sludge in Zimbabwe: implications for human health. Agriculture, Ecosystems & Environment 112 (1), 41–48.

Nies, D.H., 1999. Microbial heavy-metal resistance. Applied Microbiology and Biotechnology 51 (6), 730–750.

Padilla-Ortega, E., Leyva-Ramos, R., Flores-Cano, J.V., 2013. Binary adsorption of heavy metals from aqueous solution onto natural clays. Chemical Engineering Journal 225, 535–546.

Patra, D.K., Pradhan, C., Patra, H.K., 2019. Chromium bioaccumulation, oxidative stress metabolism and oil content in lemon grass *Cymbopogon flexuosus* (Nees ex Steud.) W. Watson grown in chromium rich over burden soil of Sukinda chromite mine, India. Chemosphere 218, 1082–1088.

Patra, D.K., Pradhan, C., Patra, H.K., 2020. Toxic metal decontamination by phytoremediation approach: concept, challenges, opportunities and future perspectives. Environmental Technology & Innovation 18, 100672.

Pendergast, M.M., Hoek, E.M., 2011. A review of water treatment membrane nanotechnologies. Energy & Environmental Science 4 (6), 1946–1971.

Peng, W., Li, H., Liu, Y., Song, S., 2017. A review on heavy metal ions adsorption from water by graphene oxide and its composites. Journal of Molecular Liquids 230, 496–504.

Pinto, E., Sigaud-kutner, T.C., Leitao, M.A., Okamoto, O.K., Morse, D., Colepicolo, P., 2003. Heavy metal–induced oxidative stress in algae 1. Journal of Phycology 39 (6), 1008–1018.
Ponder, S.M., Darab, J.G., Mallouk, T.E., 2000. Remediation of Cr (VI) and Pb (II) aqueous solutions using supported, nanoscale zero-valent iron. Environmental Science & Technology 34 (12), 2564–2569.
Quintana, A., Altube, A., García-Lecina, E., Suriñach, S., Baró, M.D., Sort, J., et al., 2017. A facile co-precipitation synthesis of heterostructured $ZrO_2|ZnO$ nanoparticles as efficient photocatalysts for wastewater treatment. Journal of Materials Science 52 (24), 13779–13789.
Ralston, N.V., Raymond, L.J., 2018. Mercury's neurotoxicity is characterized by its disruption of selenium biochemistry. Biochimica et Biophysica Acta (BBA)-General Subjects 1862 (11), 2405–2416.
Rodrigues, S.M., Henriques, B., Reis, A.T., Duarte, A.C., Pereira, E., Römkens, P.F.A.M., 2012. Hg transfer from contaminated soils to plants and animals. Environmental Chemistry Letters 10 (1), 61–67.
Sadeek, S.A., Negm, N.A., Hefni, H.H., Wahab, M.M.A., 2015. Metal adsorption by agricultural biosorbents: adsorption isotherm, kinetic and biosorbents chemical structures. International Journal of Biological Macromolecules 81, 400–409.
Sahmoune, M.N., 2018. Performance of *Streptomyces rimosus* biomass in biosorption of heavy metals from aqueous solutions. Microchemical Journal 141, 87–95.
Salem, H.M., Eweida, E.A., Farag, A., 2000. September. Heavy metals in drinking water and their environmental impact on human health. In *The International Conference For Environmental Hazards Mitigation, Cairo University Egypt* (pp. 542–656).
Sang, W., Xu, J., Bashir, M.H., Ali, S., 2018. Developmental responses of Cryptolaemus montrouzieri to heavy metals transferred across multi-trophic food chain. Chemosphere 205, 690–697.
Saravanan, A., Jayasree, R., Hemavathy, R.V., Jeevanantham, S., Hamsini, S., Yaashikaa, P.R., et al., 2019. Phytoremediation of Cr (VI) ion contaminated soil using Black gram (*Vigna mungo*): assessment of removal capacity. Journal of Environmental Chemical Engineering 7 (3), 103052.
Sarwar, N., Malhi, S.S., Zia, M.H., Naeem, A., Bibi, S., Farid, G., 2010. Role of mineral nutrition in minimizing cadmium accumulation by plants. Journal of the Science of Food and Agriculture 90 (6), 925–937.
Satoh, M., Kaji, T., Tohyama, C., 2003. Low dose exposure to Cadmium and its health effects (3) toxicity in laboratory animals and cultured cells. Nippon Eiseigaku Zasshi (Japanese Journal of Hygiene) 57 (4), 615–623.
Sharma, S., Nagpal, A.K., Kaur, I., 2018. Heavy metal contamination in soil, food crops and associated health risks for residents of Ropar wetland, Punjab, India and its environs. Food Chemistry 255, 15–22.
Shen, X., Chi, Y., Xiong, K., 2019. The effect of heavy metal contamination on humans and animals in the vicinity of a zinc smelting facility. PLoS One 14 (10), e0207423.
Sherlala, A.I.A., Raman, A.A.A., Bello, M.M., Asghar, A., 2018. A review of the applications of organo-functionalized magnetic graphene oxide nanocomposites for heavy metal adsorption. Chemosphere 193, 1004–1017.
Shyam, R., Puri, J.K., Kaur, H., Amutha, R., Kapila, A., 2013. Single and binary adsorption of heavy metals on fly ash samples from aqueous solution. Journal of Molecular Liquids 178, 31–36.
Singh, M.R., 2007. Impurities-heavy metals: IR perspective. http://www.usp.org/pdf/EN/meetings/asMeetingIndia/2008Session4track1.pdf.
Stefaniuk, M., Oleszczuk, P., Ok, Y.S., 2016. Review on nano zerovalent iron (nZVI): from synthesis to environmental applications. Chemical Engineering Journal 287, 618–632.
Suganya, S., 2019. An investigation of adsorption parameters on ZVI-AC nanocomposite in the displacement of Se (IV) ions through CCD analysis. Journal of Industrial and Engineering Chemistry 75, 211–223.
Sun, H., Xu, K., Lu, G., Lv, H., Liu, Z., 2014. Graphene-supported silver nanoparticles for pH-neutral electrocatalytic oxygen reduction. IEEE Transactions on Nanotechnology 13 (4), 789–794.
Tariq, M., Ali, M., Shah, Z.J.S.E., 2006. Characteristics of industrial effluents and their possible impacts on quality of underground water. Soil Environment 25 (1), 64–69.
Teh, C.Y., Budiman, P.M., Shak, K.P.Y., Wu, T.Y., 2016. Recent advancement of coagulation–flocculation and its application in wastewater treatment. Industrial & Engineering Chemistry Research 55 (16), 4363–4389.
Tušar, N.N., Maučec, D., Rangus, M., Arčon, I., Mazaj, M., Cotman, M., et al., 2012. Manganese functionalized silicate nanoparticles as a Fenton-type catalyst for water purification by advanced oxidation processes (AOP). Advanced Functional Materials 22 (4), 820–826.
Waisberg, M., Joseph, P., Hale, B., Beyersmann, D., 2003. Molecular and cellular mechanisms of cadmium carcinogenesis. Toxicology 192 (2–3), 95–117.

Weast, R.C., 1984. CRC Handbook of Chemistry and Physics, sixty-fourth ed. CRC Press, Boca Raton.

WHO, 2019. http://www.who.int/water_sanitation_health/publications/en/.

Widhiastuti, F., Lin, J.Y., Shih, Y.J., Huang, Y.H., 2018. Electrocoagulation of boron by electrochemically co-precipitated spinel ferrites. Chemical Engineering Journal 350, 893–901.

Wong, M.H., 2012. Environmental Contamination: Health Risks and Ecological Restoration. Taylor & Francis Group.

Wuana, R.A., Okieimen, F.E., 2011. Heavy metals in contaminated soils: a review of sources, chemistry, risks and best available strategies for remediation. International Scholarly Research Notices 2011.

Xu, J., Cao, Z., Zhang, Y., Yuan, Z., Lou, Z., Xu, X., et al., 2018. A review of functionalized carbon nanotubes and graphene for heavy metal adsorption from water: preparation, application, and mechanism. Chemosphere 195, 351–364.

Yang, X.J., Fane, A.G., MacNaughton, S., 2001. Removal and recovery of heavy metals from wastewaters by supported liquid membranes. Water Science and Technology 43 (2), 341–348.

Yola, M.L., Eren, T., Atar, N., 2014. A novel efficient photocatalyst based on TiO_2 nanoparticles involved boron enrichment waste for photocatalytic degradation of atrazine. Chemical Engineering Journal 250, 288–294.

Zazycki, M.A., Tanabe, E.H., Bertuol, D.A., Dotto, G.L., 2017. Adsorption of valuable metals from leachates of mobile phone wastes using biopolymers and activated carbon. Journal of Environmental Management 188, 18–25.

Zhou, D., Kim, D.G., Ko, S.O., 2015. Heavy metal adsorption with biogenic manganese oxides generated by *Pseudomonas putida* strain MnB1. Journal of Industrial and Engineering Chemistry 24, 132–139.

Zou, Y., Wang, X., Khan, A., Wang, P., Liu, Y., Alsaedi, A., et al., 2016. Environmental remediation and application of nanoscale zero-valent iron and its composites for the removal of heavy metal ions: a review. Environmental Science & Technology 50 (14), 7290–7304.

CHAPTER

15

Indigenous techniques to remove metals from contaminated water

Preetismita Borah[1], *Vaishali Sharma*[2], *Deepak Kashyap*[1], *Manish Kumar*[1] *and Biswa Mohan Sahoo*[3]

[1]CSIO-Central Scientific Instruments Organization, Chandigarh, Chandigarh, India [2]Panjab University, Chandigarh, Chandigarh, India [3]Roland Institute of Pharmaceutical Sciences, Berhampur, Odisha, India

15.1 Introduction

The environment is used as an amalgamated term for settings in which creatures live. It consists of air, water, land, sunlight and food, which are the elementary necessities of every living organism in order to process their life functions. By the way of explanation, the environment comprises of equally biotic in addition to abiotic components. It provides supportive surroundings in lieu of survival and improvement of corporeal existences at the present time. Altogether living creatures are primarily influenced openly or otherwise incidentally due to ecological adulteration or contagion. The pollution has presently become the global problem. Pollution is an unwanted and unwelcomed alteration in the physical, chemical as well as biological features of our air, water and land that is unsafe to humanoid life, industry based processes, living situations and cultural possessions (Malik and Saha, 2003).

Not more than 1% of the entire water on the globe's surface is accessible for drinking purposes and 40% rivers are polluted which indicates that after 10 years, water scarcity will be the gospel truth of life for most of the people (Supply et al., 2005). Water is life and it is at the heart of crisis. The tri-partite relation among the contamination, health and mother nature demands requisite understanding (Taghipour et al., 2019). Multiplication in population, haphazard industrialization and progression in anthropogenic activities servers' prime role in channelizing heavy metal into the water bodies as well as soil (Vareda et al., 2019). The water assessment gauges may perhaps be largely regarded as natural (bacteria, algae); physical (color, hotness, clearness, salinity, dissolved and suspended solids), chemical [acidity/basicity, dissolved oxygen (DO), biological oxygen demand

Metals in Water
DOI: https://doi.org/10.1016/B978-0-323-95919-3.00016-1

(BOD)], nutritional worth, noxious carbon-based/inorganic complexes, visual (odors, taste, color, moving matter) and radioactive (alpha, beta and gamma radiation emitters). Amounts of these signs can be used to observe alterations in water value, and decide whether it is appropriate for the well-being of the natural surroundings in addition to the applications for which the water is requisite (Sivaranjani and Rakshit, 2016).

The heavy metals possess larger atomic mass, metallic nature and approximately five times denser than water. Several heavy metals are non-degradable and hence their minute escape into the environment stays in circulation for long. Due to their abundance and persistence in water bodies, they tend to accumulate in the human and aquatic beings via water medium. At amplified concentrations beyond the acceptable maximum, these heavy metals result in lethal and prolonged ailments in humans and may also disturb the metabolism of animal and plant (Srivastava et al., 2017) (Table 15.1).

Hydrolysis, precipitation, ion-exchange, electrolytic tools, chemical abstraction, leaching, polymeric micro-encapsulation, adsorption, coagulation, photo-catalysis, precipitation, flocculation, fenton process, sedimentation, chlorination, reverse osmosis, membrane filtration, activated carbon technologies, photo electrochemical techniques, ozonation and ultraviolet light, sono-degradation and biodegradation etc. (physical/chemical methods) have not ascertained noteworthy heavy metal uptake. Vapor extraction, stabilization, solidification, soil washing and flushing, vitrification, incineration and thermal desorption have also been corroborated way back then. However, these are also quite high priced at macro scale and call for constant watch and control, many times even not fully eliminating the toxins (Borah et al., 2020).

Bioengineering is an innovative subdivision of civil manufacturing which incorporates alive constituents, predominantly floras and microorganisms, in order to sermonize the teething troubles of environmental administration and sustainable improvement. This technology was initiated in Germany in the middle of 1930s, but expanded its prominence near the end of 1980s, as soon as groundwork in the field of environmental bio-technology brought into light the environmental plusses of certain particularly improved plants and microorganisms. Bioengineering stands as the "green" or "soft" that is low priced substitute to the "hard" that is high priced civil engineering workings for environmental refurbishment. Therefore, green/bioremediation approaches could work as a substitute method for reversing the damage of heavy metals (Fig. 15.1).

TABLE 15.1 Guidelines given by World Health Organization for heavy metals uptake limit.

Heavy metal	Uptake limit (mg/L) (Mohod and Dhote, 2013)	Heavy metal	Uptake limit (mg/L) (Momodu and Anyakora, 2010)
Antimony (Sb)	0.005	Lead (Pb)	0.01
Arsenic (As)	0.05	Mercury (Hg)	0.001
Cadmium (Cd)	0.005	Nickel (Ni)	0.07
Chromium (Cr)	0.05	Silver (Ag)	—
Cobalt (Co)	0.9	Zinc (Zn)	4.0
Copper (Cu)	2.0	Selenium (Se)	0.04

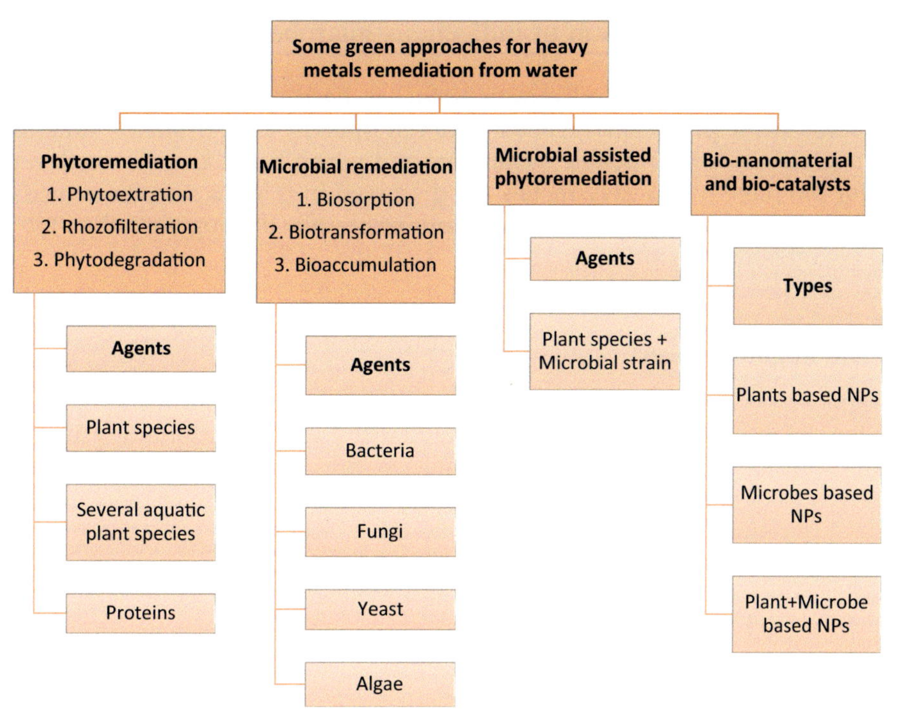

FIGURE 15.1 Some green approaches for heavy metal remediation from water.

15.1.1 Why to utilize indigenous techniques?

Indigenous refers to the species which are naturally growing, originating in or characteristics of inhabitant of particular region or country. Native materials have been utilized in curing water since olden times but dearth of knowledge on the meticulous nature and mechanism via which they work, has slowed down their widespread applications. As a result, they have been unable to race with the generally used chemicals. In latest years there has been a renaissance of interest in order to practice natural and native materials due to its budget, allied health and environmental disquiets of man-made organic polymers and inorganic chemicals (Kalibbala, 2007). Indigenous microorganisms and plant species usually finds application in on-site treatment whereas off site remediation prerequisites quarry of tainted soil and water from location trailed by handling at another site possibly by means of genetically engineered species. The morphological as well as biochemical reactions by microbes in the manifestation of metals may be cast-off in advancement of biosensors in order to identify heavy metals in foodstuff, liquid and topsoil. The exclusive individualities of genetically engineered beings in addition to the indigenous entities be able to utilize in ameliorating numerous metal contaminated spots.

Therefore, indigenous natural plants and microorganisms own high possibility in wiping out unwanted contaminates from polluted water while being easy on the pocket. In this chapter, the application of indigenous techniques which have been exploited in India in order to bio remediate dense metals out of the water is presented.

15.2 Potential solicitation of indigenous microbes in water remediation

Heavy metal acceptance in microorganisms has been comprehensively premeditated wide-reaching. Most of the indigenous bacterial species, fungal strains and their consortiums that stance to be probable contenders and which have been reported in previous writings for bio remedy of heavy metals are discussed in this chapter. Bio remedy of heavy metals can be accomplished via diverse approaches like bio augmentation, bio stimulation, and bio attenuation.

15.2.1 Bio augmentation

Bio augmentation is a methodology relevant for bio-removal of heavy metal polluted spots by the aid of particular microbes or hereditarily fixed up bacteria that are efficient in scrimmage with the specific heavy metal pollutant. Bio augmentation and bio stimulation are suggested to be the most adequate tactics for biocleaning of metal contaminated locations. The application of consortium of diverse strains over the single strain provides more favorable outputs. Bio augmentation is an extremely effective approach with sustainability and superior substrate specificity. It is an in situ method and therefore pretentious to different biotic as well as abiotic factors.

15.2.2 Bio attenuation

Bio attenuation involving the indigenous microbial population in the natural remediation of heavy metal pollutants is known as natural attenuation. This approach is tedious and economical sometimes needs to be accelerated by using bio stimulation or bio augmentation. The research group has isolated three indigenous bacterial strains from discharges of paper and pulp industry for the heavy metals remediation (Sharma and Bhattacharya, 2017). Out of the three, Pseudomonas sp. was effective in removal of manganese cadmium and mercury. However, *Staphylococcus* sp. and *Streptococcus* sp. were capable in removing copper. Another research group has exploited the indigenous bacterial strains *Lactobacillus* sp. and *Bacillus* sp. which were also out-of-the-way from the pulp and paper treating effluents which can tolerate the Cr (VI) (Das et al., 2018).

15.2.3 Bio stimulation

Bio stimulation is the method where metal tolerance potential of indigenous microbe is stimulated by using suitable conditions thus enhancing bioremediation. The addition of stimulants such as nutrients such as nitrogen, phosphorus, oxygen and carbon heightens growth of bacterial strains in soil for biodegradation (Kavitha et al., 2013).

The research group scrutinizes bioremediation probability of bio stimulated microbial culture secluded out of the heavy metals waste dumping tainted spot sited at Bhayander (east), Mumbai, India (Ali et al., 2019). The bio stimulated microbial consortium has been bring into being operational for the treatment of Cd, Cu, and Fe at elevated levels, that is,

100 mg/L up to 98.5%, 99.6%, and 100%, one-to-one. Fe present as a micro nutrient was remitted entirely in relation to Cu and Cd.

15.2.4 Discussion

The literature of the indigenous microbial strains which are indigenous to India and have been employed for remedying heavy metals is discussed. The detailed discussion includes—site of the isolation of strains and the optimum conditions for instance pH conditions, temperature, initial levels and adsorption efficiency.

The researchers procured samples from heavy metal adulterated industrialized area in Delhi national capital region (NCR) (Sadhasivam et al., 2020). Out of 118 indigenous bacterial strains, *Enterobacter cloacae* established on phylogenetic investigation of 16S rDNA sequence deciphered to be best arsenic resistive microbe. It could tolerate concentrations of up to 6000 ppm of arsenic. Additionally, it has multi-resistance nature toward numerous heavy metals for instance Cd, Zn, Se and Ni boosts desirability and finds applicability in developing an in situ bioremediation technology.

In this work, the research group collected specimens of industrial effluent, sludge of sewage management plants and ponds (Talukdar et al., 2020). Three fungal segregates viz., *Aspergillus flavus* (Cr (VI) tolerant), *Aspergillus fumigatus* (Cd (II)) tolerant), and *A. fumigatus* (Cr (VI) tolerant) isolated from different district of Punjab, Haryana (Karnal, Mullana, Faridabad, Sonepat, Panipat and Yamunanagar), Chandigarh, Jammu and Kashmir, Assam and Delhi were harnessed independently and as consortium for their metal tolerance capacity. By means of the fungi namely *A. flavus* (FS4) and *A. fumigatus* (FS6), in the aqueous phase around 70% of exclusion for Cr (VI) was seen. Also, up to 74% of elimination is exhibited by the Cd (II) accepting fungal strain FS9, which is acknowledged by the title of *A. fumigatus*. The finding of this work has evicted the remarkable adsorption capacity for even minute amount of heavy metals and encompassed novel insight into the bioremediation utilizing filamentous fungal.

Arsenic (As III) enduring strains of fungi were secluded and screened from topsoil specimens procured from numerous places of industrialized clearance situates of Davangere District, Karnataka, India (Frankel and Bazylinski, 2003). Out of five fungal isolates, *Aspergillus* spp. 1 (APR-1) and *Aspergillus* spp. 1 (APR-2) displayed better confrontation towards As-III at the test site settings. With the purpose to enlarge the superficial capacity of for biosorption, APR-1 and APR-2 isolates were restrained on sponge gourd *Lufa aegyptiaca* which is an agricultural unwanted surplus performing as biosorbent. Thru 250 mM arsenic solution, APR-1 and APR-2 indicated biosorption of 53.94% and 52.54%, one-to-one, on inductively coupled plasma-optical emission spectrometry analysis (ICP-OES) and the sorption of the fungal separates on sponge gourd was established via scanning electron microscopy (SEM). APR-1 isolate was additionally characterized by 18s rDNA typing and recognized as *Aspergillus niger*. Manufacturing scale solicitation may imaginably be occupied by means of carefully chosen fungal strains and lessens the arsenic effluence in city sewage.

In the another study, research group has isolated cadmium tolerant strain of 58 bacterial isolates from 26 diverse tainted liquid specimens procured out of 20 villages in addition to some cities positioned nearby manufacturing and mining pretentious regions of Chhattisgarh

(India) (Bazylinski and Frankel, 2000). Out of 58 bacterial isolates, finally five certain isolates (BSWC3, RgCWC2, RgUWC1, RpSWC3, KDWC1) were recognized by 16S rRNA gene sequencing belonged to the genus *Serratia liquefaciens, Klebsiella quasipneumoniae* subsp., *similipneumoniae, Klebsiella pneumoniae, Pantoea dispersa* and *Enterobacter tabaci,* respectively. Amongst these two best culture *S. liquefaciens* and *K. pneumoniae* were verified for their biocleaning efficacy independently in addition to miscellaneous culture. Cadmium (Cd) tolerance of bacterial isolates were significant in controlling and reducing Cd concentration by 44.46%, 40% and 50.92%, respectively. Therefore, the finding of this study discovered the usage of miscellaneous group or consortium of home-grown separates which are the healthier alternative for biocleaning of heavy metals.

In a study conducted by group, assessed the biocleaning ability of native bacteria sequestered from two dissimilar tannery tainted waters poised from Kanpur and Chennai (Bazylinski and Frankel, 2000). Among all the separates, *Citrobacter freundii* was capable to decrease the level of toxic agents for instance chromium and sulfate by 73% and 68% individually. The carbon-based contents described by BOD and chemical oxygen demand (COD) also curtailed by 86% and 80%, correspondingly. Findings suggested the growth of add-ons of *Artemia nauplii* in the remediated wastewater which thus additionally established decline in noxiousness of the contaminants.

In another study, wastewater illustrations were procured from a steel industry sited neighboring to the Rohtak city of Haryana, India (Sharma et al., 2019). Isolated fungal strains, *A. awamori, A. flavus and A. niger* were found to be resistant to Cu(II) and Ni(II) and be contingent on native strains confirmed, its spot of seclusion, metal beneath concern and its level in the medium.

The research group isolated total of 48 morphologically divergent arsenite enduring bacteria from central Gangetic plain (Li and Pan, 2012). Two isolates namely AK1 and AK9 belonging to genus Pseudomonas were competent in oxidizing As (III) in addition to reducing As (V) at high rate. Likewise, procured isolate presented hefty metal confrontation alongside Ag (I), Cr (IV), Ni (II), Co (II), Pb (II), Cu (II), Hg (II) and Cd (II).

The findings of another research group provided evidence insights into As mobilization by aerobic *Pseudomonas aeruginosa* bacterium in dominating and decreasing arsenate in the sub-oxic groundwater situations in Brahmaputra floodplain, India (Byrne et al., 2015). Thus reductive dissolution of Fe oxy-hydroxides together with the bio-geochemical round of As (III) and As (V) facts the occurrence of sub-oxic to decreasing alluvial aquifer settings.

The researchers investigated strong endurance headed for lead (Pb) by altogether four separates which could be accredited to the existence of Pb within the soil/water from whichever they were partitioned (Moon et al., 2010). Some unit of endurance was also presented in the direction of arsenic, symptomatic of their capability to endure and familiarize under strained circumstances. Washed live biomass indicated around 90% elimination efficacy for lead (II) solution.

A cadmium probable bacterium B. cereus was isolated by a group from the electroplating trade at Coimbatore, Tamil Nadu, India. Under the precise conditions of; primary pH 6.0, temperature 35°C with preliminary metal levels of 200 mg/L Cd (II), the 82% removal of Cd (II) was observed (Ahmad et al., 2003).

In an another study by Rajaram research group isolated 12 resistant from Cuddalore coastal zone, Tamil Nadu, India; four isolates displayed great resistance and better absorption capacity for copper (Rajaram et al., 2013). The isolates namely *Halococcus* sp., and *Microbacterium* sp. presented supreme opposition equal to 80% copper and *Staphylococcus* sp. and *Bacillus* sp. were removing up to 99.9% copper. Similarly, in another study, strain B9 of *Acinetobacter* sp. was secluded from the waste water of native discharge management plant sited in New Delhi, India (Parial and Pal, 2015). Significance of the strain for handling of heavy metals opulent industrial wastewater ensued in 93.7%, 55.4%, and 68.94% elimination of initial 30 mg/L Cr (VI), 246 mg/L total Cr (VI), and 51 mg/L Ni (II), respectively, subsequently 144 h of management in a batch manner. Heavy metal take-up strength of bacterial strain isolated from industry discharge spots of river Nagavali, Srikakulam Andhra Pradesh was conducted by the research group (Qu et al., 2017). *Staphylococcus saprophyticus* showed more resistance toward Cu metal whereas both *Staphylococus aures* and *Staphylococus epidermidis* displayed supreme resistance to cobalt, mercury and zinc metals.

The biosorption of Cr (VI) using *Acinetobacter junii*, native bacteria sequestered out of chromite mine situates of Sukinda valley, Orissa, India was reported by (Patil and Chandrasekaran, 2020). The findings discovered the lessening of Cr (VI) to Cr (V) taking place at the surface of the biomass upon Cr (VI) absorption at most favorable sorption settings (pH: 2.0, interaction time: 120 min, temperature: 27°C, preliminary Cr (VI) concentration: 100 mg/L, biosorbent amount: 2 g/L).

In another work, liquid/ solid waste specimens were procured away from landfills and production sites of Doon Valley, Uttarakhand (Kumar et al., 2010). The acclimated microorganisms viz. *Pseudomonas* sp. and *Bacillus* sp. minimized Cu and Ni concentrations respectively. *A. niger* reduced Cd and Zn whereas *Staphylococcus* sp. reduced Cr, Cu and Pb. The outcomes disclosed that *Pseudomonas* sp. removed heavy metals as compared to other microbes on the other hand *Staphylococcus* sp. removed 93% of lead which was surprising (Table 15.2).

15.3 Potential solicitation of indigenous plants in water remediation

Phytoremediation (Greek term *phyton* referring plant; Latin term *remediare* referring remedy) is emanating "bioengineering technique" which employ floras in order to cleanup environmental glitches. A large numeral of indigenous green trees, plants, grasses, herbs, and shrubs, both marine and earthly, have been revealed to have been gifted by means of the extraordinary and phenomenal possessions for environmental restoration. These plants are appealingly pleasing, acquiescent, sunlight driven, pollution remitting green technology and tolerant to very tough environmental settings working for the same ideas as that of other conventional technology. There are numeral variety of herbal kinds which possess the capability to take up considerably greater amounts of heavy metals in various portions of the body, for instance a leaf, stems and root, without displaying any indication of noxiousness (Bennett et al., 2003; Reeves et al., 2018). The plants perform mutually as "collectors," and "excluders." Accumulators stay alive even if their aerial tissues are concentrated with contaminants. They bio-degrade or bio-transform the toxins into inactive types in their tissues. The excluders limit pollutant acceptance into their biomass. Plant biomass ultimately turn into the valued biological source for the plant based industries or community.

TABLE 15.2 List of potential microbial strains available in literature which are indigenous to India, the site from where these strains are isolated and the heavy metals which can be tolerated are shown in this tabular representation.

Microbial strain/consortia		Site of isolation of strain	Heavy metal tolerated	References
Bacterium	Enterobacter cloacae	Industrial area in Delhi NCR	Cd, Zn, Se and Ni It can tolerated As upto 6000 ppm	Bhati et al. (2021)
Bacterium	Miscellaneous indigenous strains *Serratia liquefaciens* BSWC3 and *Klebsiella pneumoniae* RpSWC3	20 villages/cities of diverse polluted water illustrations from manufacturing and mining pretentious regions of Chhattisgarh (India)	*Serratia liquefaciens*, Klebsiella pneumonia and their mixed culture can reduce Cd concentration by 44.46%, 40% and 50.92%, respectively	Kumar et al. (2019)
Bacterium	*Citrobacter freundii*	Tannery wastewaters collected from Kanpur and Chennai	*Citrobacter freundii* decreased the levels of toxicants for instance chromium and sulfate by 73% and 68% respectively	Vijayaraj et al. (2018)
Bacterium	AK1 and AK9 fitting to genus *Pseudomonas*	Middle Gangetic plain, Bihar, India	Displayed heavy metal confrontation against Cr (IV), Ni (II), Co (II), Pb (II), Cu (II), Hg (II), Ag (I) and Cd (II) Oxidizes As (III) and reduce As (V) at high rate	Satyapal et al. (2018)
Bacteriium	Aerobic *Pseudomonas aeruginosa*	Groundwater situations in Brahmaputra floodplain, India	As (III) and As (V) immobilization	Sathe et al. (2018)
Bacterium	*B. cereus*	Electroplating industry at Coimbatore, Tamil Nadu, India	82% removal of Cd (II) was observed with ideal conditions; initial pH 6.0, temperature 35°C with initial metal concentration 200 mg/L Cd (II)	Arivalagan et al. (2014)
Bacterium	*Halococcus* sp., *Microbacterium* sp., *Staphylococcus* sp. and *Bacillus* sp.	Cuddalore coastal area, Tamil Nadu, India	*Halococcus* sp., and *Microbacterium* sp. indicated supreme confrontation up to 80% copper and *Staphylococcus* sp. and *Bacillus* sp. were resistant up to 99.9% copper	Rajaram et al. (2013)
Bacterium	Strain B9 of *Acinetobacter* sp.	Wastewater of local common effluent treatment plant situated in New Delhi, India	93.7%, 55.4%, and 68.94% removal of initial 30 mg/L Cr (VI), 246 mg/L total Cr (VI), and 51 mg/L Ni (II), respectively, after 144 h of treatment	Bhattacharya and Gupta (2013)

(Continued)

TABLE 15.2 (Continued)

Microbial strain/consortia		Site of isolation of strain	Heavy metal tolerated	References
Bacterial	*Staphylococcus saprophyticus, Staphylococus aures* and *Staphylococus epidermidis*	Industry effluent sites of river Nagavali, Srikakulam Andhra Pradesh	*Staphylococcus saprophyticus* showed resistance toward Cu metal whereas *Staphylococus aures* and *Staphylococus epidermidis* showed removal of cobalt, mercury and zinc metals	Bisht et al. (2012)
Bacterial	*Acinetobacter junii*	Chromite mine spots of Sukinda valley, Orissa, India	Biosorption of Cr (VI) and then reduction of Cr (VI) to Cr (V)	Paul et al. (2012)
Bacterial	*Pseudomonas sp., Bacillus sp., Aspergillus niger* and *Staphylococcus sp.*	Industries of Doon Valley, Uttarakhand	*Pseudomonas* sp. and *Bacillus* sp. reduced Cu and Ni respectively. *Aspergillus niger* reduced Cd and Zn whereas *Staphylococcus* sp. reduced Cr, Cu and 93% of Pb	Kumar et al. (2010)
Fungal	Four live biomasses (Isolate-1, Isolate-2, Isolate-3 and Isolate-4)	Dumpsite existing amidst an industrial area (Mangalore, India)	90% removal efficiency for lead (II) solution and some degree of tolerance was also shown towards arsenic	Gururajan and Belur (2018)
Consortium of Fungi	*A. flavus, A. fumigatus,* and *A. fumigatus*	Different district of Punjab, Haryana (Karnal, Mullana, Faridabad, Sonepat, Panipat and Yamunanagar), Chandigarh, Jammu and Kashmir, Assam and Delhi	*Aspergillus flavus* (FS4) and *Aspergillus fumigatus* (FS6) for the removal of 70% of Cr (VI) *Aspergillus fumigatus* (FS9) for the removal of 74% of Cd (II)	Talukdar et al. (2020)
Fungal	*Aspergillus awamori, Aspergillus flavus* and *Aspergillus niger*	Steel industry situated close to the Rohtak city of Haryana, India	Resistant to Cu (II) and Ni (II).	Rose and Devi (2018)
Fungal	*Aspergillus* spp. *1* (APR-1), *Aspergillus* spp. *1* (APR-2) and agro-waste *Lufa aegyptiaca* as a biosorbent	Industrialized disposal situates of Davangere District, Karnataka, India	APR-1 and APR-2 showed biosorption of 53.94% and 52.54% As (III) respectively	Tanvi et al. (2020)

Phytoremediation is a green technique known by various terms such as agro based treatment, green treatment, vegetative treatment and botanic treatment (Sarwar et al., 2017; Kushwaha et al., 2018). Plants basedremedial technology works largely through different ways.

15.3.1 Phytoextraction and phytoaccumulation

Herbal extracts metal toxins from the polluted soil and the tainted water, and gather them in their roots. These roots can captivate organics in addition to inorganics toxins. The bio-availability of a certain compounds be governed by the lipophilicity and the top-soil or water settings for instance pH and clay content. Significant volume of the pollutants may be migrated directly overhead via the xylem and gathered in the shoots and leaves of the plants. The roots, shoots and leaves are reaped and cremated to decay the pollutants. Occasionally phytoremediation and phytoextraction are accustomed interchangeably, which is a mistaken belief; phytoextraction is a cleaning technique however phytoremediation is the term used for a notion (Prasad et al., 2005). Phytoextraction is an apt phytoremediation technique cast-off for the removal of heavy metals from wastewater, deposits and soil (Kocoń and Jurga, 2017).

15.3.2 Phytodegradation

The pollutants are absorbed and broken down by some plant species. This work is accomplished with the help of enzyme catalyzed metabolic procedure occurring in root or shoot cells of the plant. Other plants breakdown the pollutants into the substrates on its own by discharging enzymes and chemical complexes. The enzymes oxygenases, dehydrogenases and reductases are usually secreted. The bio-degraded components are transformed towards inexplicable and inactive constituents which are stockpiled in the lignin or freed as exudates. Certain plants degrades toxins using the support of microbes that animate in symbiotic connotation on their roots (Trapp and Karlson, 2001).

15.3.3 Phytostabilisation

Definite plant species detain noxious waste in the land and water via sorption thru and sorption taking place on the roots or precipitation in the interior the root sector viz. rhizosphere.

15.3.4 Phytovolatilisation

Plants uses aerial organs in order to absorb and leak out the impurities from soil and water. Certain contaminants resembling mercury (Hg), selenium (Se) in addition to volatile organic compounds (VOCs), can be allowed thru the leaves towards the environment (Cunningham and Ow, 1996).

15.3.5 Phytotransformation

Numerous inorganic and organic pollutants after getting absorbed inside the root, might turn out to be bio-chemically bound toward cellular materials, in the arrangements which are lesser active or inert in nature (Trapp and Karlson, 2001).

15.3.6 Rhizofiltration

It is established on an alliance of norms of phytoremoval as well as phytostabilization specifically suitable to remediate metals and radio-nuclides out of contaminated water. Pollutants are taken up and gathered by plant roots, and therefore bring on as their carbonates and phosphates. Hydroponically grownup terrestrial plants similar to sunflower (*Helianthus annuus*) and vetiver (*Vetiveria zizanioides*) which possess huge root schemes and bigger biomass, are especially appropriate. Species which do not willingly relocate toxins on or after the roots to stem are chosen, meanwhile the gathered metals and radionuclides can be distant via modestly reaping the roots. Rhizofiltration mechanize in the efficient exclusion of organics such as trichloroethylene, tetrachloroethane, metachlor, dioxins, atrazine, nitrotoluenesanilines, and numerous petroleum hydrocarbons (Srivastava and Giri, 2021).

15.3.7 Discussion

The literature of the indigenous plant species which are native to India and have been exploited for the remediation of heavy metals remediation in waste water is discoursed below. The site from where the plants species were procured, the translocation factor and adsorption efficiency displayed for the acceptance of particular heavy metals has been discussed in detail.

The research group explored the bioaccumulation endurance of 23 plant species, in the neighborhood of ponds at Sahlon: site 1, Chahal Khurd; site 2, and Karnana; site 3, in Shaheed Bhagat Singh Nagar, Punjab, India (Parihar et al., 2021). The highest bio accumulation factor for Zn, Cd, Co, Fe, Cu, Mn, and Cr was displayed in *F. infectoria* (shoots), *D. sissoo* (shoots), *M. polymorpha* (shoots), *F. religiosa* (shoots), *A. conyzoides* (shoots), *A. viridis* (shoots), and *M. rotundifolia* (roots) respectively.

The naturally budding flora in a native wastewater recipient lake, Laxmi Taal, in Jhansi was studied by Pandey and his group (Pandey et al., 2019). Amongst the six dense metals deliberated, Mn, Cu, Zn, Ni, Pb and Fe, the declining arrangement in *T. angustifolia* were resulted to be Ni > Pb > Fe > Zn > Cu in the root scheme, and in the shoot, gathering of heavy metals was in the sequence of Fe > Ni > Pb > Zn > Cu however in *E. crassipus* levels in descendent sequence of Fe > Pb > Zn > Ni > Cu was perceived in the root scheme and Ni > Fe > Zn > Pb > Cu in the shoot scheme. Heavy metals movement from root to shoot was operational. The heavy metal learning discovered that uptake of numerous metal ions by *Typha angustifolia* and *Echhornia crassipus* was advanced in shoots as compared to further portions of the plant. For all heavy metals, rhizofiltration and phytoextraction processes have been established to be operative. Phytoremediation by *T. angustifolia* and *E. crassipus* can be utilized effectually to support safeguard of water bodies are freed from noxious heavy metals discarded into waterchannels.

The research group collected *T. cordifolia* biomass from the diverse areas of Chhattisgarh in India (Sao et al., 2017). Even at a much elevated levels of 450 mg/L, the biomass displayed nearby 92% removal of lead. *T. cordifolia* turned out to be a remarkable bio-adsorbent for the Pb (II) elimination from polluted water. The optimum conditions viz. workable pH was 4.0 with a contact time of 60 min at room temperature.

The study of phytoextraction patterns in 15 enduring indigenous plants growing on sludge samples procured from Unnao, located in Uttar Pradesh revealed that the *Parthenium hysterophorous, Blumea lacera, Setaria viridis, Cannabis sativa, Basella alba, Chenopodium album, Tricosanthes dioica, Achyranthes* sp., *Amaranthus spinosus L., Dhatura stramonium, Croton bonplandianum* and *Sacchrum munja* were prominent as root collector for Fe, Zn and Mn despite the fact that *P. hysterophorous, S. munja, C. sativa, T. dioica, C. album, D. stramonium, B. alba, B. lacera, Achyranthes* sp. and *Kalanchoe pinnata* were noticed as shoot gatherer for Fe. Furthermore, *A. spinosus L.* was perceived as shoot collector for Zn and Mn. Likewise, entirely all plants were noticed as leaf collector for Fe, Zn and Mn excluding *A. spinosus L.* and *Ricinus communis* (Chandra and Kumar, 2017).

Mature Aloe Vera leaves which belongs to the family of *Liliaceae* were procured from different locations in northern region of India and these were exploited by a group for their remediation properties (Malik et al., 2015). The minimal amount of the adsorbent could remove 74.6% of Pb in 30 min from a solution. Adsorption capacity was also increased up to 96.2% by amending the aloe vera and treating it with H_3PO_4. The adsorption uninterruptedly maximized in the pH range of 1.0–4.0.

Three plant types namely *Typha latifolia, Eichornia crassipes* and *Monochoria hastate* were seen budding in the samples poised from tainted water reservoirs across Ranchi, India (Hazra et al., 2015). *M. hastata* has the supreme bio-concentration factor (BCF) for root equals to 4.32 as well as shoot equals to 2.70 (for Manganese). For *T. latifolia*, BCF of maximum was perceived for root (163.5) and corresponding shoot 86.46 (for Iron), trailed by 7.3 and 5.8 for root and shoot (for Manganese) individually. *E. crassipes* was ensued to own an extreme BCF of 278.6 (for Manganese) and 151 (for Iron) and shoot as 142 (for Manganese) and 36.13 (for Iron).

Another study focused on assessment of indigenous plant species growing in National Thermal Power Corporation (NTPC), Kahalgaon, Bihar, India. Finding of the study evidenced *Typha latifolia* as effectual metal wager of Fe (927), Cu (58), Zn (87), Ni (57), Al (67), Cd (95), and Pb (69) and *Azolla pinnata* was found to be Cr (93) higher collector amongst marine species in μg/g. *Croton bonplandium* exhibited the concentrated levels of Fe (998), Zn (81), Ni (93), Al (121) and Si (156) in terrestrial species (Kumari et al., 2016).

In a laboratory based study, *Withania somnifera* has been a significant herb which develops as weeds in the inhospitable surroundings and receives polluted water from community and industrial sources. *W. somnifera* plants endured concentration of As (III) and As (V) that were higher than generally existing in polluted regions and these plants stored high amount of arsenic in roots followed by leaves (Siddiqui et al., 2015). Consequently, *W. somnifera* can grow in arsenic contaminated surroundings for heavy metal remediation tenacities under stringent regulation. Though, the usage of arsenic holding plants in medicinal research is not recommended due to its lethal effect and health risk.

The five indigenous macrophytes developing certainly in a sewer accepting tannery emission namely *Eichhornia crassipes, Bacopa monnieri, Hydrilla verticillata, Marsilea minuta,* and *Ipomoea aquatica* were assessed for their heavy metal accumulation potential in field situations at Unnao, U.P., India (Kumar et al., 2012). According to the finding, Eichhornia crassipes was acclaimed for phytoremediation of water bodies polluted by copper and nickel however Bacopa monnieri and Hydrilla verticillata were capable of removing of chromium and lead respectively.

Two species of *Portulaca* plants gathered from two ground spots in Vadodara, Gujarat, India; grows at location contaminated with multi-metals and moreover revealed speedy accretion of heavy metals and their proficient conveyance to shoots (Dwivedi et al., 2012). *Portulaca oleracea* observed to be superior collector classes for numerous metals than *P. tuberosa* and as a result, it is efficaciously engaged in phytoremediation programs.

In a study, specimens of topsoil and wastewater from the well-known situations were poised from manufacturing spots of Okhla, New Delhi, India (Ahmad et al., 2011). Amongst water plants *Hydrilla verticillata, Marsilea quadrifolia,* and *Ipomea aquatica* have shown utmost metals gathering likelihood, *Eclipta alba* and *Sesbania cannabina* amongst terrestrial plant were supreme accumulator of metals. Amongst the algal strain *Spirulina platensis* and *Phormidium papyraceum* were the best operational in accruing cadmium and mercury. The superior bio-concentration aspect was noted in *Hygroryza aristata* for the metals mercury and cadmium, in *M. quadrifolia* for the metals cadmium and chromium, in *E. alba* for the metals chromium and copper, and in *S. platensis* for the metals mercury and lead.

In a study of aquatic macrophytes, seven species were procured from Pariyej Community Reserve, Gujarat, India (Kumar et al., 2008). *Ipomoea aquatica, Eichhornia crassipes Solms, Typha angustata Bory & Chaub, Echinochloa colonum, Hydrilla verticillata, Nelumbo nucifera Gaerth* and *Vallisneria spiralis* alongside with surface sediments and water, were examined for Cd, Co, Cu, Ni, Pb and Zn pollution. Conclusively, *T. angrustata, E. crassipes* and *I. aquaitca* stored heavy metals in plentiful levels. The accumulation of heavy metals settled in the direction as Zn > Cu > Pb > Ni > Co > Cd and component systems of aquatic macrophytes have gathering capability of heavy metals for instance Sediment > Root scheme > Stem scheme > Leaf scheme.

In order to remove As (III) from aqueous solutions, an indigenous plant namely *Garcinia cambogia* found in many parts of India was investigated (Kumar et al., 2008). The pH array of 6–8 was brought into being to deliver most favorable As (III) removal. The presence of interference ions for instance calcium and magnesium at levels up to 100 mg/L and iron at 10 mg/L displayed no noteworthy influence on As (III) elimination.

The role of planktons and macrophytes were studied in phytoremediation of heavy metals from the Nainital lake U.P., India. Water of the lake is polluted with metals such as Fe, Cr, Ni, Cu, Mn, Pb and Zn. Permissible levels of a number of them such as Pb, Fe and Ni were greater in contrast to the suggested highest permitted confines (Ali et al., 1999). Plants and algae emerging in that lake mount up significant amount of metals in addition to it, water roots of Salix were extra effectual as compared to others. Extraordinary metal eliminating possibility of the above said plants would turn out be noteworthy for bio monitoring studies and may possibly be a used as phytoremediation technology in refurbishing water quality (Table 15.3).

Taking into account the numerous research examinations, microorganisms particularly bacteria be appropriate to different genera are established to potentially remediate toxic heavy metals under different bioprocesses. Their environmental compatibility and economic aspects are of foremost interest for their utilization in bio treatment. Additional advantage of bio remedial techniques is that they assure to engage smoothly next to minor levels of heavy metals, where conventional practices prove inadequate. Microbial-assisted phytoremediation practices are likewise testified to be proficient of summoning and restraining the heavy metals underneath various contrivances. Beneath several microbial

TABLE 15.3 List of potential plant species available in literature which are indigenous to India, the site from where these species are isolated and the heavy metals which can be accumulated in order to remediate the waste water are shown in this tabular representation.

Plant species	Site/location	Heavy metal tolerance	References
F. infectoria, D. sissoo, M. polymorpha, F. religiosa, A. conyzoides, A. viridis, and *M. rotundifolia*	Vicinity of ponds at Sahlon: site 1, Chahal Khurd; site 2, and Karnana; site 3, in Shaheed Bhagat Singh Nagar, Punjab, India	Highest bio accumulation factor for Zn, Cd, Co, Fe, Cu, Mn, and Cr was shown in *F. infectoria* (shoots), *D. sissoo* (shoots), *M. polymorpha* (shoots), *F. religiosa* (shoots)M *A. conyzoides* (shoots), *A. viridis* (shoots), and *M. rotundifolia* (roots) one-to-one	Parihar et al. (2021)
T. angustifolia, E. crassipus	Domestic wastewater receiving lake, Laxmi Taal, in Jhansi	The heavy metal accumulation order for *T. angustifolia,* in root system: Ni> Pb> Fe> Zn> Cu and in shoot system: Fe> Ni> Pb> Zn> Cu; and for *E. crassipus,* in root system: Fe> Pb> Zn> Ni> Cu and in shoot system: Ni> Fe> Zn> Pb> Cu respectively	Pandey et al. (2019)
T. cordifolia biomass	Diverse areas of Chhattisgarh in India	Nearby 92% removal of lead and at optimum conditions are workable pH = 4.0, contact time = 60 minutes, temperature = 25°C	Sao et al. (2017)
15 potential indigenous plant species namely *Parthenium hysterophorous, Blumea lacera, Setaria viridis, Cannabis sativa, Basella alba, Chenopodium album, Tricosanthes dioica, Achyranthes sp., Amaranthus spinosus* L., *Dhatura stramonium, Croton bonplandianum, Sacchrum munja, P. hysterophorous, S. munja, C. sativa, T. dioica, C. album, D. stramonium, B. alba, B. lacera, Achyranthes* sp., *Kalanchoe pinnata,* and *A. spinosus* L.	Sludge samples procured from Unnao, located in Uttar Pradesh	*Parthenium hysterophorous, Blumea lacera, Setaria viridis, Cannabis sativa, Basella alba, Chenopodium album, Tricosanthes dioica, Achyranthes sp., Amaranthus spinosus* L., *Dhatura stramonium, Croton bonplandianum* and *Sacchrum munja* (root collector) for Fe, Mn and Zn whereas *P. hysterophorous, S. munja, C. sativa, T. dioica, C. album, D. stramonium, B. alba, B. lacera, Achyranthes sp.* and *Kalanchoe pinnata* (shoot collector) for Fe. Furthermore, *A. spinosus L.* (shoot collector) for Mn and Zn. Likewise, all plants actions as leaf collector excepting *A. spinosus L.* and *Ricinus communis*	Chandra and Kumar (2017)
Mature Aloe Vera leaves belonging to the family of *Liliaceae*	Different locations in northern region of India	Removes 74.6% of Pb in 30 min at pH range of 1.0 to 4.0	Malik et al. (2015)
Three plant species namely *Typha latifolia, Eichornia crassipes* and *Monochoria hastate*	Contaminated water bodies across Ranchi, India	*M. hastata* has the maximum bio-concentration factor for root equals 4.32 and shoot equals 2.70 (Mn). For *T. latifolia,* BCF for root 163.5 and shoot 86.46 (Fe), 7.3 and 5.8 on behalf of root plus shoot (Mn) respectively. *E. crassipes* hold a supreme BCF for root 278.6 (Mn) and 151 (Fe) in addition to shoot as 142 (Mn) and 36.13 (Fe)	Hazra et al. (2015)

T. latifolia, A. pinnata and *C. bonplandium*	National Thermal Power Corporation (NTPC), Kahalgaon, Bihar, India	*T. latifolia* as Fe (927), Ni (57), Cu (58), Cd (95), Zn (87), Al (67), and Pb (69) and *A. pinnata* as Cr (93) active collector amongst water species in μg/g. *C. bonplandium* as Fe (998), Zn (81), Al (121), Ni (93) and Si (156) accumulator in terrestrial species	Kumari et al. (2016)
Withania somnifera	Laboratory based study	As (III) and As (V) accumulators	Siddiqui et al. (2015)
E. crassipes, B. monnieri, H. verticillata, M. minuta, and *I. aquatica*	Field situates at Unnao, U.P., India	Eichhornia crassipes for phytoremediation of water bodies polluted by Cu (II) and Ni (II). However *Bacopa monnieri* and *Hydrilla verticillata* were used to eliminate Cr (VI) and Pb (II) respectively	Kumar et al. (2012)
Two species of Portulaca plants namely *P. oleracea* and *P. tuberosa*	Two field sites in Vadodara, Gujarat, India	Multi metal accumulation	Dwivedi et al. (2012)
Aquatic plant species- *H. verticillata, M. quadrifolia,* and *I. aquatic*; terrestrial plant species- *E. alba* and *S. cannabina* and algal strains- *S. platensis* and *P. papyraceum*	Okhla industrialized sites, Okhla, New Delhi, India	Cd, Hg, Pb, Cr and Cu	Ahmad et al. (2011)
Seven aquatic macrophytes species namely *I. aquatica, E. crassipes Solms, T. angustata Bory & Chaub, E. colonum, H. verticillata, N. nucifera Gaerth* and *V. spiralis*	Pariyej Community Reserve, Gujarat, India	The acceptance of heavy metals is in the direction as Zn > Cu > Pb > Ni > Co > Cd and component systems of aquatic macrophytes have acceptance capacity of heavy metals as Sediment > Root system > Stem system > Leaf system	Kumar et al. (2008)
Garcinia cambogia	Found in many parts of India	As (III) removal at pH array of 6–8	Kamala et al. (2005)
planktons and macrophytes	Nainital lake U.P., India	Fe, Cr, Cu, Mu, Pb, Ni, and Zn	Ali et al. (1999)

methodologies, selection of the well-organized remediation procedure is prepared depending upon levels of heavy metals at the contaminated spots, whole remediating potential as well as applicability, plant flexibility, and consistency.

In contrast to conventional techniques, utmost biologically mediated processes are generally protracted and complex. Low inevitability often declines trustworthiness of their presentation in the field. Indigenous physicochemical conditions generally influence the native microbial strains at altered contaminated sites, which poses diverse effects. Additionally, many bio techniques call for supplementary assistance for a productive removal process. Owing to these shortcomings, far-reaching treatments are limited. Even though ample of laboratory examination on bioremediation has been finished up to now to pilot/demonstration scales merely, likelihoods that relay to money-making commercial operation are still deceptively not adequately reflected. All the microbial processes can subsidize to operative deduction of heavy metals merely beneath very inadequate circumstances. At these occasions, studies are actuality motivated in such directions, which employ these practices for a huge-scale solicitation.

15.4 Conclusion

Bioremediation techniques portrays a crucial part in the water remediation in addition to water treatment. Because of its easy operation, availability, wide applicability, sustainability, economical and remarkable remediation, it gains noble preference in contrast to traditional techniques. Indigenous techniques use several microbial means for instance bacteria, fungi, yeast, algae and refined plants which are indigenous to that place, as key tools in heavy metals cleanup from the environment. On basis of this, in the present chapter we have discussed various indigenous bioremediation techniques which are indigenous to India for the removal of metal from water.

References

Ahmad, A., Ghufran, R., Zularisam, A., 2011. Phytosequestration of metals in selected plants growing on a contaminated Okhla industrial areas, Okhla, New Delhi, India. Water, Air, & Soil Pollution 217, 255–266.

Ahmad, A., Senapati, S., Khan, M.I., Kumar, R., Sastry, M., 2003. Extracellular biosynthesis of monodisperse gold nanoparticles by a novel extremophilic actinomycete, Thermomonospora sp. Langmuir 19, 3550–3553.

Ali, I., Peng, C., Khan, Z.M., Naz, I., Sultan, M., Ali, M., et al., 2019. Overview of microbes based fabricated biogenic nanoparticles for water and wastewater treatment. Journal of Environmental Management 230, 128–150.

Ali, M.B., Tripathi, R., Rai, U., Pal, A., Singh, S., 1999. Physico-chemical characteristics and pollution level of lake Nainital (UP, India): role of macrophytes and phytoplankton in biomonitoring and phytoremediation of toxic metal ions. Chemosphere 39, 2171–2182.

Arivalagan, P., Singaraj, D., Haridass, V., Kaliannan, T., 2014. Removal of cadmium from aqueous solution by batch studies using Bacillus cereus. Ecological Engineering 71, 728–735.

Bazylinski, D.A., Frankel, R.B., 2000. Biologically controlled mineralization of magnetic iron minerals by magnetotactic bacteria. Environmental Microbe-Metal Interactions 109–144.

Bennett, L.E., Burkhead, J.L., Hale, K.L., Terry, N., Pilon, M., Pilon-Smits, E.A., 2003. Analysis of transgenic Indian mustard plants for phytoremediation of metal-contaminated mine tailings. Journal of Environmental Quality 32, 432–440.

Bhati, R., Sreedharan, S.M., Singh, R., 2021. Deciphering the Multi-Dimensional Abilities of Indigenous Bacteria Enterobacter Cloacae Isolated from Arsenic Contaminated Industrial Sites. Available from: http://doi.org/10.21203/rs.3.rs-524502/v1.

Bhattacharya, A., Gupta, A., 2013. Evaluation of Acinetobacter sp. B9 for Cr (VI) resistance and detoxification with potential application in bioremediation of heavy-metals-rich industrial wastewater. Environmental Science and Pollution Research 20, 6628–6637.

Bisht, S.S., Praveen, B., Rukmini, M., Dhillon, H., 2012. Isolation and screening of heavy metal absorbing bacteria from the industry effluent sites of the river Nagavali. International Journal of Pharmaceutical Sciences and Research 3, 1448.

Borah, P., Kumar, M., Devi, P., 2020. Types of inorganic pollutants: metals/metalloids, acids, and organic forms. Inorganic Pollutants in Water. Elsevier.

Byrne, J.M., Muhamadali, H., Coker, V., Cooper, J., Lloyd, J., 2015. Scale-up of the production of highly reactive biogenic magnetite nanoparticles using Geobacter sulfurreducens. Journal of the Royal Society Interface 12, 20150240.

Chandra, R., Kumar, V., 2017. Phytoextraction of heavy metals by potential native plants and their microscopic observation of root growing on stabilised distillery sludge as a prospective tool for in situ phytoremediation of industrial waste. Environmental Science and Pollution Research 24, 2605–2619.

Cunningham, S.D., Ow, D.W., 1996. Promises and prospects of phytoremediation. Plant Physiology 110, 715.

Das, S., Chakraborty, J., Chatterjee, S., Kumar, H., 2018. Prospects of biosynthesized nanomaterials for the remediation of organic and inorganic environmental contaminants. Environmental Science: Nano 5, 2784–2808.

Dwivedi, S., Mishra, A., Kumar, A., Tripathi, P., Dave, R., Dixit, G., et al., 2012. Bioremediation potential of genus Portulaca L. collected from industrial areas in Vadodara, Gujarat, India. Clean Technologies and Environmental Policy 14, 223–228.

Frankel, R.B., Bazylinski, D.A., 2003. Biologically induced mineralization by bacteria. Reviews in Mineralogy and Geochemistry 54, 95–114.

Gururajan, K., Belur, P.D., 2018. Screening and selection of indigenous metal tolerant fungal isolates for heavy metal removal. Environmental Technology & Innovation 9, 91–99.

Hazra, M., Avishek, K., Pathak, G., 2015. Phytoremedial potential of Typha latifolia, Eichornia crassipes and Monochoria hastata found in contaminated water bodies across Ranchi City (India). International Journal of Phytoremediation 17, 835–840.

Kalibbala, H.M., 2007. Application of Indigenous Materials in Drinking Water Treatment. KTH Royal Institute of Technology.

Kamala, C., Chu, K., Chary, N., Pandey, P., Ramesh, S., Sastry, A., et al., 2005. Removal of arsenic (III) from aqueous solutions using fresh and immobilized plant biomass. Water Research 39, 2815–2826.

Kavitha, K., Baker, S., Rakshith, D., Kavitha, H., Yashwantha Rao, H.C., Harini, B.P., et al., 2013. Plants as green source towards synthesis of nanoparticles. International Research Journal of Biological Sciences 2, 66–76.

Kocoń, A., Jurga, B., 2017. The evaluation of growth and phytoextraction potential of Miscanthus x giganteus and Sida hermaphrodita on soil contaminated simultaneously with Cd, Cu, Ni, Pb, and Zn. Environmental Science and Pollution Research 24, 4990–5000.

Kumar, A., Bisht, B.S., Joshi, V.D., 2010. Biosorption of heavy metals by four acclimated microbial species, Bacillus spp., Pseudomonas spp., Staphylococcus spp. and Aspergillus niger. Journal of Biological and Environmental Sciences 4 (12) , 0–0.

Kumar, J.N., Soni, H., Kumar, R.N., Bhatt, I., 2008. Macrophytes in phytoremediation of heavy metal contaminated water and sediments in Pariyej Community Reserve, Gujarat, India. Turkish Journal of Fisheries and Aquatic Sciences 8.

Kumar, N., Bauddh, K., Dwivedi, N., Barman, S., Singh, D., 2012. Accumulation of metals in selected macrophytes grown in mixture of drain water and tannery effluent and their phytoremediation potential. Journal of Environmental Biology 33, 923.

Kumar, P., Gupta, S., Soni, R., 2019. Bioremediation of cadmium by mixed indigenous isolates *Serratia liquefaciens* BSWC3 and *Klebsiella pneumoniae* RpSWC3 isolated from industrial and mining affected water samples. Pollution 5, 351–360.

Kumari, A., Lal, B., Rai, U.N., 2016. Assessment of native plant species for phytoremediation of heavy metals growing in the vicinity of NTPC sites, Kahalgaon, India. International Journal of Phytoremediation 18, 592–597.

Kushwaha, A., Hans, N., Kumar, S., Rani, R., 2018. A critical review on speciation, mobilization and toxicity of lead in soil-microbe-plant system and bioremediation strategies. Ecotoxicology and Environmental Safety 147, 1035–1045.

Li, J., Pan, Y., 2012. Environmental factors affect magnetite magnetosome synthesis in Magnetospirillum magneticum AMB-1: implications for biologically controlled mineralization. Geomicrobiology Journal 29, 362–373.

Malik, P., Saha, S., 2003. Oxidation of direct dyes with hydrogen peroxide using ferrous ion as catalyst. Separation and Purification Technology 31, 241–250.

Malik, R., Lata, S., Singhal, S., 2015. Removal of heavy metal from wastewater by the use of modified aloe vera leaf powder. International Journal of Basic and Applied Chemical Sciences 5, 6–17.

Mohod, C.V., Dhote, J., 2013. Review of heavy metals in drinking water and their effect on human health. International Journal of Innovative Research in Science, Engineering and Technology 2, 2992–2996.

Momodu, M., Anyakora, C., 2010. Heavy metal contamination of ground water: the Surulere case study. Research Journal of Environmental and Earth Sciences 2, 39–43.

Moon, J.-W., Rawn, C.J., Rondinone, A.J., Love, L.J., Roh, Y., Everett, S.M., et al., 2010. Large-scale production of magnetic nanoparticles using bacterial fermentation. Journal of Industrial Microbiology and Biotechnology 37, 1023–1031.

Pandey, S.K., Upadhyay, R.K., Gupta, V.K., Worku, K., Lamba, D., 2019. Phytoremediation potential of macrophytes of urban waterbodies in Central India. Journal of Health and Pollution 9.

Parial, D., Pal, R., 2015. Biosynthesis of monodisperse gold nanoparticles by green alga Rhizoclonium and associated biochemical changes. Journal of Applied Phycology 27, 975–984.

Parihar, J.K., Parihar, P.K., Pakade, Y.B., Katnoria, J.K., 2021. Bioaccumulation potential of indigenous plants for heavy metal phytoremediation in rural areas of Shaheed Bhagat Singh Nagar, Punjab (India). Environmental Science and Pollution Research 28, 2426–2442.

Patil, S., Chandrasekaran, R., 2020. Biogenic nanoparticles: a comprehensive perspective in synthesis, characterization, application and its challenges. Journal of Genetic Engineering and Biotechnology 18, 1–23.

Paul, M.L., Samuel, J., Chandrasekaran, N., Mukherjee, A., 2012. Comparative kinetics, equilibrium, thermodynamic and mechanistic studies on biosorption of hexavalent chromium by live and heat killed biomass of *Acinetobacter junii* VITSUKMW2, an indigenous chromite mine isolate. Chemical Engineering Journal 187, 104–113.

Prasad, M., Greger, M., Aravind, P., 2005. Biogeochemical cycling of trace elements by aquatic and wetland plants: relevance to phytoremediation. Traces Elements in the Environment: Biogeochemi Stry, Biotechnology and Bioremediation. CRC Press, pp. 451–482.

Qu, Y., Pei, X., Shen, W., Zhang, X., Wang, J., Zhang, Z., et al., 2017. Biosynthesis of gold nanoparticles by Aspergillum sp. WL-Au for degradation of aromatic pollutants. Physica E: Low-Dimensional Systems and Nanostructures 88, 133–141.

Rajaram, R., Banu, J.S., Mathivanan, K., 2013. Biosorption of Cu (II) ions by indigenous copper-resistant bacteria isolated from polluted coastal environment. Toxicological & Environmental Chemistry 95, 590–604.

Reeves, R.D., Baker, A.J., Jaffré, T., Erskine, P.D., Echevarria, G., Van Der Ent, A., 2018. A global database for plants that hyperaccumulate metal and metalloid trace elements. New Phytologist 218, 407–411.

Rose, P.K., Devi, R., 2018. Heavy metal tolerance and adaptability assessment of indigenous filamentous fungi isolated from industrial wastewater and sludge samples. Beni-Suef University Journal of Basic and Applied Sciences 7, 688–694.

Sadhasivam, S., Vinayagam, V., Balasubramaniyan, M., 2020. Recent advancement in biogenic synthesis of iron nanoparticles. Journal of Molecular Structure 1217, 128372.

Sao, K., Pandey, M., Pandey, P.K., Khan, F., 2017. Highly efficient biosorptive removal of lead from industrial effluent. Environmental Science and Pollution Research 24, 18410–18420.

Sarwar, N., Imran, M., Shaheen, M.R., Ishaque, W., Kamran, M.A., Matloob, A., et al., 2017. Phytoremediation strategies for soils contaminated with heavy metals: modifications and future perspectives. Chemosphere 171, 710–721.

Sathe, S.S., Mahanta, C., Mishra, P., 2018. Simultaneous influence of indigenous microorganism along with abiotic factors controlling arsenic mobilization in Brahmaputra floodplain, India. Journal of Contaminant Hydrology 213, 1–14.

Satyapal, G.K., Mishra, S.K., Srivastava, A., Ranjan, R.K., Prakash, K., Haque, R., et al., 2018. Possible bioremediation of arsenic toxicity by isolating indigenous bacteria from the middle Gangetic plain of Bihar, India. Biotechnology Reports 17, 117–125.

Sharma, D., Kanchi, S., Bisetty, K., 2019. Biogenic synthesis of nanoparticles: a review. Arabian Journal of Chemistry 12, 3576–3600.

Sharma, S., Bhattacharya, A., 2017. Drinking water contamination and treatment techniques. Applied Water Science 7, 1043–1067.

Siddiqui, F., Tandon, P., Srivastava, S., 2015. Analysis of arsenic induced physiological and biochemical responses in a medicinal plant, Withania somnifera. Physiology and Molecular Biology of Plants 21, 61–69.

Sivaranjani, S., Rakshit, A., 2016. Indigenous materials for improving water quality. Nature Environment and Pollution Technology 15, 171.

Srivastava, P., Giri, N., 2021. Role of plants in phytoremediation of industrial waste. Bioprospecting of Plant Biodiversity for Industrial Molecules 73–89.

Srivastava, V., Sarkar, A., Singh, S., Singh, P., De Araujo, A.S., Singh, R.P., 2017. Agroecological responses of heavy metal pollution with special emphasis on soil health and plant performances. Frontiers in Environmental Science 5, 64.

Supply, W.U.J.W., Programme, S.M., Organization, W.H., Supply, W.U.J.M.P.F.W., Sanitation, Unicef, 2005. Water for Life: Making It Happen. World health organization.

Taghipour, S., Hosseini, S.M., Ataie-Ashtiani, B., 2019. Engineering nanomaterials for water and wastewater treatment: review of classifications, properties and applications. New Journal of Chemistry 43, 7902–7927.

Talukdar, D., Jasrotia, T., Sharma, R., Jaglan, S., Kumar, R., Vats, R., et al., 2020. Evaluation of novel indigenous fungal consortium for enhanced bioremediation of heavy metals from contaminated sites. Environmental Technology & Innovation 20, 101050.

Tanvi, D.A., Pratam, K., Lohit, R., Vijayalakshmi, B., Devaraja, T., Vasudha, M., et al., 2020. Biosorption of heavy metal arsenic from Industrial Sewage of Davangere District, Karnataka, India, using indigenous fungal isolates. SN Applied Sciences 2, 1–7.

Trapp, S., Karlson, U., 2001. Aspects of phytoremediation of organic pollutants. Journal of Soils and Sediments 1, 37–43.

Vareda, J.P., Valente, A.J., Durães, L., 2019. Assessment of heavy metal pollution from anthropogenic activities and remediation strategies: a review. Journal of Environmental Management 246, 101–118.

Vijayaraj, A., Mohandass, C., Joshi, D., Rajput, N., 2018. Effective bioremediation and toxicity assessment of tannery wastewaters treated with indigenous bacteria. 3 Biotech 8, 1–11.

CHAPTER

16

Application of low-cost adsorbents for metals remediation

Mritunjay and Abdur Quaff (Rahman)

Department of Civil Engineering, National Institute of Technology Patna, Patna, India

16.1 Introduction

Beside fresh air, safe drinking water is the most essential requirement for human and civilization. Providing safe drinking water to every human being is the most challenging issue in the current scenario throughout the world. Safe water is a matter of concern as water demand is increasing globally due to increasing population. In a report of the World Health Organization (WHO), 47% of world's residents do not have reach for safe drinking water and this article also stated that this percentage will go high up to 57% by 2050. Many researchers have investigated on the degradation of water quality of natural sources and concluded that influences of climate changes such as global warming, change in the pattern of water cycle, increase in case of floods, severe drought are the causes of compromised potential water sources (Milly et al., 2002; Noyes et al., 2009; Xia et al., 2017). Rapid growth in urbanization and industrialization also responsible for the water contamination due to continuous discharge of wastewater with various pollutants including metal contamination. The main sources of metal pollution includes coal based thermal power plants (Demirak et al., 2006), electroplating industries, refineries, waste disposal and recycling activities (Herat and Agamuthu, 2012; Olafisoye et al., 2013; Perkins et al., 2014; Wu et al., 2015), agricultural runoff (Kambole, 2003), mining (Archundia et al., 2017), emissions from vehicle and additional metropolitan activities. In underdeveloped nations, according to a recent research, 80% of all industrial and municipal wastewater is disposed of without first undergoing any kind of basic treatment. These wastewater containing serious pollutants like heavy metal, organic or inorganic matter, dyes, pesticides and other harmful matter which is directly discharged to rivers, lakes, ponds or drainage land. These natural water sources are getting polluted with these pollutants making this unfit for the drinking or other household purposes. Therefore, removal of these pollutant from discharged water or polluted water is a must for both wastewater and other sources of drinking water. Flocculation (Sun et al., 2020), ultrafiltration (Lam et al., 2018), bio-filtration (Majumder

DOI: https://doi.org/10.1016/B978-0-323-95919-3.00020-3

et al., 2015), electrocoagulation (De Mello Ferreira et al., 2013), chemical precipitation (Gurmen et al., 2009), ion exchange (Veli and Pekey, 2004), and reverse osmosis are some of the active methods used for metal remediation from water or wastewater. However all these techniques have some limitations such as high initial cost, high maintenance and typical operating system making such methods not feasible for treatment of water or wastewater especially in developing countries. But recently due to extensive interest of researchers in the field of adsorption technique using low cost adsorbent, many of the above mentioned limitations were mitigated. Recently many researchers used adsorption methods for the metal remediation, which include low cost adsorbent like agricultural waste (orange peel, banana peel, rice husk, peanut shells, etc.), industrial waste (black liquor lignin, blast furnace slag, fly ash, red mud, and waste slurry), residual waste (tea waste, tobacco dust, wool, etc.), soil and mineral deposit, aquatic and terrestrial biomass. Therefore in this chapter, we concentrate on the application of low cost adsorbent for metal remediation which includes adsorption methodology, synthesis of adsorbent, mechanism involved in the remediation of metals and efficiency of different adsorbents. The objective of this chapter is to investigate the materials used for the metal remediation which should be most easily available, economic and efficient. The materials mentioned in this chapter are categorized under different aspects namely agricultural waste, industrial waste, residual waste, soil and mineral deposits, aquatic and terrestrial biomass and other waste materials that are found easily. Here the efficiency of the adsorbents concerned with the targeted metals are also discussed.

16.2 Synthesis and characterization of adsorbent

16.2.1 Synthesis of adsorbent

Generally raw materials used for the synthesis of adsorbents require some modification or activation for the application of metal remediation. The process of activation depends upon the nature of raw materials, targeted pollutants and requirements of work. Bio-waste materials are comparatively easy to prepare as it requires simple ordinary treatment. The synthesis of a bio-waste material as adsorbent include collection of materials, washing, drying, size reduction (if necessary) or some simple activation if further required. The synthesis process of adsorbents directly influences the cost and efficiency of the adsorbent. There is various synthesis techniques are reported by different researchers. The typical and most often used procedures for the production of bioadsorbents are addressed in further detail further down this page.

16.2.1.1 Activation with physical treatment

In the synthesis of adsorbents, this is the most straightforward and often used technique. This process is used as pre-requirement for any type of adsorbent preparation. Physical treatment includes washing, drying and size reduction analysis. Out of the all the treatment available for synthesization of adsorbent, this process is the most economical as well as efficient (in many cases). Many researchers prepared bio-adsorbent simply using physical treatment (Bhatti et al., 2016; Park et al., 2008). The efficiency of such type of adsorbent prepared from this method depends on bio-materials and synthesis parameters.

Many researcher reported successful removal of metals from the adsorbent prepared from simply physical treatment. For example, Semerjian (2018) used sawdust for the synthesis of adsorbent for which it was continuously washed under distilled water and then dried for 48 h at 100°C in order to adsorb metals. No other activation method was used in this case. Similarly Potato peels was also useful as an adsorbent for the adsorption of copper which was prepared merely by washing followed by drying at 50°C for 7 days (Guechi and Hamdaoui, 2016). In the same way Pb(II) was removed with cucumber peel and achieved high adsorption around 133.60 mg/g. To achieve the efficiency cucumber peel was only washed properly and then heated at 70°C followed by required sieving.

16.2.1.2 Activation with heat treatment

In this method, bio-materials are subjected to a high temperature (more than 500°C) in order to work as an adsorbent. Since this requires a lot of energy to produce heat so it is not a cost effective method. But heating the bio-materials up to such huge temperature results into much higher surface area. The higher surface area increases the efficiency of the materials as it provides more adsorption sites to get metals adsorbed (Guo et al., 2019; Li et al., 2010). For the purpose walnut shells were used as a raw material for the adsorbent. To achieve higher efficiency walnut shells were pre-treated under a nitrogen atmosphere with temperature of 600°C. The efficiency of this test was very high up to 792 mg/g for Cr(III) (Zbair et al., 2019). In another study, lemon peel was activated by keeping it in an environment of 500°C. The activated adsorbent removed cobalt from water (Bhatnagar et al., 2010). Another researcher investigated adsorption ability of oyster shells by activating it under different temperature and compared the result. For this the powdered form of oyster shells were subjected at 650°C, 700°C and 900°C of temperature for 2 h. The recorded capacity was maximum 1666.67 mg/g for 900°C (Alidoust et al., 2015).

16.2.1.3 Activation with simultaneous heat and chemical treatment

Though above mentioned unit process for activation of bio-sorbent have decent efficiency to remove metal from water but some researcher also investigated the result of simultaneous activation through more than one technique. One of them is simultaneous heat and chemical treatment of bio-sorbent. The is done due to the fact that high temperature increase the surface area of the sorbent while chemical treatment enhance the pore present in the materials. The order is not strict but generally, chemical treatment was done after heating the materials to high temperature (Changmai et al., 2018). In a study, tobacco was investigated as bio-sorbent by simultaneous treatment of the material. Powdered form of tobacco biomass was chemically modified with $ZnCl_2$ and 1 M NaOH after the pretreatment. The obtained biomass was then heated at 750°C under inert environment (Conradi et al., 2019). In another study, rice straw was used as precursor of the carbon. After essential pretreatment like washing and drying, the material was impregnated with H_2SO_4 and again dried for 24 h at 110°C. Then it was activated by keeping the material in microwave cavity under power of 700 W for 20 min. Lastly, it was washed to make the sorbent free from acid residue if present. Thus final adsorbent was ready to remove Hg(II) from liquid phase (Mashhadi et al., 2016).

16.2.1.4 *Activation with co-precipitation method*

This is a unique method of synthesis of adsorbent, in which sorbents can be regenerated (with up to 90% regeneration efficiency) by simple method as it can be magnetized in the filtration process (Rhaman et al., 2020). With this method, very effective adsorbents were synthesized but also have some its own limitation like requirement of high speed centrifugation. The advantages of this technique over other are increased surface area, crystalline structure and magnetic properties. In a research, orange peel was modified to a new magnetic adsorbent with co-precipitation using Fe_3O_4. Some static quantity of ferric chloride ($FeCl_30.6H_2O$) and hydrated ferrous sulfate ($FeSO_40.7H_2O$) were mixed in distilled water before adding orange peel powder into it, which results into formation of black precipitate after washing and drying (Gupta and Nayak, 2012). In similar manner, another magnetic nanocomposite adsorbent (pectin-iron oxide) was developed by using $FeCl_30.6H_2O$ and $FeCl_30.4H_2O$ (Gong et al., 2012).

All the above mentioned techniques have their own advantages and disadvantages. For example, physical treatments are easy and economical but comparatively lower efficiency. On other hand adsorbents prepared by heat treatment method have higher surface area but it require very high energy making it quit costly. While in chemical treatment, higher efficiency can be achieved with only pores activation but produces more amount sludge as it can't be regenerated. The adsorbent prepared can be regenerated but it requires more skill and also become laborious. Therefore, it still requires a development of simple physical techniques having more efficiency using low cost material for metal remediation.

16.2.2 Characterization of adsorbent

16.2.2.1 *Physicochemical characteristics of low cost adsorbents*

The physicochemical features of a low-cost adsorbent have a big impact on its adsorption capability. The parameters that are important for governing characteristic of adsorbent for metal remediation are surface area (higher the surface area, higher the adsorption), porosity (higher porosity favorable for adsorption), surface functional groups (depends upon the targeted metal), pore distribution and cation exchange capability (Almendros et al., 2015; Cârdei et al., 2021). Out of these characteristics, surface area and porosity plays most significant part in deciding the adsorption capacity of any adsorbent. The specific area of any adsorbent is decided by its pore structure (Pan et al., 2020). Surface area of adsorbents is responsible for holding the metals. The presence of functional group decides the metals that are getting adsorbed. Actually it is functional group like amino, carbonyl, carboxyl, hydroxyl, which interact with the metals by providing binding sites for metals. Zeta potential is another physicochemical property of synthesized adsorbent which depends upon the pH of the aqueous phase. Surface charge density at different pH of a developed adsorbent can be defined by its zeta potential (Gupta et al., 2019).

16.2.2.2 *Tools for the characterization of adsorbent*

The characteristic of adsorbents mentioned in above section can only be determined by some dedicated tools for respective characteristic. SEM (scanning electron microscopy) is a powerful tool with beam of electrons to produce high magnified high resolution images of samples (up to 15 nanometers) which tells about the morphology of the adsorbents. It helps in the surface characterization of the adsorbents. The chemical description and elemental investigation of numerous adsorbents can be determined with the help of EDS (energy dispersive X-ray spectroscopy). The key factor of the adsorption process that is functional group present in the adsorbent can be analyzed by FTIR (Fourier transform infrared spectroscopy). It also gives the idea of surface properties of the adsorbent. The XRD (X-ray diffraction) has its application in finding out the crystallinity of the materials. It also gives all the detailed information about chemical and physical properties of an adsorbent. It is the most important tool to investigate the properties of adsorbents. The surface area which plays significant role in adsorption of metals can be determined by the application of BET (Brunauer-Emmett-Teller) (Bilal et al., 2022).

16.3 Efficacy of low cost adsorbent for metal remediation

Adsorption is a process which is in application since a long time. For adsorption, it requires a adsorbent. Recently researchers are taking keen interest in developing natural adsorbent, keeping environmental as well as economic point of view. Adsorbent derived from low cost waste material (natural or industrial) are best alternative to replace costly traditional materials (commercial activated carbon, inorganic materials). Natural materials are also easily and abundantly available. Many researchers verified that some natural and industrial waste have potential as low cost sorbents for the metal adsorption. In Fig. 16.1, adsorbent derived from various low cost materials have been discussed with their efficiency any the targeted pollutant (metal in this case).

16.3.1 Low cost adsorbent derived from agricultural waste

There is a large scope in agricultural waste as raw materials for the synthesis of adsorbent to remove pollutant from aqueous phase. Many researchers all across the world have used agricultural waste for the preparation of bio-sorbent. Due the mass availability and high organic content (favorable to convert into activated carbon or biochar) agricultural waste gaining popularities among other materials. Some of the most used agricultural

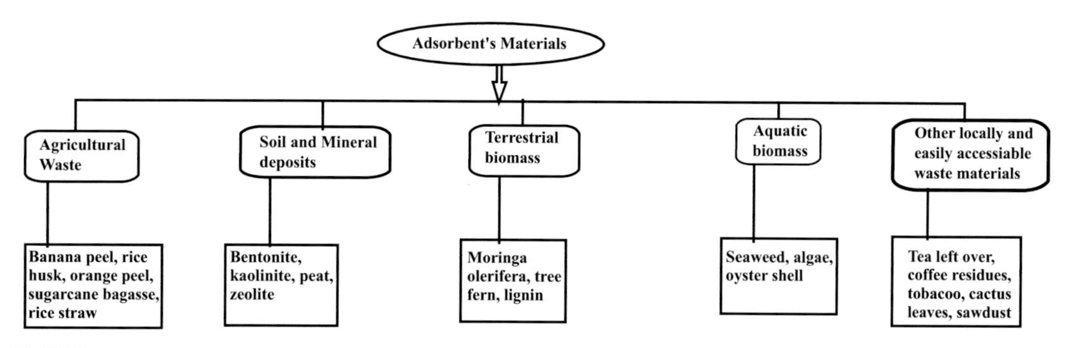

FIGURE 16.1 Systematic diagram to describe the various types of materials used in the synthesis of adsorbent.

wastes for bio-sorbent are banana peel, lemon peel, orange peel, rice husk and some other materials like coconut shells, dairy manure compost, rice straw etc.

Numerous experiments have been conducted to determine the fate of orange peel as an adsorbent. In terms of Cd (II) adsorption capability, orange peel had the highest value measured at 293 mg/g (Feng et al., 2011). Ajmal et al. (2000) removed Ni(II) achieving 97.5% of removal efficiency. Physical treatment (washing followed by heating at 100°C) was used for the activation of the orange peel. Aside from nickel, other metals were also removed, although their removal efficiency was lower than that of nickel. Nickel was the most efficient metal removed (Ajmal et al., 2000).

Banana peel is another very effective low cost adsorbent material having affinity towards adsorbing metals. Many researchers modified banana peel to work as adsorbent. Thirumavalavan et al. (2010) performed test on banana peel to investigate the adsorption of Zn(II), Pb(II), Cd(II), Cu(II) and Ni(II) and found 21.9, 25.9, 34.1, 52.4 and 54.4 mg/g of maximum adsorption capacity respectively. Another researcher conducted test to know the adsorption of different metals like Pb(II), Ni(II), Zn(II), Cu(II), and Co(II). The highest recorded capacity of 7.9 mg/g was found for lead (Annadurai et al., 2018). DeMessie et al. (2015) conducted two experiment to absorb Cu(II) from aqueous phase. First experiment was done using banana peel (activated by simply physical treatment) and found 7.4 mg/g of adsorption. While in other experiment banana peel was pyrolyzed at 500°C and the adsorption capacity recorded in this case was 38.3 mg/g, which was much more than the first experiment. Thirumavalavan et al. (2010) demonstrated lemon peel to investigate its adsorption capacity towards different metals with maximum recorded value of 80.0 mg/g for Ni(II).

Other types of agricultural waste like mushroom residues were also used in the application of metal remediation. In a study three different types of mushroom was studied and the maximum uptake (34.96 mg/g) of Cd(II) metal was taken by oyster mushroom while maximum adsorption (33.78 mg/g) of Pb(II) was done by button mushroom (Vimala and Das, 2009). Mushroom has also removed lower concentration metals like Cu(II), Zn(II) and Hg(II) with highest adsorption capacity of 81.7% among all (Li et al., 2018). Investigation has been done to find the adsorption capacity of corncob towards metals. A comparative study has been done to investigate the adsorption capacity of corncob. In first experiment corncob was activated by simple physical treatment and successfully removed lead with adsorption of 5.1 mg/g. While in other experiment when corncob was activated chemically with help of nitric acid and citric acid, the adsorption capacity increased with 4–10 times (Leyva-Ramos et al., 2005). In other experiment Pb(II) was successfully removed by corncob having adsorption of 16.2 mg/g. In this case adsorbent was modified with methanol and hydrochloric acid (Tan et al., 2010). However, after treating the same corncob with NaOH, its adsorption capacity increased to 43.4 mg/g (Tan et al., 2010).

Rice cultivation produces different agricultural waste like rice husk, rice bran and rice straw. All of these were used by researcher as an adsorbent. Rice husk is one of them that was applied for the metal remediation. A study has been investigated on rice husk to remove several metals. Many metals were targeted of which the maximum recorded adsorption was 58.1 mg/g for lead and the minimum recorded value was 5.5 for nickel. Here the rice husk was treated with 1.5% alkali treatment (300 g in 1 L) followed by heating at 121°C for 30 min (Krishnani et al., 2008). The maximum adsorption capacity recorded by

Singha and Das (2013) was 17.9 mg/g for Cu(II) using rice husk. A significant removal of Cr (VI) (more than 95%) over black rice husk was noted at pH less than 3.0 (Georgieva et al., 2015). Bansal et al. (2009) prepared rice husk bio-sorbent by two different process in order to adsorb Cr (VI).

Other agricultural residue like peanut shells, peanut hull, and cashew nut shells, pistachio hull, almonds shells were also demonstrated by researchers across the globe for remediation of metals. Peanut was used in two different forms as peanut shell and peanut husk. Taşar et al. (2014) used peanut shells as bio-sorbent (activated by physical treatment) and removed Pb(II) from aqueous solution with 39 mg/g of adsorption capacity. Using peanut shells Ahmad et al. (2017) removed Cr(VI) and Cu(II) at low pH with recorded adsorption capacity of 27.9 and 25.4 mg/g respectively. Another researcher used hull of peanut to remove Cu(II) and the extreme recorded removal was 21.3 mg/g from aqueous solution (Zhu et al., 2009). Similarly cashew nut also have capacity to remove metals from aqueous phase but the observed adsorption capacity was less in comparison to other. In a research, 85% of Cu(II) removal was achieved with cashew nut shells having adsorption of 20 mg/g (SenthilKumar et al., 2011). Adsorbent derived from cashew nut shells also removed Ni(II) with 60%–75% efficiency and 18.9 mg/g of recorded adsorption capacity (Kumar et al., 2011). In a study pecan shell was activated for the removal of metals from water. Activation was done by a method which includes activation through acid, steam and carbon dioxide. For this activated bio-sorbent, the maximum adsorption was recorded with acid activated pecan shell for Pb(II) (Bansode et al., 2003). Some other nuts like almond shells, hazelnut shells, groundnut shells were also demonstrated for the metal remediation. An experiment was again repeated with chemical modification of the groundnut shells using reactive dye, and this experiment resulted in 40%–70% increase in removal efficiency (Shukla and Pai, 2005). Metal remediation through agricultural waste has been summarized in Table 16.1.

16.3.2 Low cost adsorbent derived from soil and mineral deposits

Activated bentonite, kaolinite, peat, phosphate rock (obtained from Tunisisan ores), different type of soils (Entisols, Mollisols, Oxisols, etc.), zeolite etc. are naturally occurring soil and mineral deposits toward metals are due to solubility of heavy metal, heterogeneous nature of deposits. The mechanisms involved in the adsorption of metals into their deposits were precipitation of metal carbonates and hydroxides, electronegativity, hydrolysis constant, change in physical and chemical properties (Appel et al., 2008). Since the natural deposits requires very few operations and time for activation, so on considering the economic and technological constraints the application of soil and other natural resources are suitable approach for metal remediation. In 2008, in an experiment three types of soils from Puerto Rico were selected to eliminate lead and cadmium. The result showed varying adsorption capacity ranging from 4.1 to 6.7 and 1.6 to 3.5 mg/g for Pb(II) and Cd(II) respectively (Appel et al., 2008). The researcher reported adsorption capacity of native bentonite as 19.2 mg/g while very low adsorption capacity was reported in case of activated bentonite (Kul and Koyuncu, 2010). Another researcher also investigated the adsorption capacity of natural kaolin towards Pb(II) and found a maximum recorded

TABLE 16.1 Metal remediation by agricultural waste.

Metal	Oxidation state	Adsorbent	Initial metal concentration (ppm)	Surface area of adsorbent (m^2/g)	Maximum adsorption capacity (mg/g)	References
Cadmium	+ 2	Banana peel	100–800	1.3	34.1	Thirumavalavan et al. (2010)
		Coconut shell	100–200	212	3.5	Paranavithana et al. (2016)
		Corncob	5–120	5	5.1	Leyva-Ramos et al. (2005)
		Grape fruit peel	50	NA	42.1	Torab-Mostaedi et al. (2013)
		Lemon peel	100–800	1.3	54.6	Thirumavalavan et al. (2010)
		Orange peel	50–1200	NA	293	Liang et al. (2013)
		Rice husk	50–200	NA	16.6	Krishnani et al. (2008)
		Sugarcane bagasse	10–30	0.49	0.96	Moubarik and Grimi (2015)
Chromium	+ 3	Peanut shells	10–1000	NA	27.9	Witek-Krowiak and Harikishore Kumar Reddy (2013)
		Peanut husk	0–50	NA	7.7	Li et al. (2007)
	+ 6	peanut shells	0–40	1.8	4.3	Ahmad et al. (2017)
		Almond	20–1000	NA	10.2	Dakiky et al. (2002)
		Coconut shells	54.5	0.5	18.7	Singha and Das (2011)
		Pistachio hull waste	50–200	1.04	116	Moussavi and Barikbin (2010)
		Rice bran	54.5	0.1	12.3	Singha and Das (2011)
		Rice husk	54.5	0.5	11.4	Singha and Das (2011)
		Rice straw	54.5	1.2	12.2	Singha and Das (2011)
Cobalt	+2	Banana peel	5–25	NA	2.6	Annadurai et al. (2018)
	+2	Orange peel	5–25	NA	1.8	Annadurai et al. (2018)
	+2	Rice husk	50–200	NA	9.6	Krishnani et al. (2008)

Copper	+2	Banana peel	100–800	1.3	52.4	Thirumavalavan et al. (2010)
		Cashew nut shell	10–50	395	20	Kumar et al. (2011)
		Coconut shell	5–300	NA	19.9	Singha and Das (2013)
		Dairy manure compost	31.8	NA	27.2	Zhang (2011)
		Grape stalk	15.3–15.3	NA	10.1	Villaescusa et al. (2004)
		Lemon peel	100–800	1.3	70.9	Thirumavalavan et al. (2010)
		Orange peel	100–800	2.0	63.3	Thirumavalavan et al. (2010)
		Peanut shells	10–1000	NA	25.4	Witek-Krowiak and Harikishore Kumar Reddy (2013)
		Peanut hull	10–400	NA	21.6	Zhu et al. (2009)
		Peanut husk	0–50	NA	10.2	Li et al. (2007)
		Rice bran	50–300	NA	21.1	Singha and Das (2013)
		Rice husk	50–200	NA	10.9	Krishnani et al. (2008)
		Rice straw	5–300	NA	18.4	Singha and Das (2013)
Lead	+2, +4	Banana peel	100–800	1.3	54.4	Thirumavalavan et al. (2010)
		Coconut shell biochar	100–2000	212	13.4	Paranavithana et al. (2016)
		Corncob	20.7–414	NA	16.2	Tan et al. (2010)
		Dairy manure compost	103.6	NA	95.3	Zhang (2011)
		Lemon peel	100–800	1.3	37.9	Thirumavalavan et al. (2010)
		Orange peel	50–1200	NA	476	Feng et al. (2011)
		Peanut shell	100–350	0.84	39.0	Taşar et al. (2014)
		Rice husk	50–200	NA	58.0	Krishnani et al. (2008)
		Sugarcane begasse	57	92.3	87.0	Abdelhafez and Li (2016)

(*Continued*)

TABLE 16.1 (Continued)

Metal	Oxidation state	Adsorbent	Initial metal concentration (ppm)	Surface area of adsorbent (m^2/g)	Maximum adsorption capacity (mg/g)	References
Mercury	+2	Rice husk	50–200	NA	36.1	Krishnani et al. (2008)
Nickel	0, +2, +3	Banana peel	100–800	1.3	54.4	Thirumavalavan et al. (2010)
		Cashew nut shell	10–50	395	18.9	Kumar et al. (2011)
		Grapefruit peel	50	NA	46.1	Torab-Mostaedi et al. (2013)
		Groundnut shell	107–554	NA	3.8	Shukla and Pai (2005)
		Lemon peel	100–800	1.3	80.0	Thirumavalavan et al. (2010)
		Orange peel	50–1200	NA	162	Feng et al. (2011)
		Orange peel	100–800	2.0	81.3	Thirumavalavan et al. (2010)
		Rice husk	50–200	NA	5.5	Krishnani et al. (2008)
Zinc	+2	Banana peel	100–800	1.3	21.9	Thirumavalavan et al. (2010)
		Dairy manure compost	32.7	NA	15.5	Zhang (2011)
		Groundnut shell	38–244	NA	7.6	Shukla and Pai (2005)
		Lemon peel	100–800	1.3	27.9	Thirumavalavan et al. (2010)
		Orange peel	100–800	2.0	24.1	Thirumavalavan et al. (2010)
		Rice husk	50–200	NA	8.1	Krishnani et al. (2008)

value of 15.1 mg/g (Tang et al., 2009). The adsorption of heavy metals was in decreasing order of Pb(II) > Cd(II) > Cu(II) with maximum recorded value of 118.7, 50.2 and 25.4 mg/g. The presence of functional group like phenolic groups, carboxylic group and hydroxylic group are responsible for the adsorption of metals on peat (Qin et al., 2006). Tiede et al. (2007) examined mineral deposits containing manganese-oxyhdroxides for the application of metal remediation. In an another experiment, Cd(II), Cu(II), Pb(II) and Zn(II) were adsorbed on phosphate rock obtained from Tunisian ores. The maximum adsorption reported for Pb(II) and minimum for Zn(II) ranging from 8.5 to 12.8 mg/g. Some researcher also achieved 25%–50% more adsorption when phosphate rock was treated with NaOH and HNO_3 (Elouear et al., 2008). Activated Ganga sand (river sand, found in plain region of India) was used to remove copper from the aqueous solution. In the study, activated Ganga sand removed almost 97% of Cu(II) at optimum condition. Activation of Ganga sand was done with sodium hydroxide and heat (Mritunjay and Quaff, 2021). In the same way, Sudha Rani et al. (2018) investigated laterite soil and found its use in adsorption of copper in aqueous phase having more than 90% efficiency. Similarly many others studies are available in context of application of soil and mineral deposits in metal remediation.

16.3.3 Low cost adsorbent derived from terrestrial biomass

Various terrestrial biomass like Moringa olerifera (MO), Tree fern, Lagerstroemia speciose (found in Inida), sawdust, Lignin etc. have the potential to remove metals from aqueous solution. Recently Moringa Oleifera has received much attention as it has showed great affinity towards heavy metals and has found application in metal remediation process (Shan et al., 2017). Tree fern is also another useful bio-sorbent having potential to remove metals. Tree fern (present in Taiwan) has application in removing metals with highest noted capacity of 39.8 mg/g for lead (Ho et al., 2002). Lagerstroemia speciose, commonly known as pride-of-India have successfully removed chromium from aqueous phase. In the experiment removal of Cr(VI) was tried with native Lagerstroemia speciose and chemically modified Lagerstroemia speciose separately. In case of native Lagerstroemia speciose, the maximum recorded adsorption was 20.41 and 24.39 mg/g in case of chemically modified Lagerstroemia speciose (Srivastava et al., 2015).

Sawdust is a waste material (hence low cost) which is abundantly and easily available having high potential to adsorb metal from aqueous phase. Many researchers have investigated the adsorption capacity of different sawdust. Sawdust of different wood behaves differently like maple sawdust has successfully removed Cr(VI) with efficiency greater than 80% (Yu et al., 2003). Similarly using teakwood sawdust the metal uptake was up to 11 mg/g (Shukla and Pai, 2005). In a study, investigation of poplar tree's sawdust was done for uptake of metals like Cr(III), Cu(II) and Pb(II) from water. The result of this study was recorded as 5.5, 6.6 and 21.1 mg/g of adsorption capacity respectively (Li et al., 2007).

Many researcher also examined Lignin as a bio-sorbent to adsorb metals from aqueous solution. Lignin is a type of polymer found in the cell walls of plants and paper mills are the largest producer of any paper mills (Wu et al., 2008). In a study, lignin as a bio-sorbent has successfully removed metals (Guo et al., 2008). Chromium and cadmium were also effectively removed by using lignin (Liang et al., 2013).

16.3.4 Low cost adsorbent derived from aquatic biomass

It has been discovered that seaweed and algae (aquatic biomass) are capable of effectively remove metals. Using *Undaria pinnatifida*, an artificially created marine algae, researchers were able to successfully remove nickel and copper from water, achieving peak adsorption capacities of 29.9 and 78.9 mg/g, respectively, using the algae's adsorption capabilities. Furthermore, Romera et al. (2008) did a thorough investigation on adsorption Cd(II), Ni(II), and Pb(II) using brown (*Ascophyllum nodosum*), red (*Chondrus crispus*) and green (*Codium vermilara*) algae and concluded that in every case the removal efficiency was more than 90% for every heavy metals. The highest recorder adsorption capacity was 26.5 mg/g in case of chromium using red algae (Sari and Tuzen, 2008). In one investigation, brown, red, and green algae were utilized to eliminate Cd(II), and the results revealed a broad variety of adsorption abilities depending on the algal species used. The adsorption capacities of the algae were found to range between 17.9 and 82.9 mg/g, with brown algae being the most effective, followed by green and red algae (Hashim and Chu, 2004).

Spirodela polyrhiza, a freshwater macrophyte, was used to study the adsorption of Cu(II), Mn(II), and Zn(II) and found that the maximum adsorption capabilities were 52.6, 35.7, and 28.5 mg/g, respectively (Meitei and Prasad, 2014). Furthermore, chromium removal has been demonstrated in a variety of aquatic weeds, including water lilies and mangrove leaves, with adsorption capacities ranging from 6.1 to 7.2 mg/g for Cr(III) and 1.75 mg/g for Cr(II) (Elangovan et al., 2008). Cu(II) removal has also been demonstrated in neem leaves and hyacinth roots, with highest adsorption capacities of 17.5 and 21.8 mg/g, respectively (Singha and Das, 2013). Heavy metals can be effectively removed from aqueous solutions using shells from diverse aquatic species. Shells from the mollusk *Anadara inaequivalvis*, which can be found mostly in the Adriatic, Aegean, and Black Seas, were shown to be very efficient in removing Cu(II) and Pb(II), with maximum adsorption capacities of 330.2 and 621.1 mg/g, respectively, from water samples (Bozbaş and Boz, 2016). Cadmium, lead and zinc were significantly removed from razor clam shells, with maximal adsorption capacities of 501.3, 656.8, and 553.3 mg/g, respectively (Du et al., 2011). Furthermore, oyster shells effectively removed Cd(II), Pb(II), and Zn(II), with maximal adsorption capacities of 118.0, 159.1, and 564.4 mg/g for Cd(II), Pb(II), and Zn(II), respectively (Du et al., 2011). In addition, the adsorption of cobalt and copper from crab shell bio-sorbent has been proven, with maximum adsorption capacities of 322.6 and 243.9 mg/g, respectively, for Co(II) and Cu(II) removal (Vijayaraghavan et al., 2006).

16.3.5 Low cost adsorbent derived from locally accessible waste materials

The efficiency of locally accessible waste material in removing metals from the aqueous phase has been studied broadly. Certain types of trash may be produced in excess of others due to differences in local ecosystems, energy sources, farming practices, and civilizations. Each year, around 857,000 tons of tea are produced in India, resulting in excessive volume of tea left-over (Wasewar et al., 2009). Over the past two decades, many studies have looked at the adsorptive qualities of tea trash. Malkoc and Nuhoglu (2005) used tea trash to attain a extreme adsorption capability of 18.4 mg/g for Ni(II). Wasewar et al. (2009) also found that tea trash effectively removed Zn(II) with efficiency greater than 98%.

The efficacy of a variety of other locally accessible waste products metal remediation from water and wastewater have been evaluated. A researcher looked into the adsorption capacities of coffee residues for eliminating heavy metals. In this study, the maximum adsorption for the metals tested varied from 11.0 to 39.5 mg/g, with the highest adsorption capacities for cadmium being 39.5 mg/g and the lowest adsorption capacities for nickel being 11.0 mg/g (Boonamnuayvitaya et al., 2004). The waste material derived from the fruit of the Neem (*Acalypha indica*) plant, known as neem oil cake, has been found to successfully remove Cu(II) and Cd (II) (Rao and Khan, 2009). Qi and Aldrich (2008) studied the utilization of tobacco dust metal remediation from aqueous phase and found maximal adsorption capacities of 24.5–39.6 mg/g. A research employing olive cake, a waste product of olive oil manufacturing industries was investigated to remove Cd(II) and found that it was successful, with a maximum adsorption capacity of 65.4 mg/g (Al-Anber and Matouq, 2008). Dakiky et al. (2002) investigated the ability of cactus leaves, olive cake and wool to remove Cr (VI). In this investigation, wool found to have maximum adsorption capacities with 41.2 mg/g of chromium intake, followed by olive cake (33.4 mg/g) and cactus leaves (7.1 mg/g) (Dakiky et al., 2002). These are the few example of bio-sorbent derived from locally available waste materials. Many other researchers have also investigated different types of other waste materials. Highest recorded metal adsorption by other different materials has been summarized in Table 16.2.

16.4 Parameters affecting the metal remediation through low cost adsorbent

The efficiency of metal remediation not only depends upon the characteristics of adsorbent but also depends on the types of targeted metals, pH of the aqueous phase, ambient temperature, adsorbent's dose, initial metal concentration of targeted metal, contact period, mixing speed and existence of competing ions. Altering any of the above parameters can affect the result of the same adsorbent.

16.4.1 Effect of the targeted/selected metal

Synthesis of an adsorbent was done according to the selected metal. Adsorbent has different affinity for different metals. Metals have different chemical and physical characteristics which influence the adsorption process. For example, adsorbent prepared by Thirumavalavan et al. (2010) have different affinity towards different metals. In this investigation, the highest recorded removal of Cd(II) was 41.8 mg/g while 24.1 mg/g for Zn(II) by the same adsorbent. In another study, the adsorption capacity of coffee residue for cadmium was 39.5 mg/g while 11.0 mg/g for nickel (Boonamnuayvitaya et al., 2004).

TABLE 16.2 Highest recorded metal remediation by several other materials.

Targeted metal	Oxidation state	Adsorbent	Surface area of adsorbent (m^2/g)	Initial metal concentration (ppm)	Maximum adsorption capacity (mg/g)	References
Arsenic	+3	Green tea waste	0.75	7–23	0.4	Yang et al. (2016)
Cadmium	+2	Oyster shell	N.A.	0–300	118.00	Du et al. (2011)
	+2	*A. nodosum* (brown algae)	N.A.	10–150	69.70	Romera et al. (2008)
	+2	*C. crispus* (red algae)	N.A.	10–150	65.20	Romera et al. (2008)
Chromium	+3	*Ceramium virgatum*	N.A.	10–400	42.10	Sari and Tuzen (2008)
	+6	Olive cake	20–1000	20–1000	33.4	Dakiky et al. (2002)
	+6	Wool	N.A.	20–1000	41.2	Dakiky et al. (2002)
Cobalt	+2	Crab shell	N.A.	500–2000	323	Vijayaraghavan et al. (2006)
Copper	+2	*Anadara inaequivalvis* shell	1.82	20–100	330	Bozbaş and Boz (2016)
	+2	Crab shell	N.A.	500–2000	244	Vijayaraghavan et al. (2006)
	+2	Lignin	21.7	13–159	22.9	Guo et al. (2008)
Lead	+2	Lignin	21.7	41–518	89.5	Guo et al. (2008)
	+2	*A. inaequivalvis* shell	1.82	20–100	621	Bozbaş and Boz (2016)
	+2	Oyster shell	N.A.	0–500	1591	Du et al. (2011)
	+2	Razor clam shell	N.A.	0–500	657	Du et al. (2011)
Mercury	+2	*Flammulina velutipes*	N.A.	10–100	7–9	Li et al. (2018)
Nickel	+2	Tabacoo dust	N.A.	0–50	25.1	Qi and Aldrich (2008)
	+2	Meranti sawdust	0.6	1–200	36.0	Rafatullah et al. (2009)
Zinc	+2	Oyster shell	N.A.	0–300	564	Du et al. (2011)
	+2	Razor clam shell	N.A.	0–300	553	Du et al. (2011)

16.4.2 Effect of pH of the solution

The pH of the aqueous medium is one of the most essential operational elements to consider while adsorbing metals. pH changes have strong capability to influence adsorbate speciation, metal ionization, and adsorbent surface chemistry (Nigam et al., 2019). Generally metals present in aqueous solution are in cationic phase and solution at low pH. The point of zero charge (PZC) or zeta potential research is crucial in the development of bioadsorbents as it indicates the surface charge of the adsorbent at different pH levels. Whenever pH goes below the PZC, acidic water contributes more protons to the bio-adsorbent than hydroxide groups, resulting in a positively charged surface on the bio-sorption material (attracting anions). When pH value is greater than the PZC, the surface is negatively charged (attracting cations and repelling anions), and when the pH value is lower than the PZC, the surface is negatively charged. Researchers discovered that iron oxide nanocomposites made from bio-waste material had PZC values ranging from 5.15 to 4.6 to 7.6 and 8.9, respectively (Lingamdinne et al., 2020), while pH values of 4.2 and 5.7 were discovered which corresponded to bio-sorbents derived from olive stone and sugarcane bagasse, respectively (Moubarik and Grimi, 2015). For most cations, increased adsorption capacity is reported at higher pH for adsorbents having lower PZC value. The majority of the adsorbents showed greater adsorption at low pH when anions were present. Furthermore, raising the pH value from 1.79 to 4.80 improved the adsorption effectiveness of Pb^{2+}, Cu^{2+}, and Co^{2+} onto arborvitae leaves. It was anticipated that the increase in proton density of the bioadsorbents under study at higher pH levels would result in improved adsorption capability at pH 4.80 (Shi et al., 2016). Furthermore, it was discovered that the percentage of adsorption was greatest at lesser pH and reduces with increasing pH, as metal impurities react with OH ions to form insoluble metal hydroxide, reduces the fraction of adsorption as pH increases because metal impurities react with OH ions to form insoluble metal hydroxide (Sethy et al., 2019). When the pH of a solution is raised, it has been shown that the number of ligands available for metal ion binding increases. This has been connected to an increase in metal adsorption in certain cases. An increase in the number of H^+ ions available for metal binding capacity is associated with upsurge in the pH of the water being tested. When exposed to moderate acidic or neutral pH values, negatively charged bioadsorbents encouraged the high absorption of metals in this environment, owing to the deprotonation of their surfaces.

16.4.3 Effect of the adsorbent dose of the solution

Another crucial factor to consider is the adsorbent dose, the metal absorption capacities of a bio-sorbent are significantly influenced. Increasing the quantity of adsorbent used is directly proportional to active exchangeable adsorption sites that are accessible. In this study, the amount of Co(II) removed from three distinct bio-sorbents (carrot, tomato, and polyethylene terephthalate waste) was increased until the adsorption equilibrium was reached for each one (Changmai et al., 2018). Addition of adsorbent after the equilibrium point is not significant as at this point number of available sites are already occupied with the metals. While in some study, it was observed that increasing the adsorbent dose beyond the equilibrium point resulted into decrease in the efficiency of the bio-sorbent. For example, when the dose of adsorbent (derived from *Sargassum tenerrium*) was increased beyond the equilibrium point to remove

Cr(VI), a reverse trend was observed (Elangovan et al., 2008). When hazardous metal concentrations in solution at high dosage were reduced, a drop in the ratio of pollutants to binding sites on the bio-sorbent occurred. In response, an opposing approach was devised to counteract the decrease in the ratio of pollutants to binding sites on the bio-sorbent (Elangovan et al., 2008).

16.4.4 Effect of initial concentration of the metal

Initial concentration of metal contaminants has a direct impact on the percentage removal of metallic ions. Overall, when heavy metal concentrations decrease, adsorption efficiency decreases as a result of an increase in the possibility of ion adsorption to the adsorbent per ion at low concentrations, according to the literature available. According to a learning, the availability of a greater number of unoccupied binding sites resulted in a quick rise in Pb(II) adsorption on chemically treated *Mangifera indica* seed shells (Moyo et al., 2017). The first increase in removal rate (up to equilibrium point) was due to unsaturated binding sites, whereas the subsequent decrease was due to saturation of sites. If the amount of metal adsorbed on the adsorbent surface is divided by the amount of pollutant initially present on the surface, the adsorption efficiency may be calculated. It may be described by the following Eq. (16.1):

$$Adsorption\ efficiency\ (\%) = (Ci - Co) \times \frac{100}{Ci} \tag{16.1}$$

Whereas the adsorption capacity (q_e) of an adsorbent is defined as amount of adsorbate adsorbed per unit mass of adsorbent. It is described using Eq. (16.2):

$$Adsorption\ capacity = (Ci - Co) \times \frac{V}{W} \tag{16.2}$$

Where C_i is initial metal concentration and C_o is the final concentration of metal still present in aqueous solution, V defines the volume of the solution, and W is the adsorbent dose.

16.4.5 Effect of contact period of the adsorbent

Contact time is yet another crucial variable that has a significant influence on the adsorption procedure. The dynamics of the adsorption process are also affected by this, which is vital to understand. In general, as contact duration increases, the percentage of metal remediation increases until equilibrium is reached. In a study, it was observed that the remediation of Co (II) increased with increase in contact time of adsorbent with the metal. In this work, the equilibrium time of different adsorbents (carrot, tomato and polyethylene terephthalate waste) were between 75 and 100 min (Changmai et al., 2018). However, after a certain contact period, the adsorbate occupy all the initially available locations. The rate of metal ion adsorption reduces as the number of accessible sites decreases. Various adsorbents have different adsorption equilibrium values. Adsorbents with a shorter equilibrium period are preferable for use in practical water treatment systems.

16.4.6 Effect of mixing or agitation speed

The stirring or agitation speed has an effect on both the adsorption rate and the capacity of the solution. When mixing, the goal is to achieve a homogeneous solution while also dispersing the pollutant and adsorbent correctly in order to minimize accumulation. Because of the increased homogeneity of the waste solution and adsorbent biomass that occurs at slower speeds, the phenomenon frequently performs better (Bilal et al., 2022). The improvement in adsorption effectiveness as compared to a restricted mixing rate is due to the proper distribution of adsorbent across the solution's surface area. In this case, harmful metal contaminants could not adsorb onto bio-sorbent surfaces as a result of the high mixing levels present there (Bilal et al., 2022).

16.4.7 Effect of existing metallic ions

A range of hazardous metallic co-existing ions may be found in wastewater, and these ions can limit the adsorption capacity of the wastewater. Because a certain quantity of an adsorbent offers a few binding sites, these sites become saturated and compete with metal ions, especially at higher concentrations, once the adsorbent has been depleted of its binding sites (Bhatti et al., 2016). Therefore, to evaluate the true effectiveness of adsorbents in real-world, design and assessment of an adsorption system must take into account the interactions between the various components. In a research, using an adsorbent derived from tea industry waste, metals were removed from an aqueous environment in both single and binary (competitive) modes (Çay et al., 2004).

16.4.8 Effect of temperature of the solution

Temperature is also a significant component that should be considered while investigating the adsorption behavior of any adsorbent. Several research have looked at the effect of temperature on adsorption process by bio-waste based adsorbents. It is a critical parameter that affects physicochemical properties and has a major impact on metal remediation. When the temperature was elevated from 25°C to 60°C, it was discovered that enhanced mobility of heavy metals and bond rupturing were responsible for the increase in the removal of metal (Malkoc and Nuhoglu, 2005). After reaching 293°C, the adsorption capacity of Ni(II) dropped linearly with increasing temperature beyond that point, reaching a minimum of 13.15 mg/g (Gupta et al., 2019). Similarly, a research discovered that when the temperature rises, the Cr (VI) absorption capacity increases, suggesting that process is endothermic (Matouq et al., 2015). High temperature reduces the boundary layer's mass transfer barrier, allowing ions to escape more easily. Furthermore, some studies have shown no substantial changes as a function of temperature fluctuations. For example, When the temperature of the adsorbent generated from cucumber peel was adjusted from 20°C to 35°C, the adsorption capacity of the adsorbent remained practically constant for the absorption of Pb(II) (Basu et al., 2017). The optimum temperature for the adsorption process is crucial when evaluating the energy demand of the adsorption process.

16.5 Regeneration of prepared adsorbent

The ability to recycle and regenerate adsorbents are critical for practical adsorption applications, and adsorbents with this capability are considered as cost-effective. In addition to keeping the process's costs low, the adsorbent renewal allows for the recovery of metals. It has been examined in several research whether or not certain regenerated bioadsorbents may be recycled again. Adsorption capacity, as well as biomass regeneration and recycling, are all factors that influence the utilization of adsorbent in wastewater treatment. Many experiments have been conducted recently in which the regeneration of the adsorbent was accomplished by simply utilizing deionized water with varying pH, either acidic or alkaline.

16.5.1 Regeneration using acid

In this technique of regeneration, the saturated adsorbent is immersed in an eluent solution prepared with a strong acid. The regenerated bioadsorbents are ready for the next adsorption cycle after being completely washed and dried. The reagents used in this method are hydrochloric acid (Zbair et al., 2019), nitric acid (Guo et al., 2019), and sulfuric acid (Chen et al., 2019). When redevelopment process is carried out at suitable acid solution concentration, despite the fact that the adsorbent's adsorption efficiency decreases linearly with repeated adsorption–desorption, a large adsorption capacity may still be attained.

16.5.2 Regeneration using alkali

The method is identical to previous one, with the exception that base is utilized to renew the adsorbents instead of acid in this case. This is carried by immersing the saturated adsorbent in a base solution such as sodium hydroxide, etc. (Shakoor et al., 2019). In reality, NaOH solution is used more often in such procedures than dilute ammonia solution, and there is little information available on the redevelopment of adsorbents other than NaOH.

16.5.3 Regeneration using deionized water

Redevelopment of bioadsorbents loaded with metal ions is comparable to the previous two in terms of performance with the exception of using deionized water. The most significant advantage of utilizing deionized water that, it is sustainable, as opposed to previous two, which are harmful and corrosive by nature and hence pollute the environment. One such work is reported here, in which hazardous Cr(VI) was removed from an adsorbent made from Acacia nilotica leaves using simply deionized water (Prasad and Thirumalisamy, 2013). As acid or alkali is more expensive than ordinary deionized water, using distilled water to regenerate harmful bioadsorbents might further cut costs.

16.6 Research gap and research scope regarding low cost adsorbent

Adsorbents based on bio-waste, have come up with sustainable and cost-effective resources for metal remediation. This happened because of their high affinity for metal ions. However, there are a number of hurdles and restrictions that must be overcome in order to fully leverage the incredible properties of those adsorbents in large scale wastewater treatment applications. Because of the complication and diversity of environmental structures, a wide range of study opportunities in the future, since there are still many problems that need to be addressed immediately. After going thoroughly to number of literature following research gap have been found:

- Currently, the majority of researches for adsorption of metals from wastewater are focused on using bio-sorbents on a batch-scale basis. The cost analysis of the prepared adsorbent has not been done extensively. As a result, the commercial viability of these low-cost bioadsorbents must be investigated. The existing synthesis procedures are primarily designed to create a small number of bioadsorbents for laboratory batch applications. Hence, application of these adsorbents must be checked for the practical application. Green environmental approaches can also be used to improve these procedures by reducing the unnecessary usage of chemicals and energy necessities.
- Most current researches are established on short period investigations having insufficient data on long period impacts.
- Before these bioadsorbents may be used commercially, their possible hazardous effects on the environment must be thoroughly assessed. There is limited information on the hazardous consequences of currently used bioadsorbents.
- Adsorption studies, in majority of cases, use a single adsorbent, with only a few studies using multiple adsorbents. As a result, more study is required to develop and examine the potential of hybrid bioadsorbents in terms of improving adsorption capabilities.
- Despite the fact that adsorption has been the dominating and well-established technique for decades, the creation of advanced, low-cost, and long-lasting adsorbents having improved selectivity and stability remains a major issue.
- Incomplete adsorption, huge operating and maintenance expenses, more energy consumption, and very few method for regeneration of adsorbent for subsequent treatments are all concerns that must be addressed in future study.

16.7 Conclusion

Access to clean drinking water is one of the major problems in developing countries with rise in water scarcity, pollution and not having proper mechanism for water treatment making it even more problematic. As a result of different forms of water pollution, metal contamination has become a significant developing problem, prompting several studies to study the application and efficacy of low-cost adsorbents. Bio-sorbents like agricultural waste and their derivatives proved to be more capable while natural soil and mineral deposits least effective. Properties of the material like specific surface chemistry, surface area, etc. along with water quality parameters like pH, temperature, and ionic strength influence effectiveness of adsorbents to remove heavy metal as chemical modification increases the overall adsorption capacities. Ion exchange and electrostatic forces along with water quality conditions are most cited mechanism to remove heavy metals. Uses of these types of adsorbents in the developing world are potentially environmentally sustainable and economically viable in reducing waste disposal and improving water quality. Further research on different aspects needed to be done to overcome various technological and other challenges for successful application of adsorbents metal remediation.

References

Abdelhafez, A.A., Li, J., 2016. Removal of Pb (II) from aqueous solution by using biochars derived from sugar cane bagasse and orange peel. Journal of the Taiwan Institute of Chemical Engineers 1–9. Available from: https://doi.org/10.1016/j.jtice.2016.01.005.

Ahmad, A., Ghazi, Z.A., Saeed, M., Ilyas, M., Ahmad, R., Muqsit Khattak, A., et al., 2017. A comparative study of the removal of Cr(VI) from synthetic solution using natural biosorbents. New Journal of Chemistry 41, 10799–10807. Available from: https://doi.org/10.1039/c7nj02026k.

Ajmal, M., Rao, R.A.K., Ahmad, R., Ahmad, J., 2000. Adsorption studies on Citrus reticulata (fruit peel of orange): removal and recovery of Ni(II) from electroplating wastewater. Journal of Hazardous Materials 79, 117–131. Available from: https://doi.org/10.1016/S0304-3894(00)00234-X.

Al-Anber, Z.A., Matouq, M.A.D., 2008. Batch adsorption of cadmium ions from aqueous solution by means of olive cake. Journal of Hazardous Materials 151, 194–201. Available from: https://doi.org/10.1016/J.JHAZMAT.2007.05.069.

Alidoust, D., Kawahigashi, M., Yoshizawa, S., Sumida, H., Watanabe, M., 2015. Mechanism of cadmium biosorption from aqueous solutions using calcined oyster shells. Journal of Environmental Management 150, 103–110. Available from: https://doi.org/10.1016/J.JENVMAN.2014.10.032.

Almendros, A.I., Martín-Lara, M.A., Ronda, A., Pérez, A., Blázquez, G., Calero, M., 2015. Physico-chemical characterization of pine cone shell and its use as biosorbent and fuel. Bioresource Technology 196, 406–412. Available from: https://doi.org/10.1016/J.BIORTECH.2015.07.109.

Annadurai, G., Juang, R.S., Lee, D.J., 2018. Adsorption of heavy metals from water using banana and orange peels (PDF download available). Water Science and Technology: A Journal of the International Association on Water Pollution Research 185–190.

Appel, C., Ma, L.Q., Rhue, R.D., Reve, W., 2008. Sequential sorption of lead and cadmium in three tropical soils. Environmental Pollution 155, 132–140. Available from: https://doi.org/10.1016/j.envpol.2007.10.026.

Archundia, D., Duwig, C., Spadini, L., Uzu, G., Guédron, S., Morel, M.C., et al., 2017. How uncontrolled urban expansion increases the contamination of the Titicaca Lake Basin (El Alto, La Paz, Bolivia). Water, Air, and Soil Pollution 228. Available from: https://doi.org/10.1007/s11270-016-3217-0.

Bansal, M., Garg, U., Singh, D., Garg, V.K., 2009. Removal of Cr(VI) from aqueous solutions using pre-consumer processing agricultural waste: a case study of rice husk. Journal of Hazardous Materials 162, 312–320. Available from: https://doi.org/10.1016/J.JHAZMAT.2008.05.037.

Bansode, R.R., Losso, J.N., Marshall, W.E., Rao, R.M., Portier, R.J., 2003. Adsorption of metal ions by pecan shell-based granular activated carbons. Bioresource Technology 89, 115–119. Available from: https://doi.org/10.1016/S0960-8524(03)00064-6.

Basu, M., Guha, A.K., Ray, L., 2017. Adsorption of lead on cucumber peel. Journal of Cleaner Production 151, 603–615. Available from: https://doi.org/10.1016/J.JCLEPRO.2017.03.028.

Bhatnagar, A., Minocha, A.K., Sillanpää, M., 2010. Adsorptive removal of cobalt from aqueous solution by utilizing lemon peel as biosorbent. Biochemical Engineering Journal 48, 181–186. Available from: https://doi.org/10.1016/J.BEJ.2009.10.005.

Bhatti, H.N., Zaman, Q., Kausar, A., Noreen, S., Iqbal, M., 2016. Efficient remediation of Zr(IV) using citrus peel waste biomass: kinetic, equilibrium and thermodynamic studies. Ecological Engineering 95, 216–228. Available from: https://doi.org/10.1016/J.ECOLENG.2016.06.087.

Bilal, M., Ihsanullah, I., Younas, M., Ul Hassan Shah, M., 2022. Recent advances in applications of low-cost adsorbents for the removal of heavy metals from water: a critical review. Separation and Purification Technology 278, 119510. Available from: https://doi.org/10.1016/j.seppur.2021.119510.

Boonamnuayvitaya, V., Chaiya, C., Tanthapanichakoon, W., Jarudilokkul, S., 2004. Removal of heavy metals by adsorbent prepared from pyrolyzed coffee residues and clay. Separation and Purification Technology 35, 11–22. Available from: https://doi.org/10.1016/S1383-5866(03)00110-2.

Bozbaş, S.K., Boz, Y., 2016. Low-cost biosorbent: *Anadara inaequivalvis* shells for removal of Pb(II) and Cu(II) from aqueous solution. Process Safety and Environmental Protection 103, 144–152. Available from: https://doi.org/10.1016/J.PSEP0.2016.07.007.

Cârdei, P., Tudora, C., Vlăduț, V., Pruteanu, M.A., Găgeanu, I., Cujbescu, D., et al., 2021. Mathematical model to simulate the transfer of heavy metals from soil to plant. Sustainability 13, 1–18. Available from: https://doi.org/10.3390/su13116157.

Çay, S., Uyanik, A., Özaşik, A., 2004. Single and binary component adsorption of copper(II) and cadmium(II) from aqueous solutions using tea-industry waste. Separation and Purification Technology 38, 273–280. Available from: https://doi.org/10.1016/J.SEPPUR.2003.12.003.

Changmai, M., Banerjee, P., Nahar, K., Purkait, M.K., 2018. A novel adsorbent from carrot, tomato and polyethylene terephthalate waste as a potential adsorbent for Co (II) from aqueous solution: kinetic and equilibrium studies. Journal of Environmental Chemical Engineering 6, 246–257. Available from: https://doi.org/10.1016/J.JECE.2017.12.009.

Chen, H., Osman, A.I., Mangwandi, C., Rooney, D., 2019. Upcycling food waste digestate for energy and heavy metal remediation applications. Resources, Conservation & Recycling X 3, 100015. Available from: https://doi.org/10.1016/J.RCRX.2019.100015.

Conradi, E., Gonçalves, A.C., Schwantes, D., Manfrin, J., Schiller, A., Zimmerman, J., et al., 2019. Development of renewable adsorbent from cigarettes for lead removal from water. Journal of Environmental Chemical Engineering 7, 103200. Available from: https://doi.org/10.1016/J.JECE.2019.103200.

Dakiky, M., Khamis, M., Manassra, A., Mer'eb, M., 2002. Selective adsorption of chromium(VI) in industrial wastewater using low-cost abundantly available adsorbents. Advances in Environmental Research 6, 533–540. Available from: https://doi.org/10.1016/S1093-0191(01)00079-X.

De Mello Ferreira, A., Marchesiello, M., Thivel, P.X., 2013. Removal of copper, zinc and nickel present in natural water containing Ca^{2+} and HCO_3-ions by electrocoagulation. Separation and Purification Technology 107, 109–117. Available from: https://doi.org/10.1016/j.seppur.2013.01.016.

DeMessie, B., Sahle-Demessie, E., Sorial, G.A., 2015. Cleaning water contaminated with heavy metal ions using pyrolyzed biochar adsorbents. Separation Science and Technology 50, 2448–2457. Available from: https://doi.org/10.1080/01496395.2015.1064134.

Demirak, A., Yilmaz, F., Levent Tuna, A., Ozdemir, N., 2006. Heavy metals in water, sediment and tissues of Leuciscus cephalus from a stream in southwestern Turkey. Chemosphere 63, 1451–1458. Available from: https://doi.org/10.1016/J.CHEMOSPHERE.2005.09.033.

Du, Y., Lian, F., Zhu, L., 2011. Biosorption of divalent Pb, Cd and Zn on aragonite and calcite mollusk shells. Environmental Pollution (Barking, Essex: 1987) 159, 1763–1768. Available from: https://doi.org/10.1016/J.ENVPOL.2011.04.017.

Elangovan, R., Philip, L., Chandraraj, K., 2008. Biosorption of chromium species by aquatic weeds: kinetics and mechanism studies. Journal of Hazardous Materials 152, 100–112. Available from: https://doi.org/10.1016/J.JHAZMAT.2007.06.067.

Elouear, Z., Bouzid, J., Boujelben, N., Feki, M., Jamoussi, F., Montiel, A., 2008. Heavy metal removal from aqueous solutions by activated phosphate rock. Journal of Hazardous Materials 156, 412–420. Available from: https://doi.org/10.1016/J.JHAZMAT.2007.12.036.

Feng, N., Guo, X., Liang, S., Zhu, Y., Liu, J., 2011. Biosorption of heavy metals from aqueous solutions by chemically modified orange peel. Journal of Hazardous Materials 185, 49–54. Available from: https://doi.org/10.1016/J.JHAZMAT.2010.08.114.

Georgieva, V.G., Tavlieva, M.P., Genieva, S.D., Vlaev, L.T., 2015. Adsorption kinetics of Cr(VI) ions from aqueous solutions onto black rice husk ash. Journal of Molecular Liquids 208, 219–226. Available from: https://doi.org/10.1016/J.MOLLIQ.2015.04.047.

Gong, J.L., Wang, X.Y., Zeng, G.M., Chen, L., Deng, J.H., Zhang, X.R., et al., 2012. Copper (II) removal by pectin–iron oxide magnetic nanocomposite adsorbent. Chemical Engineering Journal 185–186, 100–107. Available from: https://doi.org/10.1016/J.CEJ.2012.01.050.

Guechi, E.K., Hamdaoui, O., 2016. Evaluation of potato peel as a novel adsorbent for the removal of Cu(II) from aqueous solutions: equilibrium, kinetic, and thermodynamic studies. Desalination and Water Treatment 57, 10677–10688. Available from: https://doi.org/10.1080/19443994.2015.1038739.

Guo, X., Zhang, S., Shan, X.quan, 2008. Adsorption of metal ions on lignin. Journal of Hazardous Materials 151, 134–142. Available from: https://doi.org/10.1016/J.JHAZMAT.2007.05.065.

Guo, J., Song, Y., Ji, X., Ji, L., Cai, L., Wang, Y., et al., 2019. Preparation and characterization of nanoporous activated carbon derived from prawn shell and its application for removal of heavy metal ions. Materials (Basel) 12. Available from: https://doi.org/10.3390/ma12020241.

Gupta, V.K., Nayak, A., 2012. Cadmium removal and recovery from aqueous solutions by novel adsorbents prepared from orange peel and Fe2O3 nanoparticles. Chemical Engineering Journal 180, 81–90. Available from: https://doi.org/10.1016/J.CEJ.2011.11.006.

Gupta, S., Sharma, S.K., Kumar, A., 2019. Biosorption of Ni(II) ions from aqueous solution using modified Aloe barbadensis Miller leaf powder. Water Science and Engineering 12, 27–36. Available from: https://doi.org/10.1016/J.WSE.2019.04.003.

Gurmen, S., Ebin, B., Stopi??, S., Friedrich, B., 2009. Nanocrystalline spherical iron-nickel (Fe-Ni) alloy particles prepared by ultrasonic spray pyrolysis and hydrogen reduction (USP-HR). J. Alloys Compd. Available from: https://doi.org/10.1016/j.jallcom.2009.01.094.

Hashim, M.A., Chu, K.H., 2004. Biosorption of cadmium by brown, green, and red seaweeds. Chemical Engineering Journal 97, 249–255. Available from: https://doi.org/10.1016/S1385-8947(03)00216-X.

Herat, S., Agamuthu, P., 2012. E-waste: A problem or an opportunity? Review of issues, challenges and solutions in Asian countries. Waste Management & Research: The Journal of the International Solid Wastes and Public Cleansing Association, ISWA 30, 1113–1129. Available from: https://doi.org/10.1177/0734242X12453378.

Ho, Y.S., Huang, C.T., Huang, H.W., 2002. Equilibrium sorption isotherm for metal ions on tree fern. Process Biochemistry 37, 1421–1430. Available from: https://doi.org/10.1016/S0032-9592(02)00036-5.

Kambole, M.S., 2003. Managing the water quality of the Kafue River. Physics and Chemistry of the Earth, Parts A/B/C 28, 1105–1109. Available from: https://doi.org/10.1016/J.PCE.2003.08.031.

Krishnani, K.K., Meng, X., Christodoulatos, C., Boddu, V.M., 2008. Biosorption mechanism of nine different heavy metals onto biomatrix from rice husk. Journal of Hazardous Materials 153, 1222–1234. Available from: https://doi.org/10.1016/J.JHAZMAT.2007.09.113.

Kul, A.R., Koyuncu, H., 2010. Adsorption of Pb(II) ions from aqueous solution by native and activated bentonite: kinetic, equilibrium and thermodynamic study. Journal of Hazardous Materials 179, 332–339. Available from: https://doi.org/10.1016/j.jhazmat.2010.03.009.

Kumar, P.S., Ramalingam, S., Kirupha, S.D., Murugesan, A., Vidhyadevi, T., Sivanesan, S., 2011. Adsorption behavior of nickel(II) onto cashew nut shell: equilibrium, thermodynamics, kinetics, mechanism and process design. Chemical Engineering Journal 167, 122–131. Available from: https://doi.org/10.1016/J.CEJ.2010.12.010.

Lam, B., Déon, S., Morin-Crini, N., Crini, G., Fievet, P., 2018. Polymer-enhanced ultrafiltration for heavy metal removal: influence of chitosan and carboxymethyl cellulose on filtration performances. Journal of Cleaner Production 171, 927–933. Available from: https://doi.org/10.1016/J.JCLEPRO.2017.10.090.

Leyva-Ramos, R., Bernal-Jacome, L.A., Acosta-Rodriguez, I., 2005. Adsorption of cadmium(II) from aqueous solution on natural and oxidized corncob. Separation and Purification Technology 45, 41–49. Available from: https://doi.org/10.1016/J.SEPPUR.2005.02.005.

Li, Q., Zhai, J., Zhang, W., Wang, M., Zhou, J., 2007. Kinetic studies of adsorption of Pb(II), Cr(III) and Cu(II) from aqueous solution by sawdust and modified peanut husk. Journal of Hazardous Materials 141, 163–167. Available from: https://doi.org/10.1016/J.JHAZMAT.2006.06.109.

Li, Z., Imaizumi, S., Katsumi, T., Inui, T., Tang, X., Tang, Q., 2010. Manganese removal from aqueous solution using a thermally decomposed leaf. Journal of Hazardous Materials 177, 501–507. Available from: https://doi.org/10.1016/J.JHAZMAT.2009.12.061.

Li, X., Zhang, D., Sheng, F., Qing, H., 2018. Adsorption characteristics of copper (II), zinc (II) and mercury (II) by four kinds of immobilized fungi residues. Ecotoxicology and Environmental Safety 147, 357–366. Available from: https://doi.org/10.1016/J.ECOENV.2017.08.058.

Liang, F.B., Song, Y.L., Huang, C.P., Zhang, J., Chen, B.H., 2013. Adsorption of hexavalent chromium on a lignin-based resin: equilibrium, thermodynamics, and kinetics. Journal of Environmental Chemical Engineering 1, 1301–1308. Available from: https://doi.org/10.1016/J.JECE.2013.09.025.

Lingamdinne, L.P., Vemula, K.R., Chang, Y.Y., Yang, J.K., Karri, R.R., Koduru, J.R., 2020. Process optimization and modeling of lead removal using iron oxide nanocomposites generated from bio-waste mass. Chemosphere 243, 125257. Available from: https://doi.org/10.1016/J.CHEMOSPHERE.2019.125257.

Majumder, S., Gangadhar, G., Raghuvanshi, S., Gupta, S., 2015. Biofilter column for removal of divalent copper from aqueous solutions: performance evaluation and kinetic modeling. Journal of Water Process Engineering . Available from: https://doi.org/10.1016/j.jwpe.2015.03.008.

Malkoc, E., Nuhoglu, Y., 2005. Investigations of nickel(II) removal from aqueous solutions using tea factory waste. Journal of Hazardous Materials 127, 120–128. Available from: https://doi.org/10.1016/J.JHAZMAT.2005.06.030.

Mashhadi, S., Sohrabi, R., Javadian, H., Ghasemi, M., Tyagi, I., Agarwal, S., et al., 2016. Rapid removal of Hg (II) from aqueous solution by rice straw activated carbon prepared by microwave-assisted H2SO4 activation: kinetic, isotherm and thermodynamic studies. Journal of Molecular Liquids 215, 144–153. Available from: https://doi.org/10.1016/J.MOLLIQ.2015.12.040.

Matouq, M., Jildeh, N., Qtaishat, M., Hindiyeh, M., Al Syouf, M.Q., 2015. The adsorption kinetics and modeling for heavy metals removal from wastewater by Moringa pods. Journal of Environmental Chemical Engineering 3, 775–784. Available from: https://doi.org/10.1016/J.JECE.2015.03.027.

Meitei, M.D., Prasad, M.N.V., 2014. Adsorption of Cu (II), Mn (II) and Zn (II) by Spirodela polyrhiza (L.) Schleiden: equilibrium, kinetic and thermodynamic studies. Ecological Engineering 71, 308–317. Available from: https://doi.org/10.1016/J.ECOLENG.2014.07.036.

Milly, P.C.D., Wetherald, R.T., Dunne, K.A., Delworth, T.L., 2002. Increasing risk of great floods in a changing climate. Nature 415, 514–517. Available from: https://doi.org/10.1038/415514a.

Moubarik, A., Grimi, N., 2015. Valorization of olive stone and sugar cane bagasse by-products as biosorbents for the removal of cadmium from aqueous solution. Food Research International 73, 169–175. Available from: https://doi.org/10.1016/J.FOODRES.2014.07.050.

Moussavi, G., Barikbin, B., 2010. Biosorption of chromium(VI) from industrial wastewater onto pistachio hull waste biomass. Chemical Engineering Journal 162, 893–900. Available from: https://doi.org/10.1016/J.CEJ.2010.06.032.

Moyo, M., Pakade, V.E., Modise, S.J., 2017. Biosorption of lead(II) by chemically modified Mangifera indica seed shells: adsorbent preparation, characterization and performance assessment. Process Safety and Environmental Protection 111, 40–51. Available from: https://doi.org/10.1016/J.PSEP.2017.06.007.

Mritunjay, Quaff, A.R., 2021. Adsorption of copper on activated Ganga sand from aqueous solution: kinetics, isotherm, and optimization. International Journal of Environmental Science and Technology. Available from: https://doi.org/10.1007/s13762-021-03651-1.

Nigam, M., Rajoriya, S., Rani Singh, S., Kumar, P., 2019. Adsorption of Cr (VI) ion from tannery wastewater on tea waste: kinetics, equilibrium and thermodynamics studies. Journal of Environmental Chemical Engineering 7, 103188. Available from: https://doi.org/10.1016/J.JECE.2019.103188.

Noyes, P.D., McElwee, M.K., Miller, H.D., Clark, B.W., Van Tiem, L.A., Walcott, K.C., et al., 2009. The toxicology of climate change: environmental contaminants in a warming world. Environment International 35, 971–986. Available from: https://doi.org/10.1016/J.ENVINT.2009.02.006.

Olafisoye, O.B., Adefioye, T., Osibote, O.A., 2013. Introduction. Polish Journal of Environmental Studies 22, 1431–1439.

Pan, J., Gao, B., Song, W., Xu, X., Yue, Q., 2020. Modified biogas residues as an eco-friendly and easily-recoverable biosorbent for nitrate and phosphate removals from surface water. Journal of Hazardous Materials 382, 121073. Available from: https://doi.org/10.1016/J.JHAZMAT.2019.121073.

Paranavithana, G.N., Kawamoto, K., Inoue, Y., Saito, T., Vithanage, M., Kalpage, C.S., et al., 2016. Adsorption of Cd^{2+} and Pb^{2+} onto coconut shell biochar and biochar-mixed soil. Environmental Earth Sciences 75. Available from: https://doi.org/10.1007/s12665-015-5167-z.

Park, D., Lim, S.R., Yun, Y.S., Park, J.M., 2008. Development of a new Cr(VI)-biosorbent from agricultural biowaste. Bioresource Technology 99, 8810–8818. Available from: https://doi.org/10.1016/J.BIORTECH.2008.04.042.

Perkins, D.N., Brune Drisse, M.N., Nxele, T., Sly, P.D., 2014. E-waste: a global hazard. Annals of Global Health 80, 286–295. Available from: https://doi.org/10.1016/j.aogh.2014.10.001.

Prasad, A.L., Thirumalisamy, S., 2013. Evaluation of the use of Acacia nilotica leaf as an ecofriendly adsorbent for Cr (VI) and its suitability in real waste water: study of residual errors. Journal of Chemistry 2013. Available from: https://doi.org/10.1155/2013/354328.

Qi, B.C., Aldrich, C., 2008. Biosorption of heavy metals from aqueous solutions with tobacco dust. Bioresource Technology 99, 5595–5601. Available from: https://doi.org/10.1016/J.BIORTECH.2007.10.042.

Qin, F., Wen, B., Shan, X.Q., Xie, Y.N., Liu, T., Zhang, S.Z., et al., 2006. Mechanisms of competitive adsorption of Pb, Cu, and Cd on peat. Environmental Pollution (Barking, Essex: 1987) 144, 669–680. Available from: https://doi.org/10.1016/J.ENVPOL.2005.12.036.

Rafatullah, M., Sulaiman, O., Hashim, R., Ahmad, A., 2009. Adsorption of copper (II), chromium (III), nickel (II) and lead (II) ions from aqueous solutions by meranti sawdust. Journal of Hazardous Materials 170, 969–977. Available from: https://doi.org/10.1016/J.JHAZMAT.2009.05.066.

Rao, R.A.K., Khan, M.A., 2009. Biosorption of bivalent metal ions from aqueous solution by an agricultural waste: kinetics, thermodynamics and environmental effects. Colloids and Surfaces A: Physicochemical and Engineering Aspects 332, 121–128. Available from: https://doi.org/10.1016/J.COLSURFA.2008.09.005.

Rhaman, M.M., Karim, M.R., Hyder, M.K.M.Z., Ahmed, Y., Nath, R.K., 2020. Removal of chromium (VI) from effluent by a magnetic bioadsorbent based on jute stick powder and its adsorption isotherm, kinetics and regeneration study. Water, Air, and Soil Pollution 231. Available from: https://doi.org/10.1007/s11270-020-04544-8.

Romera, E., González, F., Ballester, A., Blázquez, M.L., Muñoz, J.Á., 2008. Biosorption of Cd, Ni, and Zn with mixtures of different types of algae. Environmental Engineering Science 25, 999–1008. Available from: https://doi.org/10.1089/ees.2007.0122.

Sari, A., Tuzen, M., 2008. Biosorption of total chromium from aqueous solution by red algae (Ceramium virgatum): equilibrium, kinetic and thermodynamic studies. Journal of Hazardous Materials 160, 349–355. Available from: https://doi.org/10.1016/J.JHAZMAT.2008.03.005.

Semerjian, L., 2018. Removal of heavy metals (Cu, Pb) from aqueous solutions using pine (Pinus halepensis) sawdust: equilibrium, kinetic, and thermodynamic studies. Environmental Technology & Innovation 12, 91–103. Available from: https://doi.org/10.1016/J.ETI.2018.08.005.

SenthilKumar, P., Ramalingam, S., Sathyaselvabala, V., Kirupha, S.D., Sivanesan, S., 2011. Removal of copper(II) ions from aqueous solution by adsorption using cashew nut shell. Desalination 266, 63–71. Available from: https://doi.org/10.1016/J.DESAL.2010.08.003.

Sethy, T.R., Pradhan, A.K., Sahoo, P.K., 2019. Simultaneous studies on kinetics, bio-adsorption behaviour of chitosan grafted thin film nanohydrogel for removal of hazardous metal ion from water. Environmental Nanotechnology, Monitoring & Management 12, 100262. Available from: https://doi.org/10.1016/J.ENMM.2019.100262.

Shakoor, M.B., Niazi, N.K., Bibi, I., Shahid, M., Saqib, Z.A., Nawaz, M.F., et al., 2019. Exploring the arsenic removal potential of various biosorbents from water. Environment International 123, 567–579. Available from: https://doi.org/10.1016/J.ENVINT.2018.12.049.

Shan, T.C., Matar, M., Al, Makky, E.A., Ali, E.N., 2017. The use of Moringa oleifera seed as a natural coagulant for wastewater treatment and heavy metals removal. Applied Water Science 7, 1369–1376. Available from: https://doi.org/10.1007/s13201-016-0499-8.

Shi, J., Fang, Z., Zhao, Z., Sun, T., Liang, Z., 2016. Comparative study on Pb(II), Cu(II), and Co(II) ions adsorption from aqueous solutions by arborvitae leaves. Desalination and Water Treatment 57, 4732–4739. Available from: https://doi.org/10.1080/19443994.2015.1089421.

Shukla, S.R., Pai, R.S., 2005. Adsorption of Cu(II), Ni(II) and Zn(II) on dye loaded groundnut shells and sawdust. Separation and Purification Technology 43, 1–8. Available from: https://doi.org/10.1016/J.SEPPUR.2004.09.003.

Singha, B., Das, S.K., 2011. Biosorption of Cr(VI) ions from aqueous solutions: kinetics, equilibrium, thermodynamics and desorption studies. Colloids Surfaces B Biointerfaces 84, 221–232. Available from: https://doi.org/10.1016/J.COLSURFB.2011.01.004.

Singha, B., Das, S.K., 2013. Colloids and surfaces B: Biointerfaces adsorptive removal of Cu (II) from aqueous solution and industrial effluent using natural / agricultural wastes. Colloids Surfaces B Biointerfaces 107, 97–106. Available from: https://doi.org/10.1016/j.colsurfb.2013.01.060.

Srivastava, S., Agrawal, S.B., Mondal, M.K., 2015. Biosorption isotherms and kinetics on removal of Cr(VI) using native and chemically modified Lagerstroemia speciosa bark. Ecological Engineering 85, 56–66. Available from: https://doi.org/10.1016/j.ecoleng.2015.10.011.

Sudha Rani, K., Srinivas, B., Gourunaidu, K., Ramesh, K.V., 2018. Removal of copper by adsorption on treated laterite. Materials Today: Proceedings . Available from: https://doi.org/10.1016/j.matpr.2017.11.106.

Sun, Y., Zhou, S., Pan, S.Y., Zhu, S., Yu, Y., Zheng, H., 2020. Performance evaluation and optimization of flocculation process for removing heavy metal. Chemical Engineering Journal 385, 123911. Available from: https://doi.org/10.1016/J.CEJ.2019.123911.

Tan, G., Yuan, H., Liu, Y., Xiao, D., 2010. Removal of lead from aqueous solution with native and chemically modified corncobs. Journal of Hazardous Materials 174, 740–745. Available from: https://doi.org/10.1016/J.JHAZMAT.2009.09.114.

Tang, Q., Tang, X., Li, Z., Chen, Y., Kou, N., Sun, Z., 2009. Adsorption and desorption behaviour of Pb(II) on a natural kaolin: equilibrium, kinetic and thermodynamic studies. Journal of Chemical Technology and Biotechnology (Oxford, Oxfordshire: 1986) 84, 1371–1380. Available from: https://doi.org/10.1002/jctb.2192.

Taşar, Ş., Kaya, F., Özer, A., 2014. Biosorption of lead(II) ions from aqueous solution by peanut shells: equilibrium, thermodynamic and kinetic studies. Journal of Environmental Chemical Engineering 2, 1018–1026. Available from: https://doi.org/10.1016/J.JECE.2014.03.015.

Thirumavalavan, M., Lai, Y., Lin, L., Lee, J., 2010. Cellulose-based native and surface modified fruit peels for the adsorption of heavy metal ions from aqueous solution: langmuir adsorption isotherms. Journal of Chemical & Engineering Data 1186–1192.

Tiede, K., Neumann, T., Stueben, D., 2007. Suitability of Mn-oxyhydroxides from Karst caves as filter material for drinking water treatment in Gunung Sewu, Indonesia. Journal of Soils and Sediments 7, 53–58. Available from: https://doi.org/10.1065/jss2006.11.195.

Torab-Mostaedi, M., Asadollahzadeh, M., Hemmati, A., Khosravi, A., 2013. Equilibrium, kinetic, and thermodynamic studies for biosorption of cadmium and nickel on grapefruit peel. Journal of the Taiwan Institute of Chemical Engineers 44, 295–302. Available from: https://doi.org/10.1016/J.JTICE.2012.11.001.

Veli, S., Pekey, B., 2004. Removal of copper from aqueous solution by ion exchange resins. Fresenius Environmental Bulletin .

Vijayaraghavan, K., Palanivelu, K., Velan, M., 2006. Biosorption of copper(II) and cobalt(II) from aqueous solutions by crab shell particles. Bioresource Technology 97, 1411–1419. Available from: https://doi.org/10.1016/J.BIORTECH.2005.07.001.

Villaescusa, I., Fiol, N., Martínez, M., Miralles, N., Poch, J., Serarols, J., 2004. Removal of copper and nickel ions from aqueous solutions by grape stalks wastes. Water Research 38, 992–1002. Available from: https://doi.org/10.1016/J.WATRES.2003.10.040.

Vimala, R., Das, N., 2009. Biosorption of cadmium (II) and lead (II) from aqueous solutions using mushrooms: a comparative study. Journal of Hazardous Materials 168, 376–382. Available from: https://doi.org/10.1016/J.JHAZMAT.2009.02.062.

Wasewar, K.L., Atif, M., Prasad, B., Mishra, I.M., 2009. Batch adsorption of zinc on tea factory waste. Desalination 244, 66–71. Available from: https://doi.org/10.1016/J.DESAL.2008.04.036.

Witek-Krowiak, A., Harikishore Kumar Reddy, D., 2013. Removal of microelemental Cr(III) and Cu(II) by using soybean meal waste – unusual isotherms and insights of binding mechanism. Bioresource Technology 127, 350–357. Available from: https://doi.org/10.1016/J.BIORTECH.2012.09.072.

Wu, Y., Zhang, S., Guo, X., Huang, H., 2008. Adsorption of chromium(III) on lignin. Bioresource Technology 99, 7709–7715. Available from: https://doi.org/10.1016/J.BIORTECH.2008.01.069.

Wu, Q., Leung, J.Y.S., Geng, X., Chen, S., Huang, X., Li, H., et al., 2015. Heavy metal contamination of soil and water in the vicinity of an abandoned e-waste recycling site: implications for dissemination of heavy metals. The Science of the Total Environment 506–507, 217–225. Available from: https://doi.org/10.1016/J.SCITOTENV.2014.10.121.

Xia, J., Duan, Q.Y., Luo, Y., Xie, Z.H., Liu, Z.Y., Mo, X.G., 2017. Climate change and water resources: case study of Eastern Monsoon Region of China. Advances in Climate Change Research 8, 63–67. Available from: https://doi.org/10.1016/j.accre.2017.03.007.

Yang, S., Wu, Y., Aierken, A., Zhang, M., Fang, P., Fan, Y., et al., 2016. Mono/competitive adsorption of arsenic (III) and nickel(II) using modified green tea waste. Journal of the Taiwan Institute of Chemical Engineers 60, 213–221. Available from: https://doi.org/10.1016/J.JTICE.2015.07.007.

Yu, L.J., Shukla, S.S., Dorris, K.L., Shukla, A., Margrave, J.L., 2003. Adsorption of chromium from aqueous solutions by maple sawdust. Journal of Hazardous Materials 100, 53–63. Available from: https://doi.org/10.1016/S0304-3894(03)00008-6.

Zbair, M., Ait Ahsaine, H., Anfar, Z., Slassi, A., 2019. Carbon microspheres derived from walnut shell: rapid and remarkable uptake of heavy metal ions, molecular computational study and surface modeling. Chemosphere 231, 140–150. Available from: https://doi.org/10.1016/J.CHEMOSPHERE.2019.05.120.

Zhang, M., 2011. Adsorption study of Pb(II), Cu(II) and Zn(II) from simulated acid mine drainage using dairy manure compost. Chemical Engineering Journal 172, 361–368. Available from: https://doi.org/10.1016/j.cej.2011.06.017.

Zhu, C.S., Wang, L.P., Chen, W.bin, 2009. Removal of Cu(II) from aqueous solution by agricultural by-product: peanut hull. Journal of Hazardous Materials 168, 739–746. Available from: https://doi.org/10.1016/J.JHAZMAT.2009.02.085.

CHAPTER

17

Metals removal by membrane filtration

Majid Peyravi and Hossein Rezaei

Department of Chemical Engineering, Babol Noshirvani University of Technology, Babol, Iran

17.1 Introduction

Reducing drinking water resources and increasing their use in the industry requires global thinking and solidarity for innovative water management practices. The metal will be considered a heavy metal if it has atomic weights between about 63 and 200 and a relatively high density (Bach et al., 2012; Srivastava and Majumder, 2008). Heavy metals are one of the most important pollutants in water because they limit their use and also cause serious damage to the environment because they are environmentally sustainable. Heavy metals are made up of more than 50 different elements, 17 of which are among the most prevalent and toxic elements. The percentage of toxicity of heavy metals can be classified by their electronegativity, which is consistent with the stability of the complexes derived from these metals. Fig. 17.1 shows some of the most important sources of heavy metals. Bivalent metals and metals with high electronegativity form insoluble sulfurs and have toxic effects due to their high affinity for proteins, especially enzymes, in high concentrations. The most frequent heavy metals that are found in factory wastewater like plating, battery manufacturing and electronic components are nickel (Ni), zinc (Zn), silver (Ag), lead (Pb), iron (Fe), chromium (Cr), mercury (Hg), arsenic (As), uranium, copper (Cu) and cadmium (Cd). The presence of some heavy metal elements like Fe, Cu, Zn is nutritionally important for humans. Other heavy metals include Cd, As, Pb, Cr and Hg are toxic even in very small amounts because they can accumulate in the major systems of the human body and are non-degradable (Zak, 2012). Under typical conditions, the human body can endure a certain amount of heavy metals (Al-Rashdi et al., 2013). As a result, the World Health Organization (WHO) established maximum contamination levels (MCLs) to assure that zero or only a certain level of heavy metals are allowed in a water supply (Algureiri and Abdulmajeed, 2016; Hua et al., 2012). In this chapter, we examine the membrane processes, their effective and interesting role in wastewater treatment.

Metals in Water
DOI: https://doi.org/10.1016/B978-0-323-95919-3.00014-8

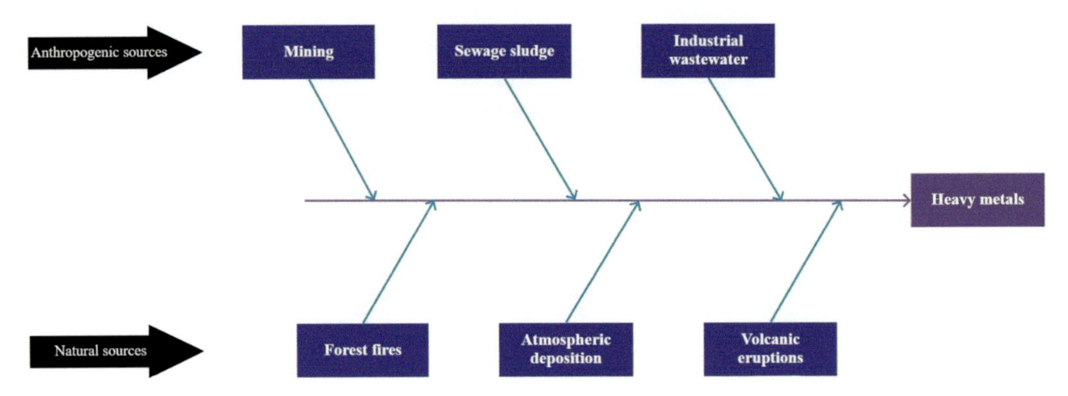

FIGURE 17.1 Shows sources of heavy metals (**Naeem et al., 2019**). Source: *Modified from Naeem, M., Mujahid, M., Umer, A., Ahmad, S., Ahmad, G., Ali, J., et al., 2019. New trends in removing toxic metals from drinking and wastewater by biomass materials and advanced membrane technologies. Journal of Biodiversity and Environmental Sciences 15, 10–17.*

17.2 Heavy metal separation methods

Most of the heavy metals released into the environment by human activities are or would be in an aqueous solution, which should be removed before surface water recycling or direct discharge. Therefore, humans need effective methods of purifying water. Common processes for this type of wastewater treatment include ion exchange, coagulation-flocculation, biosorption, adsorption, flotation, electrochemical methods and in recent years membrane filtration (Kurniawan et al., 2006; Seader and Henley, 2006). In the following, we will review some techniques for separating heavy metals and then describe the membrane filtration methods.

17.2.1 Solvent extraction

In the extraction method, limitations in the dissolution of metal salts in organic solvents, the high cost of solvents and the need for ancillary facilities such as solvent recycling are the disadvantages of this method. The precipitation process as a purification method has several advantages. This method is relatively inexpensive and the construction of new units does not require large investments. However, the usage of sediment in the elimination of heavy metals shows a large volume of hydrogen sludge that must be disposed of and is very expensive (Mukherjee, 2019).

17.2.2 Ion exchange

In this process, undesirable ions replace some of the neutral ions. This method is not technologically complex and has good efficiency and speed for ion removal even in small quantities. In this process, by passing a heavy metal contaminated stream under pressure from a column containing a resin bed, the metal ions are trapped and thus removed from the environment. Once the resin capacity has reached its saturation level, the trapped solids are removed and the column is regenerated by backwashing. Both natural and synthetic resins have the ability to exchange their cations with the metal ions in the environment, but synthetic resins are usually preferred over the elimination of heavy metals owing to better efficiency and lower cost. The advantages of this process over chemical deposition include recovery of precious metals, selectability, less sludge production and better observance of output flow standards. The most important disadvantages of this method are such as specificity (lack of suitable resins to remove all metal ions), the need for a pre-treatment unit to remove suspended solids and incomplete removal of metals (due to substrate saturation). However, this method, unlike sedimentation, has high management costs It does not contain sludge, it should not be unaware of its high start-up and operating costs (Shafaei et al., 2007).

17.2.3 Biosorption

Low cost and adaptation to the environment are important factors that the use of biological materials (methods) for the treatment of heavy metals has been accepted and the focus of various communities. Biological methods have advantages such as low operating costs, selective removal for specific metals, minimal use of chemicals (resulting in sludge) and high efficiencies for low metal concentrations. Biosorption and bioaccumulation as effective methods have the potential to replace common processes in metal removal. In biosorption of dead (inactive) residues of microorganisms has been used to eliminate heavy metals and based on cellular metabolism, this method is known as a passive process. Biosorption is a simple physicochemical phenomenon that bears a strong resemblance to surface adsorption and ion exchange methods. The only difference is the use of biosorbents such as bacteria, yeasts, algae, fungi, plant wastes and polysaccharides. These biodegradable materials, due to their inherent properties in the formation of complexes with metal ions, cause the elimination of heavy metals from the environment. The choice of biosorbents depends on factors such as origin, availability and cost-effectiveness. These materials can be collected directly from the environment or extracted from the activities of the food and pharmaceutical industries. By modifying biosorbents by physical, chemical or genetic methods, their adsorption properties can be improved. Physical methods such as heating, by removing impurities from the surface and generating active sites to form metal bonds (by changes in the cell wall of the protein) increase the adsorption capacity. In this process, the metal adheres to the surface of the molecules like s-layer protein. This bonding occurs based on one or a combination of physical adsorption processes, electrostatic interactions, complex formation, or microprecipitation (Wang and Chen, 2009).

17.2.4 Adsorption

Adsorption is the process of adsorption of atoms and molecules in a fluid by the solid surface away from equilibrium, which the solid tends to capture some of these ions to reach equilibrium. The adsorption process usually begins with a long-range van der Waals force and ends with a short-range force such as metal and ionic bonds. The surface adsorption method is one of the most generally used methods for the removal of heavy metals due to its high efficiency and ease of application. In this procedure, contaminated water is passed via a bed or filter media, and arsenic and heavy metal are separated from the water by bonds with the adsorbent. Activated alumina, activated carbon and iron oxide are among the ethane solids that are widely used as adsorbents. In general, the adsorption phenomenon occurs due to the properties on the solid surface (Babel and Kurniawan, 2003). Recently, the use of iron and zero-capacity iron nanocomposites has been considered due to their abundance, cheapness, non-toxicity, availability, high capacity and efficiency in the decomposition of pollutants and elimination of arsenic and heavy metals from water. Findings show that the reactivity of zero-capacity iron nanoparticles increases significantly with decreasing size at the nanoscale. Important cheap adsorbents include powdered activated carbon, zero-capacity iron, vegetable waste, sawdust, starch, etc. Advantages of the adsorption method include obtaining the removed metal in case of recovery from the adsorbent surface, a suitable method for removing heavy metals in dilute solutions, low cost and simple process design. Disadvantages of this method include the loss of adsorbent efficiency over time and the impossibility of active contaminants from the adsorbent in many cases (Lofrano, 2012). As noted by Hidalgo-Vázquez et al. (2011), the adsorption technique due to its simplicity of design and ability to spend less in terms of initial cost offers many advantages over other separation methods.

17.3 Membrane filtration processes

Nowadays, membranes for producing drinking water from seawater using reverse osmosis (RO) phenomenon, treatment and recycling of valuable materials from industrial plant effluents by electrodialysis (ED) and nanofiltration (NF), treatment of industrial effluents using filtration processes [microfiltration (MF), ultrafiltration (UF), and NF] and membrane bioreactors (MBRs) process, separation of macromolecular solutions in pharmaceutical and food industries using UF, elimination of urea and other toxic substances from the blood system by dialysis, separation of gases for nitrogen production, desalination of natural gas, recycling of valuable gases used in petrochemical processes. A membrane is a physical layer that selectively separates two phases and inhibits the transport of particular substances in solution through the membrane surface due to chemical potential differences (Cardew, 2007). The membrane filtering technique is comprised of five primary treatment processes: ED, NF, RO, MF and UF which are the main pressure membranes, are the most common membrane processes used to isolate endocrine disorders (Seader and Henley, 2006). Different driving forces cause the mass to move across the membrane. There are membrane processes that include equilibrium-based, non-equilibrium-based, pressure-driven, and non-pressure-driven (Jhaveri and Murthy, 2016). UF membrane is a porous membrane that can operate at low pressures and has pore sizes between 0.001 and 0.1 μm and an operating pressure of about 10–500 kpa. MF membranes are porous like UF and the pore size of MF membranes covers the range of 0.0001–0.1 mm and pressures of about 10–200 kpa (Leibundgut et al., 2011). RO membranes are non-porous and operate under membranes with low nanometer pores about 10–20 times lower operating pressures (Koyuncu et al., 2015; Leibundgut et al., 2011). NF has a pore size of 0.001 μm and operates

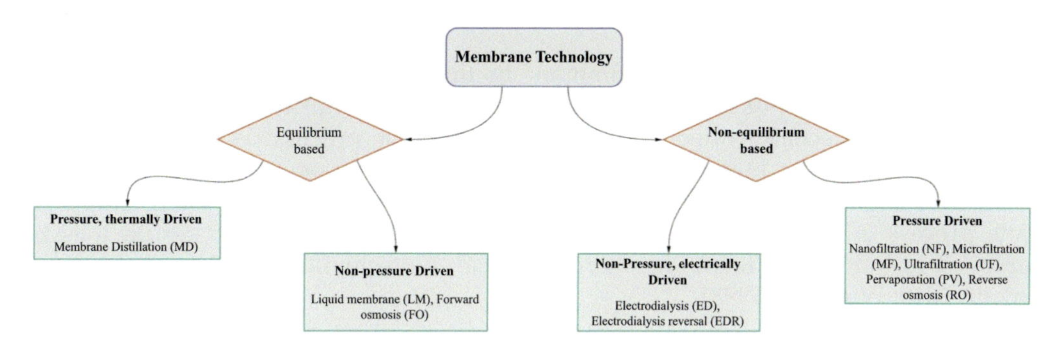

FIGURE 17.2 Some membrane processes are depicted schematically. Source: *Modified from Jhaveri, J.H., Murthy, Z., 2016. A comprehensive review on anti-fouling nanocomposite membranes for pressure driven membrane separation processes. Desalination 379, 137–154.*

at pressures ranging from 10 to 20 bar (Obotey Ezugbe and Rathilal, 2020). Fig. 17.2 summarizes some of these methods based on their driving forces.

The first use of membranes on a laboratory scale in the 1920s was used to isolate bacteria in MF operations. Membrane separation processes has been developed and used in recent decades owing to its advantages like higher removal efficiency, less contamination load and lower energy consumption than usual separation methods (Rezakazemi et al., 2018a). Quist-Jensen et al. (2015) determined that membrane technology offers many prospects for better wastewater treatment, with significant reductions in equipment size, energy requirements and low capital costs. Scientists working in the field of nanotechnology by researching membranes and achieving results in the separation of gases, separation of some compounds from contaminated water, have stated that the use of nanotechnology in the manufacture of membranes can increase the capabilities of membranes. Research in this area is research to significantly reduce costs, higher efficiency with the growing trend of nanotechnology and membranes in the world. Therefore, the development of membrane processes in promoting their efficiency depends on obtaining suitable membrane materials and how to produce them.

17.3.1 Modes of membrane

17.3.1.1 Continuous

Continuous membrane processes include the continuous input of feed and permeate removal from a membrane system. The major feature of this mode is its ability to handle enormous amounts of feed (Singh and Purkait, 2019). This makes it popular in the industry since it decreases total membrane operation time. The sole disadvantage of this mode is fouling, which develops owing to the nature of its continuous operation and results in a progressive decrease in total process efficiency. If membranes with superior antifouling characteristics can be produced soon, it may be acceptable and have the potential for wider acceptance of membrane separation techniques (Ganiyu et al., 2015; Petala and Zouboulis, 2006).

17.3.1.2 *Batch*

The batch system is the most common. In this system, the membrane receives a predetermined amount of feed. The procedure continues until the feed is exhausted or replenished. It is recommended as a virtuous membrane process system since it allows for the cleaning of the membranes also the overall membrane set up in between two cycles. This method of operation benefits the membrane process by increasing membrane process efficiency, lowering costs, and reducing fouling. As a result, it is best employed in industries including pharmaceutical, biotechnology, and food (Singh and Purkait, 2019; Wu et al., 1998).

17.3.1.3 *Semi-batch*

The semi-batch process of membrane operation is identical to the batch system of membrane operation with the exception which it allows for removal of feed, addition, recirculation. This configuration aids in the enhancement of selectivity and the overall management of the membrane process. It as well as assists in the lowering of the overall load on downstream permeate and retentate processing. Because of its minimal fouling and high efficiency, this method of membrane operation is frequently utilized in process industries (Ebrahimi et al., 2016; Singh and Purkait, 2019).

17.3.2 Classification based on the geometric shape of the membrane

During the last 20 years, membrane processes have made great progress compared to other common methods of separation from distillation, adsorption, desorption, extraction, etc. Membranes are divided according to different parameters (Chen et al., 2005; Maddah et al., 2018). These parameters are:

Driving force, membrane material, membrane geometry, membrane structure

In this part, the membrane is explained based on its geometric shape.

17.3.2.1 *Rotating disk*

This module is a kind of membrane module used in pharmaceutical, food, and biotechnology applications. This technique covers a large area while reducing membrane fouling and boosting membrane flow (Plum et al., 2003). Several disks are placed on a single shaft in this membrane module, and the system rotates between fixed circular membranes. Multi shaft systems have also been developed and are in widespread usage across the world (Waqas et al., 2021).

17.3.2.2 *Tubular*

In this system, the membranes are placed in parallel batches in stainless steel, ceramic or porous plastic tube. The diameter of the tube that covers this set is about 11 mm and the number of these groups is between four and eight (Issaoui et al., 2017; Singh and Purkait, 2019). The inlet current enters from the center of these pipes. The retentate exits parallel to the pipe and the permeate exits in a direction perpendicular to it. Tubular membranes are used in high concentrations of solids (Yan et al., 2009).

17.3.2.3 *Hollow fiber*

This type of membrane mold consists of 45–3000 intermediate fibers that are placed in parallel with each other and are embedded in the resin from the end, and finally, the smooth fibers are placed inside the effluent collecting pipe (Peng et al., 2021). The outer surface of the hollow fiber wall is not similar to the inner wall and has a rough structure that has a protective role. The filtration inlet is directed into the fibers and the effluent is collected outside the fibers and exits from the top of the pipe. In the cross-flow filtration process, the intermediate fibers can also be backwashed (Chen et al., 2020; Purchas and Sutherland, 2002).

17.3.2.4 *Plate and frame*

Various modules are used to increase the surface of the membrane. The simplest design includes smooth plates in the filter press membrane; This is called plate and frame. In this system, several layers of sandwich-shaped membranes are located between the protective wall and the input current is driven into the narrow channels of this system. The filter is usually divided into sections and the material flows through these parallel membranes. These parts are separated by walls and each leads to impenetrable walls that reverse the flow and establish a kind of row flow in this system. Packing densities for plate and frame modules range from 100 to 400 m^3/m^3 (Chen et al., 2006; Im and Jang, 2020).

17.3.2.5 *Spiral wound*

In RO, this type of old module is used. In this design, the main body consists of membranes that are wrapped around a porous collecting tube. Each of these membranes is separated by two layers of porous and permeable coating, and between these layers, there are separating membranes that allow easy flow of effluent through the membrane. The membrane surface in this case is usually 0.6–1 m^2. Commercial types are usually 100–150 cm long and 10, 15, 20 and 35 cm in diameter. Spiral membranes can be used as needed at high temperatures and pressures (Rektor and Vatai, 2004).

17.3.3 Types of membrane filtration

Membrane processes are divided into different types depending on the driving force used in the system. Of course, the actual separation mechanism can be based on the particle size difference. Membrane processes whose driving force is pressure difference are used to separate insoluble and suspended particles of different sizes.

17.3.3.1 *Microfiltration*

MF is a pressure separation process that is widely used to concentrate, purify, or separate colloids, macromolecules and suspended particles from solution (Zielińska et al., 2017). In the biotechnology industry, MF is related to applications such as cell recycling and extraction, process stream purification, and separation of recombinant proteins from cell detritus. There are two main designs in the MF process, the older and more common type of which is parallel or dead-end filtration. In this design, the entire fluid flow is passed vertically through the membrane surface using a compressive force. The particles accumulate on the surface of the membrane, so we need more pressure to hold the flow flux steady until we finally get to the point where we need to change the membrane (Azimi et al., 2017; Singh and Purkait, 2019). Since 1970, another design has been proposed called transverse flow filtration in which the feed solution (FS) flows transversely across the membrane surface. In this design, two currents are created: the flow through the membrane, which is free of particles, and the concentrated flow, which contains particles. The equipment required for cross-flow filtration is more complex, but the life of the membranes used in this design is longer than that of parallel filtration. In recent years, a type of MF operating system has been developed, which is called semi-dead-end filtration. In this system, the MF process first works in the dead-end state until the operating pressure of the system creates an acceptable output current for the membrane. As soon as the output current from the membrane decreases, the system starts operating with cross-flow filtration mode. At this time, both rinsing with the solution leaving the membrane and rinsing with air flow are applied in the opposite direction of the flow on the membrane. After the deposited fouling has been removed from the membrane, the system returns to the dead-end state. This method is especially useful for MF units used to remove bacteria and viruses from municipal water units. Vinduja and Balasubramanian (2013) investigated the electrical coagulation (EC) performance and subsequent MF process for the elimination of heavy metals in synthetic wastewater containing Zn^{2+}, Ni^{2+} and Cd^{2+} ions. They concluded that using the EC process, the optimum metal removal rate is 95% for Cd and 98% for Ni and Zn and to achieve a higher purity of up to 99%, they used MF, which shows the special role of this process for separation. Srisuwan and Thongchai (2002) investigated the separation of metal ions by MF membrane (made of cellulose acetate and pore size 200 nm) and UF (polysulfone with molecular weight cut-off (MWCO) = 30 kDa) and observed that the separation efficiency of the two processes was approximately 80% and as expected, MF is more than the flow rate.

17.3.3.2 Ultrafiltration

UF is a potential membrane technique for metal ion removal and suspended solids with a molecular weight (1000–100,000 Da) from water or wastewater at low membrane pressures (Fu and Wang, 2011; Gunatilake, 2015). In addition to compounds that can be removed using microfilters, UF also blocks proteins (Gunatilake, 2015). Owing to its high packing density, UF has various benefits, including reduced driving force and a smaller space (Igunnu and Chen, 2014; Xu et al., 2021). Because the size of dissolved metal ions in the form of hydrates or low molecular weight complexes is smaller than the size of the membrane pores used in this technique, these ions easily pass via the membrane. To correct this problem and increase the elimination efficiency of metal ions, "Micellar enhanced UF (MEUF) and Polymer enhanced UF (PEUF)" have been proposed. MEUF was first introduced in 1980 for the elimination of soluble organic compounds and multi-capacity metal ions from wastewater. The basis of this method is the addition of surfactants to the wastewater, which, after passing the critical concentration, the molecules of the surfactant accumulate in the micelle and can form bonds with metal ions and form larger structures. These ions containing metal ions are easily trapped in the cavities of the UF membrane. To improve the removal efficiency, surfactants with electrical charges contrary to metal ions can be used. The most important advantages of this process are high elimination efficiency, simplicity of equipment and easy operation in recovery of heavy metal ions. PEUF has also been used as an operational process to remove various types of metal ions. In this method, a water-soluble polymer is used to form a complex with high molecular weight metal ions, and as a result, it remains behind the membrane. This current is refined in the next step for the recovery of metals and the reuse of polymeric materials. The most important concern about using this method is to find proper polymers to form complexes with metal ions. Owing to the high efficiency and selectivity of this method and despite numerous articles in this field, this process has not yet been widely used in industry (Peyravi et al., 2012; Renu et al., 2017).

17.3.3.3 NF

NF is a pressure-driven membrane process which owing to its advantages like high permeate flux and high retention of polyvalent ionic salts, etc. The MWCO of NF is in the range of 200–2000 Da, which belongs to the range between UF and RO and can operate at operating pressures as low as 0.3–1.5 MPa. Currently, the global expansion and various applications of NF technology are the results of the introduction of thin-film composite membranes (TFCs) by interfacial polymerization in the 1971 year. TFC composite membranes are supported by selectively ultra-thin separating layers on a porous membrane. In addition, the top selective layer and the lower porous substrate can be independently selected, studied, or modified to maximize membrane performance. Polymer NF membranes are usually charged owing to the binding of specific functional groups to the membrane network. In recent decades, the use of NF to remove heavy metals is increasing rapidly, as it offers important solutions to some of the problems associated with conventional removal methods. NF is utilized in energy-efficient and ecologically sustainable applications like groundwater, surface water, and wastewater treatment. Since NF membranes contain looser polyamide structures, naturally have a selective loose thin film structure, small pore size and somewhat high charge density, which has higher water permeability. Therefore, the NF membrane technique is regarded as an energy-efficient and highly effective method of eliminating heavy metal ions. Mehdipour et al. (2015) utilized a polyamide NF membrane to remove Pb. It was also shown that increasing the pressure and starting input concentration enhances Pb removal by 97.5%. Al-Rashdi et al. (2013) showed which the NF membrane could recover about 100% of the Cu ions in Cu solution at pH values ranging from 1.5 to 5 and pressures ranging from 3 to 5 bar, demonstrating the applicability of NF membranes for Cu ion rejection. Even so, when the concentration of Cu increases to 2000 mg/L, the ability of NF membranes to reject Cu ions decreases. Table 17.1 reports Cu(II) removal utilizing suitable NF systems at various operating circumstances.

17.3.3.4 RO

RO was originally researched in the 1920s, but it was not put into practice for another 30 years (Singh, 2014). Similar to MF, UF and NF, RO also is a pressure-driven process the dense barrier layer in the polymer matrix of RO membranes is where the majority of separation happens. The RO process may be used to eliminate different types of ions and molecules from various types of polluted water, as well as in industries including food processing, desalination, textiles, biotechnology, pulp and paper, pharmaceuticals, and

TABLE 17.1 Cu(II) removal by utilizing NF, RO and NF + RO (Al-Saydeh et al., 2017).

Type of membrane	Initial concentration (M)	Removal efficiency (%)	Operation pressure (bar)	References
NF	0.01	47–66	1–3	Ahmad and Ooi (2010), Al-Rashdi et al. (2011)
NF	0.47	96–98	20	Oh-sik (2011)
RO	7.86×10^{-3}	98–99.5	5	Tran et al. (2012)
RO	$4.7 \times 10^{-4} - 1.57 \times 10^{-3}$	70–90	Low pressure RO	Zhang et al. (2009)
NF + RO	2	> 95	35	Cséfalvay et al. (2009)
NF + RO	0.015	95–99	3.8	Sudilovskiy et al. (2008)

Modified from Al-Saydeh, S.A., El-Naas, M.H., Zaidi, S.J., 2017. Copper removal from industrial wastewater: a comprehensive review. Journal of Industrial and Engineering Chemistry 56, 35–44.

water reclamation (Malaeb and Ayoub, 2011). To comprehend the mechanism of the RO process, first must comprehend the mechanism of natural osmosis. Osmosis is a type of diffusion in which water molecules diffuse over a permeable membrane from low to high solute concentration, similar to how a plant takes water from the soil through its roots. In this method, impurities are separated from water (solvent) by using a semi-permeable membrane and applying hydrostatic pressure higher than osmotic pressure. The RO process has a good efficiency in a wide range of pH and pressure. Unlike chemical deposition, where pH is the most important parameter, in this method, pressure has a great effect on the removal of heavy metals. More pressure will improve the efficiency of metal removal and instead increase energy consumption. RO has limitations, the presence of heavy metal ions like Cd and Cu aggravates the fouling of the membrane so that it can not be regenerated and replaced, resulting in increased operating costs. Another issue is the decrease in membrane performance over time, which reduces the flow rate. High energy consumption, calcium deposits and the need for experienced technical force to run the system are other disadvantages of this method (Dialynas and Diamadopoulos, 2009). Blair et al. (2017) used a dual-stage RO system to attain a high water-recovery rate. They employed an intermediate NF stage to minimize silica and aluminosilicate fouling impacts on the performance of the second RO stage. They also claimed that coupling a transitional NF phase with a coagulation stage aids in the concentration of coal seam brines.

17.3.3.5 Forward osmosis

Forward osmosis (FO) is a new membrane technology that can compete with RO and NF in water desalination. FO, like NF and RO, needs TFC membranes to attain a reasonable balance between selectivity and water flux (Darabi et al., 2017). One significant distinction between FO and pressure-driven membranes is that FO doesn't need any hydraulic pressure. The FO process may be carried out in two modes: "active layer-facing-feed solution (AL-FS) and active layer-facing-draw solution (AL-DS)." Both approaches differ in terms of solute buildup, which produces internal concentration polarization

(ICP). FO is a growing technology, because of its benefits, like low energy requirements and a lower fouling tendency, this has become a popular and common method of separation (Chian et al., 2007). Certainly, in addition to the advantages, FO has drawbacks that need attention. Aside from limited uses of DS in which the draw solute is included in the end product, further separation is required to recover freshwater. Another disadvantage of FO is low permeate flux owing to concentration polarization (CP). This CP reduces permeate flux by influencing net osmotic pressure. Heavy metal elimination utilizing the FO method was first shown in 2012 by Jin et al. (2012), who investigated the removal rates of membrane versus boron and As in various modes. According to the study, 58% boron and 90% As were rejected in the AL-FS mode, whereas 10% boron and 50% As were rejected in the AL-DS mode. The metal rejection in the FO process utilizing the AL-DS mode was lower than in the AL-FS orientation. Mondal et al. (2014) used a commercial FO membrane to remove As(V) from groundwater. They found that utilizing glucose as the draw solute resulted in 90% As(V) elimination. Nonetheless, when phosphate, Mg(II), silicate, Ca(II), and sulfate were present in the solution, metal ion rejection was reduced.

17.3.3.6 *Pervaporation*

Pervaporation (PV) is another developing technology that eliminates pollutants, especially volatile organic compounds (VOC), using a non-porous or dense hydrophobic membrane while simultaneously cleaning up the contaminated stream (Ji et al., 1994). This method separates liquid mixtures based on preference using membrane permeation and evaporation. In the PV process, the output phase of the membrane is vapor. An important feature of the PV process is that the separation rate is proportional to the permeation rate of the components in the mixture. Therefore, this process is able to separate mixtures with azeotropic point or mixtures whose components have a boiling point close to each other and can replace the distillation method or other methods that are not capable of separation. The most important application of PV is dehydration of organic solvents (Baysak, 2021; Castro-Muñoz, 2020; Zhang and Drioli, 1995). A suitable PV membrane must be produced using appropriate materials and methods, such as phase inversion. PV membranes are particularly constructed to have a stronger affinity for the component to be separated due to their specialized application. This suggests that the chemical composition and structure of the membrane are important in attaining the desired separation (Cao et al., 2021; Crespo and Brazinha, 2015). The main benefits of PV include decreased energy usage in comparison to conventional purifying techniques, as well as cheaper capital and maintenance expenses (Sadrzadeh et al., 2018). For these reasons, PV is a potential technology for treating wastewater with various components and obtaining fresh water. However, there are certain disadvantages to this technology. Due to its very sensitive operating conditions, large industrial applications have yet to be realized (Nagai, 2010). Quiñones-Bolaños et al. (2005) used PV in wastewater micro-irrigation of plants. A thick hydrophilic PV membrane was strategically placed in the soil throughout the experiment. Synthetic wastewater was circulated over the membranes in a feed tank, and the permeate flow and contaminant rejection have been measured. The findings indicated that this method has applications in the treatment of brackish groundwater or wastewater for micro-irrigation. Rezakazemi et al. (2018b) reported cyclohexane purification utilizing polydimethylsiloxane (PDMS) membranes in a PV process. To optimize the process, they also studied the influence of feed concentration on the separation factor. Dehydration of an 80 wt.% cyclohexane mixture at 27°C and a vacuum pressure of 0.0133 bar was demonstrated to be effective, with a high separation factor of 2500 attained.

17.3.3.7 ED

Other electrochemical methods include ED, which uses membrane aid in addition to electrical current. It was developed in the 1950s as a result of desalination studies. The basis of this method is according to the transfer of ions from ion exchange membranes under the effect of electric potential difference at the electrodes. The contaminant stream enters a set of ED cells for purification, including a series of anion-exchange membranes and cation-exchange membranes (Tian et al., 2021). Due to the potential difference between the electrodes, the cations move towards the cathode and the anions move towards the anode. During this movement, the ions pass via the membranes with the opposite charge and precipitate in the membranes with the like charge, and as a result, the concentration of ions decreases in one part and increases in the side part. In addition to the general advantages mentioned for electrochemical methods, selective removal and high water recovery should be considered in ED (Babilas and Dydo, 2018; Fidaleo and Moresi, 2006). Much research has been done on the successful use of this method to eliminate heavy metals. Another electrochemical method is electrodionization, which is a modified form of the ED and ion exchange process from an economic point of view. This method can be used both in continuous and batch conditions. The generality of this method, such as ED, is the use of electric current and ion transfer using an ion exchange resin located in the diluted compartment. But the main advantage of this method is the reduction of ion exchange resins by electrical energy, which eliminates the need to use chemicals to reduce the resins (Benvenuti et al., 2014; Nemati et al., 2017).

Generally, the advantages of ED over RO include:

1. More temperature tolerant, resulting in reduced viscosity, better conductivity, and minor resistance as a result of the higher temperature of generated water.
2. The grate configuration of the ED plate and frame allows for more convenient cleaning and retention procedures.
3. The ability of ED membranes to be recovered. For treat and regenerate the membranes, weak acid might be used.

The replacement of membranes and the corrosion process are significant disadvantages for (ED.) According to studies conducted by researchers increasing voltage, temperature and using membranes with higher ion exchange capacity enhanced cell performance and separation percentage reduced with an increasing flow rate (Bunani et al., 2017; Mohammadi and Pironneau, 2004). Frioui et al. (2017) created ED combined with an in situ complexation process as a hybrid (ED.) In their study, they used selective ED to be treated a mixture of metal ions that included Cu (II), Ag(II), and Zn(II). All metal ions were nearly fully removed (99.9%) by employing ethylene diamine tetraacetic acid (EDTA) as the complexing agent. When the polarity of ED electrodes is reversed, the procedure is known as electrodialysis reversal (EDR), and it can somewhen be more effective than (ED.) EDR reduces fouling and scaling while increasing recovery rates. However, EDR requires additional plumbing and electrical controllers (Valero and Arbós, 2010).

17.3.3.8 *Membrane distillation*

Membrane distillation (MD) is a developing membrane technique which studied for treating salty water with high rejection rates that conventional technologies could not achieve. This hybrid membrane technique has been reported to exist for more than 50 years, with little progress toward commercial application (Ardeshiri et al., 2018; Singh and Hankins, 2016). Owing to the porous hydrophobic membrane and phase-change thermal distillation, MD is a thermal membrane separation technique. This method is mostly useful for separating FSs with high water content. However, because MD is a thermally driven process, unlike RO, it isn't significantly impacted by FS osmotic pressure. Because only vaporized water can be transported to the membrane during this procedure, MD is an useful method for treating hypersaline solutions like coal seam gas RO brine, seawater RO brine and draws solution for FO treatment. In order to allow free flow of mass, membranes for MD must have minimal mass transfer resistance. The membrane material should have low thermal conductivity and, more importantly, the membrane pore size should be as small as feasible to reduce wetting of the membrane pores and improve heat retention in the system. MD permeate flux, on the other hand, will decrease. Pore diameters typically vary between 0.1 and 1 μm. As a result, the optimal membrane pore size for each MD usage and feed type to be treated must be established (Alkhudhiri and Hilal, 2018). MD offers a lot of promising potentials. Renewable energy sources like sun or wind can power it. Also, waste heat from industrial operations could be used. In terms of pressure needs, MD is less demanding hydrostatic pressure than RO (Deshmukh and Elimelech, 2017). MD was not frequently used for heavy metal removal; for instance, As (40–2000 ppm) polluted groundwater has been reduced to 10 ppm using direct contact membrane distillation (DCMD). Without wetting the membrane pore, over 100% As removal was achieved.

To generate the necessary driving power for MD, four major setups were often considered, as illustrated in Fig. 17.3. "DCMD, air gap membrane distillation (AGMD), vacuum membrane distillation (VMG), and sweeping gas membrane distillation (SGMD)" are the four methods (Khayet and Matsuura, 2011).

17.3.3.9 *Liquid membrane*

Liquid membrane (LM) processes are those processes that have a selective liquid phase as the membrane. The separation and retrieval process occurs continuously and in this phase. LMs are divided into three basic categories: bulk liquid membranes (BLM), Supported liquid membrane (SLM), emulsion liquid membrane (ELM). Of the above three categories, emulsion membranes have a higher mass transfer surface. This type of membrane was used to separate hydrocarbons. The dual emulsion system is of two types. The first type is the emulsion of water in oil which is in a continuous oil phase, and the second type is the emulsion of oil in water which is immersed in a water phase (Jean et al., 2018). The separation of chemicals by means of LM is an efficient and different method in the field of membrane separation processes. Compared to conventional methods, the use of ELMs has its own advantages and advantages such as process simplicity, high efficiency, single-stage extraction and recovery, common surface and larger contact surface and the possibility of sequential processes. Also, these types of membranes have a high ability to

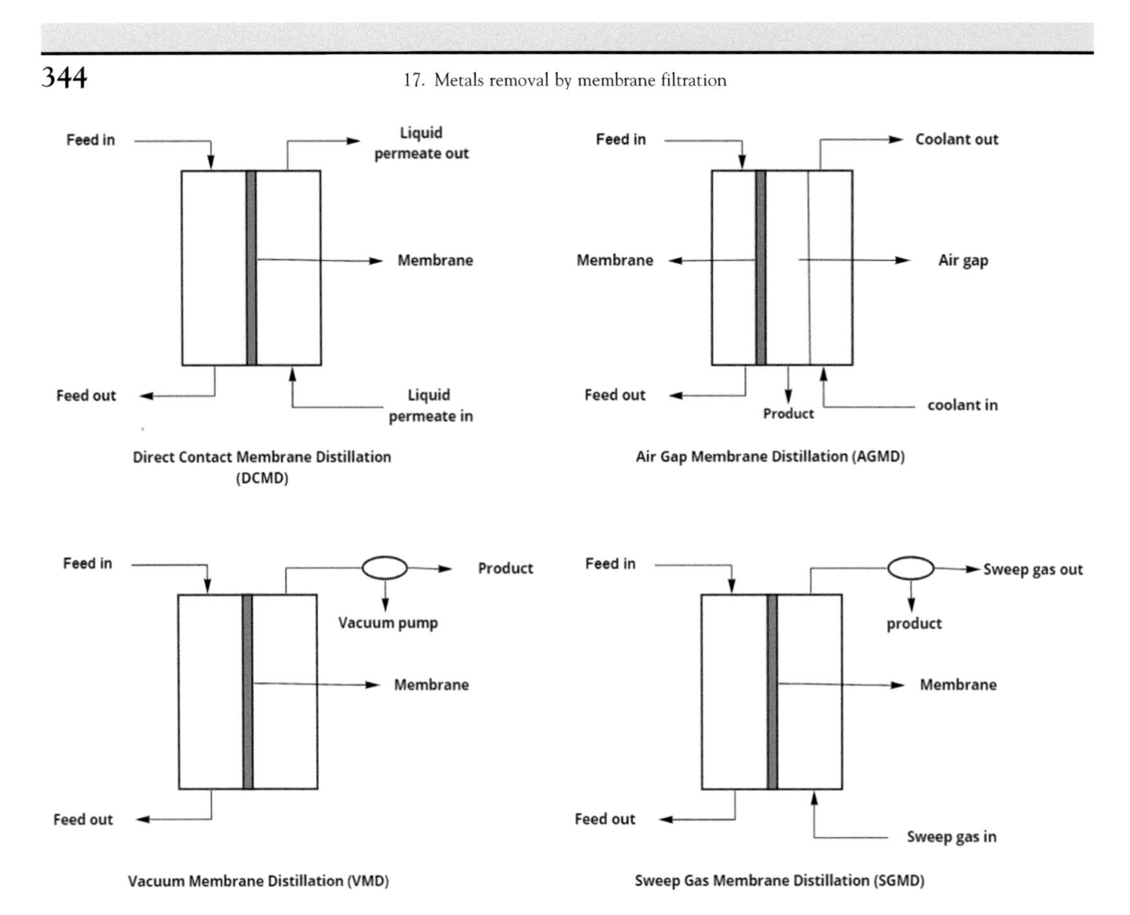

FIGURE 17.3 Schematic representations of the many sorts of MD setups. Source: *Modified from Obotey Ezugbe, E., Rathilal, S., 2020. Membrane technologies in wastewater treatment: a review. Membranes, 10, 89.*

recover and remove metal ions and various hydrocarbons from industrial effluents and have higher efficiency compared to conventional methods. The application of ELMs in wastewater treatment is when removing ionizable contaminants such as phenols or reducing or eliminating heavy or toxic metals. When metals are purified from water, metal cations are dispersed on the interface between the feed inlet and the emulsion membrane. These cations then bind to another side molecule and travel through the membrane to the droplets of the recovered material. When the metal cation combines with the side molecule and the recovered droplets collide, the metal cation separates from the compound and enters the liquid phase, which is then washed and separated by a strong acid solution (Abdullah et al., 2019; Hosseini et al., 2016). The driving force of this gradient process is the difference in chemical potential of the H^+ ion between the extracted phase and the input FS, by which the metal ions can advance against the internal concentration gradient of the aqueous phase. When separating ionizable organic compounds (such as phenolic compounds), the transfer phenomenon is based on the concentration gradient of non-ionized organic compounds between the input feed phase and the extracted phase. The pH of the inlet feed phase is adjusted so that the non-ionized particles can dissolve in the diluent solution. Inside the emulsion membrane, the non-ionized state of the impurities enters the inner space of the membrane as monoatomic or multi-atomic and moves towards the extracted droplets. Non-ionized organic molecules penetrate into the extraction phase and within this phase, the resulting microdroplets are ionized. As a result, the activity of non-ionized impurities will be lost, which makes the reverse transfer of impurities impossible. In contrast, SLM is essentially a three-phase liquid system in which the liquid phase is held together by the capillary force of the pores in the inorganic film and polymeric membrane.

17.3.3.10 Membrane bioreactor

MBR is a hybrid technology because it combines a physical membrane process such as MF or UF with a wastewater biodegradation process such as conventional activated sludge treatment. Because of their lower energy consumption needs, low-pressure membranes with a greater MWCO were commonly used in the process. MBR is a low-cost, reliable technique with less sludge generation, a smaller footprint, and frequently greater elimination efficiency than conventional activated sludge (Vieira et al., 2020). On the basis of the physicochemical characteristics of the compounds and the process parameters, the elimination of micro-contaminants by MBR is ascribed to several methods. The primary methods for the elimination of these micro-contaminants include biosorption, biodegradation, photo-transformation, volatilization, and physical retention. For instance, in the case of apolar pollutants, the major methods are sorption and membrane physical retention, whereas biodegradation governs the removal of polar pollutants. The removal process for hydrophobic compounds is biosorption onto active sludge; also, for hydrophilic pollutants, biosorption is followed by biodegradation (Goswami et al., 2018). The drawbacks of MBRs are the high cost of installation and operation, the need for frequent maintenance, pH, temperature, and pressure limitations necessary to fulfill membrane tolerances. Membrane fouling is a major drawback of MBRs, as it is with other filtering techniques. To mitigate this disadvantage, MBR is typically used in conjunction with NF or RO processes, which further improves the rate of micro-contaminant removal.

17.3.3.11 Polymeric nanocomposite membranes

Polymers are yet the basic materials used in membrane technology due to their properties such as ease of film formation, flexibility and low cost. At the same time, limitations in chemical and mechanical limit the use of the polymer. Although mineral membranes have better thermal and chemical resistance and longer life, issues such as high cost, fragility and difficulty of film formation are considered obstacles to the development of their application. Among these, composite materials with a combination of the main properties of organic and inorganic materials, are suitable options for making membranes (Castro-Muñoz et al., 2021; Elimelech and Phillip, 2011). Polymer nanocomposites obtained by dispersing mineral nanoparticles at the nanometer scale in the polymer field have been considered in recent years. Polymer nanocomposites have the advantages of polymers and nano minerals such as flexibility, ease of polymer formation and selectivity and thermal stability of mineral particles. For example, surface modification of inorganic nanoparticles using organic functional groups not only leads to better dispersion of nanoparticles in the polymer field but also in some cases improves selectivity permeability. Among the types of nanoparticles used, the application of silica is very important due to its easy process and known chemical properties. Although the effects of nanoparticles that the membrane structure depends on the components of the polymer nanoparticle system, there are some common features that Taurozzi et al. (2008) have listed: (1) increase the thickness of the shell layer, (2) increase the porosity of the shell, (3) prevent the formation and growth of large cavities, (4) increase the permeability of the membrane without flux change and (5) improve thermal and mechanical stability and reduce sediment. Some researchers have used the addition of mineral nanoparticles to modify hydrophobic polymers. It has been reported that this method causes the membrane to become hydrophilic and increases the passage of water through the membrane (Mulder, 2012). Yurekli (2016) succeeded in removing Pb and Ni ions by nanocomposite polymer membrane at low pressure of one atmosphere. The material of this membrane is polysulfone, which has been improved with zeolite nanoparticles. The maximum adsorption ability measured by hybrid membranes for Pb and Ni metals was 682 and 122 mg/g, respectively.

17.3.4 Membrane fouling

One of the biggest problems of advanced membrane filtration technology used in industry is the premature fouling of membranes by the deposition of organic and inorganic materials and compounds in the feed on their surfaces and pores. The phenomenon of concentration polarization reduces the efficiency of the membrane, although the amount of this phenomenon can occur relatively differently. Therefore, in the MF and UF process, the actual flux through the membrane can be only a fraction of the pure water flux, since this effect is less in PV. With all polarization phenomena, the flux is less than the original value in a time range (Speth et al., 1998). Consecutive flux reduction is the result of fouling, which is defined as the reversible deposition of colloids, emulsions, macromolecules and salts inside or on the membrane. This includes adsorption, blockage of cavities, sedimentation and cake layer formation. Membrane fouling usually occurs in UF and MF processes that have porous membranes. Therefore, the type of separation and the type of membrane used in the process can determine the amount of fouling. In general, three types of fouling can be diagnosed (Kang and Cao, 2012): Organic sediments (macromolecules, biological), inorganic sediments (calcium salts), particles

17.4 Conclusion

The presence of heavy metals in water resources is one of the important environmental problems that various methods have been considered to remove these metals. Separation methods, especially membrane technologies for water treatment produced in this chapter were reviewed. Membrane processes can be more efficient than other conventional separation methods due to special features such as high selectivity, simpler operating conditions like lower temperatures, lower energy requirements and therefore higher efficiencies in the elimination of heavy metals. Replacing conventional separation processes with membrane processes makes it possible to conserve large amounts of energy. This requires mass production of membranes with high mass transfer flux, good permeability, high selectivity with long life, as well as conversion of membranes into dense and cost-effective units with high specific levels per unit volume. Therefore, the development of membrane processes is essential to improve their performance.

References

Abdullah, N., Yusof, N., Lau, W., Jaafar, J., Ismail, A., 2019. Recent trends of heavy metal removal from water/wastewater by membrane technologies. Journal of Industrial and Engineering Chemistry 76, 17–38.

Ahmad, A.L., Ooi, B.S., 2010. A study on acid reclamation and copper recovery using low pressure nanofiltration membrane. Chemical Engineering Journal 156, 257–263.

Al-Rashdi, B., Johnson, D., Hilal, N., 2013. Removal of heavy metal ions by nanofiltration. Desalination 315, 2–17.

Al-Rashdi, B., Somerfield, C., Hilal, N., 2011. Heavy metals removal using adsorption and nanofiltration techniques. Separation & Purification Reviews 40, 209–259.

Al-Saydeh, S.A., El-Naas, M.H., Zaidi, S.J., 2017. Copper removal from industrial wastewater: a comprehensive review. Journal of Industrial and Engineering Chemistry 56, 35–44.

Algureiri, A.H., Abdulmajeed, Y.R., 2016. Removal of heavy metals from industrial wastewater by using RO membrane. Iraqi Journal of Chemical and Petroleum Engineering 17, 125–136.

Alkhudhiri, A., Hilal, N., 2018. Membrane distillation—principles, applications, configurations, design, and implementation. Emerging Technologies for Sustainable Desalination Handbook Elsevier.

Ardeshiri, F., Salehi, S., Peyravi, M., Jahanshahi, M., Amiri, A., Rad, A.S., 2018. PVDF membrane assisted by modified hydrophobic ZnO nanoparticle for membrane distillation. Asia-Pacific Journal of Chemical Engineering 13, e2196.

Azimi, A., Azari, A., Rezakazemi, M., Ansarpour, M., 2017. Removal of heavy metals from industrial wastewaters: a review. ChemBioEng Reviews 4, 37–59.

Babel, S., Kurniawan, T.A., 2003. A research study on Cr (VI) removal from contaminated wastewater using natural zeolite. Journal of Ion Exchange 14, 289–292.

Babilas, D., Dydo, P., 2018. Selective zinc recovery from electroplating wastewaters by electrodialysis enhanced with complex formation. Separation and Purification Technology 192, 419–428.

Bach, C., Dauchy, X., Chagnon, M.-C., Etienne, S., 2012. Chemical compounds and toxicological assessments of drinking water stored in polyethylene terephthalate (PET) bottles: a source of controversy reviewed. Water Research 46, 571–583.

Baysak, F.K., 2021. A novel approach to Chromium rejection from sewage wastewater by pervaporation. Journal of Molecular Structure 1233, 130082.

Benvenuti, T., Krapf, R.S., Rodrigues, M., Bernardes, A., Zoppas-Ferreira, J., 2014. Recovery of nickel and water from nickel electroplating wastewater by electrodialysis. Separation and Purification Technology 129, 106–112.

Blair, D., Alexander, D.T., Couperthwaite, S.J., Darestani, M., Millar, G.J., 2017. Enhanced water recovery in the coal seam gas industry using a dual reverse osmosis system. Environmental Science: Water Research & Technology 3, 278–292.

Bunani, S., Arda, M., Kabay, N., Yoshizuka, K., Nishihama, S., 2017. Effect of process conditions on recovery of lithium and boron from water using bipolar membrane electrodialysis (BMED). Desalination 416, 10–15.

Cao, X., Wang, K., Feng, X., 2021. Removal of phenolic contaminants from water by pervaporation. Journal of Membrane Science 623, 119043.

Cardew, P.T., 2007. Membrane processes: a technology guide. Royal Society of Chemistry .

Castro-Muñoz, R., 2020. Breakthroughs on tailoring pervaporation membranes for water desalination: a review. Water Research 116428.

Castro-Muñoz, R., González-Melgoza, L.L., García-Depraect, O., 2021. Ongoing progress on novel nanocomposite membranes for the separation of heavy metals from contaminated water. Chemosphere 270, 129421.

Chen, Z., Mahmud, S., Cai, L., He, Z., Yang, Y., Zhang, L., et al., 2020. Hierarchical poly (vinylidene fluoride)/active carbon composite membrane with self-confining functional carbon nanotube layer for intractable wastewater remediation. Journal of Membrane Science 603, 118041.

Chen, J.P., Mou, H., Wang, L.K., Matsuura, T., 2006. Membrane filtration. Advanced Physicochemical Treatment Processes. Springer.

Chen, X., Shen, Z., Zhu, X., Fan, Y., Wang, W., 2005. Advanced treatment of textile wastewater for reuse using electrochemical oxidation and membrane filtration, Water SA, 31. pp. 127–132.

Chian, E.S., Chen, J.P., Sheng, P.-X., Ting, Y.-P., Wang, L.K., 2007. Reverse osmosis technology for desalination. Advanced Physicochemical Treatment Technologies. Springer.

Crespo, J., Brazinha, C., 2015. Fundamentals of pervaporation. Pervaporation, Vapour Permeation and Membrane Distillation. Elsevier.

Cséfalvay, E., Pauer, V., Mizsey, P., 2009. Recovery of copper from process waters by nanofiltration and reverse osmosis. Desalination 240, 132–142.

Darabi, R.R., Peyravi, M., Jahanshahi, M., Amiri, A.A.Q., 2017. Decreasing ICP of forward osmosis (TFN-FO) membrane through modifying PES-Fe 3 O 4 nanocomposite substrate. Korean Journal of Chemical Engineering 34, 2311–2324.

Deshmukh, A., Elimelech, M., 2017. Understanding the impact of membrane properties and transport phenomena on the energetic performance of membrane distillation desalination. Journal of Membrane Science 539, 458–474.

Dialynas, E., Diamadopoulos, E., 2009. Integration of a membrane bioreactor coupled with reverse osmosis for advanced treatment of municipal wastewater. Desalination 238, 302–311.

Ebrahimi, M., Busse, N., Kerker, S., Schmitz, O., Hilpert, M., Czermak, P., 2016. Treatment of the bleaching effluent from sulfite pulp production by ceramic membrane filtration. Membranes 6, 7.

Elimelech, M., Phillip, W.A., 2011. The future of seawater desalination: energy, technology, and the environment. Science 333, 712–717.

Fidaleo, M., Moresi, M., 2006. Electrodialysis applications in the food industry. Advances in Food and Nutrition Research 51, 265–360.

Frioui, S., Oumeddour, R., Lacour, S., 2017. Highly selective extraction of metal ions from dilute solutions by hybrid electrodialysis technology. Separation and Purification Technology 174, 264–274.

Fu, F., Wang, Q., 2011. Removal of heavy metal ions from wastewaters: a review. Journal of Environmental Management 92, 407–418.

Ganiyu, S.O., Van Hullebusch, E.D., Cretin, M., Esposito, G., Oturan, M.A., 2015. Coupling of membrane filtration and advanced oxidation processes for removal of pharmaceutical residues: a critical review. Separation and Purification Technology 156, 891–914.

Goswami, L., Kumar, R.V., Borah, S.N., Manikandan, N.A., Pakshirajan, K., Pugazhenthi, G., 2018. Membrane bioreactor and integrated membrane bioreactor systems for micropollutant removal from wastewater: a review. Journal of Water Process Engineering 26, 314–328.

Gunatilake, S., 2015. Methods of removing heavy metals from industrial wastewater. Methods 1, 14.

Hidalgo-Vázquez, A., Alfaro-Cuevas-Villanueva, R., Márquez-Benavides, L., Cortés-Martínez, R., 2011. Cadmium and lead removal from aqueous solutions using pine sawdust as biosorbent. Journal of Applied Sciences in Environmental Sanitation 6.

Hosseini, S.S., Bringas, E., Tan, N.R., Ortiz, I., Ghahramani, M., Shahmirzadi, M.A.A., 2016. Recent progress in development of high performance polymeric membranes and materials for metal plating wastewater treatment: a review. Journal of Water Process Engineering 9, 78–110.

Hua, M., Zhang, S., Pan, B., Zhang, W., Lv, L., Zhang, Q., 2012. Heavy metal removal from water/wastewater by nanosized metal oxides: a review. Journal of Hazardous Materials 211, 317–331.

Igunnu, E.T., Chen, G.Z., 2014. Produced water treatment technologies. International Journal of Low-Carbon Technologies 9, 157–177.

Im, S.-J., Jang, A., 2020. Long-term performance and initial fouling evaluation of an open-loop plate and frame forward osmosis element using wastewater treatment plant secondary effluent as a feed solution. Journal of Water Process Engineering 33, 101077.

Issaoui, M., Limousy, L., Lebeau, B., Bouaziz, J., Fourati, M., 2017. Manufacture and optimization of low-cost tubular ceramic supports for membrane filtration: application to algal solution concentration. Environmental Science and Pollution Research 24, 9914–9926.

Jean, E., Villemin, D., Hlaibi, M., Lebrun, L., 2018. Heavy metal ions extraction using new supported liquid membranes containing ionic liquid as carrier. Separation and Purification Technology 201, 1–9.

Jhaveri, J.H., Murthy, Z., 2016. A comprehensive review on anti-fouling nanocomposite membranes for pressure driven membrane separation processes. Desalination 379, 137–154.

Jin, X., She, Q., Ang, X., Tang, C.Y., 2012. Removal of boron and arsenic by forward osmosis membrane: influence of membrane orientation and organic fouling. Journal of Membrane Science 389, 182–187.

Ji, W., Sikdar, S.K. & Hwang, S.-T. 1994. Modeling of multicomponent pervaporation for removal of volatile organic compounds from water. Journal of Membrane Science, 93, 1–19.

Kang, G.-D., Cao, Y.-M., 2012. Development of antifouling reverse osmosis membranes for water treatment: a review. Water Research 46, 584–600.

Khayet, M., Matsuura, T., 2011. Membrane distillation: principles and applications.

Koyuncu, I., Sengur, R., Turken, T., Guclu, S., Pasaoglu, M., 2015. Advances in water treatment by microfiltration, ultrafiltration, and nanofiltration. Advances in Membrane Technologies for Water Treatment. Elsevier.

Kurniawan, T.A., Chan, G.Y., Lo, W.-H. & Babel, S. 2006. Physico–chemical treatment techniques for wastewater laden with heavy metals. Chemical Engineering Journal, 118, 83–98.

Leibundgut, C., Seibert, J., Wilderer, P., 2011. In Treatise on Water Science. Elsevier.

Lofrano, G., 2012. Green Technologies for Wastewater Treatment: Energy Recovery and Emerging Compounds Removal. Springer Science & Business Media.

Maddah, H.A., Alzhrani, A.S., Bassyouni, M., Abdel-Aziz, M., Zoromba, M., Almalki, A.M., 2018. Evaluation of various membrane filtration modules for the treatment of seawater. Applied Water Science 8, 1–13.

Malaeb, L., Ayoub, G.M., 2011. Reverse osmosis technology for water treatment: state of the art review. Desalination 267, 1–8.

Mehdipour, S., Vatanpour, V., Kariminia, H.-R., 2015. Influence of ion interaction on lead removal by a polyamide nanofiltration membrane. Desalination 362, 84–92.

Mohammadi, B., Pironneau, O., 2004. Shape optimization in fluid mechanics. Annual Review of Fluid Mechanics 36, 255–279.

Mondal, P., Tran, A.T.K., Van Der Bruggen, B., 2014. Removal of As (V) from simulated groundwater using forward osmosis: effect of competing and coexisting solutes. Desalination 348, 33–38.

Mukherjee, S., 2019. Isolation and purification of industrial enzymes: advances in enzyme technology. Advances in Enzyme Technology. Elsevieru.

Mulder, J., 2012. Basic Principles of Membrane Technology. Springer Science & Business Media.

Naeem, M., Mujahid, M., Umer, A., Ahmad, S., Ahmad, G., Ali, J., et al., 2019. New trends in removing toxic metals from drinking and wastewater by biomass materials and advanced membrane technologies. Journal of Biodiversity and Environmental Sciences 15, 10–17.

Nagai, K., 2010. Fundamentals and perspectives for pervaporation. Membrane Operations in Molecular Separations. Elsevier Inc.

Nemati, M., Hosseini, S., Shabanian, M., 2017. Novel electrodialysis cation exchange membrane prepared by 2-acrylamido-2-methylpropane sulfonic acid; heavy metal ions removal. Journal of Hazardous Materials 337, 90–104.

Obotey Ezugbe, E., Rathilal, S., 2020. Membrane technologies in wastewater treatment: a review. Membranes 10, 89.

Oh-Sik, K., 2011. Removal of heavy metals in wastewater using a nano-filtration membrane technology. Korea Environmental Industry and Technology Institute .

Peng, N., Chung, T.-S., Wang, K.Y., 2021. Macrovoid evolution and critical factors to form macrovoid-free hollow fiber membranes. Hollow Fiber Membranes. Elsevier.

Petala, M., Zouboulis, A., 2006. Vibratory shear enhanced processing membrane filtration applied for the removal of natural organic matter from surface waters. Journal of Membrane Science 269, 1–14.

Peyravi, M., Rahimpour, A., Jahanshahi, M., Javadi, A., Shockravi, A., 2012. Tailoring the surface properties of PES ultrafiltration membranes to reduce the fouling resistance using synthesized hydrophilic copolymer. Microporous and Mesoporous Materials 160, 114–125.

Plum, A., Braun, G., Rehorek, A., 2003. Process monitoring of anaerobic azo dye degradation by high-performance liquid chromatography–diode array detection continuously coupled to membrane filtration sampling modules. Journal of Chromatography A 987, 395–402.

Purchas, D., Sutherland, K., 2002. Handbook of Filter Media. Elsevier.

Quist-Jensen, C.A., Macedonio, F., Drioli, E., 2015. Membrane technology for water production in agriculture: desalination and wastewater reuse. Desalination 364, 17–32.

Quiñones-Bolaños, E., Zhou, H., Parkin, G., 2005. Membrane pervaporation for wastewater reuse in microirrigation. Journal of Environmental Engineering 131, 1633–1643.

Rektor, A., Vatai, G., 2004. Membrane filtration of Mozzarella whey. Desalination 162, 279–286.
Renu, Agarwal, M., Singh, K., 2017. Methodologies for removal of heavy metal ions from wastewater: an overview. Interdisciplinary Environmental Review 18, 124–142.
Rezakazemi, M., Khajeh, A., Mesbah, M., 2018a. Membrane filtration of wastewater from gas and oil production. Environmental Chemistry Letters 16, 367–388.
Rezakazemi, M., Marjani, A., Shirazian, S., 2018b. Organic solvent removal by pervaporation membrane technology: experimental and simulation. Environmental Science and Pollution Research 25, 19818–19825.
Sadrzadeh, M., Rezakazemi, M., Mohammadi, T., 2018. Fundamentals and measurement techniques for gas transport in polymers. Transport Properties of Polymeric Membranes. Elsevier Amsterdam.
Seader, J., Henley, E., 2006. Separation Process Principles. United States: Jhon WIley & Sons. Inc.
Shafaei, A., Ashtiani, F.Z., Kaghazchi, T., 2007. Equilibrium studies of the sorption of Hg (II) ions onto chitosan. Chemical Engineering Journal 133, 311–316.
Singh, R., 2014. Membrane Technology and Engineering for Water Purification: Application, Systems Design and Operation. Butterworth-Heinemann.
Singh, R., Hankins, N., 2016. Emerging Membrane Technology for Sustainable Water Treatment. Elsevier.
Singh, R., Purkait, M.K., 2019. Microfiltration membranes. Membrane Separation Principles and Applications. Elsevier.
Speth, T.F., Summers, R.S., Gusses, A.M., 1998. Nanofiltration foulants from a treated surface water. Environmental Science & Technology 32, 3612–3617.
Srisuwan, G., Thongchai, P., 2002. Removal of heavy metals from electroplating wastewater by membrane. Songklanakarin Journal of Science and Technology 24, 965–976.
Srivastava, N., Majumder, C., 2008. Novel biofiltration methods for the treatment of heavy metals from industrial wastewater. Journal of Hazardous Materials 151, 1–8.
Sudilovskiy, P., Kagramanov, G., Kolesnikov, V., 2008. Use of RO and NF for treatment of copper containing wastewaters in combination with flotation. Desalination 221, 192–201.
Taurozzi, J.S., Arul, H., Bosak, V.Z., Burban, A.F., Voice, T.C., Bruening, M.L., et al., 2008. Effect of filler incorporation route on the properties of polysulfone–silver nanocomposite membranes of different porosities. Journal of Membrane Science 325, 58–68.
Tian, H., Alkhadra, M.A., Bazant, M.Z., 2021. Theory of shock electrodialysis II: mechanisms of selective ion removal. Journal of Colloid and Interface Science 589, 616–621.
Tran, A.T., Zhang, Y., Jullok, N., Meesschaert, B., Pinoy, L., Van Der Bruggen, B., 2012. RO concentrate treatment by a hybrid system consisting of a pellet reactor and electrodialysis. Chemical Engineering Science 79, 228–238.
Valero, F., Arbós, R., 2010. Desalination of brackish river water using electrodialysis reversal (EDR): control of the THMs formation in the Barcelona (NE Spain) area. Desalination 253, 170–174.
Vieira, W.T., De Farias, M.B., Spaolonzi, M.P., Da Silva, M.G.C., Vieira, M.G.A., 2020. Removal of endocrine disruptors in waters by adsorption, membrane filtration and biodegradation. A review. Environmental Chemistry Letters 18, 1113–1143.
Vinduja, V., Balasubramanian, N., 2013. Removal of heavy metals by hybrid electrocoagulation and microfiltration processes. Environmental Technology 34, 2897–2902.
Wang, J., Chen, C., 2009. Biosorbents for heavy metals removal and their future. Biotechnology Advances 27, 195–226.
Waqas, S., Bilad, M.R., Man, Z.B., Suleman, H., Nordin, N.A.H., Jaafar, J., et al., 2021. An energy-efficient membrane rotating biological contactor for wastewater treatment. Journal of Cleaner Production 282, 124544.
Wu, J., Eiteman, M.A., Law, S.E., 1998. Evaluation of membrane filtration and ozonation processes for treatment of reactive-dye wastewater. Journal of Environmental Engineering 124, 272–277.
Xu, Z., Gu, S., Rana, D., Matsuura, T. & Lan, C.Q. 2021. Chemical precipitation enabled UF and MF filtration for lead removal. Journal of Water Process Engineering, 41, 101987.
Yan, L., Hong, S., Li, M.L., Li, Y.S., 2009. Application of the Al2O3–PVDF nanocomposite tubular ultrafiltration (UF) membrane for oily wastewater treatment and its antifouling research. Separation and Purification Technology 66, 347–352.
Yurekli, Y., 2016. Removal of heavy metals in wastewater by using zeolite nano-particles impregnated polysulfone membranes. Journal of Hazardous Materials 309, 53–64.

Zak, S., 2012. Treatment of the processing wastewaters containing heavy metals with the method based on flotation. Ecological Chemistry and Engineering 19, 433.

Zhang, S., Drioli, E., 1995. Pervaporation membranes. Separation Science and Technology 30, 1–31.

Zhang, L., Yanjun, W., Xiaoyan, Q., Zhenshan, L., Jinren, N., 2009. Mechanism of combination membrane and electro-winning process on treatment and remediation of Cu^{2+} polluted water body. Journal of Environmental Sciences 21, 764–769.

Zielińska, M., Cydzik-Kwiatkowska, A., Bułkowska, K., Bernat, K., Wojnowska-Baryła, I., 2017. Treatment of bisphenol a-containing effluents from aerobic granular sludge reactors with the use of microfiltration and ultrafiltration ceramic membranes. Water, Air, & Soil Pollution 228, 1–9.

CHAPTER

18

Phytoremediation of inorganic contaminants from the aquatic ecosystem using *Eichhornia crassipes*

Khushbu Kumari and Kuldeep Bauddh

Department of Environmental Sciences, Central University of Jharkhand, Ranchi, India

18.1 Introduction

Water pollution, a serious environmental concern, is caused by industrial and mining activities, pesticides and fertilizers used in agriculture sector, waste discharged from the industries. These activities have resulted a high level of metal contamination and non-biodegradable pollutants posing a serious threat to the environment. Heavy metal (HM) contamination such as arsenic (As), lead (Pb), chromium (Cr), silver (Ag), cadmium (Cd), nickel (Ni), cupper (Cu), cobalt (Co), zinc (Zn), barium (Ba) and molybdenum (Mo) along with some ionic contaminants (ICs) like nitrate (NO_3^-), phosphate (PO_4), chloride (Cl^-) poses a negative impact on aquatic ecosystems and their functions. Heavy metals (HMs) are difficult to remove from the contaminated ecosystem because they are non-degradable in naturecome in a variety of chemical forms and therefore, they can easily pass through several trophic (Gall et al., 2015; Zhu et al., 2016). The removal of HMs from the contaminated aquatic ecosystems has become an important concern. HMs can be removed using a variety of conventional techniques such as chemical precipitation, ion exchange, adsorption, osmosis and solvent extraction (Al-Alawy and Salih, 2017; Huang et al., 2017; Levchuk et al., 2018; Burakov et al., 2018). However, the above-mentioned treatment methods are not more efficient, expensive, time consuming and usually not environment friendly. An ecologically acceptable and cost-effective treatment technique is required for the remediation of aquatic ecosystem contaminated with heavy metal and several contaminants (Shahid et al., 2018).

The plant species is used to clean, transfer, stabilize, degrade, or extract contaminants from soil or water is commonly popularized as phytoremediation, which is a low-cost,

Metals in Water
DOI: https://doi.org/10.1016/B978-0-323-95919-3.00001-X

solar-energy-driven process that aids in the removal of contaminants from water and wastewater. Some tree species, terrestrial grasses, herbs, shrubs and aquatic plants are extremely effective at adsorbing, absorbing and accumulating toxic metals or contaminants (Muradov et al., 2014; Bauddh et al., 2015; Muthusaravanan et al., 2018; Sudiarto et al., 2019; Ergönül et al., 2020; Auchterlonie et al., 2021). It has been found that HMs can be efficiently removed from highly contaminated aquatic ecosystems by using variety of free-floating and submerged-root aquatic macrophytes including *E. crassipes, Salvinia minima, Lemna minor, Phragmites australis, Pistia stratoites, Nasturtium officinale, Cyperus malaccensis, Typha latifolia*, etc., (Iha and Bianchini, 2015; da-Silva et al., 2017; Leung et al., 2017; Daud et al., 2018; Gunathilakae et al., 2018; Yasar et al., 2018; Abbas et al., 2019; Sudiarto et al., 2019; Kumar and Deswal, 2020; Ergönül et al., 2020; Auchterlonie et al., 2021; Hasani et al., 2021).

E. crassipes, popularly known as water hyacinth, has attracted a lot of interest with the potential to absorb contaminants from aquatic environments. *E. crassipes* bears several distinguish features like fast growth rate, high biomass, pollutant absorption efficiency, minimal operating cost and renewability etc. These properties indicate that *E. crassipes* is a viable aquatic plant species for phytoremediation of contaminated aquatic ecosystem. The study conducted by Ajayi and Ogunbayio (2012), observed that the *E. crassipes* have high Bioconcentration Factors (BCF) for Cd, Cu and Fe with the values of 583.83, 734.41 and 2982.95 respectively. Ali et al. (2020) reviewed the use of aquatic plants such as water hyacinth (*E. crassipes*), water lettuce (*Pistia stratiotes*) and duck weed (*L. minor*) to remove inorganic pollutants like Hg, Cu, Cr, Pb, Ni, Zn, Mn, Cd, etc., and they found that using aquatic plants for contaminant removal is both cost-effective and beneficial to the long-term viability of entire ecosystems.

The goal of this chapter is to investigate *E. crassipes'* phytoremediation potential for removing inorganic contaminants from aquatic ecosystems.

18.2 Ecology of *Eichhornia crassipes*

Eichhornia crassipes is a species of pontederiaceae family which comprises of free-floating perennial aquatic flowering plants. It possesses fibrous, feathery, long and adventitious roots which has about 50% biomass of a single *E. crassipes*. These free-floating roots of height from 10 to 300 cm appear shiny dark purple black in color. Its flower has dark blue/purple upper petal centered by a yellow diamond shaped patch in a spike. These are about 50 mm in diameter and tristylous, that is, arrangement of six stamens and one style are possible in three ways; short style, medium style and long style (Sullivan and Wood, 2012; Aboul-Enein et al., 2014). *E. crassipes* grows in water sources like ponds, reservoirs, small and large lakes and also in wetlands, marshes and rivers. It forms free floating mat which are monospecific and dense in still and sluggish flowing water. This plant has significant proliferation potential and it doubles in span of 5–15 days (Dar et al., 2011; Patil et al., 2011). *E. crassipes* is naturally present in region lying in between 330 S and 330 N of equator. It grows swiftly from 220 to 600 kg/ha day in pond with density from 224 to 412 tons/ha (Matai and Bagchi, 1980).

18.3 Contamination of inorganic substances in the water bodies

Anthropogenic and some geological activities are the primary sources of inorganic contaminants in aquatic ecosystems (Li et al., 2019). The unprocessed wastewater from industrial and domestic sources contained cyanides, suspended solids, NO_3^-, PO_4, SO_4, Cl^- and most importantly HMs (Ali et al., 2020). Furthermore, commercial activities such as tanning, mining, paper, automotive, pesticide manufacturing, metal processing, dying and geothermal energy plants contribute wide range of contaminants (Ali et al., 2011). Some other activities like municipal waste, fuel production wastewater, vehicle exhausts, discharge of urban effluent, waste incineration and smelting are produced the metal contaminants including nitrates, sulfates, chloride, cyanide, As, Cd, Pb, Ni, Cu, Ag, Hg (Ashraf et al., 2019). In addition, use of such agrochemicals in agricultural activities has increased nitrate, phosphate, Zn, Cd, Pb, Ni, Cu, and As concentrations, which can have a negative impact on the aquatic ecosystem due to agricultural runoff in water bodies (Elgallal et al., 2016). Apart from these, there are several natural causes of HM pollution (As, Ag, Hg, Cd, Zn, Cu, etc.) in water bodies such as rock weathering, erosion and volcanic eruption. One of the most challenging environmental concerns for scientists is the removal of toxic HMs and ICs from the aquatic ecosystems. Due to their persistence and abundance, as well as their potential toxicity, inorganic contaminants in aquatic ecosystems are of global concern (Raval et al., 2016; Goncalves et al., 2017; Ashraf et al., 2019; Kumar et al., 2019). However, freshwater ecosystems are also contaminated by inorganic substances that are eventually absorbed into sediments and may get bioaccumulated in benthic organisms, and from where they can enter into the food chain (Karaouzas et al., 2021).

18.4 Removal of inorganic contaminants present in water bodies using *E. crassipes*

E. crassipes has been used for phytoremediation due to its ease of cultivation and ability to produce high biomass in aquatic environments (Kumari et al., 2021). The *E. crassipes* biomass is often used as an absorbent to remove the inorganic contaminants as shown in Fig. 18.1 (Feng et al., 2017). Kumar et al. (2016) investigated the HMs like Cd, Co, Cu, Ni, Pb, and Zn accumulation in *Ipomoea aquatica*, *Typha angustata*, *E. crassipes*, *Hydrilla verticillata*, *Vallisneria spiralis*, *Echinochloa colonum*, and *Nelumbo nucifera*. They reported that *E. crassipes* had highest Zn accumulation potential. Mishra et al. (2008) investigated the phytoremediation of Hg and As in a tropical open-cast coalmine effluent using *L. minor*, *E. crassipes*, and *Spirodela polyrrhiza* and they observed that the *E. crassipes* showed highest reduction potential in comparison with *L. minor* and *S. polyrrhiza*. Similarly, Hazra et al. (2015) observed that the phytoremediation potential of *E. crassipes*, *T. latifolia* and *Monochoria hastate* against Cu, Ni, Pb, Fe, Mn, and Zn and they found that *E. crassipes* have maximum bioconcentration factor (BCF) for removal of metal content than *T. latifolia* and *Monochoria hastate*. Sung et al. (2015) reported the *Ceratophyllum demersum* and *E. crassipes* effectively removed the nutrient from wet soil and water ecosystems and they found *E. crassipes* had the highest removal capacity of N and P when compared to *C. demersum*.

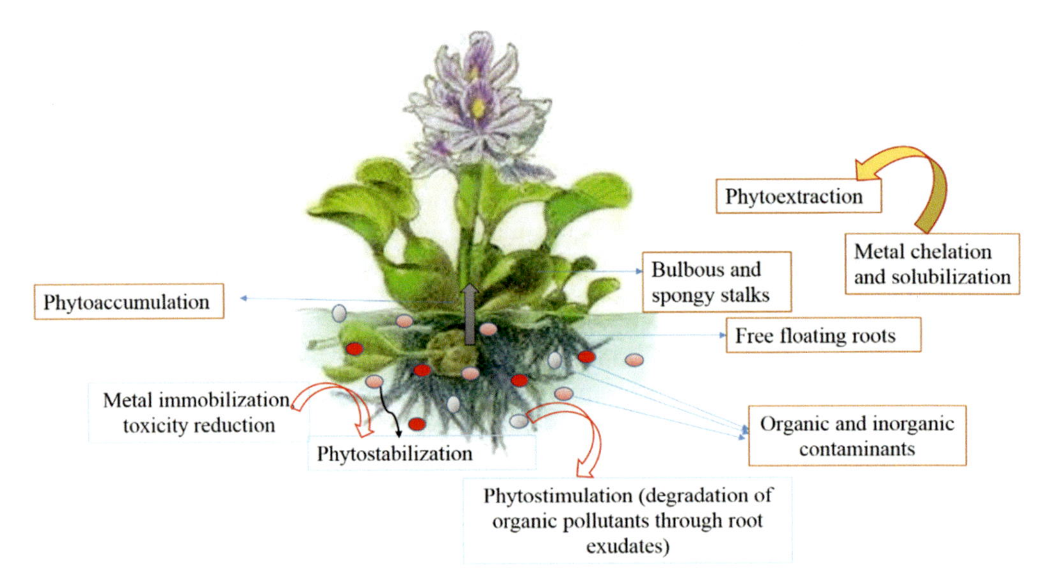

FIGURE 18.1 Mechanism of phytoremediation process of *E. crassipes*.

18.4.1 Bioaccumulation of heavy metals and ionic contaminants

Phytoaccumulation is the process by which contaminants are taken up by the plant's roots and get precipitated inside the root zone and translocated within the shoots (Mishra and Maiti, 2017). Romanova et al. (2016) have done an experiment to check the efficiency of *E. crassipes* to uptake Ag, Ba, Cd, Mo, and Pb from gold mine tailing wastewater. They found *E. crassipes* have high bioaccumulation factor (BAF) of Mo, Pb and Ba by 24,360, 18,800 and 10,040, respectively. Swain et al. (2014), investigated on the *E. crassipes* for the removal of Cd and Cu ion and they found that the highest accumulation of Cd by 230.39 mg/kg in the roots and Cu by 2314.2 mg/kg in stems. Similarly, Lu et al. (2014) studied the removal of tetracyclines (TCs) and Cu using *E. crassipes* and they found that Cu/TCs were effectively removed by 80%. Smolyakov (2012) examined efficiency of *E. crassipes* for the reduction of Cu, Pb, Zn and Cd and found that plants had a high BCF for Cu 8% and 24%, Pb 11% and 26%, Cd 24% and 50%, and Zn 18% and 57% at pH 8 and 6, respectively. Fazal et al. (2015) conducted an experiment to check the effects of *E. crassipes* in contaminated industrial wastewater. Authors found that *E. crassipes* removed metals Cd, Ni, Hg and Pb with values of 97.5%, 95.1%, 99.9% and 83.4% respectively. According to Nuraini and Felani (2015), *E. crassipes* is suitable for the reclamation of tapioca wastewater. They found that 28 days old plant was able to grow at concentration of 25% in tapioca wastewater and have ability to reduce the cyanide concentration by 0.92 mg/L. After that the treated wastewater were used for the maize plantation and results showed good impact on the plant height, leaf number and dry weight of the maize plant. Table 18.1 showed inorganic and ionic contaminant removal efficiency using this aquatic plant (Hadad et al., 2006; Su et al., 2019; Xu et al., 2020).

TABLE 18.1 Bioaccumulation of inorganic contaminants using *E. crassipes*.

Sl. No.	Contaminant(s)	Concentration Level	Removal Efficiency	Findings	References
1.	Fe, Cu and Zn (mg/L)	0%, 5%, 10%, 15%, 20%	Removal of Fe (54.15%), Cu (70.17%) at 15% and Zn (85.97%) at 20%	The metal removal percentage was highest of *E. crassipes* than that of *Typha latifolia*	Abbas et al. (2021)
2.	Cd	0, 5, 10, 20, 50 ppm	Removal from shoot and root were 0.22 and 0.25 ppm respectively	*E. crassipes* is an effective phytoremediator for Cd removal	Gunathilakae et al. (2018)
3.	Cr (VI)	0.5, 0.75, 1, 2, 5 mg/L	99.5% of Cr(VI)	*E. crassipes* had ability to restore the wastewater	Saha et al. (2017)
4.	Mn, Cu, and Cd	25 mL	Bioaccumulation factors with respect to sediment values was highest for Mn (0.20–0.27), followed Pb (0.03–0.20) & by Cu (0.13–0.20) and with respect to water was highest for Cu (428–3205), followed by Pb (242–506), Mn (285–1100) and Cd (7–130)	*E. crassipes* removed the metals effectively from the contaminated pond water	Prasad and Maiti (2016)
5.	Pb	100, 200, 400, 600, 800, 1000 mg/L	The maximum accumulation of Pb was higher in root (5.45%) than in shoot (2.72%) and leaf tissues (0.66%) at 400–800 mg/L	*E. crassipes* removed Pb from polluted wetland.	Malar et al. (2016)
6.	Total N, NH4 and NO_3	8.4–23.3, 2.65–13.96 and 4.04–7.09 mg/L respectively	Total removal efficiency of N 63.9%, NH_4 81.0% and NO_3 22.8%	The results suggested that in absence of *E. crassipes*, the efficiency of N removal decrease by 29.4%	Mayo and Hanai (2017)
7.	Cu (II) and Cd (II)	50 mL Cu (II) or Cd (II) solution at different initial concentration	Cu (II) decreased by 15.9% and Cd (II) equilibrium sorption amount decreased by 91.4%	*E. crassipes* roots efficiently absorbed Cd (II) and Cu (II)	Zheng et al. (2016)
8.	Cd	5, 10, 15 and 20 mg/L CdCl2	Plant accumulated 1927.8 μg/g dry wt. of Cd at 15 mg/L	*E. crassipes* stored the high amounts of Cd	Goswami and Das (2018)
9.	Cd, Cr, Cu, Fe, Mn, Ni, Pb and Zn	—	Removal 1.34%, 0.69%, 2.94%, 4.86%, 1.42%, 0.73%, 0.36%, 3.10% respectively	*E. crassipes* effectively restore the paper mill effluent	Kumar et al. (2016)
10.	Pb, Zn, Cd, Cu and (Cr)	—	The removal of Zn increased from 745.16 to 1304.29 of mg/kg, Cu 78.12 to 147.51 mg/kg and highest metals concentration in roots (1792.66 mg/kg for Zn, 655.25 mg/kg for Cr, 225.19 mg/kg for Cu and 104.54 mg/kg for Pb and 957.14 for Cd), leaves were found to be 817.91 mg/kg for Zn and 69.83 mg/kg for Cu	*E. crassipes* had significant HM removal efficiency than water lettuce	Victor et al. (2016)

(*Continued*)

TABLE 18.1 (Continued)

Sl. No.	Contaminant(s)	Concentration Level	Removal Efficiency	Findings	References
11.	Cr (VI)	100 mg/L	89.9% removal of Cr (VI)	The green synthesis of Ec-Fe-NPs with *E. crassipes* leaf extract was effective for removal of Cr (VI)	Wei et al. (2017)
12.	Ni (II)	1, 2, 3 and 4 mg/L	Highest Ni accumulation in aerial parts and roots were 0.29 and 3.34 mg/g respectively, when plants were exposed to 4 mg/L Ni after 72 h of treatment	*E. crassipes* removed Ni during short exposure times by stimulation with antioxidant enzymatic defense system	González et al. (2015)
13.	Cu and Cd	0.35, 0.70 and 1.05 mg/L for Cu and 0.27, 0.54 and 0.81 mg/L for Cd	Removal efficiency was more than 90% of Cu and Cd	The accumulation of Cu and Cd in their roots and stems was observed during the study	Swain et al. (2014)
14.	Cd and Ni	10 mg/L	*E. crassipes* root removal efficiency for Cd was 70% and Ni was 52% at a time of 240 min	*E. crassipes* showed excellent capacity to remove Cd and Ni from artificially contaminated water	Elfeky et al. (2013)
15.	Cu (II) and Zn (II)	250–1250 ppm	Sorption capacities for Cu (II) 99.42 mg/g and Zn (II) 83.01 mg/g respectively	*E. crassipes* was reported as a low-cost alternative biosorbent for removal of HMs from aqueous solutions	Buasri et al. (2012)
16.	Cu, Cd, Pb and Zn	5–10 mg/L	98% for Cd, 99% for Cu, 98% for Pb and 84% for Zn	*E. crassipes* is the best alternative treatment for removal of metal polluted wastewater	Sekomo et al. (2012)
17.	Cd, Cu, Ni, Zn	2152, 3372, 1850 and 2764 mg/kg respectively	Metal reduction with values of 66%, 68%, 64%, 70% for Cd, Cu, Ni and Zn respectively	*E. crassipes* significantly removed the HM	Valipour et al. (2010)
18.	(Cr) and (Zn)	Cr and Zn, i.e., 1.0, 5.0, 10.0 and 20.0 mg/L respectively	95% of Zn at 10.0 mg/L and 84% of Cr at 1.0 mg/L	*E. crassipes* showed good removal capacity for both metals	Mishra and Tripathi (2009)

18.4.2 Adsorption of heavy metals and ionic contaminants

Adsorption is a viable process of eliminating metal-contaminated water (Basu et al., 2014). Several researchers have been used the plant biomass as a natural adsorbent for the removal of HMs (Gunatilake, 2015; Qambrani et al., 2017; Tatarchuk et al., 2019; Singh et al., 2020). The plant biomass is used as an adsorbent to activate the adsorption capacity and oxidation of HM (Kwak et al., 2019).

18.4.2.1 Adsorption using E. crassipes *biomass*

Adsorption using *E. crassipes* biomass is a feasible and effective method for the removal of HM. The application of *E. crassipes* biomass as adsorbent is low-cost and environment-friendly for removing contaminants from municipal wastewater, industrial effluents, and another wastewater (Sarkar et al., 2017; Kumar and Chauhan, 2019). For adsorption, the adsorbent materials used in the form of dried powder of root, shoot, or entire plant biomass as described in below Table 18.2. Lin et al. (2018) evaluated the feasibility of *E. crassipes* root powder (RP) for the removal of As (III) and As (V) from wastewater and they found that RP removed As (III) 78%, 61%, 54%, and 38% and As(V) 89%, 70%, 63%, and 51% at initial concentrations of 1, 5, 10, and 20 mg/L, respectively. Sarkar et al. (2017) observed that *E. crassipes* shoot powder used for the reduction of Cr and Cu and they revealed that Cr and Cu ion concentrations in standard solution (SS) were 99.98% and 99.96%, respectively and in tannery effluent (TE) were 98.83% and 99.59%, respectively. Wei et al. (2017) examined on *E. crassipes* (water hyacinth) leaf extracts synthesized with Fe nanoparticles (Ec-Fe-NPs) as reductants and stabilizing agents used to remove Cr (VI) and they found extracts have potential to adsorbed Cr (VI) up to 89.9%. Ibrahim et al. (2012) investigated on the dried powered of roots and shoots of *E. crassipes* biomass reduce concentration of Cd and Pb by 75% and 90% respectively. Elfeky et al. (2013) studied the biosorption capacity of dried root of *E. crassipes* for Ni and Cd removal and they have reported that the plant root biomass significantly removed the metals by 55%–70%.

18.4.2.2 Adsorption using biochar produced from E. crassipes

Biochar made from plants' biomass can also be used as an adsorbent or bio-sorbent. Biochar is a carbon-rich products, which is derived from pyrolysis of plant biomass in the absence of oxygen (Kumari et al., 2021). Adsorption through biochar have potential to decontaminate the metal contaminated water as well as soil and increase the metal uptake capacity. Because, biochar has large surface area, existing more surface functional group and interlocking patterns in their particles that can reduce the phytotoxicity (Jin et al., 2018). Currently, several factors influence biochar adsorption efficiency, including raw materials, biochar preparation and adsorption conditions, all of which have a significant impact on adsorption capacity.

The biochar derived from *E. crassipes* for the removal of inorganic contaminants and HM has been reported by several researchers as mentioned in Table 18.3 (Cai et al., 2017; Nyamunda et al., 2019; Hashem et al., 2020). Consequently, during the biochar process of production, cellulosic carbons in *E. crassipes* biomass can be converted into stable aromatic carbons, which may aid in increasing the sorption capacity towards the negatively charged particles. According to a study conducted by Liu et al. (2020), *E. crassipes* biochar was

TABLE 18.2 Adsorption of inorganic contaminants using adsorbent materials in the form of dried powder of root, shoot and entire plant biomass of *E. crassipes*.

Sl. No.	Utilization of *E. crassipes*	Contaminants	Sources	Adsorption capacity	References
Shoot powder					
1.	*E. crassipes* shoot powder	Cr and Cu	Eluate solutions	Highest Cr removal was 99.8% in 13 min but after 13 min it decreased upto 68.9% and maximum Cu removal was 99.9% in 14 min after 14 min it decreased upto 64.8%	Sarkar et al. (2017)
2.	Dried shoot and root of *E. crassipes*	Cd (II) and Pb (II)	Aqueous solution	Adsorption by root and shoot 75% for Cd and more than 90% for Pb respectively	Ibrahim et al. (2012)
3.	Dried powder of stems and leaves of *E. crassipes* biomass	Cr6 +	Batch sorption process	Maximum adsorption capacity was 91.5181 mg/g	Hasan et al. (2010)
Root powder					
1.	Dried *E. crassipes* roots	Cr (VI)	Synthetic aqueous solution	Maximum removal was 95.43%	Kumar and Chauhan (2019)
2.	*E. crassipes* root powder (RP)	Arsenic (AS)	Aqueous solution	Maximum removal was 8 μg of arsenic/g	Gogoi et al. (2017)
3.	*E. crassipes* root powder	Pb, Zn, Cu, and Cd	Aqueous solution	Highest removal was Pb (98.33%), Cd (95.16%), Zn (93.5%) at 25 g/L solutions	Li et al. (2016a,b)
4.	*E. crassipes* roots	Pb (II) ions	Aqueous solutions	Removal concentration was 30 mg/L	Mitra et al. (2014)
5.	Dry *E. crassipes* roots (DHR)	As	Spiked drinking water sample	90% As removal	Govindaswamy et al. (2011)
Entire plant biomass					
1.	*E. crassipes* biomass	Cu, Cd, Pb and Zn	*E. crassipes* ponds and anaerobic up flow packed bed reactor	Removal percentage was 99% for Cu, 98% for Cd 98% for Pb and 84% for Zn	Sekomo et al. (2012)
2.	*E. crassipes* fiber	Cu and Zn	Aqueous solution	Adsorption capacity was found to be 99.42 mg Cu^{2+} and 83.01 mg Zn^{2+} per 1 g biomass	Buasri et al. (2012)

3.	*E. crassipes*	Zn (II) and Cd (II)	Batch binary system	Maximum metal uptake was 0.65 meq/g	Módenes et al. (2011)
4.	*E. crassipes* biomass	Pb (II), Cd (II) and Zn (II)	Aqueous solution	Sorption capacities (qm) for Pb (II), Cd (II) and Zn (II) by 26.32, 12.60 and 12.55 mg/g respectively	Mahamadi and Nharingo (2010)
5.	*E. crassipes* biomass	Cd and Zn	Aqueous metal solution	Adsorption of Cd in shoots 148.0 μg/g & roots 2006.0 μg/g and Zn in shoots 1899.0 μg/g & roots 9646.0 μg/g respectively	Mohamad and Latif (2010)
6.	*E. crassipes* biomass	Zn and Cr	Aqueous solution	Removal capacity was 95% for Zn and 84% for Cr	Mishra and Tripathi (2009)
7.	*E. crassipes* biomass	Fe	Fe rich wastewaters. In constructed wetlands	Adsorption capacity was 6707 Fe mg/kg dry weight	Jayaweera et al. (2008)
8.	*E. crassipes* biomass	Fe, Zn, Cu, Cr and Cd	Aqueous solution	The highest metal removal percentage was >90% during 15 days experiment	Mishra and Tripathi (2008)
9.	*E. crassipes* biomass	Al	Al rich waste water in constructed wet lands	Highest adsorption efficiency was 63%	Jayaweera et al. (2007)
10.	*E. crassipes* biomass	Cd and Zn	Aqueous solution	The concentration of the metal as well as the duration of the exposure time influenced the plant's metal uptake	Hasan et al. (2007)
11.	Activated carbon derived from *E. crassipes*	Hg^{2+}	Aqueous solution	Adsorption capacity was 28.4 mg/g at pH 5	Kadirvelu et al. (2004)

TABLE 18.3 *E. crassipes* biochar used as adsorbent for the removal of metals.

Sl. No.	Biochar	Heavy metals	Amount/dose	Remarks	References
1.	*E. crassipes* biochar	Cr (III)	4 g biochar for 70 mL wastewater	Cr (III) removal efficiency was 99%	Hashem et al. (2020)
2.	Magnetic biochar (Fe2O3-EC) made from *E. crassipes* biomass	Cu^{+2} and Zn^{+2}	1.2 and 6 g	Fe_2O_3-EC, which was synthesized by coprecipitating ferric and ferrous salts on water hyacinth powder, has shown high potential reduction of aqueous metal ion	Nyamunda et al. (2019)
3.	Activated carbon made from *E. crassipes* leaves	Co (II)	1:4 (w/w)	Maximum adsorption capacity was 140.725 mg/g	Riyanto and Prabalaras (2019)
4.	*E. crassipes* biochar treated with ZnO nanoparticles	Cr (VI)	ZnO/BC as 10, 20, 30, 40 and 50 wt.%	ZnO/BC showed 95% removal efficiency of Cr (VI) at 30 wt.% level	Yu et al. (2018)
5.	Magnetic biochar derived from *E. crassipes*	Phosphate (P)	1, 5, 15 mg/ L	More than 90% removal of P at 1 mg/ L concentration	Cai et al. (2017)
6.	Magnetic biochar chemically co-precipitated by Fe^{2+}/Fe^{3+} on *E. crassipes* biomass	As (V)	2.5 g	Magnetic biochar produced from *E. crassipes* through slow pyrolysis in 250°C (MW250) showed high sorption capacity for the removal of As (V) with 7.41 mg/g	Zhang et al. (2016)
7.	*E. crassipes* biochar	Cd^{2+} and Pb^{2+}	0.1 g	Biochar produced from water hyacinth through slow pyrolysis in 450°C (BC450) showed highest adsorption capacity for Cd^{2+} and Pb^{2+} in comparison to BC300 and BC600	Ding et al. (2016)
8.	*E. crassipes* biochar	Cd	0.01 to 50.00 mg/L	The maximum adsorption capacities (90% of Cd) were calculated to be 49.837, 36.899 and 25.826 mg/L	Li et al. (2016a,b)

significant for aqueous solution at a low density if modified as magnetic biochar. Chen et al. (2019) reported that biochar obtained from *E. crassipes* biomass and treated with nanoscale zero-valent ion (nZVI) for removal of Cd (II). Authors found that nZVI/biochar showed good adsorption of Cd (II) at the level of 56.62 mg/g. Nyamunda et al. (2019) done an experiment to check the efficiency of magnetic biochar (Fe_2O_3-EC) derived from *E. crassipes* biomass for the reduction of Cu^{+2} and Zn^{+2} and they found that composition of magnetic biochar (Fe_2O_3-EC) significantly enhanced metal ion adsorption efficiency upto 80%. Cai et al. (2017) checked the efficiency of magnetic biochar made from water hyacinth for phosphate (P) removal and the results showed that biochar had the greatest capacity for removing P adsorption at the MW450 level. Zhang et al. (2015) conducted an experiment with *E. crassipes* biochar for Cd removal and the results showed that biochar had the highest Cd sorption capacity at the BC450 level.

18.5 Conclusion and future prospects

Phytoremediation has been seemed to be a less disruptive, less expensive and more environment friendly clean-up technology. The selection of a suitable plant species is the most important aspect of phytoremediation. Effective aquatic plants, terrestrial and hyperaccumulator plants, play an important role in HM remediation from the contaminated water bodies. The findings in this chapter show that *E. crassipes* can be used for inorganic contaminants removal in both bioaccumulation (entire biomass) and bio-sorption (dried plant) forms. In addition, it has been found that *E. crassipes* has significant metal tolerant capacity as well as ability to remove a wide range of other inorganic contaminants. It is recommended that more field-based studies to be conducted to explore the potential of *E. crassipes* for the removal of inorganic and ionic contaminants particularly HMs from the contaminated aquatic ecosystems.

References

Abbas, N., Butt, T., Ahmad, M., Deeba, F., Hussain, N., 2021. Phytoremediation potential of *Typha latifolia* and water hyacinth for removal of heavy metals from industrial wastewater. Chemistry International 7 (2), 103–111.

Abbas, Z., Arooj, F., Ali, S., Zaheer, I.E., Rizwan, M., Riaz, M.A., 2019. Phytoremediation of landfill leachate waste contaminants through floating bed technique using water hyacinth and water lettuce. International Journal of Phytoremediation 21 (13), 1356–1367.

Aboul-Enein, A.M., Shanab, S.M., Shalaby, E.A., Zahran, M.M., Lightfoot, D.A., El-Shemy, H.A., 2014. Cytotoxic and antioxidant properties of active principals isolated from water hyacinth against four cancer cells lines. BMC Complementary and Alternative Medicine 14 (1), 1–11.

Ajayi, T.O., Ogunbayio, A.O., 2012. Achieving environmental sustainability in wastewater treatment by phytoremediation with water hyacinth (*Eichhornia crassipes*). Journal of Sustainable Development 5 (7), 80–90.

Al-Alawy, A.F., Salih, M.H., 2017. Comparative study between nanofiltration and reverse osmosis membranes for the removal of heavy metals from electroplating wastewater. Journal of Engineering 23, 1–21.

Ali, Q., Ahsan, M., Khaliq, I., Elahi, M., Ali, S., Ali, F., et al., 2011. Role of rhizobacteria in phytoremediation of heavy metals: an overview. International Research Journal of Plant Science 2, 220–232.

Ali, S., Abbas, Z., Rizwan, M., Zaheer, I.E., Yavaş, İ., Ünay, A., et al., 2020. Application of floating aquatic plants in phytoremediation of heavy metals polluted water: a review. Sustainability 12 (5), 1927. Available from: https://doi.org/10.3390/su12051927.

Ashraf, S., Ali, Q., Zahir, Z.A., Ashraf, S., Asghar, H.N., 2019. Phytoremediation: environmentally sustainable way for reclamation of heavy metal polluted soils. Ecotoxicology and Environmental Safety 174, 714–727.

Auchterlonie, J., Eden, C.L., Sheridan, C., 2021. The phytoremediation potential of water hyacinth: a case study from Hartbeespoort Dam, South Africa. South African Journal of Chemical Engineering 37, 31–36.

Basu, A., Saha, D., Saha, R., Ghosh, T., Saha, B., 2014. A review on sources, toxicity and remediation technologies for removing arsenic from drinking water. Research on Chemical Intermediates 40 (2), 447–485.

Bauddh, K., Singh, K., Singh, B., Singh, R.P., 2015. *Ricinus communis*: a robust plant for bio-energy and phytoremediation of toxic metals from contaminated soil. Ecological Engineering 84, 640–652.

Buasri, A., Chaiyut, N., Tapang, K., Jaroensin, S., Panphrom, S., 2012. Biosorption of heavy metals from aqueous solutions using water hyacinth as a low cost biosorbent. Civil and Environmental Research 2 (2), 17–25.

Burakov, A.E., Galunin, E.V., Burakova, I.V., Kucherova, A.E., Agarwal, S., Tkachev, A.G., et al., 2018. Adsorption of heavy metals on conventional and nanostructured materials for wastewater treatment purposes: a review. Ecotoxicology and Environmental Safety 148, 702–712.

Cai, R., Wang, X., Ji, X., Peng, B., Tan, C., Huang, X., 2017. Phosphate reclaim from simulated and real eutrophic water by magnetic biochar derived from water hyacinth. Journal of Environmental Management 187, 212–219.

Dar, S.H., Kumawat, D.M., Singh, N., Wani, K.A., 2011. Sewage treatment potential of water hyacinth (*Eichhornia crassipes*). Research Journal of Environmental Sciences 5 (4), 377. Available from: https://doi.org/10.3923/rjes.2011.377.385.

Chen, L., Li, F., Wei, Y., Li, G., Shen, K., He, H.J., 2019. High cadmium adsorption on nanoscale zero-valent iron coated *Eichhornia crassipes* biochar. Environmental Chemistry Letters 17 (1), 589–594.

da-Silva, C., Canatto, R., Cardoso, A.A., Ribeiro, C., Oliveira, J., 2017. Arsenic-hyperaccumulation and antioxidant system in the aquatic macrophyte *Spirodela intermedia* W. Koch (Lemnaceae). Theoretical and Experimental Plant Physiology 29 (4), 203–213.

Daud, M.K., Ali, S., Abbas, Z., Zaheer, I.E., Riaz, M.A., Malik, A., et al., 2018. Potential of duckweed (*Lemna minor*) for the phytoremediation of landfill leachate. Journal of Chemical 1–9.

Ding, Y., Liu, Y., Liu, S., et al., 2016. Competitive removal of Cd (II) and Pb (II) by biochars produced from water hyacinths: performance and mechanism. RSC Advances 6 (7), 5223–5232.

Elfeky, S.A., Imam, H., Alsherbini, A.A., 2013. Bio-absorption of Ni and Cd on *Eichhornia crassipes* root thin film. Environmental Science and Pollution Research 20 (11), 8220–8226.

Elgallal, M., Fletcher, L., Evans, B., 2016. Assessment of potential risks associated with chemicals in wastewater used for irrigation in arid and semiarid zones: a review. Agricultural Water Management 177, 419–431. Available from: https://doi.org/10.1016/j.agwat.2016.08.027.

Ergönül, M.B., Nassouhi, D., Atasağun, S., 2020. Modeling of the bioaccumulative efficiency of *Pistia stratiotes* exposed to Pb, Cd, and Pb and Cd mixtures in nutrient-poor media. International Journal of Phytoremediation 22 (2), 201–209.

Fazal, S., Zhang, B., Mehmood, Q., 2015. Biological treatment of combined industrial wastewater. Ecological Engineering 84, 551–558.

Feng, W., Xiao, K., Zhou, et al., 2017. Analysis of utilization technologies for *Eichhornia crassipes* biomass harvested after restoration of wastewater. Bioresource Technology 223, 287–295.

Gall, J.E., Boyd, R.S., Rajakaruna, N., 2015. Transfer of heavy metals through terrestrial food webs: a review. Environmental Monitoring and Assessment 187, 201. Available from: https://doi.org/10.1007/s10661-015-4436-3.

Gogoi, P., Adhikari, P., Maji, T.K., 2017. Bioremediation of arsenic from water with citric acid cross-linked water hyacinth (*E. crassipes*) root powder. Environmental Monitoring and Assessment 189 (8), 1–11.

Goncalves, A.L., Pires, J.C., Simões, M., 2017. A review on the use of microalgal consortia for wastewater treatment. Algal Research 24, 403–415.

González, C.I., Maine, M.A., Cazenave, J., Hadad, H.R., Benavides, M.P., 2015. Ni accumulation and its effects on physiological and biochemical parameters of *Eichhornia crassipes*. Environmental and Experimental Botany 117, 20–27.

Goswami, S., Das, S., 2018. *Eichhornia crassipes* mediated copper phytoremediation and its success using catfish bioassay. Chemosphere 210, 440–448.

Govindaswamy, S., Schupp, D.A., Rock, S.A., 2011. Batch and continuous removal of arsenic using hyacinth roots. International Journal of Phytoremediation 13 (6), 513–527.

Gunatilake, S.K., 2015. Methods of removing heavy metals from industrial wastewater. Methods (San Diego, Calif.) 1 (1), 14.

Gunathilakae, N., Yapa, N., Hettiarachchi, R., 2018. Effect of arbuscular mycorrhizal fungi on the cadmium phytoremediation potential of *Eichhornia crassipes* (Mart.) Solms. Groundwater for Sustainable Development 7, 477–482.

Hadad, H.R., Maine, M.A., Bonetto, C.A., 2006. Macrophyte growth in a pilot-scale constructed wetland for industrial wastewater treatment. Chemosphere 63 (10), 1744–1753.

Hasan, S.H., Talat, M., Rai, S., 2007. Sorption of cadmium and zinc from aqueous solutions by water hyacinth (*Eichchornia crassipes*). Bioresource Technology 98 (4), 918–928.

Hasan, S.H., Ranjan, D., Talat, M., 2010. Water hyacinth biomass (WHB) for the biosorption of hexavalent chromium: optimization of process parameters. Bioresource 5 (2), 563–575.

Hasani, Q., Pratiwi, N.T., Wardiatno, Y., Effendi, H., Martin, A.N., Pirdaus, P., et al., 2021. Phytoremediation of iron in ex-sand mining waters by water hyacinth (*Eichhornia crassipes*). Biodiversitas Journal of Biological Diversity 22 (2). Available from: https://doi.org/10.13057/biodiv/d220238.

Hashem, M.A., Hasan, M., Momen, M.A., Payel, S., Nur-A-Tomal, M.S., 2020. Water hyacinth biochar for trivalent chromium adsorption from tannery wastewater. Environmental Sustainability Indicators 5, 100022. Available from: https://doi.org/10.1016/j.indic.2020.100022.

Hazra, M., Avishek, K., Pathak, G., 2015. Phytoremedial potential of *Typha latifolia, Eichornia crassipes* and *Monochoria hastata* found in contaminated water bodies across Ranchi City (India). International Journal of Phytoremediation 17 (9), 835–840.

Huang, H., Liu, J., Zhang, P., Zhang, D., Gao, F., 2017. Investigation on the simultaneous removal of fluoride, ammonia nitrogen and phosphate from semiconductor wastewater using chemical precipitation. Chemical Engineering Journal 307, 696–706.

Ibrahim, H.S., Ammar, N.S., Soylak, M., Ibrahim, M., 2012. Removal of Cd (II) and Pb (II) from aqueous solution using dried water hyacinth as a biosorbent. Spectrochimica Acta Part A, Molecular and Biomolecular Spectroscopy 96, 413–420.

Iha, D.S., Bianchini Jr, I., 2015. Phytoremediation of Cd, Ni, Pb and Zn by *Salvinia minima*. International Journal of Phytoremediation 17 (10), 929–935.

Jayaweera, M.W., Kasturiarachchi, J.C., Kularatne, R.K., Wijeyekoon, S.L., 2007. Removal of aluminium by constructed wetlands with water hyacinth (*Eichhornia crassipes* (Mart.) Solms) grown under different nutritional conditions. Journal of Environmental Science and Health, Part A 42 (2), 185–193.

Jayaweera, M.W., Kasturiarachchi, J.C., Kularatne, R.K., Wijeyekoon, S.L., 2008. Contribution of water hyacinth (*Eichhornia crassipes* (Mart.) Solms) grown under different nutrient conditions to Fe-removal mechanisms in constructed wetlands. Journal of Environmental Management 87 (3), 450–460.

Jin, J., Li, S., Peng, X., Liu, W., Zhang, C., Yang, Y., et al., 2018. HNO_3 modified biochars for uranium (VI) removal from aqueous solution. Bioresource Technology 256, 247–253.

Kadirvelu, K., Kanmani, P., Senthilkumar, P., Subburam, V., 2004. Separation of mercury (II) from aqueous solution by adsorption onto an activated carbon prepared from *Eichhornia crassipes*. Adsorption Science & Technology 22 (3), 207–222.

Karaouzas, I., Kapetanaki, N., Mentzafou, A., Kanellopoulos, T.D., Skoulikidis, N., 2021. Heavy metal contamination status in Greek surface waters: a review with application and evaluation of pollution indices. Chemosphere 263, 128192.

Kumar, P., Chauhan, M.S., 2019. Adsorption of chromium (VI) from the synthetic aqueous solution using chemically modified dried water hyacinth roots. Journal of Environmental Chemical Engineering 7 (4), 103218. Available from: https://doi.org/10.1016/j.jece.2019.103218.

Kumar, S., Deswal, S., 2020. Phytoremediation capabilities of Salvinia molesta, water hyacinth, water lettuce, and duckweed to reduce phosphorus in rice mill wastewater. International Journal of Phytoremediation 22 (11), 1097–1109.

Kumar, V., Chopra, A.K., Singh, J., Thakur, R.K., Srivastava, S., Chauhan, R.K., 2016. Comparative assessment of phytoremediation feasibility of water caltrop (*Trapa natans* L.) and water hyacinth (*Eichhornia crassipes* Solms.) using pulp and paper mill effluent. Archives of Agronomy and Soil Science 1 (1), 13–21.

Kumar, S., Prasad, S., Yadav, K.K., et al., 2019. Hazardous heavy metals contamination of vegetables and food chain: role of sustainable remediation approaches—a review. Environmental Research 179, 108792. Available from: https://doi.org/10.1016/j.envres.2019.108792.

Kumari, K., Swain, A.A., Kumar, M., Bauddh, K., 2021. Utilization of Eichhornia crassipes biomass for production of biochar and its feasibility in agroecosystems: a review. Environmental Sustainability 1–13.

Kwak, J.H., Islam, M.S., Wang, S., Messele, S.A., Naeth, M.A., El-Din, M.G., et al., 2019. Biochar properties and lead (II) adsorption capacity depend on feedstock type, pyrolysis temperature, and steam activation. Chemosphere 231, 393–404.

Leung, H.M., Duzgoren-Aydin, N.S., Au, C.K., Krupanidhi, S., Fung, K.Y., Cheung, K.C., et al., 2017. Monitoring and assessment of heavy metal contamination in a constructed wetland in Shaoguan (Guangdong Province, China): bioaccumulation of Pb, Zn, Cu and Cd in aquatic and terrestrial components. Environmental Science and Pollution Research International 24, 9079–9088.

Levchuk, I., Màrquez, J.J.R., Sillanpää, M., 2018. Removal of natural organic matter (NOM) from water by ion exchange—a review. Chemosphere 192, 90–104.

Li, F., Shen, K., Long, X., et al., 2016a. Preparation and characterization of biochars from *Eichhornia crassipes* for cadmium removal in aqueous solutions. PLoS One 11 (2), 0148132. Available from: https://doi.org/10.1371/journal.pone.0148132.

Li, Q., Chen, B., Lin, P., Zhou, J., Zhan, J., Shen, Q., et al., 2016b. Adsorption of heavy metal from aqueous solution by dehydrated root powder of long-root *Eichhornia crassipes*. International Journal of Phytoremediation 18 (2), 103–109.

Li, Y., Liu, X., Nie, X., Yang, W., Wang, Y., Yu, R., et al., 2019. Multifunctional organic–inorganic hybrid aerogel for self-cleaning, heat-insulating, and highly efficient microwave absorbing material. Advanced Functional Materials 29 (10), 1807624. Available from: https://doi.org/10.1002/adfm.201807624.

Lin, S., Yang, H., Na, Z., Lin, K., 2018. A novel biodegradable arsenic adsorbent by immobilization of iron oxy-hydroxide (FeOOH) on the root powder of long-root *Eichhornia crassipes*. Chemosphere 192, 258–266.

Liu, C., Ye, J., Lin, Y., Wu, J., Price, G.W., Burton, D., et al., 2020. Removal of Cadmium (II) using water hyacinth (*Eichhornia crassipes*) biochar alginate beads in aqueous solutions. Environmental Pollution 264, 114785.

Lu, X., Gao, Y., Luo, J., Yan, S., Rengel, Z., Zhang, Z., 2014. Interaction of veterinary antibiotic tetracyclines and copper on their fates in water and water hyacinth (*Eichhornia crassipes*). Journal of Hazardous Materials 280, 389–398.

Mahamadi, C., Nharingo, T., 2010. Competetive adsorption of Pb2 + Cd2 + and Zn2 + ions onto *Eichhornia crassipes* in binary and ternary systems. Bioresource Technology 101, 859–864.

Malar, S., Vikram, S.S., Favas, P.J., Perumal, V., 2016. Lead heavy metal toxicity induced changes on growth and antioxidative enzymes level in water hyacinths [*Eichhornia crassipes* (Mart.)]. Botanical Studies 55 (1), 1–11.

Matai, S., Bagchi, D.K., 1980. Water hyacinth: a plant with prolific bioproductivity and photosynthesis. In: Gnanam, A., Krishnaswamy, S., Kahn, J.S. (Eds.), Proceedings of the International Symposium on Biological Applications of Solar Energy. SG Wasani for the Macmillan Co. of India, Madras, India, 1–5 December 1978.

Mayo, A.W., Hanai, E.E., 2017. Modeling phytoremediation of nitrogen-polluted water using water hyacinth (*Eichhornia crassipes*). Physics and Chemistry of the Earth Parts A/B/C 100, 170–180.

Mishra, V.K., Tripathi, B.D., 2008. Concurrent removal and accumulation of heavy metals by the three aquatic macrophytes. Bioresource Technology 99 (15), 7091–7097.

Mishra, V.K., Tripathi, B.D., 2009. Accumulation of chromium and zinc from aqueous solutions using water hyacinth (*Eichhornia crassipes*). Journal of Hazardous Materials 164, 1059–1063.

Mishra, S., Maiti, A., 2017. The efficiency of *Eichhornia crassipes* in the removal of organic and inorganic pollutants from wastewater: a review. Environmental Science and Pollution Research 24 (9), 7921–7937.

Mishra, V.K., Upadhyay, A.R., Pathak, V., Tripathi, B.D., 2008. Phytoremediation of mercury and arsenic from tropical opencast coalmine effluent through naturally occurring aquatic macrophytes. Water, Air, & Soil Pollution 192 (1), 303–314.

Mitra, T., Singha, B., Bar, N., Das, S.K., 2014. Removal of Pb (II) ions from aqueous solution using water hyacinth root by fixed-bed column and ANN modeling. Journal of Hazardous Materials 273, 94–103.

Módenes, A.N., Espinoza- Quinones, F.R., Trigueros, D.E.G., et al., 2011. Kinetic and equilibrium adsorption of Cu (II) and Cd (II) ions on *Eichhornia crassipes* in single and binary systems. Chemical Engineering Journal 168, 44–51.

Mohamad, H.H., Latif, P.A., 2010. Uptake of cadmium and zinc from synthetic effluent by water hyacinth (*Eichhornia crassipes*). Environmental Asia 3 (Special issue), 36–42.

Muradov, N., Taha, M., Miranda, A.F., et al., 2014. Dual application of duckweed and azolla plants for wastewater treatment and renewable fuels and petrochemicals production. Biotechnology for Biofuels 7 (1), 1–17.

Muthusaravanan, S., Sivarajasekar, N., Vivek, J.S., et al., 2018. Phytoremediation of heavy metals: mechanisms, methods and enhancements. Environmental Chemistry Letters 16 (4), 1339–1359.

Nuraini, Y., Felani, M., 2015. Phytoremediation of tapioca wastewater using water hyacinth plant (*Eichhornia crassipes*). Journal of Degraded and Mining Lands Management 2 (2), 295.

Nyamunda, B.C., Chivhanga, T., Guyo, U., Chigondo, F., 2019. Removal of Zn (II) and Cu (II) ions from industrial wastewaters using magnetic biochar derived from water hyacinth. Journal of Engineering 5656983. Available from: https://doi.org/10.1155/2019/5656983.

Patil, J.H., AntonyRaj, M., Gavimath, C.C., 2011. Study on effect of pretreatment methods on biomethanation of water hyacinth. International Journal of Advanced Biotechnology and Research 2 (1), 143–147.

Prasad, B., Maiti, D., 2016. Comparative study of metal uptake by *Eichhornia crassipes* growing in ponds from mining and nonmining areas—a field study. Bioremediation Journal 20 (2), 144–152.

Qambrani, N.A., Rahman, M.M., Won, S., Shim, S., Ra, C., 2017. Biochar properties and eco-friendly applications for climate change mitigation, waste management, and wastewater treatment: a review. Renewable & Sustainable Energy Reviews 79, 255–273.

Raval, N.P., Shah, P.U., Shah, N.K., 2016. Adsorptive removal of nickel (II) ions from aqueous environment: a review. Journal of Environmental Management 179, 1–20.

Riyanto, C.A., Prabalaras, E., 2019. The adsorption kinetics and isoterm of activated carbon from Water Hyacinth Leaves (*Eichhornia crassipes*) on Co (II). Journal of Physics: Conference Series 1307 (1), 012002. Available from: https://doi:10.1088/1742-6596/1307/1/012002.

Romanova, T.E., Shuvaeva, O.V., Belchenko, L.A., 2016. Phytoextraction of trace elements by water hyacinth in contaminated area of gold mine tailing. International Journal of Phytoremediation 18 (2), 190–194.

Saha, P., Shinde, O., Sarkar, S., 2017. Phytoremediation of industrial mines wastewater using water hyacinth. International Journal of Phytoremediation 19 (1), 87–96.

Sarkar, M., Rahman, A.K.M.L., Bhoumik, N.C., 2017. Remediation of chromium and copper on water hyacinth (*E. crassipes*) shoot powder. Water Resources and Industry 17, 1–6.

Sekomo, C.B., Kagisha, V., Rousseau, D., Lens, P., 2012. Heavy metal removal by combining anaerobic upflow packed bed reactors with water hyacinth ponds. Environmental Technology 33 (10–12), 1455–1464.

Shahid, M.J., Arslan, M., Ali, S., Siddique, M., Afzal, M., 2018. Floating wetlands: a sustainable tool for wastewater treatment. Clean-Soil Air Water 46, 1800120. Available from: https://doi.org/10.1002/clen.201800120.

Singh, S., Kumar, V., Datta, S., Dhanjal, D.S., Sharma, K., Samuel, J., et al., 2020. Current advancement and future prospect of biosorbents for bioremediation. The Science of the Total Environment 709, 135895. Available from: https://doi.org/10.1016/j.scitotenv.2019.135895.

Smolyakov, B.S., 2012. Uptake of Zn, Cu, Pb, and Cd by water hyacinth in the initial stage of water system remediation. Applied Geochemistry: Journal of the International Association of Geochemistry and Cosmochemistry 27 (6), 1214–1219.

Su, H., Chen, J., Wu, Y., Chen, J., Guo, X., Yan, Z., et al., 2019. Morphological traits of submerged macrophytes reveal specific positive feedbacks to water clarity in freshwater ecosystems. The Science of the Total Environment 684, 578–586.

Sudiarto, S.I.A., Renggaman, A., Choi, H.L., 2019. Floating aquatic plants for total nitrogen and phosphorus removal from treated swine wastewater and their biomass characteristics. Journal of Environmental Management 231, 763–769.

Sullivan, P.R., Wood, R., 2012. Water hyacinth (*Eichhornia crassipes* (Mart.) Solms) seed longevity and the implications for management. In: Eighteenth Australasian Weeds Conference. pp. 37–40.

Sung, K., Lee, G.J., Munster, C., 2015. Effects of *Eichhornia crassipes* and *Ceratophyllum demersum* on soil and water environments and nutrient removal in wetland microcosms. International Journal of Phytoremediation 17 (10), 936–944.

Swain, G., Adhikari, S., Mohanty, P., 2014. Phytoremediation of copper and cadmium from water using water hyacinth, *Eichhornia crassipes*. International Journal of Agricultural Science and Food Technology 2 (1). Available from: https://doi.org/10.14355/ijast.2014.0301.01.

Tatarchuk, T., Bououdina, M., Al-Najar, B., Bitra, R.B., 2019. Green and ecofriendly materials for the remediation of inorganic and organic pollutants in water. A new generation material graphene. Applied Water Technology 69–110.

Valipour, A., Raman, V.K., Motallebi, P., 2010. Application of shallow pond system using water hyacinth for domestic wastewater treatment in the presence of high total dissolved solids (tds) and heavy metal salts. Environmental Engineering and Management Journal 9 (6), 853–860.

Victor, K.K., Ladji, M., Adjiri, A.O., Cyrille, Y.D.A., Sanogo, T.A., 2016. Bioaccumulation of heavy metals from wastewaters (Pb, Zn, Cd, Cu and Cr) in water hyacinth (*Eichhornia crassipes*) and water lettuce (*Pistia stratiotes*). International Journal of ChemTech Research 9 (2), 189–195.

Wei, Y., Fang, Z., Zheng, L., Tsang, E.P., 2017. Biosynthesized iron nanoparticles in aqueous extracts of *Eichhornia crassipes* and its mechanism in the hexavalent chromium removal. Applied Surface Science 399, 322–329.

Xu, L., Cheng, S., Zhuang, P., Xie, D., et al., 2020. Assessment of the nutrient removal potential of floating native and exotic aquatic macrophytes cultured in Swine Manure Wastewater. International Journal of Environmental Research and Public Health 17 (3), 1103. Available from: https://doi.org/10.3390/ijerph17031103.

Yasar, A., Zaheer, A., Tabinda, A.B., Khan, M., Mahfooz, Y., Rani, S., et al., 2018. Comparison of reed and water lettuce in constructed wetlands for wastewater treatment. Water Environment Research: A Research Publication of the Water Environment Federation 90, 129–135.

Yu, J., Jiang, C., Guan, Q., Ning, P., Gu, J., Chen, Q., et al., 2018. Enhanced removal of Cr(VI) from aqueous solution by supported ZnO nanoparticles on biochar derived from waste water hyacinth. Chemosphere 195, 632–640.

Zhang, F., Wang, X., Yin, D., Peng, B., Tan, C., Liu, Y., et al., 2015. Efficiency and mechanisms of Cd removal from aqueous solution by biochar derived from water hyacinth (*Eichhornia crassipes*). Journal of Environmental Management 153, 68–73.

Zhang, F., Wang, X., Xionghui, J., Ma, L., 2016. Efficient arsenate removal by magnetite-modified water hyacinth biochar. Environmental Pollution 216, 575–583.

Zheng, J.C., Liu, H.Q., Feng, H.M., Li, W.W., Lam, M.H.W., Lam, P.K.S., et al., 2016. Competitive sorption of heavy metals by water hyacinth roots. Environmental Pollution 219, 837–845.

Zhu, C., Tian, H., Cheng, K., Liu, K., Wang, K., Hua, S., et al., 2016. Potentials of whole process control of heavy metals emissions from coal-fired power plants in China. Journal of Cleaner Production 114, 343–351.

CHAPTER

19

Application of nanotechnology for heavy metals remediation from contaminated water

Sweety Nath Barbhuiya[1,2], Dharmeswar Barhoi[1,3] and Sarbani Giri[1]

[1]Laboratory of Cell and Molecular Biology, Department of Life Science and Bioinformatics, Assam University, Silchar, Assam, India [2]Department of Zoology, Patharkandi College, Patharkandi, Assam, India [3]The Assam Royal Global University, Guwahati, Assam, India

19.1 Introduction

19.1.1 Heavy metal

In the current world, rapid industrialization has led to a massive increase in the environmental toxicants. Environmental toxicants are toxic substances that cause dangerous effects to the living beings as well as the surroundings. Among the environmental toxicants, heavy metals have immense potential to cause damage to the pattern and rationale of the entire ecosystem (Alengebawy et al., 2021). Metals with atomic weight and a density larger than 5 g/cm^3 are termed as "heavy metals" (Zhang et al., 2019). In this book chapter, we have discussed the toxic effects of heavy metals and their bioremediation using nanotechnology techniques. Besides the heavy metals, we have also included arsenic which not a perfect heavy metal. It is a metalloid present in the atmosphere. Whenever, we discuss the noxious profile of heavy metals, the toxicity of arsenic could not be overlooked. Arsenic is a highly toxic metalloid present in trivalent (As^{3+}) and pentavalent (As^{5+}) form. Several scientist and researchers consider use the term metal/metalloid in case of arsenic and in this chapter also, we have used both metal and metalloid synonymously for arsenic. However, the readers should keep it in mind that arsenic is not a perfect heavy metal, instead it is a metalloid. Arsenic (As), Lead (Pb), Cadmium (Cd), Nickel (Ni), Cobalt (Co) etc., are some common heavy metals, inducing extensive toxicity, both at

DOI: https://doi.org/10.1016/B978-0-323-95919-3.00010-0

small and higher concentrations in the environment (Abdelaal et al., 2021). Heavy metals are ubiquitously present all around the earth's crust. Some metalloids (e.g., arsenic) and light weight metals (e.g., selenium, aluminum) are extremely toxic and thus recognized as heavy metals. Gold, a heavy metal is however typically nontoxic (Briffa et al., 2020). It is reported that the heavy metals are elucidated depending in numeral criteria, such as complex generation, hard-soft acids and bases, their cationic-hydroxide development, range of specific gravity larger than 5 g/mL, and in recent times, alliance with eutrophication and environmental toxicity (Rajendran and Gunasekaran, 2007). The toxicity even gets enhanced due to the property of heavy metals to persist in nature and accumulate in plant and animal tissues (Kinuthia et al., 2020).

19.1.2 Sources of heavy metal

The sources of origin of heavy metals in the surroundings include both normal and human activities. The natural sources consisted of rock weathering, forest fires, soil-forming processes and volcanic eruptions (Bradl, 2005). The anthropogenic activities include coal mining, urban activities, industrial discharges, etc. (Morkunas et al., 2018). The other sources of heavy metal pollution include rain drops containing heavy metal, sedimentation of aerosol particles, and agrochemicals (Alengebawy et al., 2021).

19.2 Effect of heavy metal

Heavy metals are cluster of substances including both metals and metalloids that have comparatively elevated density and are noxious even at little level. Heavy metals produce hazardous effect on the environment and living organisms. Examples of heavy metals include Cadmium, Lead, Chromium, Nickel, Arsenic (metalloid), Zinc, Iron etc.

19.2.1 Effect of heavy metals in the environment

Heavy metals can induce toxicity in the atmosphere and thus, an issue of global disquiet. Heavy metals can enter water bodies from both natural and human activities and continuously pollute the aqueous environment. Heavy metals could dissolve in water and thus can be easily absorbed by the different fish species. The heavy metals upon entering the food chain, leads to bioaccumulation or biomagnification and threatens human health (Kinuthia et al., 2020). However, the harmful outcome caused by heavy metals depends upon the dose and time of exposure. It also has hazardous consequence on plants and soil quality. High concentration of lead in soils causes decreased soil fertility while a very low concentration of lead inhibits the fundamental plant processes *viz.* mitosis, photosynthesis, drooping of older leaves, diminutive foliage and brown tiny leaves (Singh and Kalamdhad, 2011). The contamination of iron in the water bodies whether directly or indirectly affects different species of fishes adversely and also other species like periphyton and benthic invertebrates (Jaishankar et al., 2014).

19.2.2 Effect in human beings induced by heavy metals

Heavy metals are usually considered as the most toxic agent to living organisms and may cause several diseases. According to a report, heavy metals exposure may damage to central nervous system, kidneys, lungs and other crucial organs (Yang et al., 2019a,b). Another study reveals that contact to heavy metals for a long time results in neurological disorders *viz.* multiple sclerosis, Alzheimer's disease, muscular dystrophy and Parkinson's disease (Järup, 2003). Hypertension and kidney related problems are caused by cadmium (Satarug et al., 2005). Exposure to lead (Pb) can generate disorders of the nervous system and reduce reproductive efficiency (Sall et al., 2020). It is reported that chromium has the potential to cause skin related problems and respiratory troubles (Mohammadi et al., 2020). Arsenic exposure via drinking water may cause cancer of kidney, lung, skin, bladder and liver (Kumar et al., 2021). Apart from cancer, arsenic exposure was found to be associated with numerous diseases like diabetes, cardiovascular diseases, nephrotoxicity, and reproductive implications of both male and female (Nath Barbhuiya et al., 2020; Barbhuiya et al., 2021). In 2015, Jan et al. (2015), reported that exposure to elevated level of metallic mercury leads to pulmonary toxicity, nausea, skin rashes, diarrhea, vomiting, hypertension, nephrotoxicity, and severe neurologic abnormalities. Heavy metals exposure *viz.* Hg (mercury) and Pb (lead) may provoke autoimmune diseases viz. rheumatoid arthritis, renal, nervous and circulatory problems (Lauwerys et al., 2007) (Table 19.1).

19.3 Bioremediation

It is already mentioned earlier that quick development in the industrial sectors and urbanization has lead to the massive raise in pollutants, causing hazardous effect to human health. Thus, the pollutants should be removed from the environment. The conventional physical and chemical remediation processes are expensive and also produces secondary contamination problems due to the transport of pollutants and chemical reagents (Yu et al., 2020). This evokes the necessity of developing some strategy to remove the pollutants without generating harmful effects.

Bioremediation is a method to facilitate the capability of life forms to condense or exterminate contaminants from aquatic and land areas, thus reducing the threat to the physical condition of human beings by reestablishing the atmosphere to its original circumstance (de Oliveira Santos et al., 2018). Adhikari et al. (2004) defined bioremediation as "the process of cleaning up hazardous wastes with microorganisms or plants and is the safest method of clearing soil of pollutants." The advantage of this technique includes minimal expenditure, no secondary pollutants are produced, and it is environment friendly and very useful in revival of the ecosystem (Igiri et al., 2018; Yu et al., 2020). Microorganisms or processes involving microbes are used in bioremediation to remove as well as alter the ecological contaminant into a lesser amount of toxic components (Girma, 2015). However, the process of bioremediation has some restrictions like heavy metals are non-degradable (Igiri et al., 2018). Bioremediation methods include: (1) degeneration of carbon-base chemicals using microbes, (2) Use of fast growing plants with intense potential to produce large biomass, (3) storage by animals found in soils and (4) utilization of plants and bacteria in

TABLE 19.1 Effects of various types of heavy metals on the animal test system.

Heavy metal	Animal model	Treatment	Observation	References
Cadmium (Cd)	Sprague-Dawley adult rat (male)	The animals were treated with different concentrations of cadmium; 1 mg/kg $CdCl_2$ (low dose), 2.5 mg/kg $CdCl_2$ (moderate dose), 5 mg/kg CdCl2 (high dose). The animals were administered orally for 60 days	The results of the study under consideration indicate that cadmium treatment induces renal injury. The mechanism behind such injury may be attributed to redox imbalance, disordered iron adsorption and apoptosis in the renal tissues	Liu et al. (2019)
Mercury (Hg)	Albino male rats (*Rattus norvegicus)*	Mercuric chloride $HgCl_2$ was administered orally at a concentration of 20 ppm for eight weeks	Exposure to mercury on the liver leads to increment in liver biomarkers such as Alaline transaminase; ALT and Aspartate transaminae; AST and decrease in serum alkaline phosphatase (ALP)	Wadaan (2009)
Chromium	Swiss albino mice (*Mus musculus*)	Mice were supplied with chromium sulfate along with normal diet at four different concentrations (2.5, 3, 3.5 and 4 g/kg) for 21 days	It was observed from the study that high concentration of chromium is associated with altered tissue structure of liver and kidney	Fatima et al. (2016)
Arsenic	Wistar albino rats (female)	The experimental animals were treated with three dissimilar concentrations of arsenic (sodium arsenite): 10, 30, 50 μg/L for two durations, 30 days and 60 days	From the study, it was observed that subchronic exposure of arsenic induces oxidave stress in the reproductive organs, which in turn leads to degeneration of the reproductive organs of the female mice	Mehta and Hundal (2016)
Manganese		Adult mice were exposed to three doses of Manganese chloride (7.5, 15 and 30 mg/kg/day of $MnCl_2$ for sixty numbers of days through oral route, while pups of mice were treated with the equal dose of manganese during developmental period. The pups parentally exposed to Mn were then administered with Mn in the form of $MnCl_2$ at three different doses (0.013, 0.13 and 1.3 mg/kg/day) through orally during 60 days of treatment	The current study indicates that chemical stress was generated in the groups treated with manganese during the early period of development in dose-dependent manner in mice	Okada et al. (2016)
Copper		The test animals were exposed to different concentrations (0, 4, 8 and 16) mg/kg of copper for a period of 21 days and 42 days.	Exposure to higher concentration of copper leads to oxidative stress due to increase in reactive oxygen species (ROS) and protein carbonyls (PC) intensity	Liu et al. (2020)
Cadmium	Balb/c mice	The animals administered with 200 ppm cadmium chloride for two time periods *viz.* 30 days water) and 60 days through orally.	The present study reveals that administration of cadmium orally induces apoptosis in the endometrium and it the apoptotic index increases with the duration of exposure that may affect the receptivity of the female mice	Sapmaz-Metin et al. (2017)

combination, known as microbe-assisted phytoremediation (Tara et al., 2014; Shehzadi et al., 2016; Fatima et al., 2018). The strategies commonly used for microbial-based bioremediation includes the method of biostimulation, that is, accumulation of particular compounds in order to arouse autochthonous microbial accumulation and the process of bioaugmentation, that is, the addition of specific microbial taxa which possess extensive biodegradation or detoxification ability (Beolchini et al., 2009; Das and Chandran, 2011; Adams et al., 2015; Fodelianakis et al., 2015; Daccò et al., 2020).

19.3.1 Bacterial bioremediation

In natural and synthetic bioremediation processes, different types of bacteria are used *viz. Pseudomonads, methanobacteria* and bacilli. Many bacteria help in the elimination of carbon, phosphorus and nitrogen compounds while others are involved in eradication of toxic metals, herbicides, aromatic compounds, pesticides and xenobiotics (Saier, 2005). The different types of bacteria associated with the process of bioremediation includes *Pseudomonas, Arthrobacter, Bacillus, Flavobacterium, Enterobacter, Acinetobacter, Alcaligens, Archromobacter* (Ojuederie and Babalola, 2017; Xu et al., 2018).

19.3.2 Phycoremediation

Phycoremediation is extensively employed to treat industrial effluents. This process involves the utilization of macroalgae and microalgae for the exclusion or biotransformation of pollutants, consisting of nutrients and other hazardous components from wastewater and carbon-dioxide from waste atmosphere (Rao et al., 2011). The use of microalgae offers an economical and valuable approach to get rid of excess nutrients and other pollutants in tertiary waste-water treatment. Moreover, it is a potentially important biomass due to its high potential to uptake inorganic nutrient (Munoz and Guieysse, 2006).

19.3.3 Mycoremediation

Mycoremediation is the method in which fungi are used to remove pollutants from contaminated soil and water effluents (Sasek and Cajthaml, 2005). Fungi has immense ecological and biochemical ability to decay organic chemicals present in the environment and diminish the menace linked with metals, radionuclides and metalloids, either by compound variation or by influencing substance bioavailability. Moreover, fungi are suitable agent for bioremediation because they can use pollutants as their growth substrate. They also have the capability to form extensive mycelial networks (Prakash, 2017).

19.3.4 Phytoremediation

It is a procedure where plants are utilized in situ to degenerate, immobilize and eliminate contaminants from the atmosphere (Murphy and Coats, 2011). Plant act as a potent remediation agent because the plant tissues can absorb the pollutants (e.g., heavy metals) from the surrounding atmosphere via phytoextraction and thus removing the pollutants.

This method is in-expensive, easy and environment friendly method for removing contaminants from contaminated sites (Itanna and Coulman, 2003).

The efficient measures for removing contaminants from contaminated region with least effect on the ecological system involve biostimulation and bioaugmentation. Even though the process of bioremediation provides an outstanding strategy for removing pollutants, but it is less effective for removal of elevated concentrations of the pollutants and xenobiotics, leading to less treatment efficiencies and revival time (Vázquez-Núñez et al., 2020). In this perspective, the integration of nanomaterials (NMs) and the development of nanotechnology seem to be the novel strategy to deal with the limitations of conventional bioremediation process (Rizwan et al., 2014). NMs have immense potential to interact with both abiotic and biotic elements, both in constructive and destructive ways; thus many researches were carried out to determine the consequence of the collective application of conventional bioremediation process as well as NMs and explicate the interactions (physical, chemical and biological) in aqueous or in soil system (Kumari and Singh, 2016; Cecchin et al., 2017).

19.4 Nanotechnology for heavy metals remediation

In the 21st century, the nanotechnology is the quickly developing area of scientific study across the globe and is known as the "Next Industrial Revolution" (Roco, 2005). In the year 1959, Richard Feynman was the first researcher who postulated the idea of nanotechnology (Feynman, 1960).

Nanotechnology is the branch of science that makes use of minute manufactured particles, less than 100 nm in size and are called nanoparticles (NPs). These are molecular aggregates that may considerably transform their physico-chemical characteristics as compared to the large quantity of materials. The important forces governing NPs are Van der Waal's forces, magnetism and electron resistance instead of forces like inertia or gravity (Bhat, 2003).

NPs are generally divided into two groups *viz.* organic and inorganic NPs (Yadav et al., 2017).

There are numerous advantages in making use of NMs over conventional bioremediation techniques. First, use of small size materials in nanoscale measure increases the surface area per unit mass; thus the bigger portion of the materials get expose to the surrounding and thus affects the reactivity greatly. Secondly, a smaller amount of activation energy is required by NMs for undergoing any chemical reaction. Thirdly, NPs exhibits another phenomenon called "Surface plasmon resonance" that can be used for uncovering toxic materials (Rizwan et al., 2014).

19.5 Mechanism of removal of heavy metals by nanomaterials

19.5.1 Adsorption

NMs have extremely elevated adsorption capacity owing to their huge surface area. There are mainly two different types of adsorption: physical and chemical adsorption.

Chemisorptions is the process of adsorption that takes place via electron transfer or sharing of electron pair to produce chemical bonds. The functional groups like amino, carboxyl and hydroxyl groups are often present on the exterior of the NMs with potential to adsorb heavy metals chemically. These functional groups can produce ionic and covalent bonds, or can generate chelates with the ions of heavy metal in order to remove heavy metals effectively. Physical adsorption however refers to the heavy metal adsorption using NMs through intermolecular forces *viz.* electrostatic attraction, van der Waals force, etc. (Ding et al., 2021).

19.5.2 Redox

The utilization of reducing NMs to eliminate multivalent heavy metals is often associated with redox reactions. Nano-zero-valent iron is zero-valent, is an excellent reducing agent. Thus, the mechanism behind heavy metal elimination by nano-zero-valent iron involves reduction in addition to adsorption. NZVI is the most studied process of elimination of contaminants using redox reaction. NZVI is generally used as electron donor for dechlorination of chlorinated organic contaminants and for alteration of high-value noxious metals (Di Palma et al., 2015; El-Temsah et al., 2016; Huang et al., 2016). FeS is an important reducing agent with immense capability to eradicate heavy metals from the polluted location; consist of Fe (II) and S (II) as electron donor.

19.5.3 Co-precipitation

Co-precipitation is also a mechanism to eliminate heavy metal from the contaminated site. A novel variety of nanomaterial was synthesized using NZVI and $Mg(OH)_2$ in order to eliminate $Pb(OH)_2$ from water (Liu et al., 2015). This study revealed that Pb(II) can form $Pb(OH)_2$ with $OH-$ and can be isolated by precipitation. Similarly, NZVI can also be produced by adsorbing inorganic ions such as As (III) and co-precipitating with them (Lackovic et al., 2000).

19.6 Different types of nanoparticles used for elimination of heavy metals

Recently, adsorbents with nanomaterial characteristics has been produced for the treatment of wastewater. These are classified on the basis of the substance used to make these nanoadsorbents. The diverse varieties of NMs used for elimination of heavy metals are discussed below:

19.6.1 Carbon-containing nanoparticles

Graphene along with its derivatives, like activated carbon, fullerenes and carbon nanotubes are the most important compounds that have been used extensively in recent times for removing heavy metals as because they are nontoxic and have larger adsorption efficiency (Madima et al., 2020). This efficiency of carbon-based NPs for eliminating heavy metals is due to the presence of active sites, functional groups at the surface that are essential for the exterior chemistry of carbon components and heavy metal accumulation (Xie et al., 2017).

19.6.1.1 Carbon nanotubes

These nanotubes have hollow structure, high porosity, huge surface area and the extremely fast transport of water, thus gained more attention for the researchers/scientists to develop as a potent adsorbent in remediation of heavy metals. The four potential active sites of carbon nanotubes (CNTs) responsible for the adsorption procedure are as follows: (1) void inside of each nanotubes; (2) presence of interstitial channels in between each nanotubes in the bundles; (3) the external surface where two adjoining parallel tubes meet and a groove that exists on the edge of a nanotube bundle; and (4) the outside face of each nanotubes. It was reported that multiwalled CNTs show lower affinity for Cr (VI) than single-walled CNTs (Dehghani et al., 2015). However, unmodified CNTs have restricted capability in the heavy metal ions' adsorption due to its reduced dispersibility, absence of functional groups and hydrophobic surfaces (Dehghani et al., 2015).

19.6.1.2 Graphene nanoparticles

Graphene and graphene-based NMs have good electrical properties, thermal conductivity and mechanical strength and thus used as adsorbents for removing heavy metal in recent times from the waste water (Santhosh et al., 2016). Graphene exists in two different forms: reduced graphene oxide and graphene oxide. The exclusion of heavy metals from wastewater by using grapheme depends on electrostatic connections and surface metal hydroxide precipitation. These connections further depends upon surface charge and area i.e., higher the surface area of graphene, higher the deletion potential (Rajabathar et al., 2017).

19.6.1.3 Nanoparticles consisting of silica

These act as a good adsorbent for removing heavy metals from contaminated water because of the unique qualities such as adjustable surface properties, large surface area, well-defined pore-size and nontoxic in nature (Mahmoud et al., 2016). The addition of amino group to silicon-based NMs has shown to increase the chelating ability of these NMs. This increases the efficiency of removing the heavy metals at a neutral pH and at a much lesser concentration of NMs, indicating the possible use of silicon based NMs for the remediation of the polluted water (Jawed et al., 2020).

It was reported that silica-based NPs, that is, the silica gel consisting of amino (NH_2-SG), non-functionalized silica nanohollow sphere and amino-functionalized silica nanohollow sphere (NH2-SNHS) showed high adsorption efficiency for lead, nickel and cadmium metal ions, among which NH2-SNHS displayed highest removal potential on comparison to other NPs (Najafi et al., 2012).

19.6.1.4 Fullerenes

Fullerenes are confined cage-like shape consisting of five-member rings (12 in numbers) and an undetermined number of six-member rings. It is a hollow globular or tubular configuration composed exclusively of carbon atoms. The conductive and electric properties of fullerenes are the basis of their distinctive characteristics on the nanoscale. Fullerenes have the affinity to generate steady crystalline NPs (25–500 nm in diameter) in a range of solutions, together with water having electrolyte concentrations and pH relevant to the environment (Fortner et al., 2005; Chen and Elimelech, 2006). The first discovered fullerene molecule is C60 (Chen et al., 2019). C60 fullerene is lubricious for which it is widely used as a super lubricant.

19.6.2 Nanoparticles consisting of metal oxide

Metal oxide based NPs have size of nanoscale range that showed to have high exclusion capacity and high selectivity for heavy metals; including manganese, zinc, titanium, zirconium, and nanosized oxides of iron, zinc, cerium and titanium. These particles are deliberately synthesized in the nature from both natural and anthropogenic activities. The following are some metal oxide based nanoparticle that could be used in removal of contaminants.

19.6.2.1 Nanoparticles consisting of iron-oxide

The iron oxide based NMs have minute dimension and greater surface vicinity also possesses good magnetic property, making it a suitable agent for the removal of heavy metals (Oh and Park, 2011; Laurent et al., 2008). There are different forms of iron oxide materials found naturally *viz.* magnetite (Fe_3O_4), hematite (alpha-Fe_2O_3) and maghemite (gamma Fe_2O_3) (Cornell and Schwertmann, 2003). It is reported that iron oxide NMs are reusable through the process of desorption (Hao et al., 2010). He et al. (2017) reported that Fe_3O_4/biochar nanocomposites showed high adsorption capacity. Loading of photosynthetic bacteria (PSB) to this composite showed more efficacy than Fe_3O_4/biochar only and removed chemical oxygen demand (83.10%), 87.5% ammonium and 91.10% phosphate ions from the wastewater. Therefore, it evident that metal oxide NPs alone or together with microorganism could efficiently and promisingly remove organic contaminants from wastewater.

19.6.2.2 Zinc oxide based nanoparticles

NMs with zinc oxide have large surface area and good absorptive capability and chemical, mechanical and electrical stability; thus a suitable candidate to eliminate heavy metals from the polluted site (Ibrahim and Asal, 2017). However, application of zinc oxide-based NPs has some disadvantages like agglomeration, colloidal stability and devoid of some properties like separation and recovery (Joshi et al., 2020).

19.6.2.3 Titanium dioxide nanoparticles

Titanium dioxide (TiO_2) is a harmless semiconductor that is steady over a pH range of 2–14, having specific physical characteristics, facilitating to produce energy (solar) and transfer the same, together with reductive electron transfer and oxidative hole transfer (Skubal et al., 2002). It is the important metal oxide substance, existing in three solid phases in environment. These phases are: anatase (3.2 eV), rutile (3.0 eV) and brookite (3.2 eV). It is reported that TiO_2 has high photo-stability, high oxidizing potential of the holes, redox selectivity, simple composition and it has elevated empathy towards arsenic, making it a very suitable candidate for ecological remediation (Balaji and Matsunaga, 2002).

19.6.3 Zerovalent metal nanoparticles

These NPs have showed their fundamental task in removing harmful pollutants from contaminated aqueous system and some of them are discussed below (Table 19.2).

19.6.3.1 Nano zerovalent iron

Nano zerovalent iron (nZVI) is the most acknowledged NMs (1–100 nm). nZVI has a stronger adsorption ability, elevated capacity of reduction and activity, and a enhanced mobility. Thus, nZVI is a very good elimination matter to treat heavy metals. nZVI can be extracted out using both chemical as well as physical process. Active hydrogen-molten metal reaction, gas-phase chemical reduction and gas-phase thermal decomposition consists the chemical methods. Physical methods mostly consist of abrasion, lithography and grinding. Among these methods, the most frequently used method in order to prepare nZVI is chemical reduction (Stefaniuk et al., 2016).

19.6.3.2 Ag and Au nanoparticles

The silver (Ag) NMs are reported to act as a suitable candidate to eliminate heavy metals from contaminated water ecosystem. This is due to their exceptional electrical conductivity that enhances their efficiency of heavy metal removal; that may be endorsed to the nano size and active sites of Ag NPs (Sultan and Mohammad, 2017). Silver nonmaterial is known to be associated with the removal of mercury (Hg^{2+}). It is reported that nanoparticle consisting of Ag and mercaptosuccinic acid revealed high elimination potential for Hg^{2+} with 800 mg/g adsorption capacity at 1:6 (Sumesh et al., 2011). However, the removal capacity of gold (Au) for Hg^{2+} is higher than Ag and the absorption capacity is recorded as 4.065 g/g (Lisha and Pradeep, 2009).

19.6.4 Nanocomposite nanoparticles

19.6.4.1 Inorganic supported nanocomposites

Nanocomposites like activated carbon, zeolite, bentonite, carbon nanotubes and montmorillonite are utilized to eliminate heavy metals.

19.6.4.2 Magnetic nanocomposites

Magnetic nanocomposites have gain attention for the heavy metals removal on account of their eagerly parting potential and they are mainly supported on iron oxides and magnetic iron. The three different mechanisms in order to engineer the magnetic nanocomposites are as follows:

(1) On the exterior surface of magnetic Fe/Fe_2O_3 NMs, functional moieties like sulfhydryl and amino groups are added, (2) Encapsulation of Fe/Fe_2O_3 NMs by substances like polyethyleneimine, humic acid, MnO_2, polypyrrole and polyrhodanine forming a core-shell association and, (3) The permeable substances like carbon nanotubes and graphene oxide are coated with Fe/Fe_2O_3 NMs (Yang et al., 2019a,b).

TABLE 19.2 Summary of application of different types of nanomaterials in heavy metal elimination from waste water.

Type of nanoparticle	Target contaminant	Parameter	Observation	References
Reduced graphene oxide based nanoscale zero-valent iron (nZVI/rGO) composites	Cadmium, Cd (II)	Adsorption isotherm study, Thermodynamic study and removal kinetics	The study revealed that the Cd(II) removal took place via monolayer physical adsorption as revealed by Langmuir isotherm. The authors reported that the process of Cd(II) removal was exothermic and spontaneous. The mechanism behind Cd(II) ions elimination was restricted by the adsorption of nZVI/rGO composites from aqueous solution.	Fan et al. (2017)
Commercially available iron oxide coated stabilized NZVI powder were used for the treatment of the aqueous system	Arsenic (As)	The amount of heavy metal in the aqueous ecosystem was determined by using Inductively coupled plasma-mass. The photometric analysis was done to determine the amount of the total dissolved iron and Fe(II).	In the study, a prominent decrease in the oxidation-reduction process along with a noticeable increase in iron concentration was observed for the first month of waste water treatment using NZVI. A considerable reduction in the arsenic concentration was noticed during the initial days of treatment, however, fluctuations in the arsenic concentrations was observed throughout the experimental period (six-months) in most of the wells under observation. The toxic effect of NZVI on the bacteria present in groundwater could not be evaluated clearly. The number of cultured bacteria fluctuated during the treatment period and at the end of sampling, the number of cultured bacteria reached below the initial number. It was observed that most of the bacteria were lenient to elevated concentrations of arsenic and showed the existence of arsenic resistant genes in it. From the present study, it was revealed that NZVI treatment stimulated the anaerobic conditions, supporting the development of sulfate-reducing bacteria, thus, increasing the removal efficiency of arsenic.	Castaño et al. (2021)

(Continued)

TABLE 19.2 (Continued)

Type of nanoparticle	Target contaminant	Parameter	Observation	References
ZnO/talc nanocomposite	Lead [Pb(II)]	The experimental data (isotherm) of lead adsorption were analyzed by means of the Freundlich and the Langmuir models.	The results showed that the highest lead adsorption capability of ZnO/talc nanocomposite was observed to be 48.3 mg/g from the waste water.	Sani et al. (2016)
B-NZVI	Chromium, Cr (VI)	Concentration of Cr (VI), pH. temperature, loading of B-nZVI and regeneration of B-nZVI.	The surface area of the iron particles increased as a consequence of decline in accretion of iron nanoparticles because of the presence of bentonite. The abolition efficiencies for lead, Copper and Chromium by B-nZVI were 90%. The capacity of B-nZVI for Cr(VI) exclusion decreased by 70% (approx) upon reuse of B-nZVI, possible only after washing it with ethylenediaminetetraacetic acid Solution.	Shi et al. (2011)
Kaolin/ZnO nanocomposites	Iron, Chromium and Chloride	Batch adsorption experiment was carried out	It was observed that Kaolin/ZnO nanocomposites were able to remove iron, chromium and chloride from tannery polluted water. The maximum removal of iron—98%, chromium—100% and Chloride—78% was obtained at 15 min by Kaolin/ZnO nanocomposites.	Mustapha et al. (2020)
NZVI, n-Fe3O4	Uranium (U VI)	Anoxic batch systems	The this study, it is shown that nano-Fe^0 is only suitable for the in situ management of Uranium in exterior surface and/or vadose zone waters if tremendously strong secondary method(s) are applied to avoid Dissolved Oxygen (DO) into the pollutant treatment sector. It is shown that adequately low Dissolved oxygen flux is required to maintain the chemically reduced groundwater environment inculcated by the nano-Fe^0 that will in turn sustain the removal of uranium on an extensive term or even semi-permanent basis	Crane and Scott (2014)

Iron doped Titanium dioxide (TiO_2)	Arsenic-As (III) and As(V)	X-ray diffraction (XRD) examination was done. Freundlich and Langmuir adsorption models were evaluated	Arsenic is removed effectively from the drinking water through adsorption process via TiO_2	Deedar and Aslam (2009)
ZnO	Lead-Pb (II)	X-ray diffraction, Transmission Electron Microscopy and UV visible studies were done to test the purity of ZnO nanoparticles. Series of experiments were done to study the various parameters that determines the removal of the lead	The maximum adsorption capacity of lead by the ZnO NPs was recorded to be 19.65 mg/g at a pH of 5 and temperature 70°C in aqueous solution	Azizi et al. (2017)
NZVI	Lead (Pb^{2+})	The sample was detected via atomic force microscope (AFM-SPM), high-resolution transmission electron microscope (HRTEM), and X-ray diffraction (XRD). Batch experiments were carried out that included effect of contact time and pH of the surrounding during the adsorption process	The current study reveals that the application of NZVI has immense potential to uptake large quantity of lead (Pb^{2+}) from waste water which accounts for 98% uptake when exposed for a short period (10 h contact time). In fact, the stored sample showed the similar efficiency on the removal of lead (Pb^{2+}) under the identical conditions	Ahmed et al. (2017)
Zeolite/Zinc Oxide Nanocomposite (Zeolite/ZnO NCs)	Lead Pb(II) and Arsenic As(V)	Langmuir model and kinetics of adsorption were studied	Study revealed that the maximum adsorption capability by Zeolite/Zinc Oxide nanocomposite was recorded as 93% for lead and 89% for arsenic ions at pH 4, 0.15 g and contact time was 30 min from 100 mg/L aqua solutions	Alswata et al. (2017)
Iron oxide (Fe_2O_3)	Cadmium, Copper, zinc, lead, iron and chromium	Physico-chemical parameters like dissolved oxygen, pH, Biological oxygen demand, turbidity and TDS of water were assessed	The study showed the improved results of dissolved oxygen (DO) and Biological oxygen demand (BOD) which in turn, indicates the high adsorption capability of the green synthesized iron oxide	Deepa et al. (2020)
Nickel oxide nanoparticles	Lead and cadmium	The synthesis of nanomaterial was carried out by one step hydrothermal method. A detailed study of metal ion adsorption onto the nickel-oxide nanoparticle was done	The study revealed that the time required to simultaneously eliminate Pb (II) and Cd (II) from an aqueous solution was 60 min and nickel-oxide nanoparticles showed maximum adsorption capacity than other adsorbents	Sardar (2021)

19.7 Conclusion

In today's world, pollutants (heavy metals) in the aqueous ecosystem are eminent due to various reasons like release of industrial effluents, use of herbicides, fertilizers and pesticides in agricultural fields that eventually goes into water bodies. It is recognized that even naturally heavy metals occur all around the Earth's crust and thus impose a risk to human health. This indicates the need to device out methods to get rid of these heavy metals from the water system. Many techniques have been developed in this regard but these are expensive and time taking. Thus, researchers across the world are constantly making efforts to find out techniques that will be cost-effective and also consumes lesser time. Recently, nanotechnology has gained researchers' attention because they are cost-effective and possess immense potential to remediate heavy metals from contaminated water bodies. The major reason behind the application of nanotechnology for the remedial process is their characteristics of large surface area, high reactivity, and strong adsorption ability. However, the risk associated with the use of NMs should not be avoided and appropriate measures are to be taken up to nullify its impact on health.

References

Abdelaal, M., Mashaly, I.A., Srour, D.S., Dakhil, M.A., El-Liethy, M.A., El-Keblawy, A., et al., 2021. Phytoremediation perspectives of seven aquatic macrophytes for removal of heavy metals from polluted drains in the Nile Delta of Egypt. Biology 10 (6), 560.

Adams, G.O., Fufeyin, P.T., Okoro, S.E., Ehinomen, I., 2015. Bioremediation, biostimulation and bioaugmention: a review. International Journal of Environmental Bioremediation & Biodegradation 3 (1), 28–39.

Adhikari, T., Manna, M.C., Singh, M.V., Wanjari, R.H., 2004. Bioremediation measure to minimize heavy metals accumulation in soils and crops irrigated with city effluent. Journal of Food Agriculture and Environment 2, 266–270.

Ahmed, M.A., Bishay, S.T., Ahmed, F.M., El-Dek, S.I., 2017. Effective Pb 2+ removal from water using nanozerovalent iron stored 10 months. Applied Nanoscience 7 (7), 407–416.

Alengebawy, A., Abdelkhalek, S.T., Qureshi, S.R., Wang, M.Q., 2021. Heavy metals and pesticides toxicity in agricultural soil and plants: ecological risks and human health implications. Toxics 9 (3), 42.

Alswata, A.A., Ahmad, M.B., Al-Hada, N.M., Kamari, H.M., Hussein, M.Z.B., Ibrahim, N.A., 2017. Preparation of zeolite/zinc oxide nanocomposites for toxic metals removal from water. Results in Physics 7, 723–731.

Azizi, S., Mahdavi Shahri, M., Mohamad, R., 2017. Green synthesis of zinc oxide nanoparticles for enhanced adsorption of lead ions from aqueous solutions: equilibrium, kinetic and thermodynamic studies. Molecules (Basel, Switzerland) 22 (6), 831.

Balaji, T., Matsunaga, H., 2002. Adsorption characteristics of As (III) and As (V) with titanium dioxide loaded Amberlite XAD-7 resin. Analytical Sciences 18 (12), 1345–1349.

Barbhuiya, S.N., Barhoi, D., Giri, S., 2021. Impact of arsenic on reproductive health. Environmental Health 89.

Beolchini, F., Dell'Anno, A., De Propris, L., Ubaldini, S., Cerrone, F., Danovaro, R., 2009. Auto-and heterotrophic acidophilic bacteria enhance the bioremediation efficiency of sediments contaminated by heavy metals. Chemosphere 74 (10), 1321–1326.

Bhat, J.S., 2003. Heralding a new future—nanotechnology? Current Science 85 (2), 147–154.

Bradl, H.B., 2005. Sources and origins of heavy metals, Interface Science and Technology, Vol. 6. Elsevier, pp. 1–27.

Briffa, J., Sinagra, E., Blundell, R., 2020. Heavy metal pollution in the environment and their toxicological effects on humans. Heliyon 6 (9), e04691.

Castaño, A., Prosenkov, A., Baragaño, D., Otaegui, N., Sastre, H., Rodríguez-Valdés, E., et al., 2021. Effects of in situ remediation with nanoscale zero valence iron on the physicochemical conditions and bacterial communities of groundwater contaminated with arsenic. Frontiers in Microbiology 12, 580.

Cecchin, I., Reddy, K.R., Thomé, A., Tessaro, E.F., Schnaid, F., 2017. Nanobioremediation: integration of nanoparticles and bioremediation for sustainable remediation of chlorinated organic contaminants in soils. International Biodeterioration & Biodegradation 119, 419–428.

Chen, K.L., Elimelech, M., 2006. Aggregation and deposition kinetics of fullerene (C60) nanoparticles. Langmuir: The ACS Journal of Surfaces and Colloids 22 (26), 10994–11001.

Chen, M., Sun, Y., Liang, J., Zeng, G., Li, Z., Tang, L., et al., 2019. Understanding the influence of carbon nanomaterials on microbial communities. Environment International 126, 690–698.

Cornell, R.M., Schwertmann, U., 2003. The Iron Oxides: Structure, Properties, Reactions, Occurrences and Uses. John Wiley & Sons.

Crane, R.A., Scott, T.B., 2014. The removal of uranium onto nanoscale zero-valent iron particles in anoxic batch systems. Journal of Nanomaterials 2014.

Daccò, C., Girometta, C., Asemoloye, M.D., Carpani, G., Picco, A.M., Tosi, S., 2020. Key fungal degradation patterns, enzymes and their applications for the removal of aliphatic hydrocarbons in polluted soils: a review. International Biodeterioration & Biodegradation 147, 104866.

Das, N., Chandran, P., 2011. Microbial degradation of petroleum hydrocarbon contaminants: an overview. Biotechnology Research International 2011.

de Oliveira Santos, J.V., Ferreira, Y.L.A., de Souza Silva, L.L., Palácio, S.B., Cavalcanti, I.M.F., 2018. Use of bioremediation for the removal of petroleum hydrocarbons from the soil: an overview. International Journal of Environment, Agriculture and Biotechnology 3 (5), 266198.

Deedar, N.A.B.I., Aslam, I., 2009. Evaluation of the adsorption potential of titanium dioxide nanoparticles for arsenic removal. Journal of Environmental Sciences 21 (3), 402–408.

Deepa, G., Jayaraj, M., Magudeswaran, P.N., 2020. Removal of heavy metals from chithrapuzha river water by iron oxide nanoparticles prepared via green synthesis methods. Oriental Journal of Chemistry 36 (6), 1154–1160.

Dehghani, M.H., Taher, M.M., Bajpai, A.K., Heibati, B., Tyagi, I., Asif, M., et al., 2015. Removal of noxious Cr (VI) ions using single-walled carbon nanotubes and multi-walled carbon nanotubes. Chemical Engineering Journal 279, 344–352.

Di Palma, L., Gueye, M.T., Petrucci, E., 2015. Hexavalent chromium reduction in contaminated soil: a comparison between ferrous sulphate and nanoscale zero-valent iron. Journal of Hazardous Materials 281, 70–76.

Ding, N., Han, M., He, Y., Wang, X., Pan, Y., Lin, H., et al., 2021. Advances in application of nanomaterials in remediation of heavy metal contaminated soil, E3S Web of Conferences, Vol. 261. EDP Sciences.

El-Temsah, Y.S., Sevcu, A., Bobcikova, K., Cernik, M., Joner, E.J., 2016. DDT degradation efficiency and ecotoxicological effects of two types of nano-sized zero-valent iron (nZVI) in water and soil. Chemosphere 144, 2221–2228.

Fan, M., Li, T., Hu, J., Cao, R., Wei, X., Shi, X., et al., 2017. Artificial neural network modeling and genetic algorithm optimization for cadmium removal from aqueous solutions by reduced graphene oxide-supported nanoscale zero-valent iron (nZVI/rGO) composites. Materials 10 (5), 544.

Fatima, I., Iqbal, R., Hussain, M., 2016. Histopathological effects of chromium (III) sulfate on liver and kidney of Swiss albino mice (Mus musculus). Asia Pacific Journal of Multidisciplinary Research 4 (3), 175–180.

Fatima, K., Imran, A., Amin, I., Khan, Q.M., Afzal, M., 2018. Successful phytoremediation of crude-oil contaminated soil at an oil exploration and production company by plants-bacterial synergism. International Journal of Phytoremediation 20 (7), 675–681.

Feynman, R.P., 1960. There is plenty of room at the bottom. California Institute of Technology. Journal of Engineering Science 4 (2), 23–36.

Fodelianakis, S., Antoniou, E., Mapelli, F., Magagnini, M., Nikolopoulou, M., Marasco, R., et al., 2015. Allochthonous bioaugmentation in ex situ treatment of crude oil-polluted sediments in the presence of an effective degrading indigenous microbiome. Journal of Hazardous Materials 287, 78–86.

Fortner, J.D., Lyon, D.Y., Sayes, C.M., Boyd, A.M., Falkner, J.C., Hotze, E.M., et al., 2005. C60 in water: nanocrystal formation and microbial response. Environmental Science & Technology 39 (11), 4307–4316.

Girma, G., 2015. Microbial bioremediation of some heavy metals in soils: an updated review. Egyptian Academic Journal of Biological Sciences, G. Microbiology 7 (1), 29–45.

Hao, Y.M., Man, C., Hu, Z.B., 2010. Effective removal of Cu (II) ions from aqueous solution by amino-functionalized magnetic nanoparticles. Journal of Hazardous Materials 184 (1–3), 392–399.

He, S., Zhong, L., Duan, J., Feng, Y., Yang, B., Yang, L., 2017. Bioremediation of wastewater by iron oxide-biochar nanocomposites loaded with photosynthetic bacteria. Frontiers in Microbiology 8, 823.

Huang, D., Xue, W., Zeng, G., Wan, J., Chen, G., Huang, C., et al., 2016. Immobilization of Cd in river sediments by sodium alginate modified nanoscale zero-valent iron: impact on enzyme activities and microbial community diversity. Water Research 106, 15–25.

Ibrahim, M.M., Asal, S., 2017. Physicochemical and photocatalytic studies of Ln3 + -ZnO for water disinfection and wastewater treatment applications. Journal of Molecular Structure 1149, 404–413.

Igiri, B.E., Okoduwa, S.I., Idoko, G.O., Akabuogu, E.P., Adeyi, A.O., Ejiogu, I.K., 2018. Toxicity and bioremediation of heavy metals contaminated ecosystem from tannery wastewater: a review. Journal of Toxicology 2018.

Itanna, F., Coulman, B., 2003. Phyto-extraction of copper, iron, manganese, and zinc from environmentally contaminated sites in Ethiopia, with three grass species. Communications in Soil Science and Plant Analysis 34 (1–2), 111–124.

Jaishankar, M., Tseten, T., Anbalagan, N., Mathew, B.B., Beeregowda, K.N., 2014. Toxicity, mechanism and health effects of some heavy metals. Interdisciplinary Toxicology 7 (2), 60.

Jan, A.T., Azam, M., Siddiqui, K., Ali, A., Choi, I., Haq, Q.M., 2015. Heavy metals and human health: mechanistic insight into toxicity and counter defense system of antioxidants. International Journal of Molecular Sciences 16 (12), 29592–29630.

Järup, L., 2003. Hazards of heavy metal contamination. British Medical Bulletin 68 (1), 167–182.

Jawed, A., Saxena, V., Pandey, L.M., 2020. Engineered nanomaterials and their surface functionalization for the removal of heavy metals: a review. Journal of Water Process Engineering 33, 101009.

Joshi, N.C., Kumar, N., Singh, A., 2020. A brief study on zinc oxide based nanosorbents and adsorptive removal of heavy metal ions. International Journal of Health and Clinical Research 3 (1), 34–38.

Kinuthia, G.K., Ngure, V., Beti, D., Lugalia, R., Wangila, A., Kamau, L., 2020. Levels of heavy metals in wastewater and soil samples from open drainage channels in Nairobi, Kenya: community health implication. Scientific Reports 10 (1), 1–13.

Kumar, A., Ali, M., Kumar, R., Kumar, M., Sagar, P., Pandey, R.K., et al., 2021. Arsenic exposure in Indo Gangetic plains of Bihar causing increased cancer risk. Scientific Reports 11 (1), 1–16.

Kumari, B., Singh, D.P., 2016. A review on multifaceted application of nanoparticles in the field of bioremediation of petroleum hydrocarbons. Ecological Engineering 1 (97), 98–105.

Lackovic, J.A., Nikolaidis, N.P., Dobbs, G.M., 2000. Inorganic arsenic removal by zero-valent iron. Environmental Engineering Science 17 (1), 29–39.

Laurent, S., Forge, D., Port, M., Roch, A., Robic, C., Vander Elst, L., et al., 2008. Magnetic iron oxide nanoparticles: synthesis, stabilization, vectorization, physicochemical characterizations, and biological applications. Chemical Reviews 108 (6), 2064–2110.

Lauwerys, R., Haufroid, V., Hoet, P., Lison, D., 2007. Toxicologie industrielle et intoxications professionnelles.

Lisha, K.P., Pradeep, T., 2009. Towards a practical solution for removing inorganic mercury from drinking water using gold nanoparticles. Gold Bulletin 42 (2), 144–152.

Liu, M., Wang, Y., Chen, L., Zhang, Y., Lin, Z., 2015. Mg (OH) 2 supported nanoscale zero valent iron enhancing the removal of Pb (II) from aqueous solution. ACS Applied Materials & Interfaces 7 (15), 7961–7969.

Liu, Q., Zhang, R., Wang, X., Shen, X., Wang, P., Sun, N., et al., 2019. Effects of sub-chronic, low-dose cadmium exposure on kidney damage and potential mechanisms. Annals of Translational Medicine 7 (8).

Liu, H., Guo, H., Jian, Z., Cui, H., Fang, J., Zuo, Z., et al., 2020. Copper induces oxidative stress and apoptosis in the mouse liver. Oxidative Medicine and Cellular Longevity 2020.

Madima, N., Mishra, S.B., Inamuddin, I., Mishra, A.K., 2020. Carbon-based nanomaterials for remediation of organic and inorganic pollutants from wastewater. A review. Environmental Chemistry Letters 18 (4), 1169–1191.

Mahmoud, M.E., Fekry, N.A., El-Latif, M.M., 2016. Nanocomposites of nanosilica-immobilized-nanopolyaniline and crosslinked nanopolyaniline for removal of heavy metals. Chemical Engineering Journal 304, 679–691.

Mehta, M., Hundal, S.S., 2016. Effect of sodium arsenite on reproductive organs of female Wistar rats. Archives of Environmental & Occupational Health 71 (1), 16–25.

Mohammadi, A.A., Zarei, A., Esmaeilzadeh, M., Taghavi, M., Yousefi, M., Yousefi, Z., et al., 2020. Assessment of heavy metal pollution and human health risks assessment in soils around an industrial zone in Neyshabur, Iran. Biological Trace Element Research 195 (1), 343–352.

Morkunas, I., Woźniak, A., Mai, V.C., Rucińska-Sobkowiak, R., Jeandet, P., 2018. The role of heavy metals in plant response to biotic stress. Molecules (Basel, Switzerland) 23 (9), 2320.

Munoz, R., Guieysse, B., 2006. Algal–bacterial processes for the treatment of hazardous contaminants: a review. Water Research 40 (15), 2799–2815.

Murphy, I.J., Coats, J.R., 2011. The capacity of switchgrass (Panicum virgatum) to degrade atrazine in a phytoremediation setting. Environmental Toxicology and Chemistry 30 (3), 715–722.

Mustapha, S., Tijani, J.O., Ndamitso, M.M., Abdulkareem, S.A., Shuaib, D.T., Mohammed, A.K., et al., 2020. The role of kaolin and kaolin/ZnO nanoadsorbents in adsorption studies for tannery wastewater treatment. Scientific Reports 10 (1), 1–22.

Najafi, M., Yousefi, Y., Rafati, A.A., 2012. Synthesis, characterization and adsorption studies of several heavy metal ions on amino-functionalized silica nano hollow sphere and silica gel. Separation and Purification Technology 85, 193–205.

Nath Barbhuiya, S., Barhoi, D., Giri, A., Giri, S., 2020. Arsenic and smokeless tobacco exposure induces DNA damage and oxidative stress in reproductive organs of female Swiss albino mice. Journal of Environmental Science and Health, Part C 38 (4), 384–408.

Oh, J.K., Park, J.M., 2011. Iron oxide-based superparamagnetic polymeric nanomaterials: design, preparation, and biomedical application. Progress in Polymer Science 36 (1), 168–189.

Ojuederie, O.B., Babalola, O.O., 2017. Microbial and plant-assisted bioremediation of heavy metal polluted environments: a review. International Journal of Environmental Research and Public Health 14 (12), 1504.

Okada, M.A., Neto, F.F., Noso, C.H., Voigt, C.L., Campos, S.X., de Oliveira Ribeiro, C.A., 2016. Brain effects of manganese exposure in mice pups during prenatal and breastfeeding periods. Neurochemistry International 97, 109–116.

Prakash, V., 2017. Mycoremediation of environmental pollutants. International Journal of ChemTech Research 10 (3), 149–155.

Rajabathar, J.R., Shukla, A.K., Ali, A., Al-Lohedan, H.A., 2017. Silver nanoparticle/r-graphene oxide deposited mesoporous-manganese oxide nanocomposite for pollutant removal and supercapacitor applications. International Journal of Hydrogen Energy 42 (24), 15679–15688.

Rajendran, P., Gunasekaran, P., 2007. Nanotechnology for bioremediation of heavy metals. Environmental Bioremediation Technologies. Springer, Berlin, Heidelberg, pp. 211–221.

Rao, P., Kumar, R.R., Raghavan, B.G., Subramanian, V.V., Sivasubramanian, V., 2011. Application of phycoremediation technology in the treatment of wastewater from a leather-processing chemical manufacturing facility. Water SA 37 (1).

Rizwan, M., Singh, M., Mitra, C.K., Morve, R.K., 2014. Ecofriendly application of nanomaterials: nanobioremediation. Journal of Nanoparticles 2014.

Roco, M.C., 2005. The emergence and policy implications of converging new technologies integrated from the nanoscale. Journal of Nanoparticle Research 7 (2), 129–143.

Saier Jr, M.H., 2005. Beneficial bacteria and bioremediation. Journal of Molecular Microbiology and Biotechnology 9 (2), 63.

Sall, M.L., Diaw, A.K.D., Gningue-Sall, D., Efremova Aaron, S., Aaron, J.J., 2020. Toxic heavy metals: impact on the environment and human health, and treatment with conducting organic polymers, a review. Environmental Science and Pollution Research 27, 29927–29942.

Sani, H.A., Ahmad, M.B., Saleh, T.A., 2016. Synthesis of zinc oxide/talc nanocomposite for enhanced lead adsorption from aqueous solutions. RSC Advances 6 (110), 108819–108827.

Santhosh, C., Velmurugan, V., Jacob, G., Jeong, S.K., Grace, A.N., Bhatnagar, A., 2016. Role of nanomaterials in water treatment applications: a review. Chemical Engineering Journal 306, 1116–1137.

Sapmaz-Metin, M., Topcu-Tarladacalisir, Y., Kurt-Omurlu, I., Karaoz Weller, B., Unsal-Atan, S., 2017. A morphological study of uterine alterations in mice due to exposure to cadmium. Biotechnic & Histochemistry 92 (4), 264–273.

Sardar, P.R., 2021. Simultaneous removal of lead and cadmium ions by nickel oxide nanoparticles. Indian Journal of Science and Technology 14 (28), 2327–2336.

Sasek, V., Cajthaml, T., 2005. Mycoremediation: current state and perspectives. International Journal of Medicinal Mushrooms 7 (3).

Satarug, S., Nishijo, M., Ujjin, P., Vanavanitkun, Y., Moore, M.R., 2005. Cadmium-induced nephropathy in the development of high blood pressure. Toxicology Letters 157 (1), 57–68.

Shehzadi, M., Fatima, K., Imran, A., Mirza, M.S., Khan, Q.M., Afzal, M., 2016. Ecology of bacterial endophytes associated with wetland plants growing in textile effluent for pollutant-degradation and plant growth-promotion potentials. Plant Biosystems—An International Journal Dealing with All Aspects of Plant Biology 150 (6), 1261–1270.

Shi, L.N., Zhang, X., Chen, Z.L., 2011. Removal of chromium (VI) from wastewater using bentonite-supported nanoscale zero-valent iron. Water Research 45 (2), 886–892.

Singh, J., Kalamdhad, A.S., 2011. Effects of heavy metals on soil, plants, human health and aquatic life. International Journal of Research in Chemistry and Environment 1 (2), 15–21.

Skubal, L.R., Meshkov, N.K., Rajh, T., Thurnauer, M., 2002. Cadmium removal from water using thiolactic acid-modified titanium dioxide nanoparticles. Journal of Photochemistry and Photobiology A: Chemistry 148 (1–3), 393–397.

Stefaniuk, M., Oleszczuk, P., Ok, Y.S., 2016. Review on nano zerovalent iron (nZVI): from synthesis to environmental applications. Chemical Engineering Journal 287, 618–632.

Sultan, A., Mohammad, F., 2017. Chemical sensing, thermal stability, electrochemistry and electrical conductivity of silver nanoparticles decorated and polypyrrole enwrapped boron nitride nanocomposite. Polymer 113, 221–232.

Sumesh, E., Bootharaju, M.S., Pradeep, T., 2011. A practical silver nanoparticle-based adsorbent for the removal of Hg2 + from water. Journal of Hazardous Materials 189 (1–2), 450–457.

Tara, N., Afzal, M., Ansari, T.M., Tahseen, R., Iqbal, S., Khan, Q.M., 2014. Combined use of alkane-degrading and plant growth-promoting bacteria enhanced phytoremediation of diesel contaminated soil. International Journal of Phytoremediation 16 (12), 1268–1277.

Vázquez-Núñez, E., Molina-Guerrero, C.E., Peña-Castro, J.M., Fernández-Luqueño, F., de la Rosa-Álvarez, M., 2020. Use of nanotechnology for the bioremediation of contaminants: a review. Processes 8 (no. 7), 826.

Wadaan, M.A., 2009. Effects of mercury exposure on blood chemistry and liver histopathology of male rats. Journal of Pharmacology and Toxicology 4 (3), 126–131.

Xie, R., Jin, Y., Chen, Y., Jiang, W., 2017. The importance of surface functional groups in the adsorption of copper onto walnut shell derived activated carbon. Water Science and Technology 76 (11), 3022–3034.

Xu, X., Liu, W., Tian, S., Wang, W., Qi, Q., Jiang, P., et al., 2018. Petroleum hydrocarbon-degrading bacteria for the remediation of oil pollution under aerobic conditions: a perspective analysis. Frontiers in Microbiology 9, 2885.

Yadav, K.K., Singh, J.K., Gupta, N., Kumar, V.J.J.M.E.S., 2017. A review of nanobioremediation technologies for environmental cleanup: a novel biological approach. Journal of Materials and Environmental Science 8 (2), 740–757.

Yang, J., Hou, B., Wang, J., Tian, B., Bi, J., Wang, N., et al., 2019a. Nanomaterials for the removal of heavy metals from wastewater. Nanomaterials 9 (3), 424.

Yang, J., Hou, B., Wang, J., Tian, B., Bi, J., Wang, N., et al., 2019b. Nanomaterials for the removal of heavy metals from wastewater. Nanomaterials 9 (3), 424.

Yu, Y., Zhang, Y., Zhao, N., Guo, J., Xu, W., Ma, M., et al., 2020. Remediation of crude oil-polluted soil by the bacterial rhizosphere community of suaeda salsa revealed by 16S rRNA genes. International Journal of Environmental Research and Public Health 17 (5), 1471.

Zhang, X., Yan, L., Liu, J., Zhang, Z., Tan, C., 2019. Removal of different kinds of heavy metals by novel PPG-nZVI beads and their application in simulated stormwater infiltration facility. Applied Sciences 9 (20), 4213.

CHAPTER

20

Arsenic contamination in water, health effects and phytoremediation

Juhi Khan[1], Himanshu Dwivedi[2], Ajay Giri[3], Ritu Aggrawal[2], Rinkey Tiwari[4] and Deen Dayal Giri[2]

[1]Department of Botany, IFTM University, Moradabad, Uttar Pradesh, India [2]Department of Botany, Maharaj Singh College, Saharanpur, Uttar Pradesh, India [3]Department of Basic Educations, Ghazipur, Uttar Pradesh, India [4]Departments of Botany, Meerut College, Meerut, Uttar Pradesh, India

20.1 Introduction

Over the last decade, arsenic (As) pollution and its toxicity to humans have become a topic of great concern. Arsenic is present in the environment, that is, soil and groundwater (Polya et al., 2019). It is classified as the most toxic compound by many organizations. Besides, this known as king of poisonous element and ranked first in the 2001 priority list of hazardous chemicals (Abdulsalam et al., 2011). Arsenic is a teratogen, mutagen and carcinogen to humans as well as other animals, which is proven through epidemiological shreds of evidence (WHO, 2003). It is class I human carcinogen as per International Agency for Research on Cancer. According to WHO, the permissible value of As concentration in drinking water should be below 10 μg/L (WHO, 2011). The arsenic can exist in different oxidation states (Henke, 2009), however the two dominant inorganic oxidation states are arsenite (As^{3+}) and arsenate (As^{5+}). Although, both arsenic species are of toxic nature, arsenite toxicity is much higher compared to arsenate (Fendorf et al., 2010). The main route of human arsenic exposure is primarily ingestion of contaminated water and food (Carbonell-Barrachina et al., 2009; EFSA., 2009). Use of arsenic-contaminated ground water for drinking, cooking and irrigation in Southeast Asia is responsible for ill effect and suffering of millions of people (Cubadda et al., 2010).

Sources of arsenic could be categorized into natural and anthropogenic. In nature As exist in three allotropic form of gray, yellow, and black color (Norman, 1998). It is 53rd most abundant element in the earth crust accounting for 4.0×10^{16} kg of arsenic (Sarkar et al., 2011). The major natural arsenic transport from the earth crust into water is

Metals in Water
DOI: https://doi.org/10.1016/B978-0-323-95919-3.00021-5

weathering of rocks (Smith et al., 1998), and erosion of As containing mineral rocks (Abbas et al., 2018). The aquifers passing through such rocks get contaminated by As (Govil and Krishna, 2018). Arsenic of the earth crust and bedrocks can leach into potable water sources (Vahter, 2008). High levels of arsenic have been reported from aquifers, especially the unconsolidated sediment aquifers around the globe (Mozumder, 2019; Smedley and Kinniburgh, 2013; Madhav et al., 2020). Most probably, the usage of arsenic-rich groundwater is the main reason for arsenic accumulation in soil and plants(Arco-Lázaro et al., 2018). Plant and crops absorb arsenic through their roots. Consumption of vegetables and cereales grown in arsenic-contaminated soil is responsible for its entry in the living animals including human (Mandal and Suzuki, 2002; Ahamad et al., 2020).

Arsenic is found in both inorganic as well as organic forms. The inorganic As ($_{i}$As) are the mineral form and found in soil and water whereas organic arsenic species are formed in organisms exposed to As contamination (Postma et al., 2007). Entry of $_{i}$As species takes place in living organisms through contaminated water and food whereas $_{org}$As species enters in animals by consumption of seafood (oysters, prawns, mussels, fish) and contaminated crops/other product (McCarty et al., 2011).

The major anthropogenic sources of arsenic in the environment consist of mining, metal smelting, combustion of fossil fuel coal, metal extraction processes, industrial exhausts, and marine aerosols (Kossoff and Hudson-Edwards, 2012; Madhav et al., 2020). The other potential anthropogenic sources include use of arsenical herbicides, fungicides, insecticides, wood preservatives, application of synthetic phosphate fertilizers, and irrigation with arsenic-contaminated groundwater (Smith et al., 1998). Global anthropogenic arsenic accounts for about 82,000 metric tons/year (Bhattacharya et al., 2002). Burning of fossil fuel emit arsenic in the environment due to the volatilization of As_4O_6 present in the fossil fuels (Bissen and Frimmel, 2003). The degree of groundwater arsenic contamination by the anthropogenic sources is very small compared to the natural one (Bhattacharya et al., 2002). Different anthropogenic and natural sources of As and mediators via which it enter in human body are depicted in Fig. 20.1.

20.1.1 Inorganic arsenic ($_{i}$As)

There are more than 300 mineral forms of arsenic, such as arsenates, sulfosalts, sulfides, arsenides, arsenites, native elements, and metal alloys (Kossoff and Hudson-Edwards, 2012). Oxide of arsenate is arsenic pentoxide (As_2O_5) whereas oxide of arsenite are Arsenolite (As_2O_3) and Claudetite (As_2O_3). In addition, it is found in the form of arsenate of iron, calcium and magnesium, Fe -suphoarsenate and arsenates of Zn, Pb, Ni and Co. Such As species contaminated aquifers are naturally found in the environment, worldwide (Sailo and Mahanta, 2014). According to a well-accepted mechanism for natural release of arsenic into the environment, takes place through desorption or dissolution of Fe-oxides and other naturally As rich minerals (Bhattacharyya et al., 2003; Eiche et al., 2008; Smith and Naidu, 2004).

20.1.2 Organic arsenic (Oars)

Entry of inorganic arsenic species into the food chain, lead to their methylation resulting in its various organic species such as monomethylarsine (MMA), dimethylarsine (DMA); and

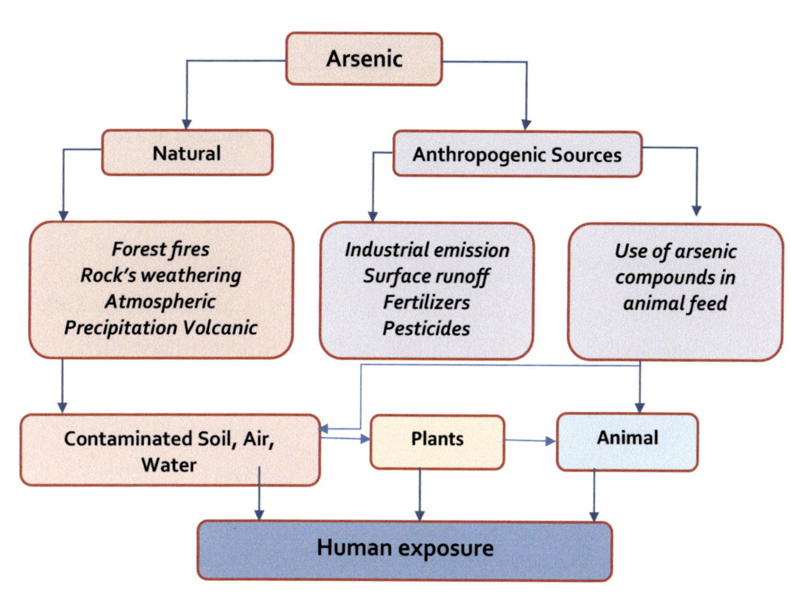

FIGURE 20.1 Flowchart showing possible human exposure to Arsenic in the environment.

trimethylarsine (TMA) (Kossoff and Hudson-Edwards, 2012). The other organic forms of arsenic are adamsite, lewisite, and cacodylic acid (Girard, 2010). Antiparasitic organic As species such as roxarsone (ROX), carbarsone, nitarsone and p-ASA are important additive in the food of chicken and swine (Chen et al., 2009; Saucedo-Velez et al., 2017). These oars species are released in through urine and feces of the animals without any alteration and enters in the environment. The small fraction of arsis retained and accumulated in animal tissues are transferred to human through their meat (Ennaceur et al., 2017). An elevated level has been recorded in tissues of animals supplemented organic As ingredient in the feed (Wu et al., 2011).

The Oars are present in the waste generated from poultry feedstuff plant, breeding facility of chicken and swine manure. Such nutrient rich manures are frequently used as fertilizer in agriculture land after composting or storage in surface ponds (Xie et al., 2016). The economic disposal of Oars rich manure distributes As compounds in the soil, surface water, sediment and groundwater sources (Fei et al., 2018; Li et al., 2013; Rani et al., 2022). Even our seas are not free from pollution of Oars. Some Oars species present in seafood are arseno-betaine and arseno-choline (Hopenhayn, 2006). The transformation of Oars into DMA and monomethyl MMA give rise to more toxic inorganic $_{i}As^{+3}$ and $_{i}As^{+5}$ species (Yin et al., 2018; Nachman et al., 2013). The highly mobile $_{i}$As species, pose greater threat to our surface waterbodies, ground water and soil (Sun et al., 2019; Tang et al., 2020). Atmospheric sources of As compounds are volcanic eruptions and wind mobilization (Patel et al., 2005).

Many countries across the world have higher groundwater concentration of As than the maximum permissible limit (10 ppb), include 32 Asian, 31 Europian, 20 African, 11 North America, and 9 South American countries (Shaji et al., 2021). In Indian subcontinent, nearly 50 million population is exposed to high level natural As in river plain of Brahmaputra in north-eastern states, Indian Gangetic plain and delta, the central Indian (Chhattisgarh), as well as several other areas of India. The data of As concentration in the ground water of different area are given in Table 20.1 (Bhattacharya et al., 2016; Rani

TABLE 20.1 As concentration (mg/L) in various regions of the India.

City/area/region	Source of water	As (mg/L)	Reference
Bhagalpur	Tube well	49.8	Singh et al. (2014)
Dehari	Tube well	110	Alam et al. (2016)
Dibang	Bore well	618	Singh (2004)
Gonda	Tube well	510	Shah (2017)
Madanpur	Tube well	700	Chakraborti et al. (2017)
Manipur	Tube well	628	Bhattacharya (2017)
Patna	Tube well	90	Chakraborti et al. (2016b)
Rajnandgaon	Tube well	506	Patel et al. (2017)
Ramnagar	Tube well	3700	Das et al. (2009)
Samastipur	Tube well Hand pump	32	Kumar et al. (2016)
Shahpur	Tube well	173	Chakraborti et al. (2016a)
Lake Chillika	Surface water	35	Konhauser et al. (1997)
Shivnath River	Surface water	100–300	Pandey et al. (2002)
Middle/Lower Ganges plain, Bihar	Groundwater	<10–1654	von Brömssen (1999)
Brahmaputra basin, Assam	Groundwater	<1–657	Mandal et al. (1996)
West Bengal basin, West Bengal	Groundwater	<1–4200	Mandal et al. (1996)
Upper Ganges plain, Uttar Pradesh	Groundwater	<1–700	Mandal et al. (1996)

et al., 2022). Use of As contaminated groundwater in irrigation results in contamination and enrichment in the soil. Thus the crops grown in these are likely to contain As in different parts (Bhattacharya et al., 2016).

In India, many states have varying level of arsenic concentration. Land area (%) of the states affected and expected population exposed to toxic As concentrations have been summarized in recent past (Podgorski et al., 2020). The state wise As-contaminated area are less than 0.1% in Rajasthan and Chhattisgarh, 1%–2.4% of Uttar Pradesh (2.5 lakh), 0.2%–0.6% of Jharkhand (2.3 lakh), 2%–12% of Bihar (4.7 lakh), 4.3%–21.6% of Arunachal Pradesh (1.6 lakhs), 12%–20% of West Bengal (10.2 lakhs) and 8.6%–22.7% of Manipur (1.2 lakhs) and 42.3–59.7% of Assam (8.8 lakhs).

In Asia around 180 million people are affected by As contamination, especially the South and Southeast Asian regions (McArthur, 2019). The countries included in the region are Bangladesh (Yang et al., 2014), India (Bhowmick et al., 2018; Bindal and Singh, 2019; Chakraborti et al., 2018; Dhillon, 2020; Mukherjee et al., 2009), China (Guo et al., 2014), Vietnam (Stopelli et al., 2020), Nepal (Pokhrel et al., 2009), Thailand (Cho et al., 2011), Myanmar (Van Geen et al., 2014) and Indonesia (Winkel et al., 2008). However, many regions of countries like the United States (Gong et al., 2014), Canada, Argentina, Chile, Pakistan, Mexico, and South Africa also suffer from issue of As toxicity (Ravenscroft et al., 2009).

TABLE 20.2 Countries affected with As contaminated groundwater in the world.

Country	Continents	Reference
Argentina	South America	Robles et al. (2016)
Australia	Australia	Appleyard et al. (2006)
Bangladesh	Asia	Edmunds et al. (2015)
Brazil	South America	Ciminelli et al. (2017)
Canada	North America	Bondu et al. (2017)
Chile	South America	Corradini et al. (2018)
China	Asia	Chen et al. (2017)
Cambodia	Asia	Richards et al. (2019)
France	Europe	Barats et al. (2014)
Ghana	Africa	Buamah et al. (2008)
India	Asia	Suhag (2019)
Italy	Europe	Vivona et al. (2007)
Japan	Asia	Even et al. (2017)
Mexico	North America	Kwong et al. (2007)
Nepal	Asia	Thakur et al. (2011)
Nigeria	Africa	Ravenscroft et al. (2009)
Pakistan	Asia	Ali et al. (2019)
Spain	Europe	Sanz et al. (2007)
Thailand	Asia	Tiankao and Chotpantarat (2018)
U.K.	Europe	Middleton et al. (2016)
The U.S.A.	North America	Flanagan et al. (2015)
Vietnam	Asia	Nguyen et al. (2020)

Many European countries especially Greece, Romania, Hungary, Serbia, Croatia, Spain, Turkey and central Italy showed elevated As in their groundwater (Katsoyiannis et al., 2015; Zuzolo et al., 2020). The reports regarding high As concentration from the countries across the world are enlisted in Table 20.2.

20.2 Diseases caused by arsenic poisonings

Worldwide groundwater As affected human population is about 500 million (Sun et al., 2019). The toxic effect of As on individual differ depending upon oxidation state of As, bioavailability, intake rate, route, frequency and exposure time (Rosen and Liu, 2009; Ying et al., 2011). The major exposure routes of As are ingestion (water and food), dust inhalation, and dermal contact (Rahman et al., 2009). The noncarcinogenic effects of As include keratosis, hyperpigmentation, hypopigmentation, peripheral vascular and cardiovascular diseases, diabetes mellitus, and disorders of peripheral and central nervous system. Arsenic poisoning causes loss of appetite, anemia, weight loss, leucopenia, weakness, and dementia (Ahamad et al., 2020). A condition called arsenicosis is related with skin, liver lung, and kidney cancer, keratosis or pigmentation caused due to ingestion of contaminated water (WHO, 2015). Long-term exposure causes severe organ damage and cancer of internal organs (WHO, 2015). The presence of low dose As in the rice was able to alter the microbiome of the mice intestine (Chen et al., 2021a,b).

The toxicity of arsenic is associated with its interference in cellular oxidative phosphorylation where it can substitute P in adenosine triphosphate (ATP) synthesis resulting in defective intracellular energy storage. In the liver, arsenic methylation takes place. Methylation process helps in the removing arsenic from the body. The product of arsenic methylation are monomethyl arsenic acid and dimethyl arsenic acid (Sharma and Sohn, 2009; Jang et al., 2016). The studies prove that all four forms of arsenic (As^{3+}, As^{5+}, DMAA, MMAA) severely effect cell metabolism causing damage to DNA resulting in formation of altered proteins and enzymes. The other effects may include sister chromatid exchanges, aneuploidy, polyploidy, chromatid aberrations, DNA amplification, and morphological changes (Rudel et al., 1996). Some of the adverse effects of arsenic on different body parts are graphically presented (Fig. 20.2) and its brief description is given below.

20.2.1 Skin

The chronic arsenic exposure results in lesions (e.g., keratosis melanosis and pigmentation) on skin (Rahman et al., 2009). The development and appearance of lesions on skin indicates long exposure duration of 5–10 years (Mazumder et al., 1998). Hyperpigmentation cause wart-like dermal lesions and hyperkeratosis on feet, hands or fingers (Guo et al., 2007). A clinical symptom of chronic exposure is hair loss (alopecia) in many cases (Amster et al., 2007).

20.2.2 Brain

Symptoms of As-induced neurotoxicity are pain, paresthesia and numbness in the soles (Vahidnia et al., 2007), oxidative stress (Mundey et al., 2013) and peripheral neuropathy caused by peripheral nerves damage (Mathew et al., 2010). Acute As exposure result in encephalopathy having symptoms of lost brain activity, headache, seizures, hallucinations, and coma (Bartolome et al., 1999). Moreover, Alzheimer's disease may develop in long-term low dose As exposure (O'Bryant et al., 2011).

20.2.3 Lungs

Respiratory complications and associated high mortality are observed in peoples drinking As-contaminated water (Argos et al., 2010; Parvez et al., 2011). Some clinical respiratory complications associated with As are chronic cough, laryngitis, chest sounds, bronchitis, and short-breath, rhinitis and presence of blood in sputum (Saha et al., 1999). Moderate to low level As toxicity may impair lung function and result in tuberculosis (Parvez et al., 2012, 2013).

20.2.4 Heart

Presence of high inorganic As in potable water causes various cardio/cerebro-vascular diseases including atherosclerosis, hypertension, cardiac arrhythmia, diabetes (Chen et al., 2011; Rahman et al., 2009) and has shown a direct link with some heart complications like myocardial injury and cardiomyopathy (Manna et al., 2008). Long-term exposure to $_i$As develop a specific type of peripheral vascular disease of legs called black foot disease (WHO, 2015). There is a strong correlation existed between As exposure and atherosclerosis (Simeonova and Luster, 2004), ischemic heart disease and hypertension (Li et al., 2013).

20.2.5 Blood

The ingested As is absorbed in the digestive tract and enters into the systemic circulation by primarily binding with oxy-hemoglobin of RBCs to induce hemolysis (Lu et al., 2004). The eventual extensive destruction of erythrocytes and reduces oxygen supply to the vital organs (Chandrasekaran et al., 2010). Thus low level of healthy erythrocytes and anemia is one of the most common effects of arsenic toxicity (Vigeh et al., 2014). Chronic As exposure causes distinct hematological effects such as leukopenia, intravascular hemolysis and thrombocytopenia (Pakulska and Czerczak, 2006).

20.2.6 Immune system

Immune cells proliferation may inhibition or induce by arsenic (Soto-Pena and Vega, 2008). Arsenic impairs function of lymphocyte by affecting their development, activation, or proliferation. Such alterations could be due to As dependent increased free radicals in cells and oxidative damage, lipid peroxidation, crosslinking of proteins, modification of DNA bases, DNA damage and apoptosis (Singh et al., 2013). In vitro studies confirmed that impaired physiology T cells can stimulate cancer of skin (Martin-Chouly et al., 2011).

20.2.7 Endocrine system

Arsenic potentially interrupt activity of endocrine gland like thyroid gland, pancreas, gonads and hypothalamic-pituitary-adrenal axis (Davey et al., 2008; Ciarrocca et al., 2012; Lu et al., 2011; Shen et al., 2013; Goggin et al., 2012). The pancreatic hormone insulin and glucagon regulate sugar level of blood glucose. Arsenic exposure for long duration decreases the secretion of insulin resulting in type II diabetes characterized by high glucose levels (Wang et al., 2014b).

20.2.8 Liver

Arsenic accumulates in the liver and show hepatotoxic effects (Jomova et al., 2011). Some frequent symptoms of As exposure are bleeding from esophagus, ascites, increased liver enzymes, jaundice, and enlarged liver. Severe arsenic toxicity result in hepatic lesions, liver cirrhosis, (Saha and Ray, 2019) noncirrhotic portal fibrosis, and lastly liver failure (Kapaj et al., 2006). Some other associated symptoms are gastrointestinal disturbances like nausea, vomiting, diarrhea, excessive salivation, problems in food swallowing, gastrointestinal cramps, abdominal pain and dehydration (Vantroyen et al., 2004).

20.2.9 Kidneys

In animals kidney filter and eliminate toxic As from body through the urine. In the elimination process, renal cells and tissues accumulate it and experience cytotoxicity (Madden and Fowler, 2000). Renal toxicity caused by elevated As manifested in form of high serum creatinine, hypo-urea and proteinuria (Sasaki et al., 2007). The symptoms appear due to acute tubular necrosis, damage of glomeruli and renal cells of the excretory system following peroxidation of lipid (El-Demerdash et al., 2009; Rahman et al., 2009).

20.2.10 Reproductive organs and growth

Teratogenic arsenic affects reproductive health and fetus developmental. It may cause growth retardation, premature delivery, birth defects (Wu et al., 2011), damage intellectual ability and motor dysfunction (Saha and Ray, 2019), and death (Chakraborti et al., 2003). It is responsible for declined testosterone synthesis and not proper functioning of resulting in male infertility whereas in females it may cause cervical cancer (Shen et al., 2013) by altering DNA methylation, apoptosis or cell proliferation (Li et al., 2009).

20.2.11 Carcinogenicity

In human, tumors and cancer of skin (Surdu, 2014), lung (Hubaux et al., 2013; Celik et al., 2008), bladder (Radosavljevic and Jakovljevic, 2008), prostates (Benbrahim-Tallaa and Waalkes, 2008; Siegel et al., 2013) and liver (Wang et al., 2014a), may be developed due to arsenic exposure. Long-term use of As contaminated water result in cancer of skin (Yu et al., 2000), and Prostate (Fig. 20.2).

20.3 Methods of arsenic remediation

20.3.1 Oxidation

The composition and chemistry of waste water is important parameter in As removal process (Singh et al., 2014). The pH of waste water affect As removal. In alkaline pH

FIGURE 20.2 Flow chart depicting various health effects of arsenic toxicity in humans.

below 9.2 arsenite is non-charged and inhibit its adsorption, precipitation or ion-exchange. The conversion of arsenite into arsenate facilitates its removal by different physical and chemical techniques (Pous et al., 2015). The arsenite is dominant arsenic species in groundwater having neutral pH (Singh et al., 2014). In aerobic habitat arsenite is oxidized into arsenate by biological entity or use of chemical like Cl_2, O_3, and $KMnO_4$, however, impurities significantly decrease rate of chemical oxidation process (Singh et al., 2014). The groundwater As oxidation is influenced by different types of competing anions and organic matter (Guan et al., 2012).

20.3.2 Coagulation and flocculation

These are the most availed and recorded practices (Choong et al., 2007) in which positively charged coagulants (aluminum sulfate, ferric chloride) are used to neutralize the negative charge for forming bigger aggregates. Arsenic removal efficiency of a coagulant is affected by the water pH. At pH below 7.6, $Al_2(SO_4)_3$, and $FeCl_3$ both significantly eliminated arsenic from wastewater (Garelick et al., 2005), however their performance may be change depending upon the pH (Garelick et al., 2005; Saha and Ray, 2019). In addition to the traditional salts, some other compounds such as titanium tetrachloride are also used in treatment of waste water.

20.3.3 Membrane filtration

Membrane filtration is a potential technique for treatment of As polluted water without use of chemicals. The technique is easily scale-up and modulated. The operation of membrane filtration is pressure dependent in nanofiltration and reverse osmosis whereas it is thermally driven in cases of membrane distillation.

20.3.4 Adsorption

Adsorption technique is frequently used for removing metal contaminants. The adsorbents used include activated carbon, activated alumina, zeolites $Fe(OH)_2$, cation-exchange resins, etc. The As present in the wastewater get sorbed on the adsorbent surface by electrostatic and the Vander Waals attraction forces. The nanosorbents with very high surface area per unit volume are suitable as adsorbent. The adsorbents prepared from copper nanoflakes very efficiently removed As from the synthetic waste water (Pal et al., 2017). Biosorbent have been prepared from variety of biomaterials ranging from bacteria, plant, fungi and algae. Seed biomass derived bio-sorbents effectively removed As from water (Giri et al., 2021). For reducing the biosorbent preparation cost, waste or low cost biomaterials are preferred. Activated carbon having large specific surface area and arsenic capture ability, however, it is inflammable at high temperature in aerobic atmosphere (Wu et al., 2021; Yang et al., 2021).

20.3.5 Bioremediation

20.3.5.1 Bacterial As remediation

Organisms contains many genes for combating the environmental challenges including toxic metal. In bacteria some genes are related to the management of As toxicity are aioA, arrA, arsC, and arsM. The genes have been frequently isolated from variety of As polluted ecosystems (Guo et al., 2019). The biotransformation and volatilization of As in algae is well documented for As uptake and redox pathways (Hussain et al., 2021). A little is known about the bacterial As sensing mechanism and chemotaxis response toward As^{+5}, however a dissimilatory metal-reducing bacterium showed energy dependent chemotaxis toward As^{+5} (Cheng et al., 2019). At the neutral pH, arsenate is transported into the bacterial cell through the phosphate transport system whereas arsenite transported via aqua-glyceroporins (Stahlberg et al., 2000). The host associated bacteria *Acinetobacter lwoffii* not only used to uptake arsenic from the environment but also help in the growth of host mung bean (Das et al., 2009). The protiobacteria associated with root of rice plant could diminish As accumulation in shoots of the rice plant. The bacterial As removal potential can be altered by changing the concentration of nutrients in the medium (Tripti and Shardendu, 2021). Some endophytic bacteria protect their host from environmental stresses and help in detoxifying pollutants. In As contaminated soils, the endophytes containing plant performed better in As phytoremediation (Mukherjee et al., 2018).

20.3.5.2 Phytoextraction of arsenic

Phytoextraction is the process of removing contaminates from soil using plant. In addition to the phytoextraction the other terms used in the process phytoremediation are phytostabilization and phytovolatilization. The phyto stabilization process plant decreases the mobility and phyto availability of pollutant by excluding their uptake by plants whereas phytovolatilization reference to the uptake of pollutant by hyper accumulating plant, its translocation in various parts followed by volatilization in the air (Vamerali et al., 2010). The related term rhizofiltration refers to adsorption and absorption of pollutant from water or wastewater by aquatic plant.

20.3.5.2.1 Arsenic uptake in plants

Uptake of As in the plant in aquatic plants takes place via energy dependent active process, passive process or by simple adsorption on the surface. The active uptake is dependent on the phosphate transporters whereas the passive uptake depends on the aquaporins or aqua-glycoporins. The uptake of As^{+5} mainly take place through the phosphate transporters (Tripathi et al., 2007; Zhao et al., 2009). The As^{+5} uptake may take place through the physochochemical sorption the surface of roots (Robinson et al., 2006). The As^{+3} and organic As species are passively transported into the plant cells by the aquaglycoporins (Zhao et al., 2009). The As hyper accumulating plants *Pteris vittata* growing in P deficient aerobic soils up take As^{+5} by use of phosphate transporter (Phts) and As^{+5} uptake is inhibited in presence of P rich soil. Similarly presence of P in the growth medium of barley seeding inhibited As^{+5} absorption and suggested a common transporter for both As and P in the root seedlings (Asher and Reay, 1979). Such transporter existed in model plant *Arabidopsis thaliana* and food crop *Oryza sativa* (Cao et al., 2017; Clark et al., 2000). The superfamily of phosphate transport (Pht) have several subfamilies P transporter 1 (Pht1) in different plants. In *A. thaliana* the transporter are (AtPht1, 1 and AtPht1, 4 whereas in *Oryza sativa* plant the transporter are OsPht1,4 and OsPht1; 8). Both the transporter in the rice plant may be used As accumulation as deletion of the OsPht1, 4 resulted 16%–32% low As^{+5} accumulation by one transporter only (Cao et al., 2017).

The soils contaminated with As can be remediated with use of intercropping of aquatic vegetables (water spinach, celery and arrowhead) cultivation before rice plant. In this case the As extracted by intercropping of was better as compared to the *Pteris vittata* fern for the same period of treatment. So, it could be a better sustainable option for remediation of As contaminated soils (Huang et al., 2021).

The phytoremediation is an attractive option for remediation of large field contaminated with pollutant. The process have been applied for remediation of As polluted soils (Jankong et al., 2007; Niazi et al., 2012). Hyper As accumulating plants are useful option for As clean up from the contaminated soils (Kraemer, 2005). However, it is necessary to do the fundamental research activity for better understanding the interaction of soil, plant with a specific contaminant. The plant may not show the presence of As in tissue under low concentration exposure but only in high concentration exposure (250 μg/L) it may be accumulated in the roots only and transported to the stems and leaves as reported in the case of *Elodea canadensis*. This plant accumulated about 63% As in the root in an hour and decreased As concentration in soil below detectable limit (Picco et al., 2019). The aquatic plant *Vallisneria natans* was able to accumulate As only at concentration of 1000 μg/L under association with plant growth promoting rhizobacterium association (Chen et al., 2021a,b). The other environmental factors such as pH, presence of ions like PO_4^{-2} NO_3^- in the waste water can affect As removed by plant. Aquatic plant *Lemna valdiviana* accumulated As only when pH ranged between 6.3–7 and PO_4^{-2} concentration of 0.0488 mmol/L, and NO_3^- concentration 7.9 mmol/L for reducing 82% As to its initial concentration (de Souza et al., 2019). Efforts are being made to optimize the As removal by different plant. The As removal by *Ludwigia octovolvis* was optimized by response surface methodology revealed maximum arsenic removal was 39 mg/kg soil in 42 days with removal about 72.6% (Titah et al., 2018).

Field crops like *Helianthus annus, Zea mays*, etc., have also been evaluated for their heavy metal remediation strategies and shown good potential. In addition to the As removal by the individual plant, association of *Glomus mosseae* with Pea not only protected host pea plant from the harmful impact of the As and other heavy metals but also enhanced growth parameters and better antioxidant defense. The symbiotic association was suggested for onsite remediation process, however crop use was not suggested for use due to high heavy metal content in the edible parts (Chaturvedi et al., 2021). The submerged aquatic plant *Vallisneria natans* showed improved As accumulation in presence of PGPR bacteria, however such effects were observed when As concentration in the water was below 1 ppm and As accumulation decreased at higher As concentration in the water. The IAA synthesis and As accumulation were correlated and decreased IAA synthesis decreased the As accumulation in plant. The PGPR association with plant used to played important role in As accumulation. In submerged aquatic plant *Vallisneria natans* showed improved As accumulation in presence of PGPR bacteria when As concentration in the water was below 1 ppm, but at higher As concentration As accumulation decreased. In this plant, auxin hormone synthesized and accumulated As content were correlated. The low IAA synthesis decreased the As accumulation in plant. Some recent updates about the As removal by the different plant have been summarized in Table 20.3 given below.

The accumulation of As in the crops pose threat to animals and human. To protect the human from the ill effect it is necessary to limit the accumulation of As in harvestable parts of the crops. Alternatively, the contaminated ecosystem may be remediated by use of hyper accumulating plants. The strategies used in remediation will be only successful when understanding about the physiology and morphology of the plant are well known under As exposure. The knowledge of plant based transporters; oxidative stresses and metabolism impairment are continuously being updated in plants including

TABLE 20.3 Plant used in phytoremediation of Arsenic from various habitats.

Plant	Comment	Reference
Lepidium sativum	This aquatic plant is consumed as vegetable in New Zealand. It accumulated As in leaves (29mg/kg and stem(15.9mg/kg) of its fresh weight.	Robinson et al. (2003)
Vallisneria natans	Submerged plant with PGPR bacteria *Rahnella aquatilis* showed better arsenic accumulation and detoxification by *V. natans* at As concentration below 1 ppm.	Chen et al. (2021a,b)
Vetiveria zizanioides	Removal of As along with fluoride and manganese from synthetic waste water	Thakur et al. (2021)
Vetiveria zizanioides, Chara vulgaris, Hyacyntus orintalis	As contaminated pot soil	Taleei et al. (2019)
Lemna valdiviana	Effect of pH, nutrient concentration on As accumulation. About 82% As removal.	de Souza et al. (2019)
Elodea canadensis	As accumulated in root only when exposed at 250 μg/L but not translocated to stem and leaves. It decreased As concentration below detectable limit in an hour.	Picco et al. (2019)
Pteris vittata, P. umbrosa P. cretica	As accumulation in leaves	Praveen and Pandey (2020)
Helianthus annus	Preferential accumulation of As and Uranium from polluted soil	Webber et al. (2021)
Zea mays	Compost dose and soil texture dependent As activity	Mehmood et al. (2021)
Pteris vitata	Recycling strategy for As rich biomass of the plant	Cai et al. (2021)
Ludwigia octovalvis	About 72.6% removal	Titah et al. (2018)

rice. There are many aspects which are still unanswered and need further study to exploit full potential of plant in an ecosystem (Kofroňová et al., 2018).

The strategic disposal of As rich biomass is necessary for the activity of phytoremediation projects. Considering such issues of As recycling from As rich biomass of *Pteris vittata,* the biomass was used for isolating bioactive compounds from it. The fern biomass (1 t) was able to produce about 48 kg phenolic compound having no acute or chronic risk to skin in long-term exposure. It could be an alternative for disposing and recycling all the fern waste cost effectively (Cai et al., 2021). In some parts of Navajo Nation, indigenous residents depending on the As and uranium polluted water sources for drinking and other routine activities. The cultivation of sunflowers accumulating these As and uranium could be useful tool for on-site phytoremediation (Webber et al., 2021). The cultivation *Zea mays* crops for fulfilling future energy need on the contaminated soil could be an alternate cost effective option for soil remediation (Mehmood et al., 2021).

20.4 Conclusion

Considering the harmful affects of As on human health, the use of arsenicals in the agriculture, animal feed, and other industry should be highly controlled. The population exposed to groundwater arsenic contamination needs to develop low cost filters to remove As from drinking water, which is still a challenge for the scientific community. The phytoremediation technique could be used for combating As contamination in the large scale on site remediation of soil and water. These techniques could supplement the presently used chemical engineering principles for remediation of polluted water from variety of sources.

References

Abbas, G., Murtaza, B., Bibi, I., Shahid, M., Niazi, N.K., Khan, M.I., et al., 2018. Arsenic uptake, toxicity, detoxification, and speciation in plants: physiological, biochemical, and molecular aspects. International Journal of Environmental Research and Public Health 15, 59.

Abdulsalam, S., Bugaje, I.M., Adefila, S.S., Ibrahim, S., 2011. Comparison of biostimulation and bioaugmentation for remediation of soil contaminated with spent motor oil. International Journal of Environmental Science and Technology 8, 187–194.

Ahamad, A., Madhav, S., Singh, A.K., Kumar, A., Singh, P., 2020. Types of water pollutants: conventional and emerging. Sensors in Water Pollutants Monitoring: Role of Material. Springer, Singapore, pp. 21–41.

Alam, M.O., Shaikh, W.A., Chakraborty, S., Avishek, K., Bhattacharya, T., 2016. Groundwater arsenic contamination and potential health risk assessment of Gangetic Plains of Jharkhand, India. Expo Health 8, 125–142.

Ali, W., Rasool, A., Junaid, M., Zhang, H., 2019. A comprehensive review on current status, mechanism, and possible sources of arsenic contamination in groundwater: a global perspective with prominence of Pakistan scenario. Environmental Geochemistry and Health 41 (2), 737–760.

Amster, E., Tiwary, A., Schenker, M.B., 2007. Case report: potential arsenic toxicosis secondary to herbal kelp supplement. Environmental Health Perspectives 115, 606–608.

Appleyard, S.J., Angeloni, J., Watkins, R., 2006. Arsenic-rich groundwater in an urban area experiencing drought and increasing population density, Perth, Australia. Applied Geochemistry: Journal of the International Association of Geochemistry and Cosmochemistry 21 (1), 83–97.

Arco-Lázaro, E., Pardo, T., Clemente, R., Bernal, M.P., 2018. Arsenic adsorption and plant availability in an agricultural soil irrigated with As-rich water: effects of Fe-rich amendments and organic and inorganic fertilisers. Journal of Environmental Management 209, 262–272.

Argos, M., Kalra, T., Rathouz, P.J., Chen, Y., Pierce, B., Parvez, F., et al., 2010. Arsenic exposure from drinking water, and all-cause and chronic-disease mortalities in Bangladesh (HEALS): a prospective cohort study. Lancet 376, 252–258.

Asher, C.J., Reay, P.F., 1979. Arsenic uptake by barley seedlings. Functional Plant Biology 6, 459–466. Available from: https://doi.org/10.1071/pp9790459.

Barats, A., Féraud, G., Potot, C., Philippini, V., Travi, Y., Durrieu, G., et al., 2014. Naturally dissolved arsenic concentrations in the Alpine/Mediterranean Var River watershed (France). The Science of the Total Environment 473, 422–436.

Bartolome, B., Cordoba, S., Nieto, S., Fernández-Herrera, J., García-Díez, A., 1999. Acute arsenic poisoning: clinical and histopathological features. The British Journal of Dermatology 141, 1106–1109.

Benbrahim-Tallaa, L., Waalkes, M.P., 2008. Inorganic arsenic and human prostate cancer. Environmental Health Perspectives 116, 158–164.

Bhattacharya, P., 2017. Assessment of arsenic accumulation by different varieties of rice (Oryza sativa L.) irrigated with arsenic-contaminated groundwater in West Bengal (India). Environmental Pollution and Protection 2, 92–99.

Bhattacharya, P., Jacks, G., Frisbie, S.H., Smith, E., Naidu, R., Sarkar, B., 2002. Arsenic in the environment: a global perspective. In: Sarkar, B. (Ed.), Handbook of Heavy Metals in the Environment. Dekker, New York, pp. 147–215.

Bhattacharya, P., Mukherjee, A., Mukherjee, A.B., 2016. Groundwater arsenic in India: source, distribution, effects and mitigation. Reference Module in Earth Systems and Environmental Sciences, Elsevier. Available from: https://doi.org/10.1016/B978-0-12-409548-9.09342-8.

Bhattacharyya, R., Chatterjee, D., Nath, B., Jana, J., Jacks, G., Vahter, M., 2003. High arsenic groundwater: mobilization, metabolism and mitigation—an overview in the Bengal Delta Plain. Molecular and Cellular Biochemistry 253, 347–355.

Bhowmick, S., Pramanik, S., Singh, P., Mondal, P., Chatterjee, D., Nriagu, J., 2018. Arsenic in groundwater of West Bengal, India: a review of human health risks and assessment of possible intervention options. The Science of the Total Environment 612, 148–169.

Bindal, S., Singh, C.K., 2019. Predicting groundwater arsenic contamination: regions at risk in highest populated state of India. Water Research 159, 65–76.

Bissen, M., Frimmel, F.H., 2003. Arsenic: a review—part I—: occurrence, toxicity, speciation, mobility. Acta Hydrochimica et Hydrobiologica 31 (1), 9–18.

Bondu, R., Cloutier, V., Rosa, E., Benzaazoua, M., 2017. Mobility and speciation of geogenic arsenic in bedrock groundwater from the Canadian Shield in western Quebec, Canada. The Science of the Total Environment 574, 509–519.

Buamah, R., Petrusevski, B., Schippers, J.C., 2008. Presence of arsenic, iron and manganese in groundwater within the gold-belt zone of Ghana. Journal of Water Supply: Research and Technology-Aqua 57 (7), 519–529.

Cai, W., Chen, T., Lei, M., Wan, X., 2021. Effective strategy to recycle arsenic-accumulated biomass of Pteris vittata with high benefits. The Science of the Total Environment 756, 143890. Available from: https://doi.org/10.1016/j.scitotenv.2020.143890.

Cao, Y., Sun, D., Ai, H., Mei, H., Liu, X., Sun, S., et al., 2017. Knocking out OsPT4 gene decreases arsenate uptake by rice plants and inorganic arsenic accumulation in rice grains. Environmental Science & Technology 51, 12131–12138. Available from: https://doi.org/10.1021/acs.est.7b03028.

Carbonell-Barrachina, A.A., Signes-Pastor, A.J., VázquezAraffljo, L., Burló, F., Sengupta, B., 2009. Presence of arsenic in agricultural products from arsenic-endemic areas and strategies to reduce arsenic intake in rural villages. Molecular Nutrition & Food Research 53, 531–541.

Celik, I., Gallicchio, L., Boyd, K., Lam, T.K., Matanoski, G., Tao, X., et al., 2008. Arsenic in drinking water and lung cancer: a systematic review. Environmental Research 108, 48–55.

Chakraborti, D., Mukherjee, S.C., Pati, S., Sengupta, M.K., Rahman, M.M., Chowdhury, U.K., et al., 2003. Arsenic groundwater contamination in Middle Ganga Plain, Bihar, India: a future danger? Environmental Health Perspectives 111, 1194–1201.

Chakraborti, D., Rahman, M.M., Ahamed, S., Dutta, R.N., Pati, S., Mukherjee, S.C., 2016a. Arsenic contamination of groundwater and its induced health effects in Shahpur block, Bhojpur district, Bihar state, India: risk evaluation. Environmental Science and Pollution Research 23, 9492–9504. Available from: https://doi.org/10.1007/s11356-016-6149-8.

Chakraborti, D., Rahman, M.M., Ahamed, S., Dutta, R.N., Pati, S., Mukherjee, S.C., 2016b. Arsenic groundwater contamination and its health effects in Patna district (capital of Bihar) in the middle Ganga plain, India. Chemosphere 152, 520–529.

Chakraborti, D., Rahman, M.M., Das, B., Chatterjee, A., Das, D., Nayak, B., et al., 2017. Groundwater arsenic contamination and its health effects in India. Hydrogeology Journal 25, 1165–1181. Available from: https://doi.org/10.1007/s10040-017-1556-6.

Chakraborti, D., Singh, S.K., Rahman, M.M., Dutta, R.N., Mukherjee, S.C., Pati, S., et al., 2018. Groundwater arsenic contamination in the Ganga River Basin: a future health danger. International Journal of Environmental Research and Public Health 15 (2), 180.

Chandrasekaran, V.R.M., Muthaiyan, I., Huang, P.C., Liu, M.Y., 2010. Using iron precipitants to remove arsenic from water: is it safe? Water Research 44, 5823–5827.

Chaturvedi, R., Favas, P.J.C., Pratas, J., Varun, M., Paul, M.S., 2021. Harnessing Pisum sativum-Glomus mosseae symbiosis for phytoremediation of soil contaminated with lead, cadmium, and arsenic. International Journal of Phytoremediation 23, 279–290. Available from: https://doi.org/10.1080/15226514.2020.1812507.

Chen, Y., Graziano, J.H., Parvez, F., Liu, M., Slavkovich, V., Kalra, T., et al., 2011. Arsenic exposure from drinking water and mortality from cardiovascular disease in Bangladesh: prospective cohort study. British Medical Journal 342, d2431.

Chen, Y., Han, Y.H., Cao, Y., Zhu, Y.G., Rathinasabapathi, B., Ma, L.Q., 2017. Arsenic transport in rice and biological solutions to reduce arsenic risk from rice. Frontiers in Plant Science 8, 268.

Chen, Y., Parvez, F., Gamble, M., Islam, T., Ahmed, A., Argos, M., et al., 2009. Arsenic exposure at low-to-moderate levels and skin lesions, arsenic metabolism, neurological functions, and biomarkers for respiratory and cardiovascular diseases: review of recent findings from the Health Effects of Arsenic Longitudinal Study (HEALS) in Bangladesh. Toxicology and Applied Pharmacology 239, 184–192.

Chen, G., Ran, Y., Ma, Y., Chen, Z., Li, Z., Chen, Y., 2021a. Influence of Rahnella aquatilis on arsenic accumulation by Vallisneria natans (Lour.) Hara for the phytoremediation of arsenic-contaminated water. Environmental Science and Pollution Research International 28, 44354–44360. Available from: https://doi.org/10.1007/s11356-021-13868-9.

Chen, F., et al., 2021b. Sub-chronic low-dose arsenic in rice exposure induces gut microbiome perturbations in mice. Ecotoxicology and Environmental Safety 227, 112934.

Cheng, L., Min, D., Liu, D.F., Li, W.W., Yu, H.Q., 2019. Sensing and Approaching Toxic Arsenate by *Shewanellaputrefaciens* CN-32. Environmental Science & Technology 17, 53(24), 14604–14611. Available from: https://doi.org/10.1021/acs.est.9b05890. Epub 2019 Dec 5. PMID: 31747260.

Cho, K.H., Sthiannopkao, S., Pachepsky, Y.A., Kim, K.W., Kim, J.H., 2011. Prediction of contamination potential of groundwater arsenic in Cambodia, Laos, and Thailand using artificial neural network. Water Research 45 (17), 5535–5544.

Choong, T., Chuah, T., Robiah, Y., Gregory, K., Azni, I., 2007. Arsenic toxicity, health hazards and removal techniques from water: An overview. Desalination 217, 139–166.

Ciarrocca, M., Tomei, F., Caciari, T., Cetica, C., Andre, J.C., Fiaschetti, M., et al., 2012. Exposure to arsenic in urban and rural areas and effects on thyroid hormones. Inhalation Toxicology 24, 589–598.

Ciminelli, V.S., Gasparon, M., Ng, J.C., Silva, G.C., Caldeira, C.L., 2017. Dietary arsenic exposure in Brazil: the contribution of rice and beans. Chemosphere 168, 996–1003.

Clark, G.T., Dunlop, J., Phung, H.T., 2000. Phosphate absorption by Arabidopsis thaliana: interactions between phosphorus status and inhibition by arsenate. Functional Plant Biology 27, 959–965.

Corradini, F., Correa, A., Moyano, M.S., Sepúlveda, P., Quiroz, C., 2018. Nitrate, arsenic, cadmium, and lead concentrations in leafy vegetables: expected average values for productive regions of Chile. Archives of Agronomy and Soil Science 64 (3), 299–317.

Cubadda, F., Ciardullo, S., D'Amato, M., Raggi, A., Aureli, F., Carcea, M., 2010. Arsenic contamination of the environment-food chain: a survey on wheat as a test plant to investigate phytoavailable arsenic in Italian agricultural soils and as a source of inorganic arsenic in the diet. Journal of Agricultural and Food Chemistry 58, 10176–10183. Available from: https://doi.org/10.1021/jf102084p.

Das, B., Rahman, M.M., Nayak, B., Pal, A., Chowdhury, U.K., Mukherjee, S.C., et al., 2009. Groundwater arsenic contamination, its health effects and approach for mitigation in West Bengal, India and Bangladesh. Water Quality, Exposure and Health 1, 5–21.

Davey, J.C., Nomikos, A.P., Wungjiranirun, M., Sherman, J.R., Ingram, L., Batki, C., et al., 2008. Arsenic as an endocrine disruptor: arsenic disrupts retinoic acid receptor-and thyroid hormone receptor-mediated gene regulation and thyroid hormone-mediated amphibian tail metamorphosis. Environmental Health Perspectives 116, 165–172.

de Souza, T.D., Borges, A.C., Braga, A.F., Veloso, R.W., Teixeira de Matos, A., 2019. Phytoremediation of arsenic-contaminated water by Lemna Valdiviana: An optimization study. Chemosphere 234, 402–408. Available from: https://doi.org/10.1016/j.chemosphere.2019.06.004.

Dhillon, A.K., 2020. Arsenic contamination of India's groundwater: A review and critical analysis. In: Fares, A., Singh, S. (Eds.), Arsenic Water Resources Contamination. Advances in Water Security. Springer, Cham, pp. 177–205. Available from: https://doi.org/10.1007/978-3-030-21258-2_8.

Edmunds, W.M., Ahmed, K.M., Whitehead, P.G., 2015. A review of arsenic and its impacts in groundwater of the Ganges–Brahmaputra–Meghna delta, Bangladesh. Environmental Science: Processes & Impacts 17 (6), 1032–1046.

EFSA, 2009. European food safety authority. Scientific Opinion on Arsenic in Food 7 (10), 1351. Available from: http://www.efsa.europa.eu/en/scdocs/scdoc/1351.htm.

Eiche, E., Neumann, T., Berg, M., Weinman, B., van Geen, A., Norra, S., 2008. Geochemical processes underlying a sharp contrast in groundwater arsenic concentrations in a village on the Red River delta, Vietnam. Applied Geochemistry: Journal of the International Association of Geochemistry and Cosmochemistry 23, 3143–3154.

El-Demerdash, F.M., Yousef, M.I., Radwan, F.M., 2009. Ameliorating effect of curcumin on sodium arsenite-induced oxidative damage and lipid peroxidation in different rat organs. Food and Chemical Toxicology: An International Journal Published for the British Industrial Biological Research Association 47, 249–254.

Ennaceur, N., Henchiri, R., Jalel, B., Cordier, M., Ledoux-Rak, I., Elaloui, E., 2017. Synthesis, crystal structure, and spectroscopic characterization supported by DFT calculations of organoarsenic compound. Journal of Molecular Structure 1144, 25–32.

Even, E., Masuda, H., Shibata, T., Nojima, A., Sakamoto, Y., Murasaki, Y., et al., 2017. Geochemical distribution and fate of arsenic in water and sediments of rivers from the Hokusetsu area, Japan. Journal of Hydrology: Regional Studies 9, 34–47.

Fei, J., Wang, T., Zhou, Y., Wang, Z., XiaoboMin, Y., Wenyong, H., Chai, L., 2018. Aromatic organoarsenic compounds (AOCs) Occurrence and remediation methods. Chemosphere 207, 665–675.

Fendorf, S., Michael, H.A., van Geen, A., 2010. Spatial and temporal variations of groundwater arsenic in South and Southeast Asia. Science 328 (5982), 1123–1127.

Flanagan, S.V., Marvinney, R.G., Zheng, Y., 2015. Influences on domestic well water testing behaviour in a Central Maine area with frequent groundwater arsenic occurrence. The Science of the Total Environment 505, 1274–1281.

Girard, J., 2010. Principles of Environmental Chemistry. Jones & Bartlett Learning. ISBN 9780763759391.

Garelick, H., Dybdowska, A., Valsami-Jones, E., Priest, N., 2005. Remediation technologies for arsenic contaminated drinking waters. Journal of Soils and Sediments 5, 182–190.

Giri, D.D., et al., 2021. Java Plum and Amaltash seed biomass based bio-adsorbents for synthetic wastewater treatment. Environmental Pollution (Barking, Essex: 1987) 280, 116890.

Goggin, S.L., Labrecque, M.T., Allan, A.M., 2012. Perinatal exposure to 50 ppb sodium arsenate induces hypothalamic-pituitary-adrenal axis dysregulation in male C57BL/6 mice. Neurotoxicology 33, 1338–1345.

Gong, G., Mattevada, S., O'Bryant, S.E., 2014. Comparison of the accuracy of kriging and IDW interpolations in estimating groundwater arsenic concentrations in Texas. Environmental Research 130, 59–69.

Govil, P.K., Krishna, A.K., 2018. Soil and water contamination by potentially hazardous elements: a case history from India, Environmental Geochemistry, second ed. Elsevier, pp. 567–597.

Guo, J.X., Hu, L., Yand, P.Z., Tanabe, K., Miyatalre, M., Chen, Y., 2007. Chronic arsenic poisoning in drinking water in Inner Mongolia and its associated health effects. Journal of Environmental Science and Health, Part A. Toxic/Hazardous Substances and Environmental Engineering 42, 1853–1858.

Guo, H., Wen, D., Liu, Z., Jia, Y., Guo, Q., 2014. A review of high arsenic groundwater in Mainland and Taiwan, China: distribution, characteristics and geochemical processes. Applied Geochemistry: Journal of the International Association of Geochemistry and Cosmochemistry 41, 196–217.

Guan, X., Du, J., Meng, X., Sun, Y., Sun, B., Hu, Q., 2012. Application of titanium dioxide in arsenic removal from water: A review. J. Journal of Hazardous Materials 27, 360–370.

Guo, T., et al., 2019. Distribution of Arsenic and Its Biotransformation Genes in Sediments from the East China Sea. Environmental Pollution (Barking, Essex: 1987) 253, 949–958.

Henke, K.R., 2009. Arsenic in natural environments. In: Henke, K. (Ed.), Arsenic: Environmental Chemistry, Health Threats and Waste Treatment. John Wiley & Sons Ltd., Chichester, UK, pp. 69–235.

Hopenhayn, C., 2006. Arsenic in drinking water: impact on human health. Elements 2, 103–107. Available from: https://doi.org/10.2113/gselements.2.2.103.

Huang, S.Y., Zhuo, C., Du, X.Y., Li, H.S., 2021. Remediation of arsenic-contaminated paddy soil by intercropping aquatic vegetables and rice. International Journal of Phytoremediation 23, 1021–1029. Available from: https://doi.org/10.1080/15226514.2021.1872485.

Hubaux, R., Becker-Santos, D.D., Enfield, K., Rowbotham, D., Lam, S., Lam, W.L., et al., 2013. Molecular features inarsenic-induced lung tumors. Molecular Cancer 12, 20.

Hussain, M.M., et al., 2021. Arsenic speciation and biotransformation pathways in the aquatic ecosystem: the significance of algae. Journal of Hazardous Materials 403, 124027.

Jang, Y.-C., Somanna, Y., Kim, H., 2016. Source, distribution, toxicity and remediation of arsenic in the environment – a review. International Journal of Applied Environmental Sciences 11 (2), 559–581.

Jankong, P., Visoottiviseth, P., Khokiattiwong, S., 2007. Enhanced phytoremediation of arsenic contaminated land. Chemosphere 68, 1906–1912. Available from: https://doi.org/10.1016/j.chemosphere.2007.02.061.

Jomova, K., Jenisova, Z., Feszterova, M., Baros, S., Liska, J., Hudecova, D., et al., 2011. Arsenic: toxicity, oxidative stress and human disease. Toxicology and Applied Pharmacology 31, 95–107.

Kapaj, S., Peterson, H., Liber, K., Bhattacharya, P., 2006. Human health effects from chronic arsenic poisoning – a review. Journal of Environmental Science and Health, Part A. Toxic/Hazardous Substances and Environmental Engineering 41, 2399–2428.

Katsoyiannis, I.A., Mitrakas, M., Zouboulis, A.I., 2015. Arsenic occurrence in Europe: emphasis in Greece and description of the applied full-scale treatment plants. Desalination and Water Treatment 54 (8), 2100–2107.

Kofroňová, M., Mašková, P., Lipavská, H., 2018. Two facets of world arsenic problem solution: crop poisoning restriction and enforcement of phytoremediation. Planta 248, 19–35. Available from: https://doi.org/10.1007/s00425-018-2906-x.

Konhauser, K.O., Powell, M.A., Fyfe, W.S., Longstaffe, F.J., Tripathy, S., 1997. Trace element chemistry of major rivers in Orissa State, India. Environmental Geology 29, 132–141.

Kossoff, D., Hudson-Edwards, K.A., 2012. Arsenic in the environment. Chapter 1. In: Santini, J.M., Ward, S.M. (Eds.), The Metabolism of Arsenite, Arsenic in the Environment, vol. 5. CRC Press, London, pp. 1–23.

Kraemer, U., 2005. Phytoremediation: novel approaches to cleaning up polluted soils. Current Opinion in Biotechnology 16 (2), 133–141.

Kumar, M., Rahman, M.M., Ramanathan, A., Naidu, R., 2016. Arsenic and other elements in drinking water and dietary components from the middle Gangetic plain of Bihar, India: health risk index. The Science of the Total Environment 539, 125–134.

Kwong, Y.T.J., Beauchemin, S., Hossain, M.F., Gould, W.D., 2007. Transformation and mobilization of arsenic in the historic Cobalt mining camp, Ontario, Canada. Journal of Geochemical Exploration 92 (2–3), 133–150.

Li, X., Li, B., Xi, S., Zheng, Q., Lv, X., Sun, G., 2013. Prolonged environmental exposure of arsenic through drinking water on the risk of hypertension and type 2 diabetes. Environmental Science and Pollution Research 20, 8151–8161.

Li, D., Lu, C., Wang, J., Hu, W., Cao, Z., Sun, D., et al., 2009. Developmental mechanisms of arsenite toxicity in zebrafish (Danio rerio) embryos. Aquatic Toxicology (Amsterdam, Netherlands) 91, 229–237.

Lu, T.H., Su, C.C., Chen, Y.W., Yang, C.Y., Wu, C.C., Hung, D.Z., et al., 2011. Arsenic induces pancreatic beta-cell apoptosis via the oxidative stress-regulated mitochondria-dependent and endoplasmic reticulum stress-triggered signaling pathways. Toxicology Letters 201, 15–26.

Lu, M., Wang, H., Li, X.F., Lu, X., Cullen, W.R., Arnold, L.L., et al., 2004. Evidence of hemoglobin binding to arsenic as a basis for the accumulation of arsenic in rat blood. Chemical Research in Toxicology 17, 1733–1742.

Madden, E.F., Fowler, B.A., 2000. Mechanisms of nephrotoxicity from metal combinations: a review. Drug and Chemical Toxicology 23, 1–12.

Madhav, S., Ahamad, A., Singh, A.K., Kushawaha, J., Chauhan, J.S., Sharma, S., et al., 2020. Water pollutants: sources and impact on the environment and human health. Sensors in Water Pollutants Monitoring: Role of Material. Springer, Singapore, pp. 43–62.

Mandal, B.K., Roychowdhury, T., Samanta, G., 1996. Arsenic in groundwater in seven districts of West Bengal, India: the biggest arsenic calamity in the world. Current Science 70 (11), 976–998.

Mandal, B.K., Suzuki, K.T., 2002. Arsenic round the world: a review. Talanta 58, 201–235.

Manna, P., Sinha, M., Sil, P., 2008. Arsenic-induced oxidative myocardial injury: protective role of arjunolic acid. Archives of Toxicology 82, 137–149.

Martin-Chouly, C., Morzadec, C., Bonvalet, M., Galibert, M.-D., Fardel, O., Vernhet, L., 2011. Inorganic arsenic alters expression of immune and stress response genes in activated primary human T lymphocytes. Molecular Immunology 48, 956–965.

Mathew, L., Vale, A., Adcock, J.E., 2010. Arsenical peripheral neuropathy. Practical Neurology 10, 34–38.

Mazumder, D.N., Das Gupta, J., Santra, A., Pal, A., Ghose, A., Sarkar, S., 1998. Chronic arsenic toxicity in west Bengal – the worst calamity in the world. Journal of the Indian Medical Association 96, 4–7.

McArthur, J.M., 2019. Arsenic in groundwater. In: Sikdar, P.K. (Ed.), Groundwater Development and Management. Springer, Cham, pp. 279–308.

McCarty, K.M., Hanh, H.T., Kim, K.-W., 2011. Arsenic geochemistry and human health in South East Asia. Reviews on Environmental Health 26, 71–78.

Mehmood, T., Liu, C., Niazi, N.K., Gaurav, G.K., Ashraf, A., Bibi, I., 2021. Compost-mediated arsenic phytoremediation, health risk assessment and economic feasibility using Zea mays L. in contrasting textured soils. International Journal of Phytoremediation 23, 899–910. Available from: https://doi.org/10.1080/15226514.2020.1865267.

Middleton, D.R.S., Watts, M.J., Hamilton, E.M., Ander, E.L., Close, R.M., Exley, K.S., et al., 2016. Urinary arsenic profiles reveal exposures to inorganic arsenic from private drinking water supplies in Cornwall, UK. Scientific Reports 6.

Mozumder, M.R.H., 2019. Impacts of Pumping on the Distribution of Arsenic in Bangladesh Groundwater. (PhD. thesis). Columbia University, p. 199.

Mukherjee, A., Bhattacharya, P., Shi, F., Fryar, A.E., Mukherjee, A.B., Xie, Z.M., et al., 2009. Chemical evolution in the high arsenic groundwater of the Huhhot basin (Inner Mongolia, PR China) and its difference from the western Bengal basin (India). Applied Geochemistry: Journal of the International Association of Geochemistry and Cosmochemistry 24 (10), 1835–1851.

Mukherjee, G., et al., 2018. An endophytic bacterial consortium modulates multiple strategies to improve arsenic phytoremediation efficacy in Solanum Nigrum. Scientific Reports 8 (1), 6979.

Mundey, M.K., Roy, M., Roy, S., Awasthi, M.K., Sharma, R., 2013. Antioxidant potential of Ocimum sanctum in arsenic induced nervous tissue damage. Brazilian Journal of Veterinary Pathology 6, 95–101.

Nachman, K.E., Raber, G., Francesconi, K.A., Navas-Acien, A., Love, D.C., 2013. Arsenic species in poultry feather meal. Science of the Total Environment 15, 417–418, 183–8. Available from: https://doi.org/10.1016/j.scitotenv.2011.12.022.

Nguyen, T., Thi, P., Yacouba, S., Pare, S., Bui, H.M., 2020. Removal of arsenic from groundwater using Lamdong laterite as a natural adsorbent. Polish Journal of Environmental Studies 29 (2), 1305–1314.

Niazi, N.K., Singh, B., Van Zwieten, L., Kachenko, A.G., 2012. Phytoremediation of an arsenic-contaminated site using Pteris vittata L. and Pityrogramma calomelanos var. austroamericana: a long-term study. Environmental Science and Pollution Research International 19, 3506–3515. Available from: https://doi.org/10.1007/s11356-012-0910-4.

Norman, N.C., 1998. Chemistry of Arsenic, Antimony and Bismuth. Springer, p. 50. Available from: http://books.google.com/?id = vVhpurkfeN4C.

O'Bryant, S.E., Edwards, M., Menon, C.V., Gong, G., Barber, R., 2011. Long-term low-level arsenic exposure is associated with poorer neuropsychological functioning: a Project FRONTIER study. International Journal of Environmental Research and Public Health 8, 861–874.

Pakulska, D., Czerczak, S., 2006. Hazardous effects of arsine: a short review. International Journal of Occupational Medicine and Environmental Health 19, 36–44.

Pal, D.B., et al., 2017. Arsenic removal from synthetic waste water by CuO nano-flakes synthesized by aqueous precipitation method. Desalination and Water Treatment 62.

Pandey, P.K., Yadav, S., Nair, S., Bhui, A., 2002. Arsenic contamination of the environment: a new perspective from central-east India. Environment International 28 (4), 235–245.

Parvez, F., Chen, Y., Yunus, M., Olopade, C., Segers, S., Slavkovich, V., et al., 2013. Arsenic exposure and impaired lung function. Findings from a large population-based prospective cohort study. American Journal of Respiratory and Critical Care Medicine 188, 813–819.

Parvez, F., Chen, Y., Yunus, M., Olopade, C., Slavkovich, V., Graziano, J., et al., 2012. E-008: arsenic induced impaired lung function and tuberculosis: findings from health effects of Arsenic Longitudinal Study (HEALS) Cohort, Bangladesh. Epidemiology (Cambridge, Mass.) 23.

Parvez, F., Chen, Y., Yunus, M., Zaman, R.-U., Ahmed, A., Islam, T., et al., 2011. Associations of arsenic exposure with impaired lung function and mortality from diseases of the respiratory system: findings from the health effects of Arsenic Longitudinal Study (HEALS). Epidemiology (Cambridge, Mass.) 22, S179.

Patel, K.S., Sahu, B.L., Dahariya, N.S., Bhatia, A., Patel, R.K., Matini, L., et al., 2017. Groundwater arsenic and fluoride in Rajnandgaon District, Chhattisgarh, northeastern India. Applied Water Science 7, 1817–1826.

Patel, K.S., Shrivas, K., Brandt, R., Jakubowski, N., Corns, W., Hoffmann, P., 2005. Arsenic contamination in water, soil, sediment and rice of central India. Environmental Geochemistry and Health 27, 131–145.

Picco, P., Hasuoka, P., Verni, E., Savio, M., Pacheco, P., 2019. Arsenic species uptake and translocation in Elodea canadensis. International Journal of Phytoremediation 21, 693–698. Available from: https://doi.org/10.1080/15226514.2018.1556588.

Podgorski, J., Wu, R., Chakravorty, B., Polya, D.A., 2020. Groundwater Arsenic Distribution in India by Machine Learning Geospatial Modeling. International Journal of Environmental Research and Public Health 17. Available from: https://doi.org/10.3390/ijerph17197119.

Pokhrel, D., Bhandari, B.S., Viraraghavan, T., 2009. Arsenic contamination of groundwater in the Terai region of Nepal: an overview of health concerns and treatment options. Environment International 35 (1), 157–161.

Polya, D.A., Sparrenborn, C., Datta, S., Guo, H., 2019. Groundwater arsenic biogeochemistry – key questions and use of tracers to understand arsenic-prone groundwater systems. Geoscience Frontiers 10 (10), 1635–1641.

Postma, D., Larsen, F., Minh Hue, N.T., Thanh Duc, M., Viet, P.H., Nhan, P.Q., et al., 2007. Arsenic in groundwater of the Red River floodplain, Vietnam: controlling geochemical processes and reactive transport modeling. Geochimica et Cosmochimica Acta 71, 5054–5071.

Pous, N., Casentini, B., Rossetti, S., Fazi, S., Puig, S., Aulenta, F., 2015. Anaerobic arsenite oxidation with an electrode serving as the sole electron acceptor: a novel approach to the bioremediation of arsenic-polluted groundwater. Journal of Hazardous Materials 283, 617–622. Process. Pres. 1, 8

Praveen, A., Pandey, V.C., 2020. Pteridophytes in phytoremediation. Environmental Geochemistry and Health 42, 2399–2411. Available from: https://doi.org/10.1007/s10653-019-00425-0.

Radosavljevic, V., Jakovljevic, B., 2008. Arsenic and bladder cancer: observations and suggestions. Journal of Environmental Health 71, 40–42.

Rahman, M.M., Ng, J.C., Naidu, R., 2009. Chronic exposure of arsenic via drinking water and its adverse health impacts on humans. Environmental Geochemistry and Health 31, 189–200.

Rani, L., Srivastav, A.L., Kaushal, J., Grewal, A.S., Madhav, S., 2022. Heavy metal contamination in the river ecosystem. Ecological Significance of River Ecosystems. Elsevier, pp. 37–50.

Ravenscroft, P., Brammer, H., Richards, K.S., 2009. Arsenic Pollution: A Global Synthesis. Wiley-Blackwell, London.

Richards, L.A., Casanueva-Marenco, M.J., Magnone, D., Sovann, C., van Dongen, B.E., Polya, D.A., 2019. Contrasting sorption behaviours affecting groundwater arsenic concentration in Kandal Province, Cambodia. Geoscience Frontiers 10 (5), 1701–1713.

Robinson, B., Duwig, C., Bolan, N., Kannathasan, M., Saravanan, A., 2003. Uptake of arsenic by New Zealand watercress (Lepidium sativum). The Science of the Total Environment 301, 67–73. Available from: https://doi.org/10.1016/s0048-9697(02)00294-2.

Robinson, B., Kim, N., Marchetti, M., Moni, C., Schroeter, L., van den Dijssel, C., et al., 2006. Arsenic hyperaccumulation by aquatic macrophytes in the Taupo Volcanic Zone, New Zealand. Environmental and Experimental Botany 58, 206–215. Available from: https://doi.org/10.1016/j.envexpbot.2005.08.004.

Robles, A.D., Polizzi, P., Romero, M.B., Boudet, L.N.C., Medici, S., Costas, A., et al., 2016. Geochemical mobility of arsenic in the surficial waters from Argentina. Environmental Earth Sciences 75 (23).

Rosen, B.R., Liu, Z.J., 2009. Transport pathways for arsenic and selenium: a minireview. Environment International 35, 512–515.

Rudel, R., Slayton, T.M., Beck, B.D., 1996. Implications of arsenic genotoxicity for dose response of carcinogenic effects. Regulatory Toxicology and Pharmacology 23, 87–105.

Saha, J., Dikshit, A., Bandyopadhyay, M., Saha, K., 1999. A review of arsenic poisoning and its effects on human health. Critical Reviews in Environmental Science and Technology 29, 281–313.

Saha, D., Ray, R.K., 2019. Groundwater resources of India: potential, challenges and management. In: Sikdar, P.K. (Ed.), Groundwater Development and Management. Springer, Cham, pp. 19–42.

Sailo, L., Mahanta, C., 2014. Arsenic mobilization in the Brahmaputra plains of Assam: groundwater and sedimentary controls. Environmental Monitoring and Assessment 186 (10), 6805–6820.

Sanz, E., Munoz-Olivas, R., Camara, C., Sengupta, M.K., Ahamed, S., 2007. Arsenic speciation in rice, straw, soil, hair and nails samples from the arsenic-affected areas of Middle and Lower Ganga plain. Journal of Environmental Science and Health 42 (12), 1695–1705.

Sarkar, D., Datta, R., Hannigan, R., 2011. Concepts and Applications in Environmental Geochemistry. Elsevier, Amsterdam.

Sasaki, A., Oshima, Y., Fujimura, A., 2007. An approach to elucidate potential mechanism of renal toxicity of arsenic trioxide. Experimental Hematology 35, 252–262.

Saucedo-Velez, A.A., Hinojosa-Reyes, L., Villanueva-Rodríguez, M., Caballero-Quintero, A., Hernández-Ramírez, A., Guzmán-Mar, J.L., 2017. Speciation analysis of organoarsenic compounds in livestock feed by microwave-assisted extraction and high performance liquid chromatography coupled to atomic fluorescence spectrometry. Food Chemistry 1 (232), 493–500. Available from: https://doi.org/10.1016/j.foodchem.2017.04.012.

Shah, B.A., 2017. Groundwater arsenic contamination from parts of the Ghaghara Basin, India: influence of fluvial geomorphology and Quaternary morphostratigraphy. Applied Water Science 7, 2587–2595.

Shaji, E., Santosh, M., Sarath, K.V., Prakash, P., Deepchand, V., Divya, B.V., 2021. Arsenic contamination of groundwater: A global synopsis with focus on the Indian Peninsula. Geoscience Frontiers 12. Available from: https://doi.org/10.1016/j.gsf.2020.08.015.

Sharma, V.K., Sohn, M., 2009. Aquatic arsenic: toxicity, speciation, transformations, and remediation. Environment International 35, 743–759.

Shen, H., Xu, W., Zhang, J., Chen, M., Martin, F.L., Xia, Y., et al., 2013. Urinary metabolic biomarkers link oxidative stress indicators associated with general arsenic exposure to male infertility in a Han Chinese population. Environmental Science & Technology 47, 8843–8851.

Siegel, R., Naishadham, D., Jemal, A., 2013. Cancer statistics, 2013. CA: A Cancer Journal for Clinicians 63, 11–30.

Simeonova, P.P., Luster, M.I., 2004. Arsenic and atherosclerosis. Toxicology and Applied Pharmacology 198, 444–449.

Singh, A., 2004. Arsenic contamination in groundwater of North Eastern India. In: Proceedings of Eleventh National Symposium on Hydrology With Focal Theme on Water Quality, National Institute of Hydrology, Roorkee, pp. 255–262.

Singh, S., Ghosh, A., Kumar, A., Kislay, K., Kumar, C., Tiwari, R., et al., 2014. Groundwater arsenic contamination and associated health risks in Bihar, India. International Journal of Environmental Research 8, 49–60.

Singh, M.K., Yadav, S.S., Gupta, V., Khattri, S., 2013. Immunomodulatory role of Emblica officinalis in arsenic induced oxidative damage and apoptosis in thymocytes of mice. BMC Complementary and Alternative Medicine 13, 193.

Smedley, P.L., Kinniburgh, D.G., 2013. Arsenic in groundwater and the environment. In: Selinus, O. (Ed.), Essentials of Medical Geology. Springer, Dordrecht. Available from: https://doi.org/10.1007/978-94-007-4375-5_12.

Smith, E., Naidu, R., 2004. Distribution and nature of arsenic along former railway corridors of South Australia. The Science of the Total Environment 363 (1–3), 175–182.

Smith, E., Naidu, R., Alston, A.M., 1998. Arsenic in the soil environment: a review. Advances in Agronomy 64, 149–195.

Soto-Pena, G.A., Vega, L., 2008. Arsenic interferes with the signaling transduction pathway of T cell receptor activation by increasing basal and induced phosphorylation of Lck and Fyn in spleen cells. Toxicology and Applied Pharmacology 230, 216–226.

Stahlberg, H., Braun, T., de Groot, B., Philippsen, A., Borgnia, M.J., Agre, P., et al., 2000. The 6.9-Å structure of GlpF: a basis for homology modeling of the glycerol channel from Escherichia coli. Journal of Structural Biology 132, 133–141.

Stopelli, E., Duyen, V.T., Mai, T.T., Trang, P.T., Viet, P.H., Lightfoot, A., et al., 2020. Spatial and temporal evolution of groundwater arsenic contamination in the Red River delta, Vietnam: interplay of mobilisation and retardation processes. The Science of the Total Environment 717, 137–143.

Suhag, R., 2019. Overview of ground water in India. PRS Legislative Research, 1–11.

Sun, G., Yu, G., Zhao, L., Li, X., Xu, Y., Li, B., et al., 2019. Endemic arsenic poisoning. In: Sun, D. (Ed.), Endemic Disease in China. *Public Health in China*, Vol 2. Springer, Singapore. Available from: https://doi.org/10.1007/978-981-13-2529-8_4.

Surdu, S., 2014. Non-melanoma skin cancer: occupational risk from UV light and arsenic exposure. Reviews on Environmental Health 29, 255–265.

Taleei, M.M., Karbalaei Ghomi, N., Jozi, S.A., 2019. Arsenic Removal of Contaminated Soils by Phytoremediation of Vetiver Grass, Chara Algae and Water Hyacinth. Bulletin of Environmental Contamination and Toxicology 102, 134–139. Available from: https://doi.org/10.1007/s00128-018-2495-1.

Tang, R., Wu, G., Yue, Z., Wang, W., Zhan, X., Zhen-Hu, H., 2020. Anaerobic biotransformation of roxarsone regulated by sulfate:Degradation, arsenic accumulation and volatilization. Enviro pollut (Barking, Essex: 1987) 267, 115–602. Available from: https://doi.org/10.1016/j.envpol.2020.115602.

Thakur, J.K., Thakur, R.K., Ramanathan, A.L., Kumar, M., Singh, S.K., 2011. Arsenic contamination of groundwater in Nepal—an overview. Water 3 (1), 1–20.

Thakur, L.S., Varma, A.K., Goyal, H., Sircar, D., Mondal, P., 2021. Simultaneous removal of arsenic, fluoride, and manganese from synthetic wastewater by Vetiveria zizanioides. Environmental Science and Pollution Research International 28, 44216–44225. Available from: https://doi.org/10.1007/s11356-021-13898-3.

Tiankao, W., Chotpantarat, S., 2018. Risk assessment of arsenic from contaminated soils to shallow groundwater in Ong Phra Sub-District, SuphanBuri Province, Thailand. Journal of Hydrology: Regional Studies 19, 80–96.

Titah, H.S., Halmi, M.I.E.B., Abdullah, S.R.S., Hasan, H.A., Idris, M., Anuar, N., 2018. Statistical optimization of the phytoremediation of arsenic by Ludwigia octovalvis- in a pilot reed bed using response surface methodology (RSM) vs an artificial neural network (ANN). International Journal of Phytoremediation 20, 721–729. Available from: https://doi.org/10.1080/15226514.2017.1413337.

Tripathi, R.D., Srivastava, S., Mishra, S., Singh, N., Tuli, R., Gupta, D.K., et al., 2007. Arsenic hazards: strategies for tolerance and remediation by plants. Trends in Biotechnology 25, 158–165. Available from: https://doi.org/10.1016/j.tibtech.2007.02.003.

Tripti, K., Shardendu, S., 2021. Efficiency of arsenic remediation from growth medium through Bacillus Licheniformis modulated by phosphate $(PO_4)_3$- and nitrate (NO_3)- enrichment. Archives of Microbiology 203 (7), 4081–4089.

Vahidnia, A., Van der Voet, G., De Wolff, F., 2007. Arsenic neurotoxicity—a review. Human & Experimental Toxicology 26, 823–832.

Vahter, M., 2008. Health effects of early life exposure to arsenic. Basic & Clinical Pharmacology & Toxicology 102, 204–211.

Vamerali, T., Bandiera, M., Mosca, G., 2010. Field crops for phytoremediation of metal-contaminated land. A review. Environmental Chemistry Letters 8, 1–17. Available from: https://doi.org/10.1007/s10311-009-0268-0.

Van Geen, A., Ahmed, E.B., Pitcher, L., Mey, J.L., Ahsan, H., Graziano, J.H., et al., 2014. Comparison of two blanket surveys of arsenic in tubewells conducted 12 years apart in a 25 km^2 area of Bangladesh. The Science of the Total Environment 488, 484–492.

Vantroyen, B., Heilier, J.F., Meulemans, A., Michels, A., Buchet, J.P., Vanderschueren, S., et al., 2004. Survival after a lethal dose of arsenic trioxide. Clinical Toxicology 42, 889–895.

Vigeh, M., Yokoyama, K., Matsukawa, T., Shinohara, A., Ohtani, K., 2014. The relation of maternal blood arsenic to anemia during pregnancy. Women & Health 55, 42–57.

Vivona, R., Preziosi, E., Madé, B., Giuliano, G., 2007. Occurrence of minor toxic elements in volcanic-sedimentary aquifers: a case study in central Italy. Hydrogeology Journal 15 (6), 1183–1196.

von Brömssen, M., (1999). Genesis of High Arsenic Groundwater in the Bengal Delta Plains, West Bengal and Bangladesh. (MSc. thesis), AMOV-EX-1999-18, Stockholm, Sweden: Royal Institute of Technology.

Wang, W., Cheng, S., Zhang, D., 2014a. Association of inorganic arsenic exposure with liver cancer mortality: a meta-analysis. Environmental Research 135, 120–125.

Wang, W., Xie, Z., Lin, Y., Zhang, D., 2014b. Association of inorganic arsenic exposure with type 2 diabetes mellitus: a meta-analysis. Journal of Epidemiology and Community Health 68, 176–184.

Webber, Z.R., Webber, K.G.I., Rock, T., St Clair, I., Thompson, C., Groenwald, S., et al., 2021. Diné citizen science: Phytoremediation of uranium and arsenic in the Navajo Nation. The Science of the Total Environment 794, 148665. Available from: https://doi.org/10.1016/j.scitotenv.2021.148665.

WHO, 2003. Environmental Health Criteria-224, Arsenic and Arsenic Compounds, second ed. World Health Organization, Geneva.

WHO, 2011. Arsenic in Drinking-water. Background Document for Preparation of WHO Guidelines for Drinking-water Quality. World Health Organization, Geneva.

WHO, 2015. World Health Organisation. International Programme on Chemical Safety. Environmental health criteria 224 Arsenic and arsenic compounds report [online]. Geneva. <http://espace.library.uq.edu.au/view/UQ:40271> (Accessed 01.08.15).

Winkel, L., Berg, M., Amini, M., Hug, S.J., Johnson, C.A., 2008. Predicting groundwater arsenic contamination in Southeast Asia from surface parameters. Nat. Geosci. 1 (8), 536–542.

Wu, J., Chen, G., Liao, Y., Song, X., Pei, L., Wang, J., et al., 2011. Arsenic levels in the soil and risk of birth defects: a population-based case–control study using GIS technology. Journal of Environmental Health 74, 20–25.

Wu, Y.W., Zhou, X.Y., Mi, T.G., Xu, M.X., Zhao, L., Lu, Q., 2021. Theoretical insight into the interaction mechanism between V2O5/TiO2 (001) surface and arsenic oxides in flue gas. Applied Surface Science 535, 147752.

Xie, X., Hu, Y., Cheng, H., 2016. Rapid degradation of p-arsanilic acid with simultaneous arsenic removal from aqueous solution using Fenton process. Water Research 89, 59–67.

Yang, T., Wu, S., Liu, C., Liu, Y., Zhang, H., Cheng, H., Wang, L., Guo, L., Li, Y., Liu, M., Ma, J., 2021. Efficient Degradation of Organoarsenic by UV/Chlorine Treatment: Kinetics, Mechanism, Enhanced Arsenic Removal, and Cytotoxicity. Environmental Science & Technology 55 (3), 2037–2047.

Yang, M.H., Zang, Y.S., Huang, H., Chen, K., Li, B., Sun, G.Y., et al., 2014. Arsenic trioxide exerts anti-lung cancer activity by inhibiting angiogenesis. Current Cancer Drug Targets 14 (6), 557–566.

Yin, Y., Wan, J., Li, S., Li, H., Dagot, C., Wang, Y., 2018. Transformation of roxarsone in the anoxic-oxic process when treating the livestock wastewater. Science of the Total Environment 616–617, 1235–1241.

Ying, S.C., Kocar, B.D., Griffis, S.D., Fendorf, S., 2011. Competitive Microbially and Mn Oxide Mediated Redox Processes Controlling Arsenic Speciation and Partitioning. Environmental Science & Technology 45, 5572–5579.

Yu, R.C., Hsu, K.-H., Chen, C.-J., Froines, J.R., 2000. Arsenic methylation capacity and skin cancer. Cancer Epidemiology, Biomarkers & Prevention: a Publication of the American Association for Cancer Research, Cosponsored by the American Society of Preventive Oncology 9, 1259–1262.

Zhao, F.J., Ma, J.F., Meharg, A.A., McGrath, S.P., 2009. Arsenic uptake and metabolism in plants. The New Phytologist 181, 777–794. Available from: https://doi.org/10.1111/j.1469-8137.2008.02716.x.

Zuzolo, D., Cicchella, D., Demetriades, A., Birke, M., Albanese, S., Dinelli, E., et al., 2020. Arsenic: geochemical distribution and age-related health risk in Italy. Environmental Research 182.

CHAPTER

21

Time series analysis of dissolved trace elements in Gomati River, the Ganga River tributary, northern India: its environmental implications

Dharmendra Kumar Jigyasu[1,2], *Priyanka Singh*[3], *Munendra Singh*[1] *and Sandeep Singh*[4]

[1]Department of Geology, University of Lucknow, Lucknow, India [2]Central Muga Eri Research and Training Institute, Jorhat, India [3]Department of Geology, Babasaheb Bhimrao Ambedkar University, Lucknow, India [4]Department of Earth Sciences, Indian Institute of Technology, Roorkee, India

21.1 Introduction

Time series analysis refers to a systematic approach that answers scientific questions posed by time correlations. It is crucial in understanding various geo-environmental processes and in elucidating the controlling mechanisms of their variability caused by biological, physical, or environmental phenomena of interest (Shumway and Stoffer, 2011). In a river system, the dissolved phase of trace elements could be derived either from natural chemical weathering processes or anthropogenic sources. Trustworthy detection and quantification of the amount of dissolved trace elements are desired for measureing oceanic chemical mass balances, considerating continental erosion and weathering, river geochemistry, and estimating human-induced chemical perturbations (Shiller et al., 2001).

The Gomati River Basin is an alluvial river basin located in the interfluve area of the Ganga and Ghaghara Rivers within the Ganga Alluvial Plain, northern India (Fig. 21.1). It is forming one of the densely populated, the most intensely irrigated, and the extremely water deficient areas of the globe. Trace elemental composition and its variability in the

Metals in Water
DOI: https://doi.org/10.1016/B978-0-323-95919-3.00008-2

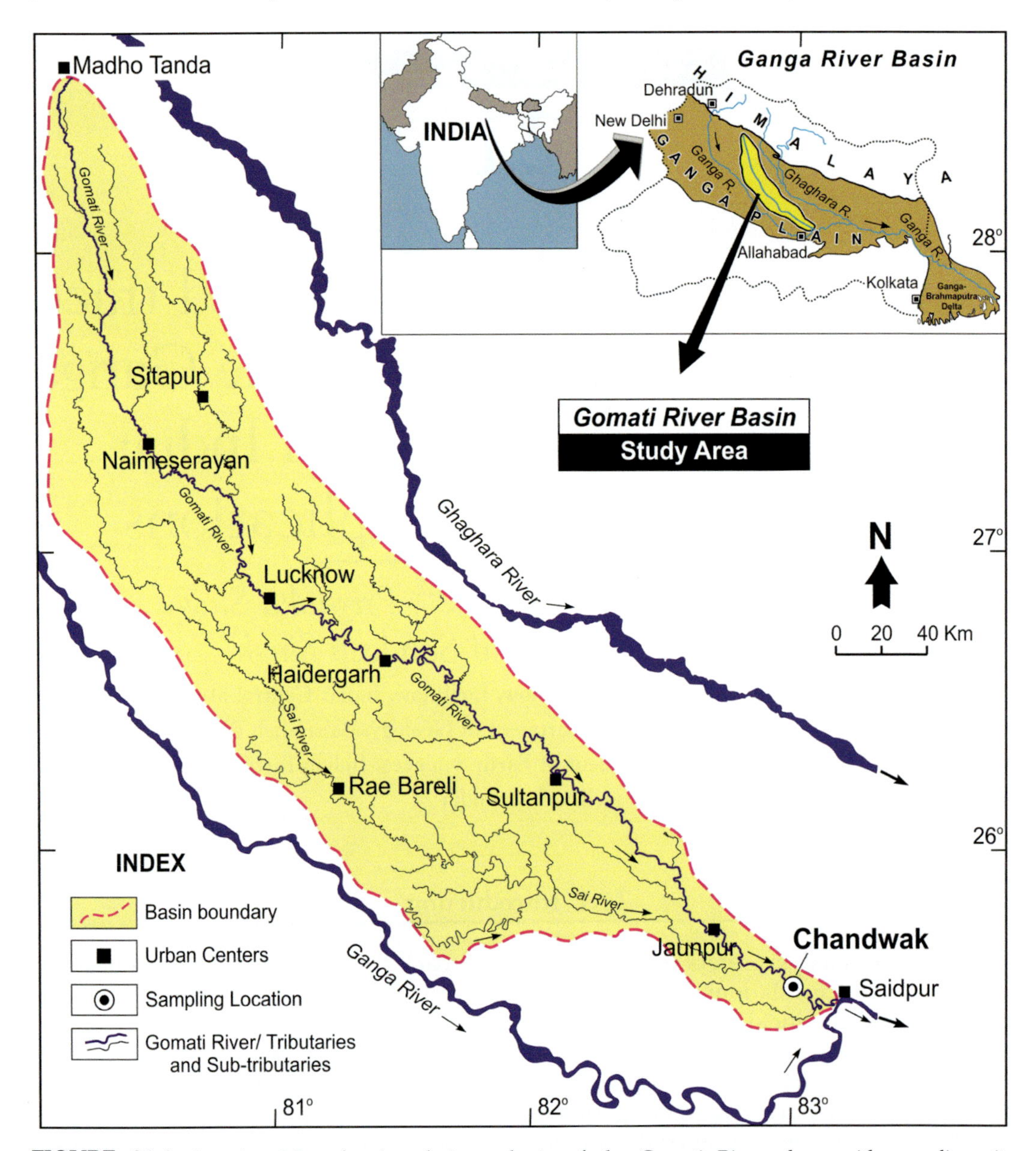

FIGURE 21.1 Location Map showing drainage basin of the Gomati River along with sampling site "Chandwak" selected for the present study.

Gomati River water are of interest for the understanding of fluvial geochemistry of alluvial river or as a source of dissolved load to the Ganga River, or for the quantification of chemical weathering progressions of the Ganga Alluvial Plain and an evaluation of anthropogenic perturbations of river's environment or their interactions.

The present study comprises time series data for 12-month's on 16 trace elements to light the preliminary indication for the environment of unevenness in dissolved trace elements quantity in the Gomati River. The study will make us understand the fluvial geochemistry of the Gomati River basin within the Ganga Alluvial Plain, acting as a source of dissolved load to the main Ganga river. This will also evaluate the anthropogenic contribution of the trace elements.

21.2 Study area: the Gomati River and its basin

The physiographic, climatic, geology, sediments uniqueness of the Gomati River and its drainage pattern watershed are of utmost significance to understand the mobilization and sources of trace elements in the Gomati River. In northern India, the 900-km long Gomati River is a offshoot of the Ganga River. Its drainage watershed is placed within the Ganga Alluvial Plain and drains 30,437 km^2 of the interfluve area of the Ganga and Ghaghara Rivers. In the Gomati River Basin, the primary land use pattern is covered by crops because the basin is made up of highly fertile alluvium and comprises about 67% of total geographical land, 13% is the settlement, 14% is barren land, 4% is forest cover, and about 2% is covered by water bodies (Dutta et al., 2007). Table 21.1 sum up the essential individuality of the Gomati tributary Basin. The regional upland surface of the river watershed has discrete rising and falling topography differentiated by low relief beside to well-established stream vally, ponds, lakes, meandering scars and alluvial edges. The maximum as well as minimum elevations of the river basin are near 186 m, and 61 metres above mean sea level. The main tributary of the Gomati River is the Sai stream which depletes almost 1/3rd of the watershed area. All existing sub-basins of the Gomati River are drained in 3rd to 5th array of alluvial waterways with drainage density from 0.44 to 1.04 km/km^2 and basin relief ranging from 10 to 44 m (Thakur, 2008).

21.2.1 Climate

The Gomati River watershed posses a humid subtropical climatic conditions with four different seasons; the wet/monsoon season (June–September), the post-monsoon season (October–November), the winter season (December–February), and the dry/summer season (March–May). The temperature difference is intense, arraign from highest (46.6°C) in the summer season to the lowest (3.2°C) in the winter season. The monsoon season is differentiate by elevated precipitation due to the SW Monsoon arrangement, and the basin receives about 84% of total annual rainfall. In this season, precipitation interrelate with the alluvium of the Ganga Alluvial Plain. It seeps into the aquifer to turn into the resource of all rivers of the Ganga Alluvial Plain. The post-monsoon season is characterized by moderate temperature and rainfall, where temperature varies from 7.6°C to 38.6°C with about 9.12% of total rainfall.

The winter season expands from December to February and is distinguished by lowest temperature and less precipitation and temperature ranging from 27.6°C to 32°C (maximum temperature) and 3.2°C to 4.4°C (maximum temperature). The rain occurs due to western disturbances and sometimes higher around 20 mm and contributes 3% of total

TABLE 21.1 General characteristics of the Gomati River draining central part of the Ganga Alluvial Plain, northern India.

Gomati River Basin:	
Human population – ~40 million (in 2010); Drainage area- 30,437 km^2; Climate Type-Humid sub-tropical climate; Annual rainfall range – 81–125 cm; Maximum and minimum elevation – 186 and 61 m (above mean sea level; Maximum and minimum temperature – 47°C and 2°C; Relief – 25 m (maximum); Sub basin – Sai River Basin; Number of micro-basins – 33; Land use pattern – agriculture (67%), settlement (13%), barren land (14%), forest (4%), water bodies (2%); Groundwater pH – 7.6; Groundwater chemistry – Ca-Mg-HCO_3 geochemical facies	
River water	
pH	8.0–8.9
Conductivity	320–350 μS/cm
Total suspended matter	18–26 mg/L
HCO_3^-	300 mg/L
Temperature	16°C in winter, 35°C in summer
Hydro-facies	Ca^{2+} – Mg^{2+} – dominant HCO_3^-
Dissolved oxygen (in urban stretch)	1–3 mg/L
Dissolved oxygen (in rural stretch)	5–15 mg/L
River sediments	
Mean grainsize (Bedload sediments)	Very fine sand
Mean grainsize (Suspended load sediments)	Medium silt
Minerals composition	(Bedload sediments) Quartz, muscovite, biotite, rock fragments, K-feldspar, plagioclase
Common clay minerals	Illite and montmorillonite
River Hydrology	
Low flow (during summer season)	2–5 m^3/s
High flow (during monsoon season)	500–900 m^3/s
Total annual discharge (Maighat, Chadwak)	5.85×10^9 m^3 (June, 2009–May, 2010)
Seasonal contributions	Monsoon (48%), Post-monsoon (30%), Winter (16%), Summer (6%)

Data from various sources.

rainfall. The summer season is dry and hot, widen from March to May and described by elevated temperatures as the northwestern winds dominate and dust storms. The maximum temperature reaches up to 40°C to 44.7°C. The rainfall variation is very low, sometimes nil. The summer season contributes nearly 3.8% of total rainfall.

21.2.2 Sub-surface geology

The Gomati River watershed is composed of unconsolidated sediments draw from the Himalayas region as the products of weathering and erosion and deposited as alluvial sediments. These alluvial deposits are multiple layers of 1–2 m wide fine sand and silty–mud deposits with widespread irregular calcrete horizons. The mineralogical characteristics of alluvial sediment of the Gomati River Basin are largely create of mica and quartz, feldspar, and plagioclase minerals, which is similar to alluvial sediments of the Ganga Plain (Pal et al., 2012). Singh (1996) recognized mainly two types of lithofacies which are muddy set downs and sandy set downs. Muddy interfluve set downs are composed of up of 0.2–1.0 m broad fine silt with widespread calcrete development and 1.0–2.0 m broad greatly mottled fine sand set downs with 5–10 cm broad bedded calcrete and shell-containing mud. Sandy set downs are consist of 0.5–2.0 m wide lenticular sand cadavers stand for meandering river set downs and 1.0–2.0 m wide well-sorted silty fine sand stand for flood deposits like a sheet with 10–50 cm wide irregular horizons of calcrete (Singh, 1996).

21.2.3 River hydrology

The Gomati River is a aquifer-fed stream of the Gangetic basin. The river has a slothful current all through the time, excluding in the wet season. Its hydrology is characterized and proscribed by the strength and period of rainfall in the wet season. The strength of precipitation and the time of the wet season cause 50–100 fold augment in the river's overflow. Thus, the Gomati River obtains its maximum water account during the wet season. According to Rao (1975), the river hydrograph of the Gomati River is recurrently forbidden and pointed with an yearly discharge of about $7390 \times 10^6\ m^3/year$.

21.2.4 River sediments

According to physical and chemical weathering processes operating under monsoon-controlled climatic conditions, the Gomati River derives its sediment and dissolved load from the Ganga Alluvial Plain. The river sediments properties are robustly restricted by the geology of alluvium in the Gomati River Basin. The Gomati River Sediments are mainly very fine sand (57%), silt, and clay. The mean grain size of the suspended load sediments was reported from coarse silt to very fine silt. Mineralogically, the sand fraction comprises quartz, muscovite, biotite, chlorite, rock fragments, k-feldspar, and plagioclase. The average composition of the sand fraction is quartz (55%), rock fragments (19%), mica (17%), and feldspars (9%) (Kumar and Singh, 1978). The silt fraction of Gomati River sediments consists of quartz, plagioclase, alkali feldspar, and mica. Clay comprises Kaolinite, Chlorite, Montmorillonite, and Illite with minor amounts of non-clay fractions like quartz and feldspar. Illite is the main content in clay and possesses good crystallinity in the Gomati River sediments (Kumar and Singh, 1978). The Gomati River annually transports total suspended materials 3.4×10^5 tonnes and total dissolved solids 3.0×10^6 tones to the Ganga River (Gupta and Subramanian, 1994).

21.3 Material and method

The Gomati River receives its water from several tributaries after draining nearly 30,000 km^2 area within the Ganga Alluvial Plain and transports its dissolved load to the Ganga River. Chandwak (25° 35′ N and 80° 00′ E) being at the distal part of the Gomati River Basin is the most important location for sampling all tributaries of the Gomati River before joining the Ganga River. Therefore, 36 water samples from the middle of the active river channel, approximately every 10-days, were collected at Chandwak from June 2008 to May 2009, covering the Monsson, the Post-monsoon, the Winter, and the Summer seasons. Throughout the sampling phase, the current of the Gomati River was illustrated by greatest discharge (700 m^3/s) in the Monsoon season and by minimum discharge (65 m^3/s) in the Summer season.

All river water samples were obtained using acid-cleaned wide mouth 250 mL polyethylene containers (©Tarsons) with sealed caps and acidified on the sampling point with HNO_3 (5 mL/L). Replicate samples were obtained when possible. Every water container was labeled with the suitable tag, cautiously carried to the laboratory without cross-contamination, and stored at 4°C in the shadowy awaiting more chemical investigation. The collected water samples were filtered by using 0.45 μm cellulose filters. They were investigated for trace elements Al, Ti, V, Cr, Mn, Fe, Co, Ni, Cu, Zn, As, Se, Rb, Zr, Mo, Cd and Pb in the laboratory by inductively coupled plasma-mass spectrometry (ICP-Ms). Every sample was measured in replica, and mean values were consider as the outcome. All the samples were determined in the laboratory go after the standard protocols (APHA, 2002).

For the excellence declaration of the ICP-Ms analysis, replicates and blank analytical solutions were formed and examined to ensure the consistency of the data. The reliability of the procedures of estimating the trace elements in water has been guaranteed by numerous actions performing such as careful standardization, appling of blanks, standardization in reproduce examination, re-examine of chosen samples and for the chemical analysis, double distilled water and chemicals were used of analytical grade. All chemicals of analytical rating and Milli Q water were utilized during the elemental analysis. The precision of ICP-Ms used at IIT Roorkee is better than 5%.

21.4 Results and discussion

21.4.1 Distribution of dissolved trace elements concentrations

The results of the trace elements analysis in the Gomati River Water are accessible in Table 21.2. The annual variation of dissolved trace elements is Al (273–77, 861 μg/L), Ti (4–996 μg/L), V (3–50 μg/L), Cr (202–27,427 μg/L), Mn (4–1097 μg/L), Fe (202–27,427 μg/L), Co (0.2–23 μg/L), Ni (2–57 μg/L), Cu (2–54 μg/L), Zn (1–125 μg/L), As (0.5–5.7 μg/L), Se (72–158 μg/L), Rb (2–128 μg/L), Zr (0.01–0.09 μg/L), Mo (0.3–4.6 μg/L), Cd (0.07–7.36 μg/L) and Pb (0.1–41 μg/L). The average concentration of dissolved trace elements is Al (5288 μg/L), Ti (60 μg/L), V (8 μg/L), Cr (5 μg/L), Mn (115 μg/L), Fe (2080 μg/L), Co (2 μg/L), Ni (7 μg/L), Cu (7 μg/L), Zn (17 μg/L), As (2.3 μg/L), Se (100 μg/L), Rb (11.6 μg/L), Zr (0.03 μg/L), Mo (1.7 μg/L), Cd (0.96 μg/L) and Pb (4.3 μg/L). Fig. 21.2 displays a Box and whisker diagram

TABLE 21.2 Dissolved metals concentration (in μg/L) in the Gomati River Water at Chandwak.

Sample code	Sampling date (dd/mm/yy)	Season	Al	Ti	V	Cr	Mn	Fe	Co	Ni	Cu	Zn	As	Se	Rb	Zr	Mo	Cd	Pb
GRW-01	05/06/09	Monsoon	1588	8	8	2.0	85	723	1.0	5	5	15	4.3	122	6.2	0.07	2.0	3.20	3.4
GRW-02	14/06/09	Monsoon	2757	10	8	2.9	115	1252	1.5	5	5	11	2.4	111	7.0	0.03	1.3	2.54	3.7
GRW-03	25/06/09	Monsoon	4185	75	10	4.0	97	1930	1.7	7	7	19	2.4	96	11.7	0.01	1.4	1.44	3.5
GRW-04	05/07/09	Monsoon	1751	11	9	2.2	69	851	1.0	9	6	10	3.8	103	6.0	0.01	1.7	1.70	3.0
GRW-05	15/07/09	Monsoon	4225	55	10	3.9	87	1777	1.7	6	7	14	3.3	110	10.9	0.07	1.8	2.88	3.7
GRW-06	25/07/09	Monsoon	77,861	996	50	49.7	1097	27,427	23.3	57	54	125	1.3	145	127.9	0.01	0.7	7.36	41.0
GRW-07	05/08/09	Monsoon	22,677	147	19	16.4	365	8072	7.2	19	20	59	1.0	93	38.9	0.03	0.6	0.98	14.4
GRW-08	16/08/09	Monsoon	6396	73	12	5.7	108	2582	2.2	7	10	32	2.1	107	13.4	0.09	1.5	0.89	4.7
GRW-09	05/08/09	Monsoon	13,100	161	15	9.9	236	4974	4.5	12	13	29	1.3	94	25.7	0.05	0.9	0.81	8.4
GRW-10	05/09/09	Monsoon	375	5	7	6.8	4	202	0.2	2	2	1	2.4	78	1.8	0.05	4.6	0.07	0.1
GRW-11	16/09/09	Monsoon	6413	4	5	6.8	388	1692	4.4	9	13	28	0.5	90	4.9	0.08	0.3	1.18	13.8
GRW-12	25/09/09	Monsoon	7637	58	11	6.7	211	3389	3.5	9	11	25	1.0	78	15.8	0.02	0.5	0.56	8.0
GRW-13	05/10/09	Post-Monsoon	5087	73	9	4.7	119	2312	2.1	8	9	21	1.3	83	12.1	0.05	0.7	0.33	4.4
GRW-14	17/10/09	Post-Monsoon	9530	81	13	8.1	270	4143	4.5	11	13	26	1.0	85	19.2	0.09	0.4	0.64	9.9
GRW-15	25/10/09	Post-Monsoon	7574	127	9	6.6	168	3381	2.9	8	11	20	0.8	99	17.5	0.09	0.4	0.42	6.2
GRW-16	05/11/09	Post-Monsoon	3095	56	6	3.2	66	1480	1.2	5	7	11	1.1	92	7.6	0.08	1.0	0.29	2.3
GRW-17	15/11/09	Post-Monsoon	1182	13	6	1.7	42	635	0.6	3	5	7	1.6	85	3.3	0.07	1.4	0.26	1.8
GRW-18	25/11/09	Post-Monsoon	1376	23	6	2.2	44	758	0.7	4	5	7	1.9	87	4.6	0.07	1.9	0.35	1.4
GRW-19	05/12/09	Winter	1453	20	5	1.9	38	741	0.6	4	4	6	1.6	90	4.4	0.02	1.8	0.25	1.1
GRW-20	15/12/09	Winter	1460	18	5	2.0	41	734	0.7	3	3	5	1.2	88	5.7	0.01	1.4	0.34	1.7
GRW-21	25/12/09	Winter	1184	15	4	1.7	26	555	0.5	3	4	20	1.4	94	3.9	0.01	1.6	0.41	1.1
GRW-22	05/01/10	Winter	1262	19	3	1.8	23	592	0.5	6	4	7	1.3	91	4.6	0.01	1.7	0.30	1.2
GRW-23	15/01/10	Winter	862	12	3	1.4	17	434	0.3	3	3	4	1.3	80	4.2	0.01	1.9	0.27	0.8

(*Continued*)

TABLE 21.2 (Continued)

Sample code	Sampling date (dd/mm/yy)	Season	Al	Ti	V	Cr	Mn	Fe	Co	Ni	Cu	Zn	As	Se	Rb	Zr	Mo	Cd	Pb
GRW-24	25/01/10	Winter	787	11	3	1.4	16	413	0.3	3	4	5	1.2	72	4.0	0.01	1.8	0.32	1.2
GRW-25	05/02/10	Winter	558	9	3	1.2	17	329	0.3	2	3	3	1.3	75	3.0	0.01	1.8	0.41	0.8
GRW-26	15/02/10	Winter	488	7	3	1.2	21	310	0.3	3	3	3	1.9	104	3.2	0.01	2.2	0.38	0.6
GRW-27	25/02/10	Winter	461	5	3	1.2	16	269	0.3	3	5	25	1.6	98	3.3	0.01	1.8	0.33	0.9
GRW-28	05/03/10	Summer	365	5	4	1.1	16	253	0.3	3	3	10	2.2	97	3.3	0.01	2.2	0.30	0.7
GRW-29	15/03/10	Summer	273	5	4	1.0	19	231	0.3	3	3	5	2.6	119	3.5	0.01	2.4	0.36	0.6
GRW-30	25/03/10	Summer	437	6	5	1.3	45	285	0.3	3	3	7	2.7	88	4.0	0.01	2.0	0.69	1.0
GRW-31	05/04/10	Summer	565	10	5	4.0	45	340	0.4	4	3	5	3.3	86	4.9	0.01	2.4	0.42	1.2
GRW-32	15/04/10	Summer	727	11	5	1.4	45	387	0.4	7	4	10	3.5	101	5.9	0.01	2.3	0.48	1.8
GRW-33	25/04/10	Summer	465	4	6	1.2	47	265	0.5	3	3	8	4.7	158	5.1	0.01	2.7	0.47	1.8
GRW-34	05/05/10	Summer	730	9	6	1.3	49	370	0.6	4	3	9	5.2	123	5.9	0.01	2.8	0.47	1.3
GRW-35	15/05/10	Summer	757	7	7	1.3	46	405	0.6	4	3	6	5.2	129	6.1	0.02	2.7	1.24	1.4
GRW-36	25/05/10	Summer	741	7	7	1.3	44	393	0.6	5	4	13	5.7	157	6.5	0.02	2.8	1.27	1.6

Refer for sampling location in Fig. 21.1. Metals are arranged in increasing order of their atomic number.

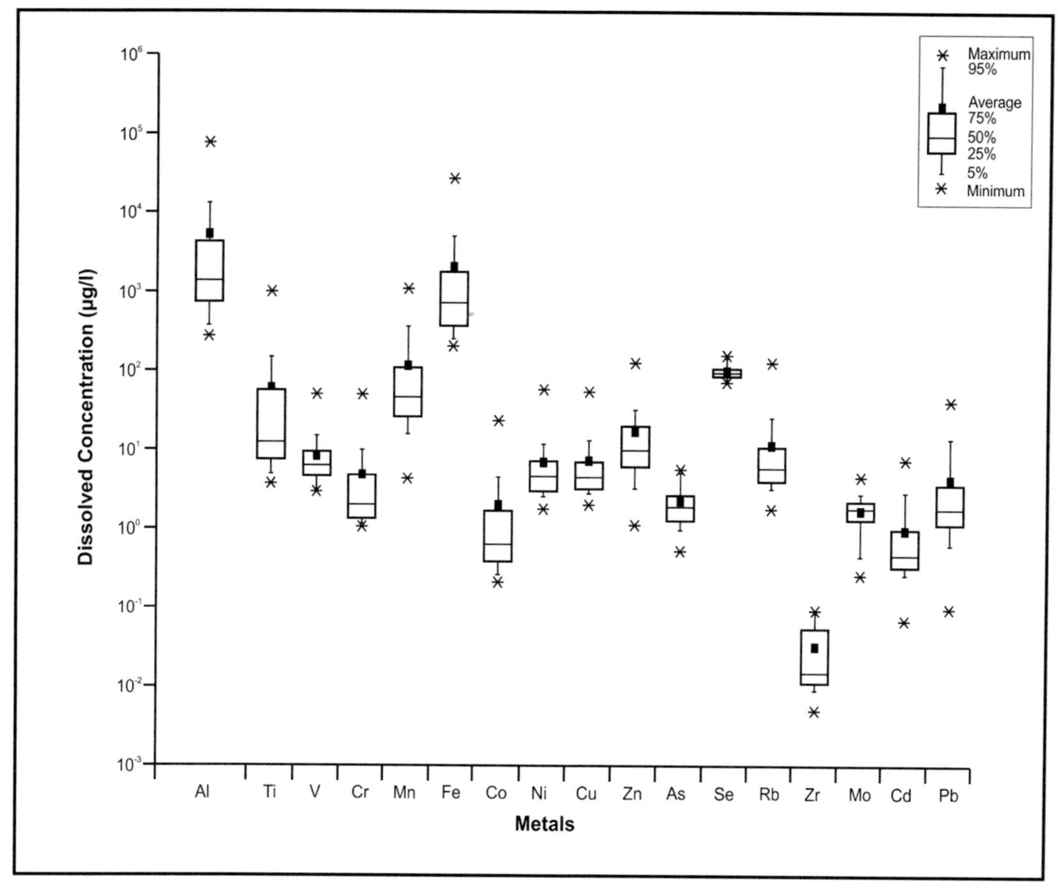

FIGURE 21.2 Box and Whisker diagram showing the distribution of dissolved concentration of metals in the Gomati River Water. Note the metal concentrations vary in the eight orders of magnitude from 0.005 µg/L for Zr to 7,7861 µg/L for Al.

showing variability of trace-elements concentration in the eight orders of magnitude in the Gomati River Water.

21.4.2 Seasonal variation in dissolved trace elements concentrations

The examination of Fig. 21.2 and Table 21.2 indicates a simple relation between variations in the dissolved concentrations of trace elements and the physical parameters. Most trace elements show a high degree of variability in the Monsoon season but relatively constant concentrations in other seasons. Dissolved concentrations of trace elements V, Cr, Pb, Cu, Mo and Cd show the prominent single peak during the monsoon season (Fig. 21.3). Most trace elements concentration in the Gomati River vary temporally, with generally increasing concentrations as discharge increases. This trend is due to the interaction of monsoon precipitation with alluvial sediments after the summer

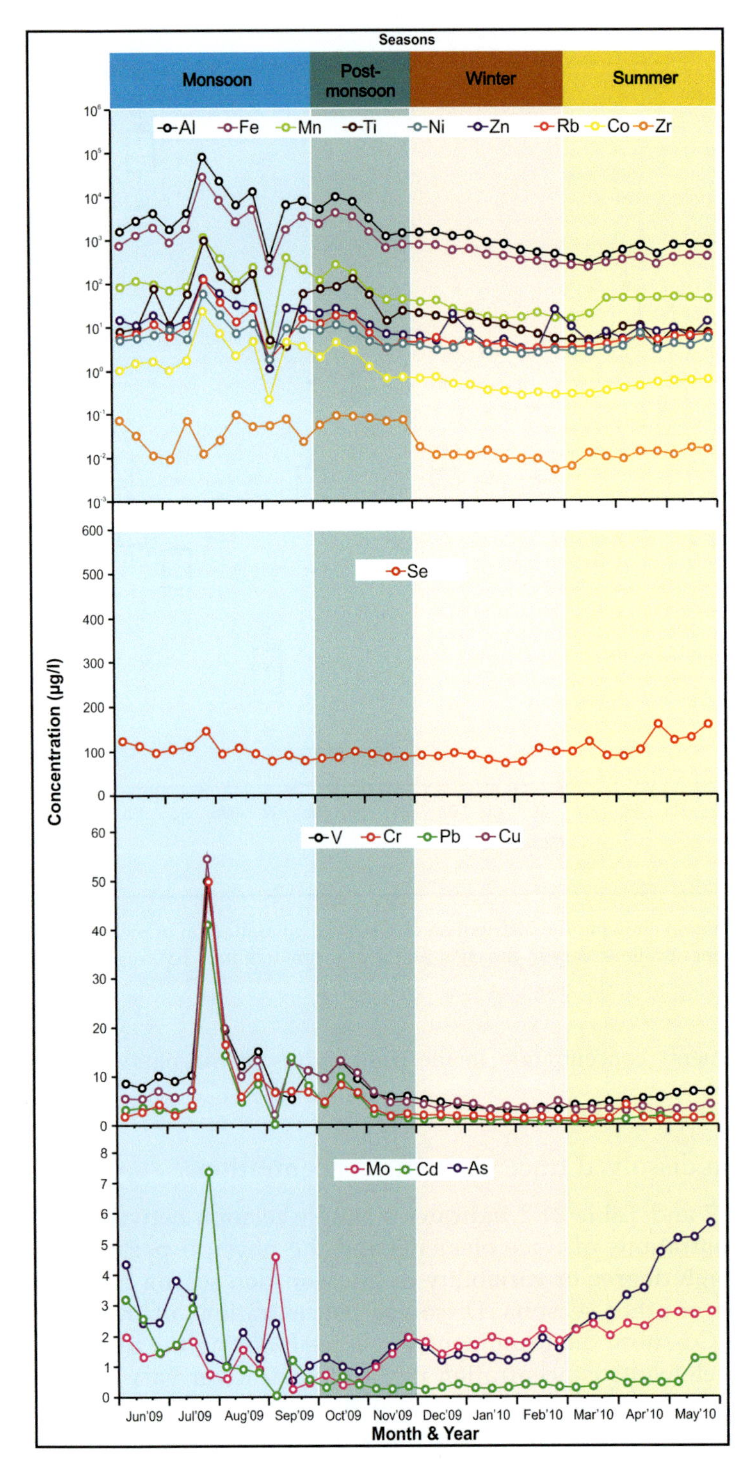

FIGURE 21.3 Seasonal variability of the dissolved metal concentration in the Gomati River. Note the metal concentrations of Al, Mn, Fe, Ti, Co and Zn vary in the three orders of magnitude during the monsoon season.

season. It is also possible that anthropogenic processes such as adding particulates and organic matter into the Gomati River might alter the distribution of natural dissolved trace elements.

21.4.3 Inter-elemental correlation

The correlation coefficient matrix of the dissolved trace elements in the Gomati River Water is presented in Table 21.3. The significant correlations are Al with Fe, Cr with Fe, Al with Ni, Ti with Fe, Al with Ti, Al with Rb, Rb with Fe, Co with Fe, Co with Fe in the Gomati River Water and suggest that these trace elements bear same characteristics like their geogenic sources and behavior. Aluminum shows the most significant correlation coefficient with Fe, Ni, Ti, and Rb elements, indicating the common mineral sources present in the Gomati River Sediments and sediments of the Ganga Alluvial Plain. Iron also shows a significant correlation with Cr, Ti, Rb, Co, and other trace elements indicating the common geogenic trace elements source in the mineral present in the river sediments.

TABLE 21.3 Correlation coefficient matrix of trace elements concentration in the Gomati River water (n = 36).

	Al	Ti	V	Cr	Mn	Fe	Co	Ni	Cu	Zn	As	Se	Rb	Zr	Mo	Cd	Pb
Al	1.00	**0.98**	**0.97**	**0.99**	**0.96**	**1.00**	**0.99**	**0.99**	**0.99**	**0.96**	−0.24	0.28	**1.00**	**0.01**	−0.39	0.80	**0.97**
Ti		1.00	**0.96**	**0.97**	**0.92**	**0.99**	**0.97**	**0.97**	**0.96**	**0.92**	−0.21	0.31	**0.99**	0.00	−0.33	0.81	**0.93**
V			1.00	**0.97**	**0.94**	**0.98**	**0.97**	**0.97**	**0.97**	**0.95**	−0.14	0.33	**0.98**	0.10	−0.37	0.82	**0.94**
Cr				1.00	**0.96**	**0.99**	**0.99**	**0.98**	**0.98**	**0.95**	−0.26	0.24	**0.99**	0.05	−0.35	0.78	**0.97**
Mn					1.00	**0.96**	**0.99**	**0.97**	**0.98**	**0.95**	−0.28	0.26	**0.95**	0.12	−0.50	0.78	**1.00**
Fe						1.00	**0.99**	**0.99**	**0.99**	**0.96**	−0.25	0.28	**1.00**	0.03	−0.40	0.80	**0.96**
Co							1.00	**0.99**	**1.00**	**0.96**	−0.27	0.27	**0.98**	0.07	−0.45	0.80	**0.99**
Ni								1.00	**0.99**	**0.96**	−0.20	0.31	**0.99**	0.02	−0.41	0.81	**0.97**
Cu									1.00	**0.98**	−0.30	0.24	**0.98**	0.12	−0.50	0.78	**0.99**
Zn										1.00	−0.26	0.28	**0.95**	0.07	−0.48	0.76	**0.96**
As											1.00	0.70	−0.20	−0.24	0.65	0.12	−0.29
Se												1.00	0.31	−0.14	0.25	0.50	0.25
Rb													1.00	0.00	−0.37	0.80	**0.95**
Zr														1.00	−0.35	0.03	0.12
Mo															1.00	−0.21	−0.51
Cd																1.00	0.78
Pb																	1.00

High correlation coefficient values are in bold. Transitional metals (Ti, V, Cr, Mn, Fe, Co, Ni and Cu) show the most significant inter-elemental correlation coefficient, indicating their common geogenic sources present in the Gomati River Sediments. [Significant at 90% probability].

21.4.4 Geogenic sources of trace elements

Common silicate minerals present in parent and weathered materials of the Ganga Alluvial Plain are quartz, muscovite, plagioclase, and biotite. Langmuir (1997) reported the lifetime span of common silicate minerals, indicating that biotite and Ca-plagioclase are the most weatherable silicate minerals. Kumar and Singh (1978) reported that the fine sand fraction of the bedload sediments of the Gomati River is composed of predominantly quartz and nearly 17% of mica minerals. Therefore, mica minerals can take par a significant role in recognizing the trace-elements distribution in river water of the Ganga Alluvial Plain.

Mica contain an octahedral (o-layer) sheet sandwich between two alike tetrahedral sheets (t-layer) constructing the feature t-o-t layer. The tetrahedral layers are engaged by Si, Fe, B, Be, and Al, whereas octahedral sheets are usually inhabited by Fe, Mg, Ti, Al, Mn, Cr, V, Mg, and other cations. The interlayer site is typically resided by K, Na, Rb, Ca, and Ba (Tischendorf et al., 2007). The phlogopite refers to Mg-rich biotite, and annite refers to Fe-rich biotite, whereas they are also known as brown and Black micas, respectively (Tischendorf et al., 2007). Biotite [$K(Mg, Fe)_3(AlSi_3O_{10})(OH)_2$] is illustrious by its alliance, character, and dark color. The two t-o-t layers of biotite are allied by K^+ ions, which occupy big sites amid them. The Fe occupies octahedral sites mixes generously with Mg, making a solid solution amid the annite and phlogopite.

Several trace elements can be incorporated in octahedral sites of the phyllo structure in mica minerals. Elements Cr, Mn, Cu, Fe, Se, Ti, Ni, Mo, Co, V, and As are incorporated with Al in VI-fold coordination. During the chemical weathering processes, trace elements exchange may be involved in releasing and depositing similar elements simultaneously from the t-o-t layer of mica minerals (Kabata-Pendias, 2011). Weathering of biotite may release significant concentrations of elements like Si, Fe, and Al from the tetrahedral sheet (-t layer), Ti, Al, Fe, Mn, Cr, V, and Mg from the octahedral sheet (-o layer) and K, Na, Rb, Ca, Ba from the interlayer (Tischendorf et al., 2007). The chemical change in biotite occurs as a concurrent process that encourages the replacement of K and the movement of cations from the octahedral layer into solution (Dong et al., 1998).

$$K_{0.65}\left(Al_{1.10}Ti_{0.15}Fe_{0.50}Mg_{0.55}\right)(Si_{3.20}Al_{0.80})O_{10}(OH)_2 + 8H^+ = (Al_{1.71}Fe_{0.31}Ti_{0.03})Si_2O_5(OH)_4$$
$$+ 0.65K^+ + 0.19Al^{3+} + 0.55Mg^{2+} + 0.19Fe^{2+} + 0.12Ti^{4+} + 1.2Si^{4+} + 3H_2O$$

Chemical weathering of mica minerals, especially biotite present in alluvial and river sediments, is involved in releasing trace elements in the Ganga Alluvial Plain (Singh et al., 2005; Jigyasu et al., 2014). The dissolved concentration of trace elements in the Gomati River of the Ganga Alluvial Plain may be influenced by the interference of anthropogenic activities.

21.5 Conclusions

The statistics stated here point out that quantity of dissolved metals in the Gomati River display high variability linked with seasonality. There are methodical connections amid dissolved metals and the rising stage of the river's discharge during the monsoon season. The variability of the dissolved amount of trace elements in the Gomati River indicates that a single sample analysis may not represent the magnitude of trace elements concentration. Biotite act as a significant mineral to understand the geogenic variability of the dissolved concentration of trace elements in the Gomati River when coupled with trace elements geochemistry of the river water and mineralogy of the river sediments.

Acknowledgments

The authors express their gratitude to Prof. Indra Bir Singh (University of Lucknow) for his guidance and encouragement in the current research. This study is devoted to Dr. Rohit Kuvar (Sagar University, India), who was involved in instigating the research plan. The current research work was funded as Rajiv Gandhi National Fellowship [Grant no. F.14-2(SC)/20109SA-III] to DKJ from University Grant Commission and as research fellowship (Grant no. MHR-02-41-106-429) to PS from Ministry of Human Resource Development, Government of India. The authors express thanks to Dr. Ratan Kar (Birbal Sahni Institute of Palaeosciences, Lucknow) for fruitful discussions.

References

APHA, 2002. Standard Methods for the Examination of Water and Wastewater, twentieth ed. *American Public Health association, Washington, DC.*

Dong, H., Peacor, D.R., Murphy, S.F., 1998. TEM study of progressive alteration of igneous biotite to kaolinite throughout a weathered soil profile. Geochimica et Cosmochimica Acta 62, 1881–1887.

Dutta, V., Srivastava, R.K., Yunus, M., Ahmed, S., Pathak, V.V., Rai, A., et al., 2007. Restoration plan of Gomti River with designated best use classification of surface water quality based on river expedition, monitoring and quality assessment. Earth Science India 4, 80–104.

Gupta, L.P., Subramanian, V., 1994. Environmental geochemistry of the River Gomti: A tributary of the Ganges River.Environmental. Geology 24, 235–243. Available from: https://doi.org/10.1007/BF00767084.

Jigyasu, D.K., Kuvar, R., Shahina, Singh, P., Singh, S., Singh, I.B., et al., 2014. Chemical weathering of biotite in the Ganga Alluvial Plain. Current Science 106 (10), 1484–1486.

Kabata-Pendias, A., 2011. fourth ed. Trace Elements in Soils and Plants, 534. *CRC Press Taylor & Francis Group,* Boca Raton, FL.

Kumar, S., Singh, I.B., 1978. Sedimentological study of Gomati River Sediments, Uttar Pradesh, India. Example of a river in alluvial plain. Senckenbergiana Marit 10, 145–211.

Langmuir, D., 1997. Aqueous Environmental Geochemistry, 590. *Prentice-Hall, Inc, Upper Saddle River, NJ.*

Pal, D.K., Bhattacharyya, T., Sinha, R., Srivastava, P., Dasgupta, A.S., Chandran, P., et al., 2012. Clay minerals record from Late Quaternary drill cores of the Ganga Plains and their implications for provenance and climate change in the Himalayan foreland. Palaeogeography, Palaeoclimatology, Palaeoecology 356–357, 27–37.

Rao, K.L., 1975. India's Water Health. Orient Longman Limited, New Delhi, 267p.

Shiller, A.M., Chen, Z., Hannigan, R., 2001. A time series of dissolved rare earth elements in the lower Mississippi River. In: Cidu, R. (Ed.), Water-Rock Interaction. Swets & Zeitlinger, Lisse, pp. 1005–1007.

Shumway, R.H., Stoffer, D.S., 2011. Time Series Analysis and Its Applications. Springer Science.

Singh, I.B., 1996. Geological evolution of Ganga plain – an overview. Journal of the Palaeontological Society of India 41, 91–137.

Singh, M., Sharma, M., Tobschall, H.J., 2005. Weathering of the Ganga alluvial plain, northern India: implications from fluvial geochemical of the Gomati River. Applied Geochemistry 20, 1–21.

Thakur, A., 2008. Morphology and Basin Characteristics of the Gomati River, the Ganga Plain, India (Unpublished Ph. D. Thesis), 125p.

Tischendorf, G., Förster, H.-J., Gottesmann, B., Rieder, M., 2007. True and brittle micas: composition and solid-solution series. Mineralogical Magazine 71 (3), 285–320.

Index

Note: Page numbers followed by "*f*" and "*t*" refer to figures and tables, respectively.

Printed in the United States
by Baker & Taylor Publisher Services